DIAGRAMMATOLOGY

SYNTHESE LIBRARY

STUDIES IN EPISTEMOLOGY,

LOGIC, METHODOLOGY, AND PHILOSOPHY OF SCIENCE

VOLUME 336

DIAGRAMMATOLOGY

AN INVESTIGATION ON THE BORDERLINES OF PHENOMENOLOGY, ONTOLOGY, AND SEMIOTICS

by

Frederik Stjernfelt
Learning Lab Denmark
The Danish University of Education
Copenhagen, Denmark

Springer

A C.I.P. Catalogue record for this book is available from the Library of Congress.

ISBN 978-1-4020-5651-2 (HB)
ISBN 978-1-4020-5652-9 (e-book)

Published by Springer,
P.O. Box 17, 3300 AA Dordrecht, The Netherlands.

www.springer.com

Printed on acid-free paper

TABLE OF CONTENTS

PREFACE

This treatise is the result of some years of work, not least because work on it was several times disrupted for university reasons, personal reasons, but probably most of all because my curiosity temporarily took me elsewhere.

Working on an interdisciplinary treatise as encompassing as the present volume would be impossible without the support from my workplaces at the institutions of the University of Copenhagen and of Learning Lab Denmark at the Danish University of Education.

For different species of help and discussions, I am indebted to Rudolf Bernet, Mikkel Bogh, Peter Borum, Mette Böll, Per Aage Brandt, Berit Brogaard, Peer Bundgaard, Peder Voetmann Christiansen, Terrence Deacon, Anne Marie Dinesen, Sten Ebbesen, Francis Edeline, Claus Emmeche, Anne Fastrup, Erik Fischer, Michael Friedman, Frans Gregersen, Thor Grünbaum, Niels Hansen, Nils Gunder Hansen, Thomas Illum Hansen, Peter Harder, Jaakko Hintikka, Michael Hoffmann, Jesper Hoffmeyer, Isak Winkel Holm, Nathan Houser, Hans Siggaard Jensen, Jørgen Dines Johansen, Keld Gall Jørgensen †, Søren Harnow Klausen, Simo Køppe, John Michael Krois, Kalevi Kull, Michael May, W.J.T. Mitchell, Dieter Münch, Winfried Nöth, Peter Øhrstrøm, Svend Østergaard, Stig Andur Pedersen, Jean Petitot, Roland Posner, João Queiroz, Thomas Riese, Wolfgang Ruttkowski, Lucia Santaella, Sune Schmidt-Madsen, Thomas Sebeok †, Martin Skov, Barry Smith, Jesper Sørensen, Søren Ulrik Thomsen, André de Tienne, Ola Tunander, Mark Turner, Dan Zahavi – as well as the library staff at the Peirce Edition Project in Indianapolis.

Thanks also to the Houghton Library, Harvard University, for permission to quote from unpublished Peirce manuscripts, and to Garland Science/Taylor & Francis for diagram reprint permission.

Thanks to Vincent Hendricks for native English advice and good suggestions, to the anonymous referee for fertile ideas, and to my parents-in-law Jens and Gerda Christoffersen for their generous placing their summer house at my disposition as an ideal writing exile.

Special thanks to Mette Stjernfelt for life-long assistance, and more than thanks to Agnete and our daughters Agnes and Karoline for never-ceasing support of any conceivable sort.

Copenhagen, September 2006

INTRODUCTION

This treatise deals with the sign types of icons and diagrams. Icons understood as those signs whose function as signs is due to some sort of similarity between them and their objects – and diagrams as that special sort of icons which represent the internal structure of those objects in terms of interrelated parts, facilitating reasoning possibilities. Thus, the treatise develops the idea that signs cannot, in general, be understood after the linguistic model of the Saussurean tradition as an arbitrary coding relation connecting pre-established expression and content, but that, quite on the contrary, the sign in most interesting cases is based on a structural similarity between signifier, signified, and the sign's object. A corollary to this idea is that signs should not, in general, be understood with the isolated linguistic word as the prototype – rather, the prototypical sign is an argument mirroring part of the structure of the states-of-affairs to which it refers.

This counterrevolution in semiotics holds important implications for the very foundations of that discipline as well as for its relation to other domains. It immediately raises the question of the sign's own foundation – which are the phenomenological prerequisites to sign use? This foundational task leads in two directions which appear not to be so diverse as may seem at a first glance: one is towards a reinterpretation of Charles Peirce's semiotics and its relation to the phenomenology, logic, and metaphysics as envisaged in his sketches of a philosophical architectonics;[1] another leads in the direction of Husserlian semiotics and phenomenology founded in the same period. In Peirce's philosophy it leads, more specifically, to an appreciation of the central role played in his mature thought by the sign type of *diagrams*, connecting the reasoning process, Peirce's logical realism, and his metaphysical doctrine of the primacy of continuity. In Husserl, it leads, more specifically, to a reinterpretation of his notions in and around *Logische Untersuchungen* of 'categorial intuition' and whole-and-parts as well as the related ideas of 'eidetic variation' and 'Wesenserschauung'.

The critical comparison of unnoticed parallels in these two classics of semiotics and phenomenology leads to a semiotic realism, emancipating semiotics from the anti-scientific tendencies inherent in the merely conventionalist sign approach and its possible affiliations with all sorts of quasi-religious vitalisms to explain which forces might be responsible for sign coding when no inner relation between sign and object are presumed to be possible.

The fact that the diagram displays the interrelation between the parts of the object it depicts is what facilitates its use in reasoning and thought processes, thus taking the concept of sign far from the idea of simple coding and decoding and to the epistemological issues of the acquisition of knowledge through signs.

Let us immediately take an example in order to illustrate this basic issue in diagram reasoning. It is well known that Pythagoras already in the fifth century BC was able to prove the insight that for right-angled triangles, the square of the hypotenuse equals the sum of the square of the two short legs, the so-called cathetes – algebraically expressed as $x^2 + y^2 = z^2$. It is interesting, however, that this proof may also be made with diagrammatical means. A well-known illustration may serve to geometrically show the size of the squares of the three sides of the triangle, respectively:

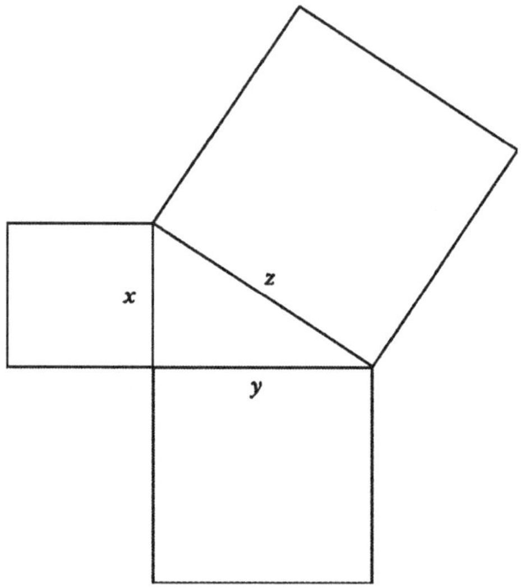

Figure 1.

The size of the large square upper right evidently corresponds to z squared, while the two lesser squares left and bottom correspond to x and y squared, respectively. Already at this stage of understanding, before any proof of the theorem is attempted, a series of important diagram reading conventions have been tacitly observed. One is the abstraction from arbitrary aspects of the figure given in order to focus upon the aspects relevant for the question discussed – an act of abstraction so ingrained and automatized that most readers perform it without thinking about it as a sort of tacit knowledge. This involves, for instance, abstraction from the fact that the triangle presented consists of ink on a sheet of paper. It goes without saying that the same relations as depicted here hold for a continuum of other triangles with different material support. It also involves the abstraction from the particular black-and-white color of the drawing – the size issues discussed are evidently independent of color. Moreover, it includes the abstraction from the fact that the drawing presented is made of lines with a physical breadth – while the figure they represent is supposed to be composed of lines with zero breadth. The same goes for the imprecision of the figure – given the granularity of paper and ink, the lines of the figure only seem straight at a macrophysical level – we know that not very strong magnification would suffice to reveal that the lines are not, as a matter of fact, really straight lines. Abstracting from these contingent aspects of the drawing in order to grasp an idealized figure, we also perform a generalization of a stunning extension: we understand that the figure does not portray only a triangle of the precise size shown, but also the same figure in all possible sizes. Even more so: we understand that the continuous deformation of the two small angles of the triangle, letting the right angle and the straight lines of its sides remain, will take it into an infinity of other rectangular triangles. Thus, the very understanding of a diagram like the given presupposes a whole series of cognitively complicated operations giving the result of an idealized figure which is not exhausted by the diagram token on the page but which involves a general continuum of other triangles satisfying the requirements given. On that basis, we may construct the following diagram:

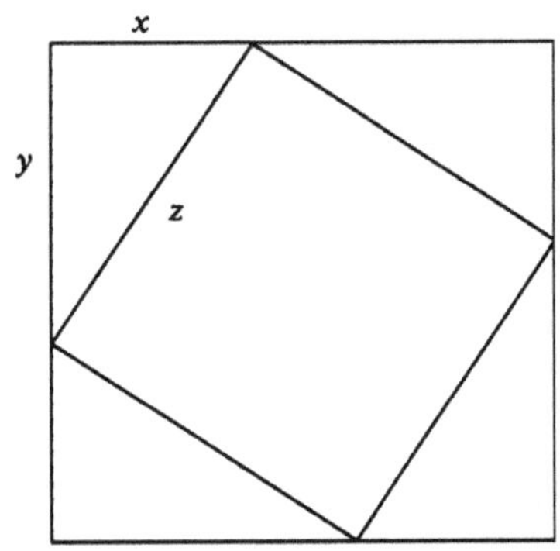

Figure 2.

The large square has as its side $x + y$, and the oblique square in the middle of the larger square is, of course, still the area corresponding to the square of the hypotenuse. The area of the total square is easy to calculate geometrically, because it corresponds to the square of the hypotenuse plus four times the area of the triangle. Now consider the next figure:

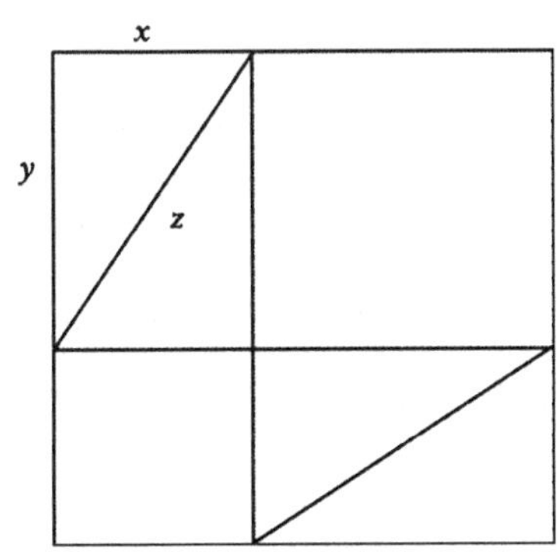

Figure 3.

The size of the large square remains the same, having the side $x + y$, and it also, just like the former square, includes four copies of the triangle. But this implies that, when comparing the two figures, we may subtract the four triangles in both of them, and now the remaining area must, in the two cases, be the same. And the two small squares remaining in the second diagram correspond exactly to the squares of the two cathetes in the triangle which is obvious from the fact that they have, each of them, one of the cathetes as their side. So, from this mental calculation with diagrams, it is evident that the large inner square in the former figure – equal to the square of the hypotenuse – must have the same area as the sum of the two small squares in the latter – equal to the sum of the squares of the hypotenuse. Thus, $x^2 + y^2 = z^2$. As was the case in the first diagram of the triangle, this diagrammatical proof also bases its generality on a possible continuous experiment with the diagram. For why is it not the case that the proof given holds only for the exact triangle represented in the drawing? This is because in the last figure, we may let the meeting point between the two small squares slide along the dotted diagonal of the large square, giving rise to a continuum of triangles with gradually changing sizes of x, y, and z:

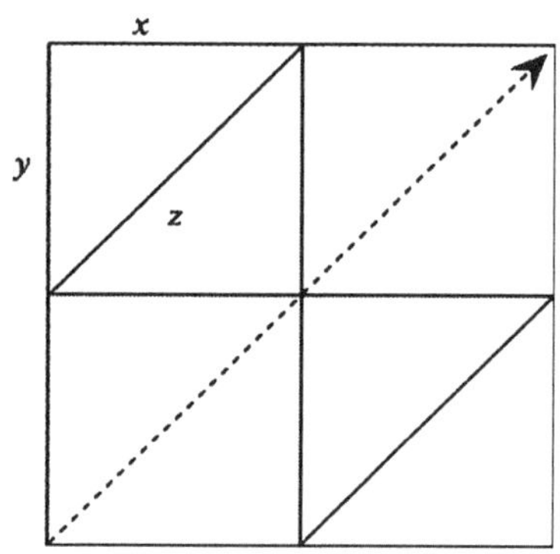

Figure 4.

With the changing of the meeting point between the two small squares, the triangles will change shape and run through a continuum of different values of the two small angles. By this means, we grasp the fact that the diagram experiment holds not only for the specific triangle represented on the page, but for all other triangles as well. If we let the meeting point approach one of the corners, the triangle will become still thinner and the square of the larger cathete will converge towards the square of the hypotenuse and indeed towards that of the whole larger square, while at the same time the area of the triangle will converge towards zero. If we let the meeting point hit the middle of the diagonal – approximately as in the diagram above – then the triangles will become isosceles and the area of the two squared cathetes will equal the area of four triangles, as can easily be seen from the figure. The continuous manipulation with the figure thus makes possible further reasoning about special cases – but not the other way around: no sum of single special cases will give us the general law which is only proved by holding for the whole continuum of possible right-angled triangles.

I have tried to go slowly through these elementary diagram reasonings – thereby boring some readers – in order to show a very important point about diagram reasoning, namely that here the generality of the states-of-affairs represented is depicted by the continuum of variations to which it is possible to subject the diagram without essentially changing its conditions. We begin this introduction with this example, thus, in order to grasp what is highlighted in Peirce's close connection between diagrams, continuity, and generality – a relation which is also intended in a related way in Husserl's idea that in order to grasp ideal objects, a continuous deformation of the object must be performed in an 'eidetic variation' thought experiment. The visually supported proof may, of course, be given further algebraical support: the area of the large square in Fig. 2 is evidently $z^2+4xy/2=z^2+2xy$. The area of the large square in Fig. 3 is $x^2+y^2+4xy/2=x^2+y^2+2xy$. But as the two large squares in Figs. 2 and 3 are equivalent, it follows that their area is the same: $z^2+2xy=x^2+y^2+2xy$. But then $z^2=x^2+y^2$. To Peirce, such a proof is no less diagrammatical. Here, the generality of the proof is granted by the fact that it holds for any values x, y determining a unique z. But a variable x in an algebraic formula is so to speak a sign that a whole continuum of different values may be inserted here. The seminal insight in Peirce is that diagrammatical reasoning representing generality by continuous shapes provides the general form of necessary and hypothetical inferences in thought. This gives the basic corollary that logical understanding is not, as it might seem to be the case, deprived of observation. It is, rather, the meticulous observation of general diagrams. As Peirce puts it:

> The first things I found out were that all mathematical reasoning is diagrammatic and that all necessary reasoning is mathematical reasoning, no matter how simple it may be. By diagrammatic reasoning, I mean reasoning which constructs a diagram according to a precept expressed in general terms, performs experiments upon this diagram, notes their results, assures itself that similar experiments performed upon any diagram constructed according to the same precept would have the same results, and expresses this in general terms. This was a discovery of no little importance, showing, as it does, that all knowledge without exception comes from observation. (MS CSP L75 C, 1902, 91–92)

We may add the psychological observation that many scientists use diagrams not only in the process of justification of their results, but also in the process of discovering them – see e.g. the classic investigations of Hadamard (1954).[2] For semiotics, this central role of the sign type of diagrams implies a basic realism. It is indeed possible to acquire and develop knowledge about different subjects by the construction and manipulation of diagrams charting those subjects – based on the fact that the structure of these diagrams are, in some respects, similar to the structure of their objects. This similarity does not have to be evident for a first glance – as the difference between the geometrical and the algebraical diagrams of the Pythagorean formula just presented shows. The recognition of the foundational role of diagrams in thinking thus immediately implies the recognition of the iconicity in thought and signs – the similarity between sign and object which is especially developed and constrained in the diagram case. The development of a semiotics based on iconical realism is the issue of this book.

The first part of the treatise thus has as its aim to articulate the basic tenets of such a semiotic realism. The central node of this part – and indeed of the whole book – is the fourth chapter, on Peirce's doctrine of diagrammatical reasoning. The three chapters preceding it prepare the ground by discussing Peirce's metaphysics of continuity, the 'extreme realism' of his sign theory, and his general doctrine of iconicity as prerequisites to his diagrammatology. The fourth chapter discusses the central idea: diagrammatical reasoning as central to the study of thought processes – encompassing far more than the usual scope of logic (as for instance represented in Peirce's logical graph systems). The four chapters following it discuss different aspects of Peircean diagrams, comparing them to other semiotic traditions with emphasis on the phenomenological tradition culminating in the early Husserl. Peirce's diagram concept is compared to Husserl's idea of 'categorial intuition', leading into a discussion of mereology – the doctrines of part and whole – in semiotics as a natural consequence of the diagram being a sign analyzing its objects in related parts. Husserl's mereology is, at the same time, a doctrine of the synthetic a priori which is discussed, based on the idea that diagrams form a major road of access to a priori structures in Husserlian formal and material ontologies.

The second part of the treatise is, in comparison, applicative – although perhaps in a special sense of the word. It consists of three major parts, one dealing with biosemiotics, one with pictures, and one with literature – in all cases referring to foundational issues with respect to these three domains. As to biology, thus, it aims at participating in the rational foundation of the emerging science of sign use in biology, the so-called 'biosemiotics'. The basic observation here is that the basic stock of concepts in biology never ceases – even in the most reductionist versions – to involve a host of semiotic concepts (genetic information, DNA code, messenger-RNA, etc.), the constant use of which can not be without any scientific reason. It thus continues the basic semiotic realism outlined in the first half of the treatise by investigating the ontological implications of this occurrence of biosemiotic vocabulary – and thus, by the same token, contributing to a biological foundation of semiotics. While the Peirco-Husserlian foundation of semotics envisaged in the former half rests on a structural, a-temporal argument, what is investigated here is the empirical and thus genetic, temporal foundation of semiotics, aiming at understanding the gradual appearance of still more complicated sign types in the course of biological evolution. In that sense, it attempts at understanding the phylogenetic instantiation, within the regional ontology of biology, of the realist semiotics developed at a principal level in the former half of the book.

To conclude, two more specific – but still general – applications follow. One deals with pictures – given that the sign type of icons figures prominently in the phenomenological as well as in the biological part, the basic iconicity of the sign provides new insights in the diagrammatical understanding of pictures proper with implication for the analysis of paintings and other pictures in art history. The experimental approach to diagram reading is exemplified in analyses of figurative and abstract art, a Husserlian picture theory is developed, and a theory of the sketch is sketched. The last application deals with literature and interpretation – the continuist and diagrammatic understanding of semiotics maintained is seen to be ripe with consequences for the understanding of the interpretation process and for the schematic aspects of the very notion of 'literary work'. The very process of literary interpretation is analyzed in terms of Peirce's abduction-deduction-induction cycle, the basic types of iconicity in the literary work are discerned on the basis of a Peircean interpretation of Ingarden's theory of literature, and finally, the spy novel is investigated as an exemplary case.

The title 'Diagrammatology' thus refers to two issues: one specific and one more general. The former deals with diagrammatical reasoning and Peirce's central idea that diagram manipulation forms the prototype of a wide class of thought processes which has not earlier been recognized as such. The latter deals with the more general status of iconicity – of which diagrams, in turn, form a central case – in sign use, which is exemplified in the three applied sections in the second part of the book, thus connecting the philosophical discussions of the first part with the special sciences of biology, of art history, and of literary theory. Let us present the overall course of the book in a little more detail to clarify the relation of the broad field discussed to the central issue of diagrammatology.

Chapter 1 briefly presents the mature Peirce's intense involvement with continuity which is taken to be the basic idea underlying his whole attempt at a philosophical architecture in the years around the turn of the century (a more thorough presentation of the detail of his mathematical and metaphysical arguments may be found in the Appendix). Thus, the chapter ends by introducing and discussing Peirce's three fundamental phenomenological categories, First-, Second-, and Thirdness, on the basis of his continuum metaphysics. Chapter 2 discusses Peirce's sign theory with reference to this continuum metaphysics which gives his semiotics its 'extreme realism'. The basic outline of Peirce's sign theory – the 'three trichotomies' – is presented with constant reference to continuity, and the Scotist inspiration for Peirce's realism is discussed. Metaphysically, Peirce's realism involves the existence of 'real possibilities', which are here compared to actual discussions of 'dispositions' and 'powers' – they form the ontological connecting link between his continuity doctrine on the one hand and diagrammatical iconicity in his epistemology and pragmatic theory of science on the other. Chapter 3 presents Peirce's theory of iconicity. In recent semiotics, a controversy on iconicity has raged, and two main anti-icon proponents, Nelson Goodman and the early Umberto Eco, are presented as contrast, while the later Eco's hesitating acceptance of basic iconicity is taken as a step in the right direction, even if still marred with culturalism. The different aspects of iconicity in Peirce are discussed – in addition to the ordinary definition of iconicity by similarity, Peirce's overlooked but non-trivial, operational icon definition is discussed: icons as signs it is possible to manipulate in order to learn more about their object than what is explicitly present in the sign – a property which may be purified for explicit and controlled use in the diagram. Chapter 4 forms the key chapter of the book. Here, Peirce's general diagram doctrine is analyzed in detail from a long quote from an overlooked 1906 paper, 'PAP'. A bouquet of central themes are discussed: the diagram as type; the diagram as the heir in Peirce's system to Kant's notion of 'schematism'; the diagram as a skeleton-like icon of parts interconnected by relations; the diagram as a vehicle for thought experiments; the diagram's connection to deduction and mathematics; the wide extension of a Peircean diagram notion, comprising algebra, logic, grammar, graphs, etc.; maps as a diagram example; a comparison of a Peircean diagrammatical reasoning process with Hilbert's superficially wholly different idea of a 'formale Redeweise', a formal reading mode; Hintikka's discussion of Peirce's distinction between two kinds of diagram reasoning, simple 'corollarial' reasoning versus more complex 'theorematical' reasoning involving auxiliary constructions not mentioned in the formulation of the theorem proved; the connection of diagrams to the pragmatic maxim, synthetic a priori structures, and 'real possibilities'.

Chapter 5 takes its departure in the idea that diagram transformations form the core of diagram use and takes an overview over types of transformations in some existing semiotic theories, taken in a broad sense. Lévi-Strauss' and Greimas' semiotics, D'Arcy Thompson's morphological biology having inspired Lévi-Strauss, Klein's Erlangen program as a deeper inspiration with its interdependent concepts of transformations and invariants, cognitive semantics, recent picture semiotics, leading up to the introduction of the Peirce-Husserl comparison. Chapter 6 introduces this comparison by making an overview of the two's mutual knowledge of each other, concluding that each of them mistook the other for falling prey to psychologism in logic, and that Peirce probably inherited his intense use of 'phenomenology' in the fertile years 1902–04 from Husserl. Husserl's doctrine of 'kategoriale Anschauung' – categorial intuition – from *Logische Untersuchungen* claims that grammar, syntax, and other ideal, categorial stuff may in some sense be directly observed in analogy to perception – evolving later into his doctrine of 'Wesensschau' – and this idea is compared to Peircean diagrammatology which in many respects are parallel, even if Peirce's favorite example, that of construction in geometrical diagrams, differs widely from Husserl's favorite example, syncategorematical expressions in language. Chapter 7 extends the comparison to mereology – doctrines of parts and wholes – where the two are seen in wider semiotic perspective by comparing them also to Jakobson's and Hjelmslev's semiotic mereologies. The foundational role of Husserl's mereology of the third investigation, also (explicitly) for Jakobsonian structuralism and maybe even (implicitly) for Hjelmslevian structuralism is discussed. The restriction to discrete dependency relations in Husserl's and especially Hjelmslev's mereologies is criticized in favor of Peirce's more generous acceptance of what may appear as diagram wholes and be experimentally used as such by the manipulation of their parts. Chapter 8 concludes the first part of the book with the placing of Husserl's mereology and its redefinition of the concept of the synthetic a priori within a larger framework, that of Austrian philosophy. Barry Smith's charting of this tradition, including the a priori concept in Austrian economics, is introduced as a prerequisite to Smith's own 'fallibilist apriorism'. The investigation of the interpretation of synthetic a priori propositions by Wojciech Zelaniec of the same tradition is presented, and these ontological a priori theories are compared to Michael Friedman's Neo-Kantian attempt at redefining the a priori from studies in recent history of science. The chapter concludes by discussing Peirce's position which is explicitly enmical towards a Kantian a priori, but the conclusion opts for diagrammatical reasoning to be a royal road to a Peircean synthetic a priori along Husserlian lines, supposedly comprising the metaphysical bases of the special sciences along with the 'theorematical' parts of mathematics. So, the chapter sets the scene for the three different regional ontologies selected for investigation in part two: biosemiotics, picture theory, and literature theory.

Chapter 9 introduces biosemiotics as a recent brand of semiotics (Thomas Sebeok, Jesper Hoffmeyer, etc.) taking the involvement of semiotic expressions (genetic information, DNA code, etc.) in biology at face value. An interpretation of Kant's critique of teleological judgment from his third Critique leads to an idea of the role of such expressions: they are indispensable, but not quantitatively formalizable. The neo-Kantian Cassirer's idea of necessary 'mediate concepts' in the special sciences are taken as an extrapolation of that idea, and the middle part of the chapter attempts at isolating a whole cluster of such 'mediate concepts', interpreted in an Austrian a priori framework – beginning with the analysis of an example, the bacterium E. coli's semiotic ability to detect sugar and act accordingly, pointing to 'categorical perception' as a primitive semiotic process in biology. The last part of the chapter discusses the extension of this biosemiotic material ontology and concludes that it may serve, also, as a formal ontology for higher domains than biosemiotics, thus giving rise to a subtypology of the ontological 'niche' concept proposed by Barry Smith. Chapter 10 investigates the basic assumptions in one of the central 'rediscoveries' of biosemiotics, the theoretical biology of the German-Estonian father of ethology Jakob von Uexküll. His theory about the 'functional circle' defining a species correlatively to the 'Umwelt' constituted by its schematic sign use in shape of perceptions and actions, is built on the idea of extrapolating Kant in two directions: towards the body, towards the animal world. This naturalized Kantianism, it is argued, implies a tension between naturalized subjects and the constitutive power supposedly inherent in transcendental subjects: you cannot preserve both. Uexküll's attempt at doing so leads him to exaggerate the perfect fit between animal and Umwelt as well as between different species – and leads him, moreover, to a pessimist idea of human culture as having broken with such perfection. Other currents in Uexküll – such as his idea that higher animals have 'neutral' objects, not exhaustively defined by their functional circle, point in sounder directions, as does the categorization idea inherent in his musical metaphor of nature as a symphony. Such categorization is proposed as a basic semiotic process in nature, supported by the notion of 'categorical perception' originating in phonetics. Chapter 11 introduces Terrence Deacon's idea that the Peircean symbol forms the main semiotic transition between animal and man. Deacon's idea, however, redefines Peircean symbols as more specific and complex than they are, and a more restricted version of Deacon's hypothesis is proposed: that the decisive semiotic 'missing link' is rather the special symbol subtype named by Peirce 'hypostatic abstraction'. Peirce's abstraction theories are introduced and discussed: while 'prescission', corresponding to Scotus' 'formal distinction', allows for isolating aspects of a phenomenon, 'hypostatic abstraction' allows for making such isolated aspects into thought objects in their own right to be investigated. These abstraction types are intimately connected to diagrammatical reasoning and the ability to construct diagrams as well as meta-diagrams taking simpler diagrams as their object. Chapter 12 takes up the alamodic notion of 'embodiment' in recent cognitive semantics and cognitive linguistics and charts the role of it in various actual semiotic currents. Husserl's influence on Merleau-Ponty is well-known: in his late notes on nature, the functional circle is definitely opened, and Uexküll is phenomenologically interpreted alongside Lorenz and other contemporary biologists. Stuart Kauffman's complexity theory interpretation of the origin of life involves a diagrammatical notion of a network of catalyzing chemicals which form the first, simple biological metabolism – his theory even makes biology itself an example of how such ontologies come in packets of interrelated concepts. The chapter ends the biosemiotic section by proposing a task for biosemiotics to construct an evolutionary ladder of still more complex sign types correlated with still more developed body types.

Chapter 13 argues that diagrammatical experiment is ubiquitous, also in picture viewing, both in the ordinary sense of the word and in the art historian's and picture theorists sense of the word. The chapter takes two examples, the Danish 'Golden Age' painter Eckersberg's altarpiece in Frederiksberg Church and Malevich's well-known white square on white – in order to show diagrams at work in figurative as well as in abstract art. Diagram experiment in Eckersberg reveals an intricate use of triple perspective in an apparently normal realist painting – connected to the biblical references of the depicted Last Supper scene. In Malevich, on the other hand, it is argued that even the understanding and interpretation of a work as seemingly 'empty' and simple as this, necessitates diagram experimentation. Chapter 14 returns to Husserl, now in a presentation of his picture theory as it appears in the many scattered papers and notes on fantasy, memory, and pictures from around 1900–20. Husserl's development is outlined, and the many connections between Husserl's picture concept and Peircean iconicity are discussed. Husserl provides, inter alia, for a foundation of picture consciousness already as a limit case *within* ordinary perception, and taking its point of departure in this observation, a new distinction between two picture types, accessible and non-accessible, respectively, is proposed – a distinction independent of that between figurative and abstract. The interplay between these two distinctions is illustrated by ten commented pictures. Chapter 15 is a short discussion on the notion of 'sketch', taking its point of departure in the Belgian semiotician gang 'Groupe μ' and their concept of transformation and the correlative concept of a pure visual type – as against the linguistic imperialism of French picture semiotics. The sketch is seen as a special sort of diagram, supporting certain transformation types.

Chapter 16 takes literary interpretation as a basic process taken as prerequisite by all the many competing sorts of more or less irrational 'theory' dominating the contemporary comparative literature domain. This process is seen as a special case of general

Peircean interpretation leading from abduction over diagrammatical deduction and to induction, and, via trial-and-error, back again. The place for the introduction of theory in that interpretation process is argued to be restricted to the diagrammatical phase, and the literary specificity of the process is argued to lie in the closedness of the literary artwork text as opposed to ordinary prose. Finally, a home-made interpretation example from Peirce's unpublished papers is discussed. Chapter 17 makes a reinterpretation of a classic piece of literary theory, Roman Ingarden's Husserlian doctrine of the four strata of the literary work. Special attention is given to the diagrammatical level of 'schematized aspects' showing the represented objectivities of the work as seen with respect to selected aspects, in turn allowing for completion because of the unsatiated character of schemata. Ingarden's doctrine is extrapolated to the extent that such 'schematized aspects' need not be restricted to structures close to perception but may also comprise many sorts of more abstract diagrammatical structures. Thus, this extension of Ingarden gives rise to the distinction between no fewer than five different sorts of schematic iconicity in the literary work – one of them being the fulfilling of certain regional ontological structures in the literary work. Chapter 18, finally, finds a specific example of this mirroring of regional ontology in the literary subgenre of the spy novel. On the basis of Greimas' semiotics, an ontological molecule of espionage is outlined, and with background in some observations in Carl Schmitt's controversial brand of political science, different regularities in spy novels are attempted explained. Drawing in a host of material from the world of espionage and spy novels, this chapter forms an example of how the charting of diagrammatical structure necessitates the involvement of both semantic presuppositions and structure abstracted from the body of empirical knowledge of the domain in question.

An Appendix presents in more detail Peirce's complicated discussion of the continuum involving both mathematical and metaphysical argumentation. It sets out from a discussion of his 1892 'Monist' papers, and leads into a discussion of Peirce's own – in some aspects fallacious – version of set theory around the turn of the century. Peirce's set theoretic formalization attempts are seen as deep down metaphysically rather than mathematically motivated: his main idea is that the continuum must be taken as a primitive, transgressing any formalization attempts. Peirce's continuum notion is compared to a series of actual philosophical and mathematical positions, to some extent supporting his idea of the inexhaustibility of the continuum. The Appendix may be read, thus, as a further substantiation of the more immediate presentation of the first chapter.

Thus, the overall course of the 18 chapters of the book leads from the more to the less general. As is evident, this book goes a long way from Peirce's continuum doctrine and to the spy novel, and along the line it involves many different continents of phenomenology, ontology, and semiotics as well as results from several different special sciences. Still, the book constantly focuses on the main topics of diagrammatical schemata and access to regional ontologies and thus attempts a synthesis of domains most often widely separated. The central notions of diagrams and diagrammatical reasoning are taken to integrate a whole series of 'schema' concepts in current philosophy, linguistics, biology, literary studies and elsewhere:

diagrammatic schemata seem to form a basic and economical tool for understanding and thought experimentation.

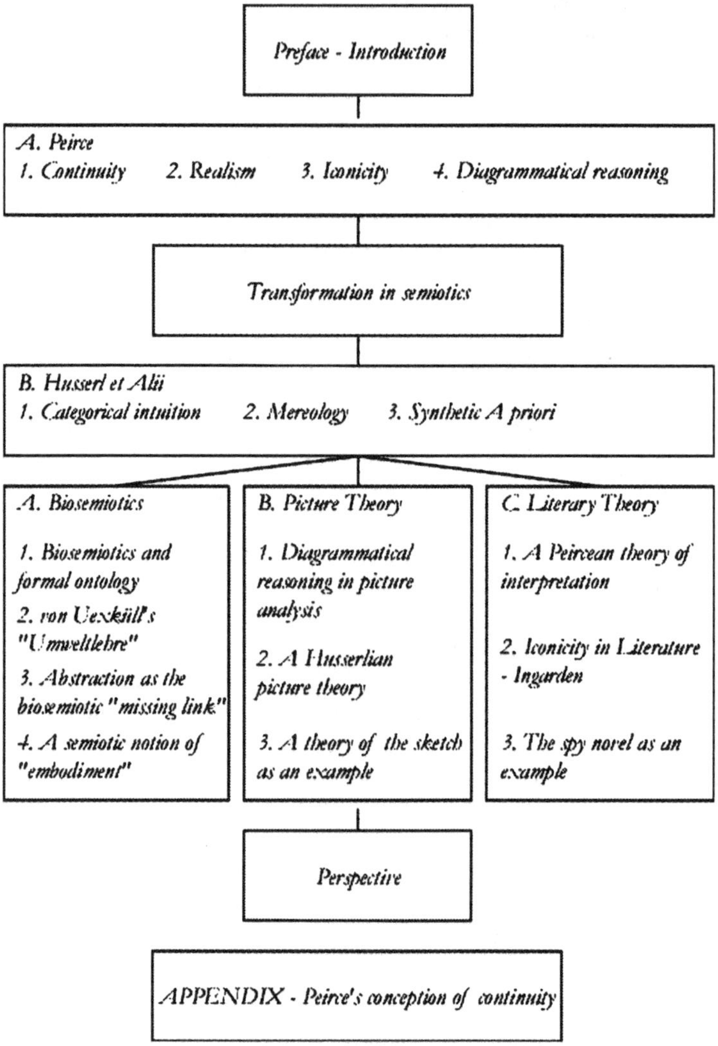

Figure 5.

The overall disposition of the book may be outlined in the diagram over the content structure (Fig. 5).

I have, for the title of the book, chosen the rarely used and maybe somewhat strange notion of 'Diagrammatology'.[3] It may seem slightly misleading, as the book is not only about diagrams. Still, diagrams form the core of the various issues covered by the book. The term 'Diagrammatology' is taken to refer to the investigation into the general status of diagrams in reasoning, in logic, theory of science, and in heuristics, on the one hand, and into the basis of diagrams in metaphysics and ontology, on the other. The notion is chosen with reference to the central role played by diagrams in Peirce's semiotics where it covers all kinds of deductive reasoning with the emphasis on the 'creative', experimental, so to speak strategic aspect of such reasoning. Peirce's idea is that the manipulation of diagrams forms the prototypical core of these areas rather than being a marginal phenomenon on its fringe – and thus goes against a strong tendency in the sciences as well as the humanities in the last 150 years for skepticism against the role of iconic representations. Thus, the central reasoning process in all these areas – as well as in the special sciences – is taken not to be best represented, necessarily, by means of formal, symbolic languages, but rather by means of diagrammatical representations governed by more or less explicit rules (of which formal, symbolic languages turn out to form a special subset). Thus, the idea is not to oppose symbolical and iconical representation with a preference for the latter – the idea is rather to reinstate iconical presentation and representation in its proper, more general position of which symbolic representations form a rich and important species.[4] If the notion of 'diagrammatology' is opposed to anything, then, it is not formal, symbolical representations, but rather 'grammatology' as a headline for all sorts of deconstructivist, vitalist, social-constructivist skepticisms against the possible reliability of iconically represented information.

As a substitute for the neologism of 'diagrammatology', the title might equally well had been 'diagrammatical reasoning'. But two things argued against such a choice: (1) a book with that title already exists (Glasgow et al. 1995, on the reemergence of diagrammatical representations in computer science), and, consequently, (2) the close association of that title with diagrams in computer science specifically. My aim here is more general. It is an attempt at reconstructing Peirce's original idea in his last large burst of philosophical creation almost exactly a hundred years ago, with its pinnacle in the years 1905–06, to appreciate and develop the many fertile parallels of that explosion in the birth of European phenomenology in the very same period – and trace the actual implications of this idea in ontology and epistemology as well as in biosemiotics, picture theory, and literary theory.

PART I

DIAGRAMS – PEIRCE AND HUSSERL

CHAPTER 1

LET'S STICK TOGETHER

Peirce's Conception of Continuity

> The entelechy and soul of the work, from which every part of its contents manifestly flows, is the principle of continuity, which has been the guiding star of exact science from the beginning, but of which novel and unexpected applications are now made.
>
> Peirce, 1893[5]

The concept of continuity is so central and basic to Peirce that it is not too much to say that he built the whole final version of his philosophy around it. Thus, the mature version of his phenomenological categories, the detailed semiotics he developed in the years after 1900 – and, a fortiori, his doctrine of icons and of diagrams – rest on a philosophy of continuity.[6] This chapter seeks to elucidate the basic reasons for the importance accorded to continuity by the mature Peirce in the years around the turn of the century.

Already in the 1880s, Peirce took part in the mathematical development concerning the continuum. Dedekind's formalization of the real numbers by means of the so-called 'Dedekind cut' only later came to Peirce's knowledge, but Cantor's foundation of set theory during that period gained Peirce's enthusiastic support and prompted his attempts to improve that theory. Most of Peirce's efforts, despite letters to the two internationally known mathematicians, were not published, and what was published only appeared later, so the mathematical side of Peirce's struggle with continuity remained with little influence on the scientific development. Dedekind had discovered that infinite sets could be described as sets containing subsets which have the same size as the sets themselves (like the even numbers are as infinite as the whole numbers, even if constituting a mere part of them), and Cantor, in his set theory, constructed a hierarchy of different infinities with increasing size, giving rise to his 'transfinite numbers' measuring them. An important step in that research was his ability to prove that one such class may be represented by the integers, while a larger class may be represented by the real numbers, that is, all numbers on the arithmetic line, defined, each of them, by an infinite decimal expansion. Cantor's proof, the famous 'diagonal' proof, showed that the former class could be represented in a list (like 1, 2, 3, 4, . . .), while the latter class could not. Even if the latter forms a linear series to the extent that any two given numbers can be ordered after size, they may not be listed nor counted one by one. The latter class can be described as containing all subsets of the former (each decimal expansion is written with integers and so can be seen as an infinite subset of integers). This provided a procedure for constructing ever higher such sets, by taking the set of subsets of the former set – the so-called power set. Seemingly large sets, like the set of all rational fractions, could be shown also to be contained in the former, countable class – which naturally gave rise to the question of whether there were any set size 'in between' these two sets. Cantor's immediate idea was that there were not – which is equivalent to the so-called continuum hypothesis (CH), claiming that the real numbers, the continuum, constitute the next transfinite number after that of the integers. As is well-known, it turned out not to be possible to prove the CH, and it was only in the twentieth century proved consistent with (Gödel) and independent of (Cohen) set theory, thus being undecidable. To our day, research continues, trying to enlarge set theory by new axioms making CH decidable.

This was the context in which Peirce intervened. Peirce subscribed to the CH to the extent that it claimed that the real numbers were in fact Aleph-1, the second transfinite number after the Aleph-0 of the integers. But he did not subscribe to the implicit addition that the real numbers, in turn, corresponded to the continuum. This forms the core of Peirce's repeated attempts to rearticulate Cantorian set theory: Peirce wanted a version of set theory which made evident that the continuum transgressed any attempt at formalizing it as a line built up from points, a set consisting of individual numbers. As Cantor's whole series of Alephs, each of them, are sets consisting of individual numbers, Peirce's claim was that the continuum must lie beyond the whole series of Alephs. This implied a distinction between the arithmetical line and the geometrical line. While the former is defined by being the linear arrangement of real numbers, the latter is most often, since Descartes' analytical geometry, identified with it. This caused Peirce's protest: the geometrical line was a true continuum, thus being larger than any point set, including the arithmetical line. This basic idea formed the reason behind Peirce's repeated attempts at defining the geometrical line and constructing a revised set theory which could prove that this real continuum exceeded all sets. Potter and Shields (1977) have given an overview over Peirce's development from a Pre-Cantorian through a Cantorian to a Kantistic and finally (maybe) a post-Kantistic period in his understanding of the continuum. In the former, he naively accepted one of Kant's attempts at describing continuity as infinite divisibility (but this also holds already for the rational numbers), but from the mid-eighties far into the nineties, he struggled with his reformulations of Cantor, adding to infinite divisibility (now nicknamed 'Kanticity') the notion that all infinite series in a continuum contain their limit ('Aristotelicity'). In the last years of the nineties, he embraced an alternative, recursive idea in Kant, the idea that the continuum was defined by having parts which had parts of the same kind. During the Cantorian and Kantistic periods, Peirce's set theoretical attempts pointed in one direction: to prove that the continuum, considered as a set, differed from all other sets in consisting of *indistinct* elements only. The parts of a continuum being 'welded' or 'merged' together into a homogeneous mass, they may not be treated as if made up of determinate individuals. True, such individuals may be *selected* in any number from a continuum, but this does not mean it consists of them. Peirce over and over attempted to find this indistinctive 'merging' which he supposed to appear gradually already within Cantor's transfinite hierarchy of Alephs, probably because he felt the transition from discontinuous sets to continuity should itself be continuous, so that in large transfinite numbers, indistinctness should somehow gradually emerge.

A recurrent way of stating Peirce's point is that the geometrical line may *contain* as large multiplicities of points as one may want – but that it, in no way, *consists* of those points. The continuum is a primitive concept of its own, and if anything, it is rather composed by infinitesimal line segments (the problem with this idea, of course, being that such segments are harder to identify or locate than points). Peirce's conception immediately entails a series of differences from the ordinary set theoretical conception of a line. In that conception, if you take away the end point of a line, what remains is an open interval without its limit point. That can not ('Aristotelicity') be allowed in Peirce's account which is why he would say that the point removed does not diminish the continuum at all. Rather, as many points as you wish (even transfinite sets of points) may 'fly off' the end of the line, leaving the original continuum unaffected. The explanation behind this idea is that the continuum 'sticks together', that points are only potentially parts of the continuum, and, as potential points only, they are indistinct. Only when, by some procedure, a point is singled out within a continuum, it gains actual existence and now forms a discontinuity within the continuum. If the continuum is broken, is crossed by another continuum, is composed by two adjoined continua, or if points by some function or other are selected within it, point sets accessible for set theoretical descriptions are actualized within it. But those points or point sets are invariably 'small' in comparison to the continuum, to the extent that they tend to vanish, or in any case, lose importance, in the overall continuous object:

> A continuum cannot be disarranged except to an insignificant extent. An instant cannot be removed. You can no more, by any decree, shorten a legal holiday by transferring its last instant to the work-day that follows that feast, than you can take away intensity from light, and keep the intensity on exhibition while the light is thrown into the ash-barrel. A limited line AB may be cut into two, AC and C'B, and its ends joined, C' to A and C to B. That is to say, all this may be done in the imagination.
> ('The Logic of Mathematics: An Attempt to Develop my Categories From Within', c. 1896, 1.499)

Such an idea of the geometrical continuum obviously breaks with Dedekind's description of the real numbers by means of Dedekind cuts: if the real line is broken in two parts, one of the parts will constitute a closed interval containing its limit point, the other part will constitute an open interval because its limit point is left at the former part. Not so in Peirce's conception which satisfies a comment by Gödel claiming that the geometrical line, if broken in two, ought to give two symmetrical parts as a result. In Peirce's account, of course, the two parts form, both of them, closed intervals, each of them containing a limit point which before the breaking was one and the same point, and which, if the two parts were put together again, would merge to one point again. Ken Ketner and Hilary Putnam have (1992) attempted, in a preface to Peirce's Harvard lectures (RLOT), to articulate Peirce's point of view within non-standard analysis as developed since the 1960s. In this account, the line consists of standard points, each of which spans an infinitesimal 'monad' of further non-standard points around itself. Using non-standard analysis vocabulary, Peirce's seemingly inconsistent talk of points having points as parts, gives more sense, even if Peirce would not agree in the distinction between standard points and points belonging to their monads. Yet, in Peirce's discussion of the continuity of time, he takes care to distinguish infinitesimal 'moments' as opposed to purely punctual 'instants' where the latter may form parts of the former, so the distinction between monads and points would not be alien to him. I refer to the Appendix for a deeper discussion of these issues and their relation to philosophy of mathematics.

Here, it is important to emphasize the roots of Peirce's constant interest in the mathematics of the continuum. It stems from Peirce's metaphysical concept of continuity, and his mathematical interest is an attempt to construe a consistent vocabulary in which to reason about metaphysical continuity. The need for continuity in metaphysics has a whole series of related reasons, and to begin with Peirce's theory of perception, the basic status of continuity is suitably summed up in Peirce's argument against the idea that continuity may be dissolved as an illusion (as he believed it would be in a Cantorian set theory):

> My notion is that we directly perceive the continuity of consciousness; and if anybody objects, that which is not really continuous may seem so, I reply, 'Aye, but it could not seem so, if there were not some consciousness that is so.' I should like to see a good criticism of that reply.
> ('A Sketch of Logical Critic,' c. 1911, 6.182)

Essentially the same argument has been made independently by René Thom, claiming that as part of our experience, the continuum has an objective existence which, like all other experiences, may be subject to illusions (cf. the 24 discontinuous pictures per second giving a continuous time flow illusion in cinema), but if no real continuity is possible within neural physiology, how then could such an illusion be explained? (Thom 1992, 140). This argument for the ineradicability of continuity in experience might be nicknamed the Peirce-Thom argument.

This argument, however, only involves the phenomenological aspects of a whole related bunch of issues. Continuity is deemed metaphysically necessary to explain: (1) the intensional meaning of general concepts (the inexhaustibility of continuous extension); (2) the embeddedness of actually existing objects and occurring events within a horizon continuum of potentiality; (3) realism as to general tendencies (as opposed to the powder of unconnected singular events without continuity); (4) continuity of research – from its infinitesimal beginnings long before science and to its converging end point in truth; (5) fallibilism as implied by the vagueness inherent in continuity; and finally: (6) diagram manipulation as basically continuous and hence able to mirror real continuity. Let us run through these main points of Peirce's continuum metaphysics.

(1) Continuity is first and foremost important, because it provides an account for the generality of concepts. Concepts with real reference are seen by Peirce as abbreviated propositions without any specific subject. 'Heavy' is an abbreviation of

'X is heavy' where X may be filled in with any subject that satisfies the proposition. The law of gravitation, one of Peirce's favourite examples of a real force functioning in Nature, may not be reduced to any finite number of heavy objects subject to attraction. This implies that the collection of subjects potentially referred to by the proposition 'X is heavy' exceeds any extension made up of recorded, actual, or even imagined single cases of gravitation attraction. It is continuous. Herein lies, Peirce argues, the reality of gravity: it has worked, and will continue to work, in a number of cases which are so vast as to be beyond the reach of any possible charting – that is, they remain vague and indistinct. The vagueness and indistinctness of extensional reference is thus, according to Peirce, a necessary flip side of that reference involving real, general forces, tendencies or patterns in reality:

> True generality is, in fact, nothing but a rudimentary form of true continuity. Continuity is nothing but perfect generality of a law of relationship. ('Synechism', Baldwin's Dictionary, 1902, 6.172)

Therefore, no positivist reduction of such laws or relationships to mutually similar aspects of single cases is possible. It might indeed be the way such laws are discovered or recorded by induction, but this does not imply that the relation at work may be reduced to such cases:

> At any rate, it is plain that no possible collection of single occasions of conduct can be, or adequately represent all conceivable occasions. For there is no collection of individuals of any general description which we could not conceive to receive the addition of other individuals of the same description aggregated to it. The generality of the possible, the only true generality, is distributive, not collective. ('Consequences of Critical Common-Sensism', 1902, 5.532)

The distinction made here between distributivity and collectivity corresponds to the normal terminology of intension and extension. The latter may give rise to probabilistic investigations, the former, however, to apodictic, universal results:

> To say that the probability that a calf will not have more than six legs is 1, is to say that in the long run, taking calves as they present themselves in experience, the ratio of the number of those with not more than six legs to the total number is 1. But this does not prevent there being any finite number of calves with more legs than six, provided that in the long run, that is, in an endless course of experience, their number remains finite, and does not increase indefinitely. A universal proposition, on the other hand, asserts, for example, that any calf which may exist, without exception, is a vertebrate animal. The universal proposition speaks of experience distributively; the probable, or statistical proposition, speaks of experience collectively. ('Predicate', Baldwin's Dictionary, 1902, 2.358)

(2) Due to the reality of (certain) concepts, these claims of semantics immediately lead into ontology. The finite number of cases recorded by any extensional investigation thus refers to actual events selected from a continuum of potential, interrrelated events. The rule, law, tendency, or pattern governing that continuum hence has the character of potentiality:

> Since Kant it has been a very wide-spread idea that it is time and space which introduce continuity into nature. But this is an anacoluthon. Time and space are continuous because they embody conditions of possibility, and the possible is general, and continuity and generality are two names for the same absence of distinction of individuals. ('Multitude and Number', 1897, 4.172)

Such possibilities do not possess the actual individuality of realized cases, but they are still real – from around 1897 Peirce begins to term them 'real possibilities' (to which we shall return in the next chapter on Peirce's 'extreme realism'):

> A true continuum is something whose possibilities of determination no multitude of individuals can exhaust. ('Synechism', Baldwin's Dictionary, 1902, 6.170)

(3) Continuity thus forms the central feature of Peirce's realism with respect to such 'real possibilities':

> That which is possible is in so far general and, as general, it ceases to be individual. Hence, remembering that the word 'potential' means indeterminate yet capable of determination in any special case, there may be a potential aggregate of all the possibilities that are consistent with certain general conditions; and this may be such that given any collection of distinct individuals whatsoever, out of that potential aggregate there may be actualized a more multitudinous collection than the given collection. Thus the potential aggregate is, with the strictest exactitude, greater in multitude than any possible multitude of individuals. But being a potential aggregate only, it does not contain any individuals at all. It only contains general conditions which permit the determination of individuals.
> ('The Logic of Continuity', 1898, 6.185)

Generality, real possibility, and indistinctness are thus connected in metaphysical continuity. It should be added that the continuity doctrine is also intimately connected to Peirce's evolutionism which extends Darwinism from biology to cover the whole of the physical evolution in a strange cosmology, taking its beginning in pure continuous, chaotic (Peirce does not quite agree with himself how ordered or unordered pure possibility is) possibility which, via a growing amount of actualization, lets still more laws and tendencies introduce in the actual world which is thus aimed at continuous perfection in a remote future ... We shall not go into these cosmological areas of Peirce's metaphysics in this book which focuses upon the basic relation between continuous realism and diagrams.

(4) Given that law- or rule-governed aspects of reality are thus continuous, the very process of acquiring knowledge must – as it is in itself such a process – be continuous. Peirce's epistemology, in which diagrams play center stage, thus rests upon an ontology of knowledge acquisition. Knowledge creation being continuous, it can never really begin nor end – in both cases continuity replaces any discontinuous beginning or halting of knowledge:

> If it is objected that there must be a first thing learned, I reply that this is like saying that there must be a first rational fraction, in the order of magnitudes, greater than zero.
> ('Some Logical Prolegomena', undated, 7.536)

Science thus is continuous with everyday knowledge which is, in turn, continuous with animal cognition and so on indefinitely down the scale of evolution. Knowledge is always already in the process of being constructed – even if this idea causes trouble for Peirce's definition of logical thought as explicitly self-controlled: already perception, entrance gate of raw knowledge, is not thus controlled, and even less so biological forms of knowledge gathering. We shall return to this. In the other end, of course, continuity defines Peirce's famous pragmatist notion of truth as that to which the scientific community will converge in the long run:[7]

But a scientific proposition is merely something you take up provisionally as being the proper hypothesis to try first and endeavor to refute. The only belief you – as a purely scientific man – have about it is that it is adopted in accordance with a method which must lead to the truth in the long run.
('Logic of Events', 1898, 6.216)

As in many rule-bound, continuous processes, there are considerable fluctuations in science, but they even out in the continuity of the long run, making necessarily science an unending, collective process involving generation after generation of scientists:

As we go on drawing inference after inference of the given kind, during the first ten or hundred cases the ratio of successes may be expected to show considerable fluctuations; but when we come into the thousands and millions, these fluctuations become less and less; and if we continue long enough, the ratio will approximate toward a fixed limit. ('The Doctrine of Chances', 1878, EPI, 146; 2.650)

(5) The continuity of knowledge thus also implies the pragmatist's fallibilism due to the ineradicable imprecision inherent in continuity:

The principle of continuity is the idea of fallibilism objectified. For fallibilism is the doctrine that our knowledge is never absolute but always swims, as it were, in a continuum of uncertainty and of indeterminacy. Now the doctrine of continuity is that all things so swim in continua.
(Untitled manus., c. 1897, 1.171)

This does not, as might be expected, wed Peirce to any irrational skepticism against the possibilities of science, quite on the contrary. General, continuous processes occur with many degrees of fluctuations, and the only minimally fluctuating parts of those may be charted with a high degree of precision.
(6) As indicated in the Introduction, continuity plays a central role for the possibility of reading off general regularities of a diagram. Diagrams may chart ideal relationships – as in the Pythagoras example of the Introduction – or they may map idealized aspects of empirical states-of-affairs of many varied sorts. In all cases, however, they involve a moment of observation. To Peirce, observation is also a process necessarily infused with continuity. There is no such thing as an observation of a completely unique event or entity: already in ordinary perception, generality and continuity play a central role – e.g. in our spontaneous recognition that this or that aspect of perception is an instantiation of some general type or process. This reliance of perception and knowledge on continuous generality is now highlighted and made an issue of explicit control in diagrams. This is why the very continuity of the sheet upon which a diagram is drawn becomes a matter of central importance:

Let the clean blackboard be a sort of diagram of the original vague potentiality, or at any rate of some early stage of its determination. This is something more than a figure of speech; for after all continuity is generality. This blackboard is a continuum of two dimensions, while that which it stands for is a continuum of some indefinite multitude of dimensions. This blackboard is a continuum of possible points; while that is a continuum of possible dimensions of quality, or is a continuum of possible dimensions of a continuum of possible dimensions of quality, or something of that sort. There are no points on this blackboard. There are no dimensions in that continuum. I draw a chalk line on the board. This discontinuity is one of those brute acts by which alone the original vagueness could have made a step towards definiteness. There is a certain element of continuity in this line. Where did this continuity come from? It is nothing but the original continuity of the blackboard which makes everything upon it continuous. What I have really drawn there is an oval line. For this white chalk-mark is not a line, it is a plane figure in Euclid's sense – a surface, and the only line there, is the line which forms the limit between the black surface and the white surface. Thus the discontinuity can only be produced upon that blackboard by the reaction between two continuous surfaces into which it is separated, the white surface and the black surface. The whiteness is a Firstness – a springing up of something new. But the boundary between the black and white is neither black, nor white, nor neither, nor both. It is the pairedness of the two. It is for the white the active Secondness of the black; for the black the active Secondness of the white.

Now the clue, that I mentioned, consists in making our thought diagrammatic and mathematical, by treating generality from the point of view of geometrical continuity, and by experimenting upon the diagram. ('The Logic of Continuity', 1898, 6.203–4)

Diagrams may, of course, use discontinuity, but it is only possible to represent discontinuity as a break of a presented continuity – like the triangle drawn on the geometer's blank slate. The continuity of the sheet is crucial, furthermore, because it is what allows us to perform the thought (or real) experiments continuously varying the diagram so as to ensure that the state-of-affairs recorded therein hold not only for the one case drawn, but for a continuum of similar cases. Which continuum of cases are in fact envisioned of course depends on the (most often) written or tacitly implied instructions relevant for the specific diagram in question – but the important thing is that the diagram always inherits the spatiotemporal continuity of the sheet in some specified respects. So, the continuity of diagrams is no accidental feature of representation, rather, it is what makes thought representing real relations possible in diagrams. Thus, diagram continuity is intimately connected to the continuity tying together semantics, realism, epistemology, and fallibilism in Peirce's mature doctrine. It goes without saying that the sketches of a philosophical architectonics which we have here briefly introduced do not form a coherent doctrine. Many obscure points indeed remain, and we shall address some of them in the chapters to come. Still, the continuity doctrine forms the basis of the most impressive results of Peirce's final endeavor. It comes as little surprise, then, that the continuum also lies beneath Peirce's mature version of his categories after the turn of the century, now considered under the headline of 'phenomenology' or 'phaneroscopy' (see Chap. 6). Peirce's categories belong to the earliest of his preoccupations, already presented, of course, in 'A New List of Categories' (1867). The continuity metaphysics of the later Peirce, however, permits him to cast them in a new light, integrating them at the basis of his thought.

CONTINUITY AS METAPHYSICAL GLUE IN PEIRCE'S SYSTEM

Continuity has wide implications in the different parts of Peirce's architectonics of theories. Time and time again, Peirce refers to his 'principle of continuity' which has not immediately anything to do with Poncelet's famous such principle in geometry (Chap. 5).[8] It is, rather, a metaphysical implication taken to follow from fallibilism: if all more or less distinct phenomena swim in a vague sea of continuity then it is no wonder that fallibilism must be accepted. And if the world is basically continuous, we should not expect conceptual borders to be definitive but rather conceive of terminological distinctions as relative to an underlying, monist continuity. We shall not go far into these ideas, but rather depict the role of continuity in Peirce's efforts to build a system. In this system, mathematics is first science. Thereafter follows philosophy which is distinguished form purely hypothetical mathematics by having an empirical basis.[9] Philosophy, in turn, has three parts, phenomenology, the normative sciences, and metaphysics. The first investigates solely 'the Phaneron' which is all what could be imagined to appear as an object for experience: '... by the word *phaneron* I mean the collective total of all that is in any way or in any sense present to the mind, quite regardless whether it corresponds to any real thing or not.' ('Adirondack Lectures', 1905, CP 1.284) As is evident, this definition of Peirce's 'phenomenology' is parallel to Husserl's phenomenological reduction in bracketing the issue of the existence of the phenomenon in question (see Chap. 6). Even if it thus is built on introspection and general experience, it is – analogous to Husserl and other Brentano disciples at the same time – conceived in a completely antipsychological manner: 'It religiously abstains from all speculation as to any relations between its categories and physiological facts, cerebral or other.' ('Logic viewed as Semeiotics', 1904, 1.287) and '... I abstain from psychology which has nothing to do with ideoscopy.' (Letter to Lady Welby, Oct 12, 1904, 8.330). The normative sciences fall in three: aesthetics, ethics, logic, in that order (and hence decreasing generality), among which Peirce does not spend very much time on the former two. Aesthetics is the investigation of which possible goals it is possible to aim at (Good, Truth, Beauty, etc.), and ethics how they may be reached. Logic is concerned with the grasping and conservation of Truth and takes up the larger part of Peirce's interest among the normative sciences. As it deals with how truth can be obtained by means of signs, it is also called *semiotics* ('logic is formal semiotics') which is thus coextensive with *theory of science* – logic in this broad sense contains all parts of philosophy of science, including contexts of discovery as well as contexts of justification. Semiotics has, in turn, three branches: *grammatica speculativa* (or stekheiotics), *critical logic*, and *methodeutic* (inspired by mediaeval trivium: grammar, logic, and rhetoric). The middle one of these three lies closest to our days' conception of logic; it is concerned with the formal conditions for truth in symbols – that is, propositions, arguments, their validity and how to calculate them, including Peirce's many developments of the logic of his time: quantifiers, logic of relations, ab-, de-, and induction, logic notation systems, etc. All of these, however, presuppose the existence of simple signs which are investigated by what is often seen as semiotics proper, the *grammatica speculativa*;[10] it may also be called formal grammar. It investigates the formal condition for symbols having meaning, and it is here we find Peirce's definition of signs and his trichotomies of different types of sign aspects. Methodeutic or formal rhetorics, on the other hand, concerns the pragmatical use of the former two branches, that is, the study of how to use logic in a fertile way in research, the formal conditions for the 'power' of symbols, that is, their reference to their interpretants; here can be found, e.g., Peirce's famous definitions of pragmati(ci)sm and his directions for scientific investigation. To phenomenology – again in analogy to Husserl – logic adds the interest in signs and their truth. After logic, metaphysics follows in Peirce's system, concerning the inventarium of existing objects, conceived in general – and strongly influenced by logic in the Kantian tradition for seeing metaphysics mirroring logic. Also here, Peirce has several proposals for subtypologies, even if none of them seem stable, and under this headline classical metaphysical issues mix freely with generalizations of scientific results and cosmological speculations.

Peirce himself saw this classification in an almost sociological manner, so that the criteria of distinction do not stem directly from the implied objects' natural kinds, but after which groups of persons study which objects: '... the only natural lines of demarcation between nearly related sciences are the divisions between the social groups of devotees of those sciences...' (CP 8.342). Science collects scientists into bundles, because they are defined by their causa finalis, a teleologial intention demanding of them to solve a central problem.[11]

Measured on this definition, one has to say that Peirce himself was not modest, not only does he continuously transgress such boundaries in his production, he frequently does so even within the scope of single papers. There is always, in his writings, a brief distance only from mathematics to metaphysics – or between any other two issues in mathematics and philosophy, and this implies, first, that the investigation of continuity and generality in Peirce's system is more systematic than any actually existing exposition of these issues in Peirce's texts, second, that the discussion must constantly rely on cross-references. This has the structural motivation that as soon as you are below the level of mathematics in Peirce's system, inspired by the Comtean system, the single science receives determinations from three different directions, each science consisting of material and formal aspects alike. First, it receives formal directives 'from above', from those more general sciences which stand above it, providing the general frameworks in which it must unfold. Second, it receives material determinations from its own object, requiring it to make certain choices in its use of formal insights from the higher sciences. The cosmological issue of the character of empirical space, for instance, can take from mathematics the different (non-)Euclidean geometries and investigate which of these are fit to describe spatial aspects of our universe, but it does not, in itself, provide the formal tools. Finally, the single sciences receive in practice determinations 'from below', from more specific sciences, when their results by means of abstraction, prescission, induction, and other procedures provide insights on its more general, material level. Even if cosmology is, for instance, part of metaphysics, it receives influences from the empirical results of physics (or biology, from where Peirce takes the generalized principle of evolution). The distinction between formal and material is thus level specific: what is material on one level is a formal bundle of possibilities for the level below; what is formal on one level is material on the level above.[12]

For these reasons, the single step on the ladder of sciences is only partially independent in Peirce, hence also the tendency of his own investigations to zigzag between the levels. His architecture of theories thus forms a sort of phenomenological theory of aspects: the hierarchy of sciences is an architecture of more and less general aspects of the phenomena, not completely independent domains. Finally, Peirce's realism has as a result a somewhat disturbing style of thinking: many of his central concepts receive many, often highly different determinations which has often led interpreters to assume inconsistencies or theoretical developments in Peirce where none necessarily exist (this does not imply, of course, that there are no developments or inconsistencies at all in Peirce; there are indeed many). When Peirce, for instance, determines the icon as the sign possessing a similarity to its object, and elsewhere determines it as the sign by the contemplation of which it is possible to learn more about its object, then they are not conflicting definitions. Peirce's determinations of concepts are rarely definitions at all in the sense that they provide necessary and sufficient conditions exhausting the phenomenon in question. His determinations should rather be seen as descriptions from different perspectives of a real (and maybe ideal) object – without these descriptions necessarily conflicting. This style of thinking can, however, be seen as motivated by metaphysical continuity. When continuous grading between concepts is the rule, definitions in terms of necessary and sufficient conditions should not be expected to be exhaustive.

A recurring skeleton on all levels, however, is provided by Peirce's famous 'triadomania', as he himself calls it, which lets most of his decisive distinctions appear in threes, following the tripartition of his list of categories, the famous triad of First, Second, and Third, or Quality, Reaction, Representation, or Possibility, Actuality, Reality – or any other of the manifold of descriptions of this triad which he gives through his work.

The probably most concise – but also very self-referential – description is found in one of the letters to Lady Welby (CP 8.327) from 1904:

> Firstness is the mode of being of that which is such as it is, positively and without reference to anything else.
>
> Secondness is the mode of being of that which is such as it is, with respect to a second but regardless of any third.
>
> Thirdness is the mode of being of that which is such as it is, in bringing a second and third into relation to each other.

The justification for this triad of possible experience can be stated briefly as follows: Firstness constitutes the *quality* of experience: in order for something to appear at all, it must do so due to a certain constellation of qualitative properties. Peirce often uses sensory qualities as examples, but it is important for the understanding of his thought that the examples may refer to phenomena very far from our standard conception of 'sensory data', e.g. forms or the 'feeling' of a whole melody or of a whole mathematical proof, not to be taken in a subjective sense but as a concept for the continuity of melody or proof as a whole, apart from the analytical steps and sequences in which it may be, subsequently, subdivided. In short, all sorts of simple and complex Gestalt qualities also qualify as Firstnesses. Firstness tend to form continua of possibilities such as the continua of shape, color, tone, etc. These qualities, however, are, taken in themselves, pure possibilities and must necessarily be incarnated in phenomena in order to appear. Secondness is the phenomenological category of 'incarnation' which makes this possible: it is the *insistency*, then, with which the individuated, actualized, existent phenomenon appears. Thus, Secondness necessarily forms discontinuous breaks in Firstness, allowing for particular qualities to enter into existence. The mind may imagine anything whatever in all sorts of quality combinations, but something appears with an irrefutable insisting power, reacting, actively, yielding resistance. Peirce's favorite example is the resistance of the closed door – which might be imagined reduced to the *quality* of resistance feeling and thus degenerate to pure Firstness so that his theory imploded into a Hume-like solipsism – but to Peirce this resistance, surprise, event, this thisness, 'haecceity' as he calls it with a Scotist term, remains irreducible in the description of the phenomenon (a Kantian idea, at bottom: existence is no predicate).[13] About Thirdness, Peirce may directly state that continuity represents it perfectly (1.337): '... continuity and generality are two names of the same absence of distinction of individuals' ('Multitude and Number', 1897, 4.173). As against Secondness, Thirdness is *general*; it *mediates* between First and Second. The events of Secondness are never completely unique, such an event would be inexperiencable, but *relates* (3) to other events (2) due to certain features (1) in them; Thirdness is thus what facilitates understanding as well as pragmatic action, due to its continuous generality. With a famous example ('Thirdness', c. 1895, 1.341): if you dream about an apple pie, then the very qualities of that dream (taste, smell, warmth, crustiness, etc.) are pure Firstnesses, while the act of baking is composed of a series of actual Secondnesses. But their coordination is governed by a Thirdness: the recipe, being general, can never specify all properties in the individual apple pie, it has a schematic frame-character and subsumes an indefinite series – a whole continuum – of possible apple pies. Thirdness is thus necessarily general and vague. Of course, the recipe may be more or less precise, but no recipe exists which is able to determine each and every property in the cake, including date, hour, place, which tree the apples stem from, etc. – any recipe is necessarily general. In this case, the recipe (3) mediates between dream (1) and fulfilment (2) – its generality, symbolicity, relationality and future orientation are all characteristic for Thirdness. An important aspect of Peirce's realism is that continuous generality may be experienced directly in perceptual judgments: 'Generality, Thirdness, pours in upon us in our very perceptual judgments ...' ('The Three Normative Sciences', 1902, EPII, 207; 5.150).[14]

All these determinations remain purely phenomenological, even if the later semiotic and metaphysical interpretations[15] clearly shine through. In a more general, non-Peircean terminology, his phenomenology can be seen as the description of minimum aspects inherent in any imaginable possible world – for this reason it is imaginability which is the main argument, and this might point in the direction that Peirce could be open to critique for subjectivism, so often aimed at Husserl's project, in some respects analogous. The concept of consciousness is invoked as the basis of imaginability: phenomenology is the study of invariant properties in any phenomenon appearing for a mind. Peirce's answer would here be, on the one hand, the research community which according to him defines reality – an argument which structurally corresponds to Husserl's reference to intersubjectivity as a necessary ingredient in objectivity (an object is a phenomenon which is intersubjectively accessible). Peirce, however, has a further argument here, namely his consequent refusal to delimit his concept of mind exclusively to human subjects (a category the use of which he obviously tries to minimize), mind-like processes may take place in nature without any subject being responsible. Peirce will, for continuity reasons, never accept any hard distinction between subject and object and remains extremely parsimonious in his use of such terms.[16] Without calling him naturalist through and through, we may at this stage claim that he tries to let the antinomy between naturalism and idealism form a circle[17] which hinders scepticist critiques of idealism as well as subjectivist critiques of naturalism to gain foothold.

The place of continuity in Peirce's category doctrine has already been hinted at by its appearance as a central predicates for Thirdness, but all has not been said here. Secondness evidently has discontinuity as one of its properties, but what about Firstness? The fact that sensory qualities – prototypical examples of Firsts – as a rule form continua, suggests that continuity is also a crucial property in Firstness. Peirce is not unanimous here: a basic idea in Firstness is that each quality appears independently of anything else, and so Firstness seems a powder of infinitesimal quality bits. But still they have a tendency to form continua, and the different quality continua may even meet in a higher dimension continuous space: 'At one end of the sequence all the qualities come together in a *zero*. But they are separate from one another as they separate from zero' ('Abstracts of 8 lectures', undated, NEM IV, 128[18]).

In the text quoted here, Peirce indulges in a larger investigation of whether the space of qualities is 'perissid' or 'artiad' – depending on whether a Hegelian-like transformation from a maximal intensity of one quality is allowed to pass directly into a maximal intensity of the opposed quality. This is refused by Peirce, and he consequently envisages the continuum of possible qualities mirror itself in a zero point where two half continua meet. The decisive idea here, however, is that qualities constitute a continuum of many dimensions, and that one type of qualities – cf. the point zero argument – may pass continuously into another. The existent sensory and other qualities of experience are thus, to Peirce, only remaining, actualized fragments of a original, basic continuum of possibility, uniting all possible qualia in one continuous manifold[19] (and thus making visual, auditory, olfactory, etc. qualities parts of the same continuum).

> But now to return to the continuum of possible quality. Every complexus of qualities is a quality, and as such, considered by itself, is all that it is in and for itself. Not only every complex of qualities but every *generalization* of such complexes is a possible quality. But in this way, the dimensions of the continuum ought to exceed every discrete multitude. In short, they should form a continuum of dimensions. It is impossible. Hence these dimensions of complex qualities are only abstractly possible. They cannot have simultaneous being in the world of potentialities. ('Abstract of 8 lectures', undated, NEM IV, 135)

It is not completely evident why a quality continuum with a continuous number of dimensions should not be possible – but the decisive issue here is the idea that

Firstness appears as a multidimensional continuum, in which the single quality yet '... in itself is absolutely severed from every other' (ibid. 133).[20]

This implies that the reaction events of Secondness may be conceived on the background of this vast space of possibilities:

But just as the qualities, which as they are for themselves, are equally unrelated to one other, each being mere nothing for any other, yet form a continuum in which and because of their situation in which they acquire more or less resemblance and contrast with one another; and then this continuum is amplified in the continuum of possible feelings of quality, so the accidents of reaction, which are waking consciousnesses of pairs of qualities, may be expected to join themselves into a continuum. (137)

Secondness is now taken to actualize these quality possibilities based on an idea that any actual event involves a clash of qualities – in the ensuing argumentation Peirce underlines that the qualities involved in actualization need not be restrained to two but may be many, if they may only be 'dissolved' into pairs and hence do not break into the domain of Thirdness. This appearance of actuality, hence, has the property of singularities, spontaneously popping up in the space of possibilities and actualizing pairs of points in it:

Since, then an accidental reaction is a combination or bringing into special connection of two qualities, and since further it is accidental and antigeneral or discontinuous, such an accidental reaction ought to be regarded as an adventitious singularity of the continuum of possible quality, just as two points of a sheet of paper might come into contact. (137)

This transition from First to Second is conceived of along Aristotelian lines: as an actualization of a possibility – and this is expressed in the picture of a discontinuous singularity in the quality continuum. The topological fact that singularities must in general be defined with respect to the neighborhood of the manifold in which they appear, now becomes the argument for the fact that Secondness can never be completely discontinuous but still 'inherits' a certain small measure of continuity from the continuum of Firstness:

But although singularities are discontinuous, they may be continuous to a certain extent. Thus the sheet instead of touching itself in the union of two points may cut itself all along a line. Here there is a continuous line of singularity. In like manner, accidental reactions though they are breaches of generality may come to be generalized to a certain extent. (137)

Singularities, being discontinuous along certain dimensions, may be continuous in others, which provides the condition of possibility for Thirdness to exist as a tendency for Secondness to conform to a general law or regularity. As is evident, a completely pure Secondness is impossible in this continuous metaphysics – it remains a conceivable but unrealizable limit case, because a completely discontinuous event would amount to nothing. Thirdness already lies as a germ in the non-discontinuous aspects of the singularity. The occurrences of Secondness seem to be infinitesimal, then, rather than completely extensionless points.

The continuity of Thirdness is, in contrast to Firstness, real – the recipe does in fact refer to a continuum of apple pies which gradually, to greater or lesser extent, are actualized. We are now well into Peirce's metaphysics – while the quality continuum of Firstness is perfect, complete, but purely potential (and thus neither universal nor particular), then that of Thirdness is imperfect and fractioned, but real – realized via the swarm of Secondness actualizations, but still transgressing these in scope, as a finite number of events may never exhaust the universal law that governs them. From the reservoir of possibility in First, still larger doses of continuity seek through the slits of actual events in Secondness out into the growing reality of Thirdness – to sum up the three categories' relation to continuity in a metaphor which Peirce does not himself use. First is potential, Second actual, and Third real, and Peirce occasionally hints at the idea that the reality of Thirdness constitutes a consistent conception of a third Aristotelian notion: entelechy.[21]

VAGUENESS, DETERMINATENESS, GENERALITY

Continuity thus appears in two different forms in Firstness and Thirdness, respectively:

> Perhaps a more scientific pair of definitions would be that anything is general in so far as the principle of excluded middle does not apply to it and is vague in so far as the principle of contradiction does not apply to it. ("Issues of Pragmaticism", 1905, EPII, 351; 5.448)

Elsewhere, Peirce connects vagueness and generality to Firstness and Thirdness, respectively. In the former, continuity thus appears as *vagueness*, implying that a quality in the Firstness quality continua can never be identified with full exactitude and constitutes an infinitesimal variation of the quality dimension in question. In the latter, continuity appears as *generality* which is underspecified in comparison to its incarnations in particular, actual events. Peirce's brief definition of these two modes of indeterminateness by means of the logical principles of excluded middle (PEM) and contradiction (PC) requires some clarifications.

When claiming that the vagueness of Firstness does not follow the principle of contradiction, Peirce's idea does not refer to the standard PC in modern propositional logic (the principle that for a proposition p, not both p and non-p hold).[22] Thus he does not mean that propositions relating to Firstness in general can have several truth values (as would be the case when referring to the normal PC being false. Peirce's non-standard PC focuses on properties in predicate logic – it is the claim that for all properties P, no subject has both of the properties P and non-P (not both "S is P" and "S is non-P") – and it explicitly applies for definite subjects only. But firstnesses are possibilities and hence not such subjects. Their metaphysical status is thus in a typical Peircean move defined by the an ontological interpretation of logical principles. They are ontologically may-bes, and a may-be does not exclude the correlated may-not-be: whether any single subject instantiates a given may-be or not can not be decided on the basis of the may-be. The fact that PC does not apply thus refers to the modal character of the entities of Firstness, and its logical expression is that "S may be P" and "S may be non-P" may both be true – Peirce's own example is that "It may rain tomorrow" and "It may not rain tomorrow" are both true.

The generality of Thirdness, similarly, is claimed not to follow the principle of the excluded middle (or, the excluded Third). This should not be understood as referring to the modern standard PEM (that for a proposition p, either p or non-p hold). claiming that propositions referring to Thirdnesses (general necessities) admit a third truth-value such as would be the case if standard logical PEM does not apply. Peirce's non-standard PEM is the idea that for all properties P, any subject has either the property P or the property non-P (either "S is P" or "S is non-P"), and it explicitly applies to individuals only. Claiming that PEM does not apply to Thirdnesses thus refers to the ontological status of Thirdness would-bes – they are not such subjects as required by PEM. Here, the logical expression is that both "S must be P" and "S must be non-P" may both be false (in Peirce's example: "It must rain tomorrow" and "It must not rain tomorrow" may both be false.[23] And even if a specific would-be holds, the general failure of PEM to apply refers to the fact that the single objects of would-bes remain undetermined as to all aspects not covered by the would-be real possibility in question – they are thus objects with lots of indeterminate aspects.[24] The following exemplifies this idea:

The *general* may be defined as that to which the principle of excluded middle does not apply. A triangle in general is not isosceles nor equilateral; nor is a triangle in general scalene. The *vague* might be defined as that to which the principle of contradiction does not apply. For it is false neither that an animal (in a vague sense) is male, nor that an animal is female. (Pragmaticism, Prag. [4], 1905; 5.505)

In generals, further equally possible specifications are indeterminate (different kinds of triangles), while in vaguenesses, contradictory properties may appear (genderless or hermaphroditic animals).

The actual individual existence of Secondness, by contrast, may then be defined by its adherence to both of the principles of contradiction and excluded middle, because individuality taken as complete determination of all properties must obey both principles. The determinateness of Secondness thus forms the third member of the triad vagueness-determinateness-generality. A completely determinate individual must possess the property P or its contradictory non-P (PEM), but not both (PC):

Although the principles of contradiction and excluded middle may be regarded as together constituting the definition of the relation expressed by "not", yet they also imply that whatever exists consists of individuals. (Baldwin's Dictionary, 1911, 537–38; 3.612)

Peirce hesitates, though, to accept full determinacy of all properties as a definition of the existence mode of actual, individual objects or events. His continuist reason for hesitating is that this would make the relation between the possibilities of Firstness and its subsets, the real possibilities of Thirdness, on the one hand, and individual Secondness, on the other, too discontinuous and insurmountable – and he points to the large and vain efforts of the Scholastics to explain the relation between general concepts and individual objects in terms of "contraction" etc. Vagueness and generality are supposed to inhere also in individual existence, albeit only infinitesimally in contrast to the two continuous possibility categories. This is why an alternative definition by *reactivity* is proposed for individuality:

> Another definition which avoids the above difficulties is that an individual is something which reacts. That is to say, it does react against some things, and is of such a nature that it might react, or have reacted, against my will. (Baldwin's Dictionary, 1911, 537–38; 3.612)

Full adherence to both PC and PEM, according to this idea, is taken to pertain to the ideal of fully determinate objects or events only, an ideal which actual existing individuals can never completely satisfy. Individuals may be singled out on all levels of reality and discourse without the ontological requirement of full determinateness, if reactivity is what grants their basic haecceity (thisness, see next chapter).

Peirce also describes the difference between the two indeterminacies, vague continuity and general continuity, in semiotic terms, based upon a dialogical notion of semiotics nowadays identified as a version of game-theoretical semantics (Pietarinen 2006). In vague signs, the possible further selection and specification of the object is left to the same source as uttered the first sign (The fortune-teller: "I see a great event..." "What is it?" "Some tall, dark man..."). In generality, the possible further selection and specification of the object is handed over to the dialogue partner (The logician: "Any man is mortal" "Which man?" "Any man you like!").[25] The normal situation being that the utterer commands the scope of signification of signs, general signs form a special, restricted subset of vague signs – a special subset where the utterer permits the dialogue partner to select examples (real generals supposedly not admitting (but few) counterexamples).

While the vagueness of Firstness has its metaphysical expression in the reality of (relative) indeterminism, the generality of Thirdness has its metaphysical expression in the real existence of laws and tendencies. The relation between the three categories and the two principles in Peirce's version may be summed up as follows (with the proviso that Secondness as defined by adherence to both principles forms an ideal limit case only).

		Principle of Contradiction	Principle of Excluded Middle
1	vagueness, possibility	–	+
2	determinateness actuality	+	+
3	generality necessity	+	–

Metaphysically speaking, the continuities of Firstness and Thirdness thus refer to the real existence of indeterminateness and laws, respectively, in addition to the reactive actuality of Secondness. Logically speaking, Peirce initiates the development of logic formalisms covering different indeterminate cases, such as modal logic, intuitionistic logic and fuzzy logic[26] – in his unfinished Gamma graphs he attempted to extend his logic representation systems to cover such cases.

This presentation of the role of continuity in Peirce's phenomenology has until now been merely reconstruction of his position and does not in any way present an argument for its validity. What is the *motivation* for this crucial role of continuity in his category doctrine? In the case of Firstness, continuity serves the aim of describing the infinite density of *possible* predicates: phenomenologically spoken it corresponds to the experience of the infinite and continuous variability of qualia. Furthermore, the 'composition' of the continuum from parts, infinitesimals, of which the single infinitesimal is but vague and escapes the law of contradiction, serves to underline a conception of the possible quality itself as vague, because only the discontinuity of Secondness makes the precise quality evident. But it is especially the rule of continuity as central to Thirdness which is controversial. Continuity here serves the phenomenological purpose of accounting for experienced, realistically conceived *regularities* in the phenomenon: the fact that processes are more or less rule-bound. Continuity in Thirdness thus supports the idea of 'real possibilities' or 'would-be's which Peirce introduces in his mature theory from around 1896–97.[27] Whether this is a necessary implication of the identification of Thirdness with real continuity is, however, an open question to which we shall return.

An even more basic critical question will ask for the legitimacy of the three categories. In our days, the legitimacy of Secondness and its ontology of particular events will probably give rise to least controversy; the actually existing world of particular entities appears to most metaphysicians as beyond any reasonable doubt. The fact that this actual world realizes certain possibilities we may conceive makes some version of Firstness easily digestible if bracketing Peirce's further metaphysical implications such as the continuity of all possible qualities. Thirdness is obviously the most problematic category, gathering in one bundle a whole set of philosophical issues in one grand solution proposal:

Real possibilities; reality of tendencies, relations, and patterns; rule-following; the iconic structure of propositions; intensional meanings

An obvious conclusion might seem the positivist reaction: regularities found in certain processes do not have any further substance and do not require their own basic phenomenological category – they are in the last account reducible to psychological or logical organizations of what is empirically given in atomist data. Such explanations, however, loosens a psychological or logical level from the phenomenon itself – in the former case with the result that phenomenology is re-psychologized and based on a preliminary subject-object distinction with all the well-known lamentable consequences – in the latter case with the result that the validity and role of logic remains unexplained, with the result that it may be conceived of as mere formal tautologies whose contribution to the organization of experience seems trivial. From a logical point of view, this argument against Thirdness (and its ensuing universalist realism) targets Peirce's so-called reduction thesis claiming that all many-sided relations may be reduced to a combination of three-sided relations, but that one- and two-sided relations may not, on the other hand, combine to form genuine three-sided relations which consequently gets a fundamental role.

THE THREE CATEGORIES IN DIAGRAMS

We already commented upon the continuous sheet as basic for diagrams. The three categories also, however, inform the signs further used as constituents of diagrams. Let us as an example take Peirce's 'existential graphs' of which the Alpha version charts propositional logic, the Beta part first order predicate logic, and the unfinished Gamma part different aspects of modal logic, temporal logic, speech act logic, etc. In the Beta Graphs, the continuous sheet admits the following sign types: predicates (typically in verbal form) may be directly written on the sheet. They form propositions by being attached to subjects which are indicated by identity lines, at the other end(s) of which an index for the subject involved may appear. The end of an identity line merely states that such a subject exists, and the line may branch and connect to various predicates. Predicates having up to three places may also connect to other identity lines. Propositions occurring side by side on the sheet are linked with a conjunction. The last sign type is the cut, a connecting line severing part of the sheet from the rest. The diagram parts inside that cut are negated.[28] Two cuts, one appearing inside the other, are called a Scroll which forms a material implication. Certain rules govern the graphs: it is allowed to write a true proposition everywhere on the blank sheet or in evenly enclosed cuts, while it is allowed to add any proposition in unevenly enclosed cuts. Graphs not transcending cuts may be iterated or deleted. Double cuts with nothing in between may be deleted or inserted.[29] The basics of the Beta Graphs are most easily understood from an example ('Prolegomena to an Apology for Pragmaticism', 1906, 4.569):

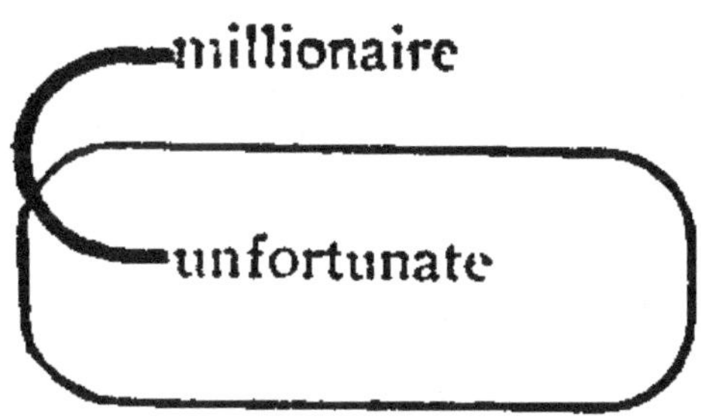

Figure 6.

This graph says that there exists a millionaire who is not unfortunate. The outermost end of the identity line functions as a quantifier claiming the existence of an object, the label identifies it as a millionaire. The innermost end of the identity line connects this subject to the predicate 'unfortunate', while the cut denies that predication. What interests us here is Peirce's category motivation for the inventory of signs used in the Graphs. In his 1898 lectures, not long after the invention of the Graphs, he says:

In the system of graphs may be remarked three kinds of signs of very different natures. First, there are the verbs, of endless variety. Among these is the line signifying identity. But second the ends of the line of identity (and *every* verb ought to [be] conceived as having such loose ends) are signs of a totally different kind. They are demonstrative pronouns, indicating existing objects, not necessarily material things, for they may be *events*, or even *qualities*, but still objects, merely designated as *this* or *that*. In the third place, the writing of verbs side by side, and the ovals enclosing graphs not asserted but subjects of assertion, which last is continually used in mathematics and makes one of the great difficulties of mathematics, constitute a third, entirely different kind of sign. Signs of the first kind represent objects in their Firstness, and give the significations of the terms. Signs of the second kind represent objects as existing, – and therefore as reacting, – and also in their reactions. They contribute the *assertive* character to the graph. Signs of the third kind represent objects as representative, that is in their Thirdness, and upon them turn all the inferential processes. In point of fact, it was considerations about the categories which taught me how to construct the system of graphs. ('Detached Ideas Continued', NEM IV, 339)

The three categories here permits Peirce to distinguish signs (1) referring to qualities and verbs, that is, polyvalent predicates, (2) referring to subjects related to those predicates, and (3) referring to logical relations like the operators of conjunction and negation, and, more implicitly, quantifiers which appear in the system due to the system of cuts modifying a basic existential quantification of the sheet. The three sign types given here relate to the term-proposition-argument triad in the way that the former are simply terms or rhemes, while the two latter are what may be added to rhemes in order to constitute propositions and arguments, respectively. We should thus expect these sign types to appear as typical instruments of different diagrams, representing qualities, subjects, and the relations between those, respectively.

To Peirce, it was an ideal to use signs representing these phenomena as iconically as possible.[30] He thus saw his logic graphs with their continuous depiction of continuous states-of-affairs as superior to his own earlier algebra of logic (which through Schröder and Peano developed into actual symbolic logic notation): all diagrams are based on the continuum, but still, the more continuous a diagram, the better.

CHAPTER 2

THE PHYSIOLOGY OF ARGUMENTS – PEIRCE'S EXTREME REALISM

The Continuum in Peirce's Theory of Signs

Peirce's mature theory of signs consists of a long series of argued distinctions between types of signs (or rather, between types of *aspects* of signs) – based on the three phenomenological categories, in turn described with reference to continuity as presented in Chap. 1. In this chapter, we shall discuss the realist bases of Peirce's sign theory with continuous reference to continuity.

Let us begin with the definition of the sign. A variety of hundreds of sign definitions are disseminated all over Peirce's work, but a certain agreement among Peirce scholars seem to focus on a few of them, among those the following: 'A *Sign*, or *Representamen*, is a First which stands in such a genuine triadic relation to a Second, called its *Object*, as to be capable of determining a Third, called its *Interpretant*, to assume the same triadic relation to its Object in which it stands itself to the same Object' ('Syllabus', 1902, EPII, 272–73; 2.274).

This sign definition from 1902 involves explicit reference to the three phenomenological categories. The sign relates a quality (1) to an actual object (2) and, by doing so, involves a more or less general regularity (3). This regularity now stands in the same relation to the object as did the initial First, that is, it is to the same degree a sign of the object. Thus, Peirce's sign definition is recursive; the fact that the interpretant is in itself a sign implies that it has its own interpretant and so on ad infinitum. In this recursivity of the sign definition lies Peirce's famous 'unlimited semiosis'. It is not determined *how much more* general the interpretant is than the initial representamen (which NB qua First is in itself general). The sign may so to speak 'run idle' when the interpretant does not express any more than the representamen; this will be the limit case of 'perfect communication' where the intended meaning is decoded identically. Both misunderstandings and more-understandings, so to speak, let the interpretant deviate from the representamen, and it is thus no necessary implication of the sign definition itself that unlimited semiosis converges towards the perfect representation of the object (this idea is added, to be sure, in Peirce's philosophy of science). Given Peirce's sign definition, it is perfectly possible to imagine unlimited semiosis circling around the object without getting neither closer nor farther away (something like what deconstruction has made out of unlimited semiosis), or, indeed, a pessimistic philosophy of culture where semiosis would gradually dwindle farther and farther away from the object and tend asymptotically towards pure nonsense; a sort of Heideggerian Peirce-interpretation could lead to such a somber *Seinsgeschichte*. Let us thus – even if the positive asymptote of knowledge indeed has certain connections to continuity – bracket Peirce's reasons for assuming it until further notice. Another very important and often highlighted property in the sign definition is its lack of reference to (human) subjectivity. Of course, it is possible to find expressions like 'mind', or '... Thirdness is the triadic relation existing between a sign, its object and the interpreting *thought*, itself a sign...' (Letter to Lady Welby, Oct 12, 1904, 8.332, our italics), but they refer to the real existence of ideas in Peirce's objective idealism, never to any necessary role of active subjectivity only.[31] The irreducible connection of the sign to the three phenomenological categories may, according to the definition mentioned, take place without any subjective or conscious intermediary – the presence in the real world of regularities *are*, in fact, according to this description already signs – because they connect objects with properties in a stable and general way. Natural regularities – 'habits' in Peirce's lingo – determine that the world is no everchanging, ungraspable chaos, and they are thus signs, stably connecting objects and properties. This consequence lies – given phaneroscopy – at the surface of Peirce's general sign definitions and becomes the condition of possibility for Peirce's 'extreme realism': the possibility that general signs created by cognizing beings (e.g., humans) may correspond to naturally occurring general signs.[32]

Peirce's logic is as mentioned triadic, and its first branch, *grammatica speculativa*, takes the form of distinctions between sign types, while the second formalizes inference structures and the third reflects upon the use of them, to put it briefly. The *argument* is thus the fundamental type of sign, it is the generic, full-blown sign, and it is defined as '... a sign which separately signifies its interpretant.' (Carnegie Application, 1902, NEM IV, 22 (1902)). The argument makes explicit – in contrast to more primitive signs – how it should be interpreted, this interpretation being normally known as its conclusion. As Peirce's definition of signs lets signs appear in nature and mind alike, there must, consequently, also exist arguments in nature: 'We all think of nature as syllogizing. Even the mechanical philosopher, who is as nominalistic as a scientific man can be, does that.' (1898, RLOT, 161) What we are supposed to understand by this seemingly radical claim, that nature not only involves processes structurally analogous to inferences, but also makes their conclusions explicit, is not immediately evident.[33] We must suppose that explicit arguments may, in some cases, mirror natural processes but without these latter having the explicit, self-controlled character of the former. Such claims belong, of course, to philosophy of nature and thus under the more special science of cosmology (see the discussion of 'real possibilities' later in this chapter) – but it is important to emphasize that the conditions of possibility for the radical claims of the latter lies already in the more general definition of signs. Whether cosmology might be better interpreted in another way is not immediately a problem for Peirce's semiotics.

Thus continuity in the form of generality is a direct consequence of Peirce's definition of signs: complicated signs imply the assumption of a continuum of possible objects as their reference as the foundation of meaning as such. The fact that the argument is the basic sign in Peirce does by no means imply that it is primitive or without internal parts; it is rather the fully developed sign, and all of Peirce's semiotics may, in fact, be seen as a physiology of arguments which investigates how the argument is composed of more primitive signs and sign aspects. But these simpler signs are not, according to Peirce, fundamental, true to his definition of sciences by their causa finalis, it is the definition of logic from the exploitation of its truth-preserving properties which makes him privilege the argument, because it is the kind of sign facilitating explicit truth-preservation. The more primitive sign types are thus to be seen as 'degenerate' arguments,[34] arguments not completed, sign types dependent on their possible role in argumentation. In presentations of Peirce it is more the rule than the exception that these radically teleological aspects of the sign are understated, so the presentation tends to suppose that Peirce sees the argument as composed 'from below' of independent primitives. But with our knowledge of the role of continuity in Peirce we can grasp how Peirce rejects the idea of building anything up from below of independent, discontinuous building blocks – in this respect, he is a sort of holist, because the simpler sign types are to be grasped as parts of the argument and the unlimited semiosis' growth towards truth. In the following presentation we shall bracket this teleology beyond the argument level with the deliberate purpose to let the question of any cosmological striving towards truth and generality remain open. We will not however, assume the opposite either, amounting to something like a sort of neo-Darwinist Peirce-interpretation taking the argument to be the result of an arbitrary concatenation of simpler signs subject to selection. We will envisage the argument and its parts as a physiologist taking the body of the argument as the upper limit of investigation until further notice.

TOKEN AND TYPE

Peirce's First Trichotomy[35]

As the sign consists of three components it comes hardly as a surprise that it may be analyzed in nine aspects – every one of the sign's three components may be viewed under each of the three fundamental phenomenological categories.[36] The least discussed of these so-called trichotomies is probably the first, concerning which property in the sign it is that functions, in fact, to make it a sign. It gives rise to the trichotomy *qualisign, sinsign, legisign*, or, in a little more sexy terminology, *tone, token, type*.[37] The latter will be preferred here, also because it makes evident the heritage philosophy has taken from it – the dichotomy token/type which stems from Peirce's terminology without its tone basis is made explicit.

The oftenmost quoted definition is from 'Syllabus' (1903, EPII, 291; 2.244):

> According to the first division, a Sign may be termed a *Qualisign*, a *Sinsign*, or a *Legisign*.
>
> A *Qualisign* is a quality which is a Sign. It cannot actually act as a sign until it is embodied; but the embodiment has nothing to do with its character as a sign.
>
> A *Sinsign* (where the syllable *sin* is taken as meaning 'being only once', as in *single, simple*, Latin *semel*, etc.) is an actual existent thing or event which is a sign. It can only be so through its qualities; so that it involves a qualisign, or rather, several qualisigns. But these qualisigns are of a peculiar kind and only form a sign through being actually embodied.

> A *Legisign* is a law that is a Sign. This law is usually [sic] established by men. Every conventional sign is a legisign. It is not a single object, but a general type which, it has been agreed, shall be significant. Every legisign signifies through an instance of its application, which may be termed a *Replica* of it. Thus, the word 'the' will usually occur from fifteen to twenty-five times on a page. It is in all these occurrences one and the same word, the same legisign. Each single instance of it is a Replica. The Replica is a Sinsign. Thus, every Legisign requires Sinsigns. But these are not ordinary Sinsigns, such as are peculiar occurrences that are regarded as significant. Nor would the Replica be significant if it were not for the law which renders it so.

In some sense, it is a strange fact that this first and basic trichotomy has not been widely discussed in relation to the continuity concept in Peirce, because it is crucial. It is evident from the second noticeable locus where this trichotomy is discussed, the letters to Lady Welby – here Peirce continues (after an introduction which brings less news):

> The difference between a legisign and a qualisign, neither of which is an individual thing, is that a legisign has a definite identity, though usually admitting a great variety of appearances. Thus, &, *and*, and the sound are all one word. The qualisign, on the other hand, has no identity. It is the mere quality of an appearance and is not exactly the same throughout a second. Instead of identity, it has *great similarity*, and cannot differ much without being called quite another qualisign.
> (Letter to Lady Welby, Oct 12, 1904, 8.334)

The legisign or type is distinguished as being general which is, in turn, defined by continuity: the type has a 'great variety of appearances'; as a matter of fact, a continuous variation of appearances. In many cases even *several* continua of appearances (as *&, and*, and the spoken sound of 'and'). Each continuity of appearances is gathered into one identity thanks to the type, making possible the repetition of identical signs. Reference is not yet discussed (it concerns the sign's relation to its object), nor is meaning (referring to its relation to its interpretant) – what is at stake is merely the possibility for a type to incarnate a continuum of possible actualizations, however this be possible, and so repeatedly appear as one and the same sign despite other differences. Thus the reality of the type is the very foundation for Peirce's 'extreme realism', and this for two reasons. First, seen from the side of the sign, the type provides the possibility of stable, repeatable signs: the type may – opposed to qualisigns and those sinsigns not being replicas of a type – be repeated as a self-identical occurrence, and this is what in the first place provides the stability which renders repeated sign use possible. Second, seen from the side of reality: because types, legisigns, are realized without reference to human subjectivity, the existence of types is the condition of possibility for a sign, in turn, to stably refer to stably occurring entities and objects. Here, the importance of the irreducible continuity in philosophy of mathematics appears for semiotics: it is that which grants the possibility of collecting a continuum in one identity, the special characteristic of the type concept.[38] The opposition to the type is the qualisign or tone lacking the stability of the type – they are not self-identical even through a second, as Peirce says – they have, of course, the character of being infinitesimal entities, about which the principle of contradiction does not hold. The transformation from tone to type is thus the transformation from unstable pre-logic to stable logic – it covers, to phrase it in a Husserlian way, the phenomenology of logic. The legisign thus exerts its law over specific qualisigns and sinsigns – like in all Peirce's trichotomies the higher sign types contain and govern specific instances of the lower types. The legisign is incarnated in singular, actual sinsigns representing the type – they are tokens of the type – and what they have in common are certain sets of qualities or qualisigns – tones – selected from continua delimited by the legisign. The amount of possible sinsigns, tokens, are summed up by a type, a stable and self-identical sign. Peirce's despised nominalists would to some degree agree here: the universal as a type *is* indeed a 'mere word' – but the strong counterargument which Peirce's position makes possible says that if 'mere words' may possess universality, then the world must contain it as well, because words are worldly phenomena like everything else.[39] Here, nominalists will typically exclude words from the world and make them privileges of the subject, but for Peirce's welding of idealism and naturalism nothing can be truly separated from the world – all what basically is in the mind must also exist in the world. Thus the synthetical continuum, which may, in some respects, be treated as one entity, becomes the very condition of possibility for the existence of types.[40]

Whether some types or legisigns now refer to existing general objects or not is not a matter for the first trichotomy to decide; legisigns may be part of any number of false or nonsensical propositions, and not all legisigns are symbols, just like arguments, in turn, are only a subset of symbols – but all of them are legisigns because they must in themselves be general in order to provide the condition of possibility of identical repetition, of reference to general objects and of signifying general interpretants.

A CONTINUUM OF SIMILARITY

Peirce's Second Trichotomy

The second trichotomy is probably the most well-known piece of Peirce's semiotics: it distinguishes three possible relations between the sign and its (dynamical) object.[41] This relation may be motivated by similarity, by actual connection, or by general habit – giving rise to the sign classes *icon, index, and symbol*, respectively. The implications of continuity for these sign classes are particularly vast. Let us as an introduction take a look at the classical locus of description of these classes:

> According to the second trichotomy, a Sign may be termed an Icon, an Index, or a Symbol.
>
> An Icon is a sign which refers to the Obect that it denotes merely by virtue of characters of its own, and which it possesses, just the same, whether any such Object actually exists or not. It is true that unless there really is such an Object, the Icon does not act as a sign; but this has nothing to do with its character as a sign. Anything whatever, be it quality, existent individual, or law, is an Icon of anything, in so far as it is like that thing and used as a sign of it.
>
> An Index is a sign which refers to the Object that it denotes by virtue of being really affected by that Object. It cannot, therefore, be a Qualisign, because qualities are whatever they are independently of anything else. In so far as the Index is affected by the Object, it necessarily has some Quality in common with the Object, and it is in respect to these that it refers to the Object. It does, therefore, involve a sort of Icon, although an Icon of a peculiar kind; and it is not the mere resemblance of its Object, even in these respects which makes it a sign, but it is the actual modification of it by the Object.

> A Symbol is a sign which refers to the Object that it denotes by virtue of a law, usually an association of general ideas, which operates to cause the Symbol to be interpreted as referring to that Object. It is thus itself a general type or law, that is, a Legisign. As such it acts through a Replica. Not only is it general in itself, but the Object to which it refers is of general nature. Now that which is general has its being in the instances it will determine. There must, therefore, be existent instances of what the Symbol denotes, although we must here understand by 'existent', existent in the possibly imaginary universe to which the Symbol refers. The Symbol will indirectly, through the association or other law, be affected by those instances; and thus the Symbol will involve a sort of Index, although an Index of a peculiar kind. It will not, however, be by any means true that the slight effect upon the Symbol of those instances accounts for the significant character of the Symbol. ('Syllabus', 1903, EPII, 291–2; 2.247)

Let us begin with the icon. As it appears from the definition, the icon refers to its object solely by means of its own properties. This implies that an icon potentially refers to an indefinite class of objects, namely all those objects which have, in some respect, a relation of similarity to it. In recent semiotics, it has often been remarked (Nelson Goodman, cf. below) that any phenomenon can be said to be like any other phenomenon in some respect, if the criterion of similarity is chosen sufficiently general, just like the establishment of any convention immediately implies a similarity relation. If Nelson Goodman picks out two otherwise very different objects, then they are immediately similar to the extent that they now have the same relation to Nelson Goodman. Goodman and others have for this reason deemed the similarity relation insignificant – and consequently put the whole burden of semiotics on the shoulders of conventional signs only. But the counterargument against this rejection of the relevance of the icon lies close at hand. Given a *tertium comparationis*, a measuring stick, it is no longer possible to make anything be like anything else. This lies in Peirce's observation that 'It is true that unless there really is such an Object, the Icon does not act as a sign ...' The icon only functions as a sign to the extent that it is, in fact, used to refer to some object – and when it does that, some criterion for similarity, a measuring stick (or, at least, a delimited bundle of possible measuring sticks) are given in and with the comparison. In the quote just given, it is of course the immediate object Peirce refers to – it is no claim that there should in fact *exist* such an object as the icon refers to. Goodman and others are of course right in claiming that as 'Anything whatever (...) is an Icon of anything ...', then the universe is pervaded by a continuum of possible similarity relations back and forth, but as soon as some phenomenon is in fact used as an icon for an object, then a specific bundle of similarity relations are picked out: '... in so far as it is like that thing.'

Just like the qualisign, the icon is a limit category. 'A possibility alone is an Icon purely by virtue of its quality; and its object can only be a Firstness.' (from 'Syllabus', c. 1902, EPII, 273; 2.276). Strictly speaking, a pure icon may only refer one possible Firstness to another. The pure icon would be an identity relation between possibilities. Consequently, the icon must, as soon as it functions as a sign, be more than iconic. The icon is typically an aspect of a more complicated sign, even if very often a most important aspect, because providing the predicative aspect of that sign. This Peirce records by his notion of 'hypoicon': 'But a sign may be *iconic*, that is, may represent its object mainly by its similarity, no matter what its mode of being. If a substantive is wanted, an iconic representamen may be termed a *hypoicon*' (ibid.). Hypoicons are signs which to a large extent makes use of iconical means as meaning-givers: images, paintings, photos, diagrams, etc. But the iconic meaning realized in hypoicons have an immensely fundamental role in Peirce's semiotics. As icons are the only signs that look-like, then they are at the same time the only signs realizing meaning. Thus any higher sign, index and symbol alike, must contain, or, by association or inference terminate in, an icon. If a symbol can not give an iconic interpretant as a result, it is empty. In that respect, Peirce's doctrine parallels that of Husserl where merely signitive acts require fulfillment by intuitive ('anschauliche') acts. As we shall consider later, this is actually Peirce's continuation of Kant's famous claim that intuitions without concepts are blind, while concepts without intuitions are empty. When Peirce observes that 'With the exception of knowledge, in the present instant, of the contents of consciousness in that instant (the existence of which knowledge is open to doubt) all our thought and knowledge is by signs' (Letter to Lady Welby, Oct 12, 1904, 8.332), then these signs necessarily involve iconic components. Peirce has often been attacked for his tendency towards a pan-semiotism which lets all mental and physical processes take place via signs – in the quote just given he, analogous to Husserl, claims there must be a basic evidence anterior to the sign – just like Husserl this evidence before the sign must be based on a 'metaphysics of presence' – the 'present instant' provides what is not yet mediated by signs. But icons provide the connection of signs, logic and science to this foundation for Peirce's phenomenology: the icon is the only sign providing evidence (PAP, 1906, in NEM IV, 317). The icon is, through its timeless similarity, apt to communicate aspects of an experience 'in the present instant'. Thus, the typical index contains an icon (more or less elaborated, it is true): the footstep on the beach as the prototypical index is an evident example of this: we are able to recognize it as a footprint only because it looks like a foot, because it is an icon of a foot. And thus any symbol intends an iconic interpretant.

Continuity is at stake in relation to the icon to the extent that the icon, while not in itself general, is the bearer of a potential generality. The footprint on the beach refers (potentially) iconically to all feet of approximately (give and take a certain margin dependent on the granularity and wetness of the sand) this shape and size, as well as to all artificial feet, etc. of the same shape and size – that is, in turn, to a continuum of possible feet. This infinitesimal generality is decisive for the higher sign types' possibility to give rise to thought: the symbol thus contains a bundle of general icons defining its meaning. A special icon providing the condition of possibility for general and rigorous thought is, of course, the diagram.

The index connects the sign directly with its object via connection in space and time; as an actual sign connected to its object, the index is turned towards the past: the action which has left the index as a mark must be located in time earlier than the sign, so that the index presupposes, at least, the continuity of time and space without which an index might occur spontaneously and without any connection to a preceding action. Maybe surprisingly, in the Peircean doctrine, the index falls in two subtypes: designators vs. reagents. Reagents are the simplest – here the sign is caused by its object in one way or another: the footprint on the beach or smoke signifying fire as arch examples. Designators, on the other hand, are more complex: the index finger as pointing to an object or the demonstrative pronoun as the subject of a proposition are prototypical examples. Here, the index presupposes an intention – the will to point out the object for some receiver. Designators, it must be argued, presuppose reagents: it is only possible to designate an object if you have already been in reagent contact (simulated or not) with it (this forming the rational kernel of causal reference theories of meaning). The closer determination of the object of an index, however, invariably involves selection on the background of continuities: did I mean the object as it is today, this year, this century; did I mean smoke as a sign of fire or as a sign of a pyroman or as a sign of the pyroman's sad childhood? Even the determination of the object depends on the local delimitation of a continuum.[42]

On the level of the symbol, continuity and generality play a main role – as always when approaching issues defined by Thirdness. The symbol is, as we noted, in itself a legisign, that is, it is a general object which exists only due to its actual instantiations. The symbol itself is a real and general recipe for the production of similar instantiations in the future. But apart from thus being a legisign, it is connected to its object thanks to a habit, or regularity. Sometimes, this is taken to mean 'due to a convention' – in an attempt to distinguish conventional as opposed to motivated sign types. This, however, rests on a misunderstanding of Peirce's doctrine in which the trichotomies record *aspects* of sign, not mutually exclusive, independent classes of signs: symbols and icons do not form opposed, autonomous sign classes; rather, the content of the symbol is constructed from indices and general icons. The habit realized by a symbol connects it, as a legisign, to an object which is also general – an object which just like the symbol itself exists in instantiations, be they real or imagined. The symbol is thus a connection between two general objects, each of them being actualized through replicas, tokens – a connection between two continua, that is:

Definition 1. Any *Blank* is a symbol which could not be vaguer than it is (although it may be so connected with a definite symbol as to form with it, a part of another partially definite symbol), yet which has a purpose.
Axiom 1. It is the nature of every symbol to blank in part. [...]
Definition 2. Any *Sheet* would be that element of an entire symbol which is the subject of whatever definiteness it may have, and any such element of an entire symbol would be a Sheet.
('Sketch of Dichotomic Mathematics' (c. 1903?), NEM IV, 292)

The symbol's generality can be described as it having always blanks having the character of being indefinite parts of its continuous sheet. Thus, the continuity of its blank parts is what grants its generality. The symbol determines its object according to some rule, granting the object satisfies that rule – but leaving the object indeterminate in all other respects.

The connection between symbol and object can take place 'by association or other law', that is, it may have natural as well as human causes. When we call a horse 'a horse', to take a semiotic arch example, the coupling between these two continua (a continuum of different pronunciations or written instances of the signifier 'horse', and a continuum of hoofed animals of a certain description), is a purely human convention. The very fact, however, that a real continuum of horses exist – however difficult it may be to describe, define, or delimit it – which is possible to connect to a continuum of spoken sounds or written letters, is not created by the word.

It is tempting to take the typical symbol to be a word, but it should rather be taken as the argument – the predicate and the proposition being degenerate versions of arguments with further continuous blanks inserted by erasure, so to speak, forming the third trichotomy of term, proposition, argument.

THE GENERALITY OF PREDICATES

Peirce's Third Trichotomy

The decisive logical role is played by continuity in the third trichotomy which is Peirce's generalization of the old distinction between term, proposition and argument in logic. In him, the technical notions are rhema, dicent and argument,[43] and all of them may be represented by symbols. A crucial step in Peirce's logic of relations (parallel to Frege) is the extension of the predicate from having only one possible subject in a proposition – to the possibility for a predicate to take potentially infinitely many subjects. Predicates so complicated may be reduced, however, to combination of (at most) three-subject predicates, according to Peirce's reduction hypothesis. Let us consider the definitions from 'Syllabus' in continuation of the earlier trichotomies:

According to the third trichotomy, a Sign may be termed a *Rheme*, a *Dicisign* or *Dicent Sign* (that is, a proposition or quasi-proposition), or an *Argument.*

A *Rheme* is a Sign which, for its Interpretant, is a Sign of qualitative possibility, that is, is understood as representing such and such a kind of possible Object. Any Rheme, perhaps, will afford some information; but it is not interpreted as doing so.

A *Dicent Sign* is a Sign, which, for its Interpretant, is a Sign of actual existence. It cannot, therefore, be an Icon, which affords no ground for an interpretation of it as referring to actual existence. A Dicisign necessarily involves, as a part of it, a Rheme, to describe the fact which it is interpreted as indicating. But this is a peculiar kind of Rheme; and while it is essential to the Dicisign, it by no means constitutes it.

An *Argument* is a Sign which, for its Interpretant, is a Sign of a law. Or we may say that a Rheme is a sign which is understood to represent its object in its characters merely; that a Dicisign is a sign which is understood to represent its object in respect to actual existence; and that an Argument is a Sign which is understood to represent its Object in its character as Sign. (...) The proposition need not be asserted or judged. It may be contemplated as a sign capable of being asserted or denied. This sign itself retains its full meaning whether it be actually asserted or not. (...) The proposition professes to be really affected by the actual existent or real law to which it refers. The argument makes the same pretension, but that is not the principal pretension of the argument. The rheme makes no such pretension.

The interpretant of the Argument represents it as an instance of a general class of Arguments, which class on the whole will always tend to the truth. It is this law, in some shape, which the argument urges; and this 'urging' is the mode of representation proper to Arguments.
('Syllabus', 1903, EPII, 292–93; 2.250)

Predicates being general is of course a standard logical notion; in Peirce's version this generality is further emphasized by the fact that the simple predicate is seen as relational and containing up to three subject slots to be filled in; each of them may be occupied by a continuum of possible subjects. The predicate itself refers to a possible property, a possible relation between subjects; the empty – or partly satiated – predicate does not in itself constitute any claim that this relation does in fact hold. The information it contains is potential, because no single or general indication has yet been chosen to indicate which subjects among the continuum of possible subjects it refers to. The proposition, on the contrary, the dicisign, is a predicate where some of the empty slots have been filled in with indices (proper names, demonstrative pronomina, deixis, gesture, etc.), and is, in fact, asserted. It thus consists of an indexical part and an iconical part, corresponding to the usual distinction between subject and predicate, with its indexical part connecting it to some level of reference reality.[44] This reality needs not, of course, be actual reality; the subject slots may be filled in with general subjects thus importing pieces of continuity into it – but the reality status of such subjects may vary, so it may equally be filled in with fictitious references of all sorts. Even if the dicisign, the proposition, is not an icon, it contains, via its rhematic core, iconical properties. Elsewhere, Peirce simply defines the dicisign as a sign making explicit its reference.[45] Thus a portrait equipped with a sign indicating the portraitee will be a dicisign, just like a charicature draft with a pointing gesture towards the person it depicts will be a dicisign. Even such dicisigns may be general; the pointing gesture could single out a group or a representative for a whole class of objects. While the dicisign specifies its object, the argument is a sign specifying its interpretant – which is what is normally called the conclusion. The argument thus consists of two dicisigns, a premiss (which may be, in turn, composed of several dicisigns and is traditionally seen as consisting of two dicisigns) and a conclusion – a dicisign represented as ensuing from the premiss due to the power of some law. The argument is thus – just like the other thirdness signs in the trichotomies – in itself general. It is a legisign and a symbol – but adds to them the explicit specification of a general, lawlike interpretant. In the full-blown sign, the argument, the more primitive degenerate sign types are orchestrated together in a threefold generality where no less than three continua are evoked: first, the argument itself is a legisign with a halo of possible instantions of itself as a sign; second, it is a symbol referring to a general object, in turn with a halo of possible instantiations around it; third, the argument implies a general law which is represented by one instantiation (the premiss and the rule of inference) but which has a halo of other, related inferences as possible instantiations.[46] As Peirce says, the argument persuades us that this lawlike connection holds for all other cases being of the same type.

Scholastic Realism and Real Possibilities

Continuity is thus in play in all corners of Peirce's sign doctrine where generality plays any role. This geometrical metaphor is thus constitutive for the relation between sign theory and metaphysics: general signs are general due to the real existence of continua which may never be exhausted by any number of actual events or objects. The existence of such real continua is decisive for the 'scholastic realism' of the mature Peirce after around 1896–97.

'I am myself a scholastic realist of a somewhat extreme stripe' (5.470) says Peirce, not only in 'A Survey of Pragmaticism' from 1906; often he declares himself a radicalizing heir to especially the realist position of John Duns Scotus. In a very early version (1868), he sums up the position as follows:

There is a real difference between man irrespective of what the other determinations may be, and man with this or that particular series of determinations, although undoubtedly this difference is only relative to the mind and not *in re*. Such is the position of Scotus. Occam's great objection is, there can be no real distinction which is not *in re*, in the thing-in-itself; but this begs the question for it is itself based only on the notion that reality is something independent of representative relation.
('Some Consequences of Four Incapacities', EPI, 53; 5.312)

As we shall see, Peirce develops his view in between these two quotes, but already in this ultra-short resume of the medieval fight over universals it is evident why Peirce must chose the side of Scotus against William of Ockham. If you claim that the distinction between universals – the human being – and single objects – one human being – is a logical distinction only, which are constructed by us, then no 'real generals' exist, and Thirdness evaporates as metaphysical-phenomenological category to leave only psychology of categorization behind. The opposite point of view is of course not tenable either, the point of view against which Ockham directs his critique – the idea that this distinction should be *in re*, interpreted as claiming that generals should possess the same mode of being as the existing particulars (which both sides agreed were the only existing objects). This would imply that universals were things on a par with particulars, and Scotus perfectly realizes this position is untenable: the distinction is 'relative to the mind', as Peirce says – but it is *still* a real difference, as he adds. This is Peirce's version of the Subtle Doctor's famous 'formal distinction' which teaches that the distinction between universals and particulars is *neither* real taken in the sense that universals exist like particulars *nor* merely logical in the sense that they are only man-made *entia rationis*. The distinction is 'formal', that is to say it is more than logical, but less than real.[47]

Boler (1963) is a thorough presentation of Peirce's relation to Scotus. Let us run through his presentation in order to place Peirce's extreme realism in relation to continuity, a relation Boler hardly takes into consideration (apart from Peirce's identification of generality with continuity).

Boler begins with the observation that Peirce's references are in general to the 'accepted' works of Scotus, but since Boler (1963), critical philology has progressed and now ascribes the crucial *Grammatica Speculativa* not to Scotus, but to Thomas of Erfurt. Boler (2004) recognizes this, but argues that the fact that Peirce never quotes from the *Grammatica Speculativa* may indicate that he was hardly very influenced by that work (64). The Scholastics generally assumed that only individual objects, 'supposita', could have existence – on the other hand, they possessed an understandable structure, a 'nature' which was common to several different particulars, consequently having a 'common nature' or essence. By abstraction, the single particular's essence could be recognized – and the crucial question now is which status these abstractions have. Are they only shorthands constructed by the mind or do they (or at least some of them) have real correlates, and if so, which reality status do these correlates possess? One solution, inherited from Aristotle, would say that the common nature is 'individuated' in the particular, but what is this more precisely supposed to mean? When things' nature or essence is referred to in a proposition, nature is often referred to as 'universalium', hence the name 'the fight over universals': which status do such universals have?

In this common conceptual frame, Scotus sought a middle way. He presented it as against two extreme positions, one claiming that universals as such existed *in* the particulars, another which found them only in the human mind (Boler 1963, 41). He took as premiss the Aristotelian definition of the universal as that which is *in multis et de multis* – that which is in many and can be said of many – a definition, of course, containing the problem rather than solving it: the two extremes which Scotus outlines can be said to lay the emphasis on each their part of that dictum. In a particular, there exists no correlate to the predicate, but we may make predications about particulars, such is the credo of nominalism. In the particular, the universal inheres in the property giving rise to the predicate, such is the corresponding credo of realism. The universal is ascribed, in these two competing interpretations, two different modes of being which lie as possibilities in Scholastic doctrine: *esse extra anima* versus *esse in anima*, being outside of the mind versus being in the mind. The corresponding objects having these forms of being are consequently *ens reale* and *ens rationis*, respectively. Particulars possess the former and may achieve the latter – if they are recognized; the fight over universals can now be expressed as the disagreement whether universals have any other mode of existence than the latter. Among *entia rationis*, Scholasticism distinguishes between two levels, grossly corresponding to our days' distinction between object language and metalanguage – first and second intentions, respectively. They both belong to the mind, but the former refers to real objects, the other refers to first-intention objects, that is, to other propositions. The basic question can thus, in turn, be rephrased as follows: are universals first or second intentions? (43).

Scotus opens his solution by distinguishing two uses of the concept of universal: in one sense, the universal is a second intention, in so far as it is predicated about a series of particular entia rationis. This goes for the 'predicabilia' of Scholasticism, inherited from Aristotle: the five canonised ways in which a predicate could be used about a subject: as genus, species, accidens, property, and differentia specifica. The subject for these logical universals is the type of relation between predicate and subject, and they are, in themselves, of course entia rationis. But the concept of universal may also be used in another sense, namely as that which provides the basis for the common predicability of several particulars – the common nature. The distinction here runs between universals as *de multis* and *in multis*, and the latter is first-intentional and refers to real common properties in nature. This distinction of course pertains to the fact that it does not follow from the existence of a universal predicate (the universal in the first sense) that there is in fact any common nature (the universal in the second sense) which corresponds to it (45). To Peirce, it is this 'real commonness' which constitutes Scholastic realism: 'The nature of the *fundamentum universalitatis* distinguishes the mediaeval realist from the nominalist' ('Unity and Plurality', Baldwin's *Dictionary* 1901, 6.377), cf. his concept of 'ground' in some of the sign definitions. With this distinction between the universal as a logical device and as a common nature, Scotus wants to insist that the question of universals can not be a merely logical issue, for then our knowledge about the world will dissolve into singular objects on the one hand and logic on the other – the equivalence between this position and twentieth century logical positivism is no coincidence.

The question to Scotus is now: which *unity* does this common nature possess? The question of unity is important here, because the nominalists – here Henry of Ghent who according to Alain de Libera played the same role as an opponent to Scotus, as he himself a generation later played for Ockham – maintained that only quantitative differences existed. If two particulars are numerically distinct, their common nature must have a 'less than numerical unity' (Boler, 47). That existing particulars are characterized by being numerically distinct does not pertain to their direct countability (of course, diffuse particulars exist), but it refers to the metaphysical fact that they are individuals. And the common nature is not unique, it is no *suppositum*. Which kind of unity may be ascribed to it, then? Common natures are possible without having to exist in any particular at all: unicorns also have a common nature. This idea, that the common nature is not in itself neither one nor many is inherited by Scotus from Avicenna's famous dictum 'Horseness is just horseness, it is not in itself one or many, universal or particular.' (50, cf. de Libera). The relation between the common nature and the particular is consequently critical, but much points to the fact that the Scotist solution was built on the introduction of the term 'haecceitas' (which Peirce inherits after his Scotus interpretation) – *thisness* as a privilege of the particular – and the idea of an operation of 'contraction' which allows the common nature to individuate itself in the particular.

The universal, understood as a logical predicate, is through and through a product of the mind, but this creation has, in certain cases, a basis – a condition of possibility, as de Libera proposes in Kantian slang – which is the common essence. Haecceity, thisness, incarnating the common nature in the particular is no further property (cf. Kant: existence is no property): this is important in order not to get the idea that it is in reality distinct from the common nature. Haecceity and common nature are so not separate in reality; they are *formally distinct*, as Scotus puts it, and this is what constitutes Scotus's famous 'formal distinction'. According to de Libera, this 'formal distinction' distinguishes three aspects in the object: the common nature, the haecceity or existence, and the *singularity* (the fully determined single object) to which only heavenly beings have direct access. Even if these three aspects may not be separated as parts of the object – not even by divine force, as Scotus says – then they may be formally distinguished. Here, it is obvious to make a parallel to the distinction in twenteith century phenomenology between Moments or Aspects on the one hand and Genuine Parts or Pieces on the other hand – after Husserl's 3rd Logical Investigation: the nominalist viewpoint would be that the only existing things are real 'countable' parts while moments (e.g. the surface, color, edges, etc.

of things) should be merely mental constructions on the basis of parts (we return to this in Chap. 7). But just like Husserl, Scotus claims that there are aspects of the object which are not really distinct in the sense that they are separable, but 'formally distinct' only. They are not distinguishable as *res et res*, but still being more than *ens rationis et ens rationis*, Scotus nicknames them with what Peirce would call a hypostatic abstraction: *realitas et realitas* – they are distinguishable *formalities*, or (these two terms are acccording to Boler used interchangeably): *realities*. This result of the formal distinction is, as is evident, not without parallels in the Brentano tradition's idea of *Sachverhalte*, states-of-affairs as the reference of logical propositions. So the common nature might be rephrased as the general state-of-affairs which the logical universals refer to or may refer to.

The argument of Ockham and his followers against this subtle distinction has been, of course, to reject the possibility of a third position: either a distinction is real or it is invented by the mind, no third possibility is at hand, and as the Scotists themselves admit that the universals are not of the former kind, then there only the latter possibility remains, and only 'numerical' differences exist in reality.

Boler now assumes that some of the difficulties in Scotus's position may be sorted out by interpreting him as an objective idealist – and de Libera's hints at Scotus as a sort of Kantian *avant la lettre* point in the same direction: the common nature as a sort of condition of possibility. Boler envisages this possibility in the fact that it is decisive for Scotus to prove the objective reality of common nature, not to prove its non-mental character (56). The objectivity of understanding is his aim, this is why he assumes metaphysical realities which grants him this – as against the Thomist doctrine which does not admit a third position between reality and logic. In de Libera's more detailed discussion of Scotus's position, the distinction between the universal as a logical device, a second intention, on the one hand, and the universal as that common nature referred to by this logical entity, is, like in Boler, taken as basic. But de Libera adds a further distinction within the universal as common nature (344–45). On the one hand primitive realism takes universals as absolute beings, not particularized, but on the other, Scotus's sophisticated realism attacks this notion for not being 'sufficiently indeterminate', thereby lacking the ability to individuate in particulars. Scotus' own alternative, the 'complete universal' is in contrast to any possible determination, 'thanks to which *man* is sufficiently indeterminate to be, by a unique intellection, conceived quidditatively in all men.'[48] We find a related indetermination in Peirce's idea of the continuity of universals.

But what about the relation between the common essence and its representation by means of logical universals? In the Aristotelian tradition it is, of course, the task of abstraction to construct universals on the basis of particulars – science only deals with what is general. Scotus turns against Thomas's version of this doctrine, according to which matter is a principle of individuation 'constraining' nature; then abstraction would face insurmountable problems when it shall restitute something thus delimited or fragmented. The process of abstraction presupposes that the common nature is not fragmentarily present, and it proceeds in two stages: (1) common nature is distinguished from the thisness, and (2) it is given numerical unity and becomes a universal which may now be predicated of many particulars (Boler, 60–61). This abstraction is second-intention, like the metalanguage of logic, but to Scotus we are now in metaphysics: abstractions of this sort are not only signs about signs, but refer to reality. Of course this difference is seminal to Peirce.

Why, now, is Peirce's continuity realism, as he himself claims, more 'extreme' than Scotus' realism? One decisive Scotist idea rejected by Peirce is the contraction hypothesis: 'Even Duns Scotus is too nominalistic when he says that universals are contracted to the mode of individuality in singulars, meaning as he does, by singulars, ordinary existing things. The pragmaticist cannot admit that.' (Letter to Signor Calderoni, c. 1905, 8.208) Why not? We have no problems in recognizing *haecceity* in Peirce's concept of Secondness, even if Peirce's version refers to events rather than to objects; is it not possible to reconstruct Peirce's theory in a manner such that generality, Thirdness, is contracted in the concrete particulars of Secondness? The reason is somewhat similar to what de Libera hints at in his discussion of Scotus's concept of common nature: it is too foreign to concretion. To Peirce any property – except for haecceity which is no property – is general, and for this reason also the property of being thinglike. Secondness and *haecceity* to Peirce is only a matter of the individuality of particular *events*. But there is a further reason of huge semiotic importance: the hypothesis of contraction allows the individual objects to be, as Kant puts it, the substance of generalities. But as individual objects are always numerical, they can never exhaust the continuity of a universal which subsumes all possible instantiations and not only those which are accidentally actualized in this world: 'Now no collection of actual individual events or other objects of any general description can amount to all possible events or objects of that description; for it is possible that an addition should be made to that collection.' ('New Elements' 1904, EPII, 316; NEM IV, 253) To put it briefly: if this were not the case, the road would be paved for an extensional semantics reducing Thirdness, generality, and continuity to the set of objects in which the universal is contracted and hence to the resurrection of nominalism. So to say that Thirdness is contracted into Secondness makes the latter 'larger' and too important in comparison to the former – in a Peircean view, it is rather the more comprehensive Thirdness which – as continuous – *contains* the actual, discontinuous events of Secondness. Finally, Peirce does not admit the real existence of the third entity distinguished by the formal distinction according to de Libera: the fully determined object. From early on in his carreer, Peirce maintained that this is an idea only and that particular objects or events are not fully determined, e.g. with respect to future states. It is not determined who my actual process of writing will influence or not. Thus Peirce does not identify the individuality of haecceity with full determinacy.

This version of scholastic realism has direct consequences for Peirce's semiotics: 'But all the realists agree in reversing the order of Aristotle's evolution by making the form come first, and the individuation of that form come later' (Lowell Lectures, 1903, 1.22). Peirce sophisticates, as we have seen, this idea by splitting the form in two: form as mere possibility in Firstness, anterior to anything actual, and form as realized possibility in Thirdness, where it governs Secondness in the shape of habits. In some of his sign definitions this becomes obvious:

> [A] Sign may be defined as a Medium for the communication of a Form. It is not logically necessary that anything possessing consciousness, that is, feeling of the peculiar common quality of all our feeling should be concerned. But it is necessary that there should be two, if not three, *quasi-minds*, meaning things capable of varied determination as to the forms communicated. As a *medium*, the Sign is essentially in a triadic relation, to its Object which determines it, and to its Interpretant which it determines [...] That which is communicated from the Object through the Sign to the Interpretant is a Form; that is to say, it is nothing like an existent, but is a power, is the fact that something would happen under certain conditions. This Form is really embodied in the object, meaning that the conditional relation which constitutes the form is true of the form as it is in the Object. In the Sign it is embodied only in a *representative* sense, meaning that whether by virtue of some real modification of the Sign, or otherwise, the Sign becomes endowed with the power of communicating it to an interpretant.
> (MS CSP 793, 1–3, c. 1905)

Here, the form – the universal – is connected to Peirce's pragmatic maxim requiring concepts to be explained conditionally: a form is equivalent to 'the fact that something would happen under certain conditions'. Under the concept hides a conditional proposition (What is a 'diamond'? – it is a very hard gemstone composed of carbon – 'hard' meaning *it would hurt other items if scratched on their surface*). This form is present in the object and represented in the sign. Thus the Scholastic universal is directly connected to the pragmatic concept definition and its ties to scientific experiments (investigating such 'would-be's) and the general laws and regularities they aim at mapping – and thereby also the prerequisite for the use of diagrams to investigate the behavior of such would-bes.

Would-Bes, or Real Possibilities

When Susan Haack discusses Peirce's scholastic realism (Haack 1992, 37–39), she makes the observation that Quine's commitment to similarity, natural kinds, dispositions, causal statements, etc. takes him close to Peirce's conception of Thirdness – even if Quine in an utopian vision imagines that all such concepts in a remote future will dissolve and vanish in favor of purely microstructural descriptions.

A crucial difference remains, however, which becomes evident when you look at Quine's brief formula for ontological commitment, the famous idea that 'to be is to be the value of a bound variable'. For even if this motto is stated exactly to avoid commitment to several different types of being, it immediately prompts the question: the equation, in which the variable is presumably bound, which status does it have? Governing the behavior of existing variable values, is that not in some sense being real?

This will be Peirce's realist idea – that regularities, tendencies, dispositions, patterns, may possess real existence, independent of any observer. In Peirce, this description of Thirdness is concentrated in the expression 'real possibility', and even it may sound exceedingly metaphysical at a first glance, it amounts, at a closer look, to what Haack calls a modest claim – namely that regularities charted by science are not mere shorthands for collections of single events but do possess reality status. In

Peirce, the idea of real possibilities thus springs (around 1897) from his philosophy of science – he observes that science, contrary to philosophy, is spontaneously realist, and is right in being so. Real possibilities are thus counterposed to mere subjective possibilities due to lack of knowledge on the part of the subject speaking: the possibility of 'not known not to be true'.

In a famous piece of self-critique from his late, realist period, Peirce attacks his earlier arguments (from 'How to Make Our Ideas Clear', 1878, in the late 1890s considered by himself the birth certificate of pragmatism after James's reference to Peirce as pragmatism's inventor). Then, he wrote

> ... let us ask what we mean by calling a thing *hard*. Evidently that it will not be scratched by many other substances. The whole conception of this quality, as of every other, lies in its conceived effects. There is absolutely no difference between a hard thing and a soft thing so long as they are not brought to the test. Suppose, then, that a diamond could be crystallized in the midst of a cushion of soft cotton, and should remain there until it was finally burned up. Would it be false to say that that diamond was soft? [...] Reflection will show that the reply is this: there would be no *falsity* in such modes of speech. (EPI, 132; 5.403)

More than twenty-five years later, however, he attacks this argument as bearing witness to the nominalism of his youth. Now instead he supports the

> ... scholastic doctrine of realism. This is usually defined as the opinion that there are real objects that are general, among the number being the modes of determination of existent singulars, if, indeed, these be not the only such objects. But the belief in this can hardly escape being accompanied by the acknowledgment that there are, besides, real *vagues*, and especially real possibilities. For possibility being the denial of a necessity, which is a kind of generality, is vague like any other contradiction of a general. Indeed, it is the reality of some possibilities that pragmaticism is most concerned to insist upon. The article of January 1878 endeavored to gloze over this point as unsuited to the exoteric public addressed; or perhaps the writer wavered in his own mind. He said that if a diamond were to be formed in a bed of cotton-wool, and were to be consumed there without ever having been pressed upon by any hard edge or point, it would be merely a question of nomenclature whether that diamond should be said to have been hard or not. No doubt this is true, except for the abominable falsehood in the word MERELY, implying that symbols are unreal. Nomenclature involves classification; and classification is true or false, and the generals to which it refers are either reals in the one case, or figments in the other. For if the reader will turn to the original maxim of pragmaticism at the beginning of this article, he will see that the question is, not what *did* happen, but whether it would have been well to engage in any line of conduct whose successful issue depended upon whether that diamond *would* resist an attempt to scratch it, or whether all other logical means of determining how it ought to be classed *would* lead to the conclusion which, to quote the very words of that article, would be 'the belief which alone could be the result of investigation carried *sufficiently far*.' Pragmaticism makes the ultimate intellectual purport of what you please to consist in conceived conditional resolutions, or their substance; and therefore, the conditional propositions, with their hypothetical antecedents, in which such resolutions consist, being of the ultimate nature of meaning, must be capable of being true, that is, of expressing whatever there be which is such as the proposition expresses, independently of being thought to be so in any judgment, or being represented to be so in any other symbol of any man or men. But that amounts to saying that possibility is sometimes of a real kind.
> ('Issues of Pragmatism', 1905, EPII, 354; 5. 453)

In the same year, he states, in a letter to the Italian pragmatist Signor Calderoni:

> I myself went too far in the direction of nominalism when I said that it was a mere question of the convenience of speech whether we say that a diamond is hard when it is not pressed upon, or whether

we say that it is soft until it is pressed upon. I *now* say that experiment will prove that the diamond is hard, as a positive fact. That is, it is a real fact that it would resist pressure, which amounts to extreme scholastic realism. I deny that pragmaticism as originally defined by me made the intellectual purport of symbols to consist in our conduct. On the contrary, I was most careful to say that it consists in our *concept* of what our conduct *would* be upon *conceivable* occasions. For I had long before declared that absolute individuals were *entia rationis*, and not realities. A concept determinate in all respects is as fictitious as a concept definite in all respects. I do not think we can ever have a logical right to infer, even as probable, the existence of anything entirely contrary in its nature to all that we can experience or imagine. (8.208)

Here lies the core of Peirce's metaphysical insistence on the reality of 'would-be's. Real possibilities, or would-bes, are vague to the extent that they describe certain tendential, conditional behaviors only, while they do not prescribe any other aspect of the single objects they subsume. They are, furthermore, represented in rationally interrelated *clusters of concepts*: the fact that the diamond is in fact hard, no matter if it scratches anything or not, lies in the fact that the diamond's carbon structure displays a certain spatial arrangement – so it is an aspect of the very concept of diamond. And this is why the old pragmatic maxim of 1878 may not work without real possibilities: it is they that the very maxim rests upon, because it is they that provide us with the 'conceived consequences' of accepting a concept. The maxim remains a test to weed out empty concepts with no conceived consequences – that is, empty a priori reasoning and superfluous metaphysical assumptions. But what remains after the maxim has been put to use, is real possibilities. Real possibilities thus connect epistemology, expressed in the pragmatic maxim, to ontology: real possibilities are what science may grasp in conditional hypotheses.

A serious discussion in Peirce scholarship attacks the question whether Peirce's 1905 revision of his old 'nominalist' beliefs form part of a more general development in Peirce from nominalism to realism. The *locus classicus* of this idea is Fisch (1967) where Fisch outlines a development from an initial nominalism (albeit of a strange kind, refusing, as always in Peirce, the existence of individuals determinate in all respects) in 1867–68 via a series of steps towards realism, culminating after the turn of the century. Fisch's first step is then Peirce's theory of the real as that which reasoning would finally have as its result (1868); the second step his Berkeley review with its anti-nominalism and the idea that the real is what is unaffected by what we may think of it (1871); the third step is his pragmatist idea that beliefs are conceived habits of action, even if he here clings to the idea that the conditionals in which habits are expressed are material implications only – like the definition of 'hard' (1878); the fourth step his reading of Abbott's realist *Scientific Theism* (which later influenced his conception of scientific universals) and his introduction of the index in his theory of signs (1885); the fifth step his acceptance of the reality of continuity (in 1891–93, the Monist series, the 'Grand Logic'); the sixth the introduction of real possibilities in January 1897, accompanied by the development of existential graphs, topology and Peirce's changing view of Hegelianism; the seventh, as we have seen, the identification of pragmatism with realism (1905); the eighth 'his last stronghold, that of Philonian or material implication' the same year (Fisch, 196). A further realist development exchanging Peirce's early frequentist idea of probability for a dispositional theory of probability was, according to Fisch, never finished. Even if accounts such as Houser's introduction to EPII and Short's sign theory development paper in Misak (2004) support large parts of Fisch's sketch, there are also dissenting voices. Especially the 'last stronghold' of material implication has given rise to some controversy, e.g. Zeman (1997), and it has certain bearings on how to conceptualize real possibilities.

The issue of implication concerns the old discussion quoted by Cicero between the Hellenistic logicians Philo and Diodorus.[49] The former formulated what we know today as material implication, while the latter objected on common-sense ground that material implication does not capture implication in everyday language and thought and another implication type should be sought. As is well known, material implication says that p $\Rightarrow$ q is equivalent to the claim that either p is false or q is true – so that p $\Rightarrow$ q is false only when p is true and q is false. The problems arise when p is false, for any false p makes the implication true, and this leads to strange possibilities of true inferences: 'If lions live in South America, Elvis is still alive'. This appears unsatisfactory for two reasons: not only are both p and q false, but if a lion was found in South America, common sense would say that the implication then would make Elvis living. The two parts of the implication have no connection with each other at all, such as would be the spontaneous idea in everyday thought. It is true that Peirce as a logician generally supports material ('Philonian') implication – but it is also true that he does express some second thoughts at around the same time as the afterthoughts on the diamond example:

> Some years ago [1903 is suggested by the editors of the CP] when ... I was led to revise [the] doctrine [that a mere possibility is an absolute nullity], in which I had already found difficulties. I soon discovered, upon a critical analysis, that it was absolutely necessary to insist upon and bring to the front, the truth that a mere possibility may be quite real. That admitted, it can no longer be granted that every conditional proposition whose antecedent does not happen to be realized is true, and the whole reasoning just given breaks down.
> (1906, 4.580, quoted with Zeman's abbreviatons and additions, Zeman 1997, 416)

Zeman convincingly argues on the basis that Peirce has, as early as 1885, when talking about hypothetical statements of what would happen, said that when p is false, then 'it is a matter of indifference whether the hypothetical be understood to be true or not, since it is useless' (3.374, quoted from Zeman 1997, 406). Thus, to Zeman, material implication was long since a second in relation to the third of hypotheticals, taking Peirce close to possible world semantics. This is probably correct, but still Zeman fails to quote what Peirce wrote shortly before the introduction of real possibilities:

> Although the Philonian views lead to such inconveniences as that it is true, as a consequence de inesse, that if the Devil were elected president of the United States, it would prove highly conducive to the spiritual welfare of the people (because he will not be elected), yet both Professor Schröder and I prefer to build the algebra of relatives upon this conception of the conditional proposition. The inconvenience, after all, ceases to seem important, when we reflect that, no matter what the conditional proposition be understood to mean, it can always be expressed by a complexus of Philonian conditionals and denials of conditionals. It may, however, be suspected that the Diodoran view has suffered from incompetent advocacy, and that if it were modified somewhat, it might prove the preferable one. (1896, 3.443)[50]

In such claims, Peirce is a forerunner of the attempts to construct alternatives such as strict implication, and the reason why is, of course, that real possibilities are not adequately depicted by material implication.[51] Peirce is in need of an implication which may somehow picture the causal dependency of q on p. The diamond example in fact derives from Peirce's implicit use of material implication in his 1878 paper: 'If x scratches all other things, then x is a diamond'. Here, if we do not know whether x scratches all other things, p may be false – but x may still be a soft diamond. Peirce's later criticism of this idea implies that the case when x does not scratch other objects must somehow be discounted. It is, of course, the interpretation of real possibilities as existing dispositions, would-bes, that makes mere material implication appear nominalist to the mature Peirce. The problem is that real possibilities can not be mapped by any merely formal logical notation, because it invariably requires certain semantic connections between p and q,[52] and real possibilities would require something like David Lewis' modeling of counterfactuals in possible world semantics to be expressed.

The basic reason for the mature Peirce's problems with the representation of real possibilities is not primarily logical, however. It is scientific. Peirce realizes that the scientific charting of anything but singular, actual events necessitates the real existence of tendencies and relations connecting singular events. Now, what kinds are those tendencies and relations? The hard diamond example seems to emphasize causality, but this probably depends on the point of view chosen. The 'conceived consequences' of the pragmatic maxim may be causal indeed: if we accept gravity as a real concept, then masses will attract one another – but they may all the same be structural: if we accept horse riders as a real concept, then we should expect horses, persons, the taming of horses, etc. to exist, or they may be teleological: if I am baking a cake, I am probably going to use an oven. In any case, the interpretation of the pragmatic maxim in terms of real possibilities paves the way for a distinction between empty a priori suppositions and real a priori structures to which we return later.

Real Possibilities and Absolute Idealism

Peirce's doctrine of would-bes nevertheless contains numerous problems. Being prime examples of Thirdness, they embody real existent continuity – but by the same token there is a tendency in Peirce to let real possibilities incarnate all the very different metaphysical issues which the category of Thirdness is expected to solve. This includes no less than habit, symbols, teleology, mind, purpose, evolution, life; sometimes even personality, love, God, etc. We should take care here to 'cleanse' the notion of real possibility as existing worldly dispositions from these other issues grouped together in Thirdness by Peirce's sometimes extravagant cosmology. The reason why real possibilities or habits involve teleology is Peirce's inclusion of all kinds of tendencies in the concept – including the goal-oriented actions of human beings or other animals, such as the example with the apple pie whose recipe forms a habit subsuming a continuity of possible singular apple pies. The inclusion of such teleological processes in the concept of real possibilities alongside the hardness of a diamond and the gravity pulling a stone is unproblematic, was it not for Peirce's tendency to claim that the *most* complex properties of a category should hold for *all* cases subsumed by that category. This is motivated, of course, by his 'synechism', the continuity doctrine excluding any sharp limits, but even given this doctrine, it is hard to swallow that gravity, e.g. should in any sense possess a teleological cosmological purpose. The error here seems to lie in Peirce's cosmology where notions like teleology are taken in a global meaning and predicated on the process of evolution as a whole and the universe as such. It is not, however, necessary to subscribe to the whole of this cosmology in order to embrace Peirce's notion of real possibility. Rather, we can see where his cosmology goes wrong: in its Hegelian tendency to globalize notions like mind, teleology, even personality onto the whole of cosmic evolution and the whole of the universe. Similarly with the notion of 'mind'. Peirce's un-psychological rendering of 'mind' as pertaining to all that is able to process signs forms a welcome divorce of mind from the more specific notion of consciousness – and even gives an first idea of how the latter might have evolved out of complex versions of the former – but still the Hegelian error enters when Peirce concludes that even atoms have mind or that all of the universe consists of signs. Here, Peirce's objective idealism of real possibilities tends to fall into the trap of absolute idealism. Even worse with the concept of personality – go to the Appendix to see how the idea emerges in Peirce's first cosmological sketches in the Monist papers of 1892: the real possibility inherent in the real referent of a symbol all of a sudden becomes a personality developing in its own right. We recognize the argumentative pattern taking Peirce to such conclusions: personality is taken to be the synthesizing force in the thought process – but when similar syntheses go on in nature, a corresponding personality must be at stake there as well...

Thus, it is extremely important to keep the notion of real possibilities apart from the quasi-religious extravaganzas of Peircean cosmology.[53] If we take final causes to be local and if we take mind to be the ability,[54] not necessarily conscious, of sign processing, we can easily include them as special cases in real possibilities – but we should not let absolute idealism lead us astray into assuming *all* real possibilities to be significant finalities, connected in any overarching, supersensible whole.

Peirce's Hegelian aberrations aside, real possibilities are continuous for the reason that they are would-bes, they have the form of 'Any state of things in which A is true *would be* a state in which B is true' (Noble 1989) – as he quotes Peirce, it '...consists in what *would be* the case under certain circumstances if they should remain unchanged throughout an endless series of actual occurrences' (Draft of a letter to Paul Carus, c. 1910, 8.226). As referring to a continuum of possibilities among which only particular instances may be actualized, real possibilities thus embody real-world continuity.

Real Possibilities as Dispositions

If real possibilities, on the other hand, are purged from the vagaries of absolute idealism, they form one of the first modern attempts at describing an ontology of dispositions. As Susan Haack emphasizes (1992), real possibilities come with a whole bundle of companion concepts: real generals – law – explanation – induction – relations, etc. To that extent, they provide a virtual reservoir of thought for the ongoing reflections on the ontology of powers (e.g. George Molnar) or dispositions (e.g. Steven Mumford). The French Peircean Claudine Tiercelin presents in a recent paper the development in actual metaphysics away from a strong tendency in early analytical philosophy to reduce or even eliminate dispositions (Quine, Goodman, Ryle) and to a tendency to reinstate or even inflate dispositions, beginning in Karl Popper ('propensities') and continuing in Rom Harré, Ian Thompson, Nancy Cartwright, Hugh Mellor and Simon Blackburn during the recent decades (Tiercelin 2002, 1). Molnar, Mumford, and David Armstrong may easily be added to that list.

The discussion about dispositions involves the issue whether dispositions should be distinguished from other, structural properties, so-called categorical properties, on which they are assumed to be dependent, or whether all properties in some sense could be said to be dispositional. It seems like some properties could very well be described as dispositional on one level (salt's being able to dissolve in water) while seeming categorical on other levels (salt's molecular structure). Armstrong, in a recent paper (undated), highlights the problems of both possibilities: property dualism makes it hard to understand how non-dispositional properties (distance between bodies) partake in the unfolding of dispositions (gravity), while the idea that all properties are dispositions faces the problem that the difference between a power and the manifestation of it gets lost: effects are then powers in their own right. He himself ends up endorsing a middle way originally proposed by C.B. Martin: all properties have two aspects, a qualitative and a dispositional. The problem left, of course, is to account for the – presumably necessary – relation, maybe even identity, between the two sides of the property. In that case, only a formal distinction will be able to isolate each of them. But what about the relation between the (universal) disposition in question and the particular incarnating it? Armstrong long held the idea that this relation was contingent, but influenced by Don Baxter, he now thinks otherwise: the universal and the particular are, by necessity, partially identical. Particulars are principles of unity, collecting universals, while universals are principles of unity across particulars, and a state-of-affairs is the intersection between these two principles.

As to this issue, the metaphysical implications of Peirce's pragmatic maxim seem clear: all properties must be dispositional. That is, in fact, the very crux of the pragmatic maxim: if a property does not, on some level, possess conceivable effects, then it may be discarded as irrelevant. Ergo, all properties must be dispositions – or, more cautiously, all properties must possess dispositional aspects. Concerning the problems raised by Armstrong, Peircean real possibilities are embedded in the three-category doctrine: the qualitative aspect of any real possibility is described by its inherent firstness (which may be prescinded), and the potency-act difference is described by the inherent secondnesses as act bundles. As to the 'incarnation' issue, we have seen Peirce support an interpretation anticipating Armstrong's: particulars are secondnesses which necessarily participate in universals via the continuity doctrine: they are actual parts of the continuity of real possibilities This necessity, we would claim, belongs to 'material necessities' in the sense of Barry Smith (see Chap. 8) and is investigated by material ontologies of the reality domain in question (such necessities connect being hard with being a diamond with being this diamond). The intersection hypothesis also seems supported by Peirce's account: both secondness and thirdness are principles of unity (referring to existence and reality, respectively) and the single particular is then the intersection between the continuum of haecceity and the continuum of real possibility, as in the idea of thisness as a sort of crack through which real possibility leaks. Thus, Peirce's real possibilities doctrine may be formulated alongside contemporary conceptions of dispositions.

Another issue is the relation between laws and dispositions as discussed by Tiercelin (2002). Is it the case that talking about dispositions instead of laws sort of locates the causal powers in the single systems where the causal process takes place – so that the relation between laws and dispositions mirrors that of universals versus tropes? Are dispositions thus trope-like in being abstract particulars and can they thus make laws superfluous or at least supervenient upon dispositions? Are laws determined by particular events rather than the other way around? Such a point of view will, of course, be much too actualist to Peirce: real possibilities transgress any actual realization of them. On the other hand, the continuity principle entails that real possibilities never govern particulars with full necessity, they remain tendencies, weaker or stronger, even if some of them may be so close to necessities that it is hard to tell the difference. On that issue, Peirce's doctrine of real possibilities will thus be inclusive: real possibility is a dispositional concept all right and thus functions as a local force here and now, but it possesses the universality of a law at the same time as it contains the single case where it holds (corresponding, in fact, to the possibility of claiming the existence of both universals and tropes, rejected by universal-supporter Armstrong for Ockham razor reasons). Real possibilities include actual instantiations among its continuum of virtual events and so do not take sides in the quarrel between universals and tropes which metaphysical Ockhamite ascetes will be sure to hold against them. If, however, laws and dispositions are seen as to interrelated aspects of real possibilities, the alleged ontological overkill might seem bearable.

Dispositional essentialism faces, according to Tiercelin, among other things the issue of substantial versus relational realism. As she argues (13), an important issue is that real possibilities are inherently *relational* on Peirce's account. And, as Dipert (2004) argues, Peirce's notion of deduction necessitates a metaphysics of relations. Peirce's relationalism is highlighted in his arguments against the Scholastics' tendency to see dispositions inhere in monadic predicates in thing-like substances. As we have seen, he of course agrees with Scotus in the existence of 'realitates' – realities – (cf. Boler 2004) and real possibilities is his account for them, but they are not identical to substances seen as monadic things. Rather, they involve the lawlike relations between singulars and thus pertain to relational events rather than objects. Being hard is a relation between scratching objects, gravity is a relation between heavy bodies, apple pies are relations between bakers, ingredients, recipes, and eaters, and so on. Dispositions as Peircean real possibilities are thus different from object universals which should rather be taken as a sort of second intention reified hypostatic abstractions from relational real possibilities. Real possibilities in Peirce thus immediately involve an ontology of relations. But it is important to note here that exactly the pragmatic maxim in Peirce makes the doctrine immune against the standard regress argument against relational ontologies: the multiplication of entities deriving from the need to posit the second order relation between any relation and its relata, and so on ad nauseam et infinitum. The pragmatic maxim immediately rules out such excesses.

As to the issue whether real possibilities are themselves contingent or necessary, Tiercelin (15–16) adopts a prudent fallibilist stance (see Chap. 8): It is probably wisest to refrain from trying to find any ultimate explanation of reality. She refers to Peirce's notion of 'conditional necessity' ('Notes on the doctrine of chances', 1910, 2.664) as more or less complex habits in nature in the context of Peirce's attempt at a dispositional theory of probability. Suffice it to refer to the ongoing discussions in physics and philosophy of physics regarding whether the basic physical constants are contingent (could the gravity constant have been different and what would the universe have looked like), or are they interconnected by some deep necessity we do not (yet) know? In single cases, we may discover that some dispositions are contingent to the extent that they depend on other, simpler dispositions, but as to any deepest level of dispositions we might be better off by biding our time and restrict ourselves to the pertinent levels of observation for the time being. Real possibilities may be located at any level of organization, and we do not need to base them in the deepest level of microphysics in order to grant their reality.

A further issue we have touched upon already is the extension of real possibilities as to causes and structural interconnectedness. In my view, it is a strength of real possibilities to comprise both. Take the hard diamond as an example. One aspect of it is immediately causal: the scratching which causes other objects to be scratched. Another aspect of it is just as immediately structural: the hardness which is conditioned by the special molecular structure of carbon in diamonds, just like the case with salt's solvability in water. Dispositions, hence, are by no means the metaphysical 'last word' in the sense that they refer to any deepest, microstructural level of reality. They may depend on structural topological systems of ever deeper causes, and we have no reason to believe that the present knowledge of fermions and bosons necessarily refers to the deepest level of physical reality. Rather, the causal-structural descriptions are complementary and refer to timeless ontological necessity on the one hand (the structural aspect) and concrete, developing tendencies in events on the other (the causal aspect). The structural description of dispositions implies that real possibilities involve what Barry Smith calls 'material necessity' (see Chap. 8), referring to the synthetic a priori necessities connecting together different aspects of the ontological structure of a given, material domain – just like hardness and molecular structure in physics. Thus, real possibilities necessarily refer to ontological interconnectedness of a domain as well as the causes, tendencies and currents characterizing that domain.

Peirce's would-bes thus cover relations, tendencies, patterns, laws, dispositions, propensities, affordances[55] – and they are motivated by the repeatability of scientific experiment (the pragmatic maxim being so to speak a thought experiment equivalent to an empirical laboratory experiment). Thus, they exemplify continuity. Real possibilities *are* Peircean continuities, and on Peirce's account they often even appear as prototypical representatives of continuity: they are general because they do not prescribe anything about the unmentioned, blank aspects of the objects they subsume, and actual instantiations, however numerous, of them may never exhaust them. Their continuity, on the other hand, determines the epistemological means we may have at our disposal for gathering knowledge about them: diagrams.

Real possibilities are therefore seminal also for Peirce's semiotics. It is they that grant that signification is more than just any actual interpretation of a sign. Short's much-discussed account for the development of Peirce's theory of signs (2004) argues that Peirce's introduction of real possibilities around 1896–97 inter alia serves to mend this lacuna in his sign theory: that there inheres an intensional meaning in (some) signs which may not be exhausted in any actual set of interpretations (cf. also Dipert 1997 on Peirce's intensional account for sets). Diagrams are the proper epistemologial means to get to know about continuous, real possibilities, because they are themselves continuous, iconic representations – they are drawn on a continuous sheet, facilitating continuous experiments with them (thus mirroring real laboratory experiment on a conceptual level), and they refer to a continuity of possible instantiations of the processes referred to by the experiments, be they real or imagined.[56] This is why diagrams as 'moving images of thought' occupies center field in the semiotic epistemology of the mature Peirce which we shall investigate in the next two chapters.

Real possibilities are thus what furnishes the connecting link between Peirce's metaphysical continuity doctrine on the one hand and his diagram epistemology on the other.[57] Real possibilities are understood by the manipulation of applied diagrams, and in Chap. 8 we shall argue diagrams thus furnish our access to the synthetic a priori of regional ontologies underlying the different special sciences.

CHAPTER 3

HOW TO LEARN MORE

An Apology for a Strong Concept of Iconicity

Iconicity is generally conceived of as the sign-relation making one phenomenon signify another by similarity in some respects. Iconicity is hence based on similarity, but it is not simply defined by it, for a very simple reason. Similarity is generally symmetrical:[58] if *a* is similar to *b*, then *b* is also similar to *a*; while sign-relations are generally asymmetrical: if *a* signifies *b*, it does not follow that *b* signifies *a*. Iconicity adds to similarity an intention aimed at *b*, using *a* as a means to this end.[59] This difference is acknowledged by most schools, for instance Peirce, who states that

> An Icon is a sign which refers to the Object that it denotes merely by virtue of characters of its own, and which it possesses, just the same, whether any such Object actually exists or not. It is true that unless there really is such an object, the Icon does not act as a sign; but this has nothing to do with its character as a sign. Anything whatever, be it quality, existent, individual, or law, is an Icon of anything, in so far it is like that thing and used as a sign of it. ('Syllabus', 1903, EPII, 291; 2.247)

Here, similarity as well as its utilization in sign reference are necessary prerequisites for the Icon, but only taken together do they become sufficient prerequisites. In the other great phenomenologist Husserl, we find a similar distinction in the sixth *Logische Untersuchung*:[60]

> The sign has in general no community of content with the thing it signifies; it can stand as readily for what is heterogeneous, as for what is homogeneous with itself. The likeness [Das Bild] on the other hand is related to the thing by *similarity*: where there is no similarity, there can be no talk of a likeness, an image. [...] It would be a descriptively wrong notion of the matter, to think of the whole difference as lying in the fact that the same intention which, in the one case is tied to the appearance of an object *like* the object referred to, is in the other case tied to the appearance of an object *unlike* it. For the sign, too, can be like what it signifies, even entirely like it: the sign-presentation is not thereby made into a presentation by way of likeness. A photograph of the sign *A* is immediately taken to be a picture of the sign. But when we use the sign *A* as a sign of the sign *A*, as when we write '*A* is a letter of the Latin written alphabet', we treat *A*, despite its representational similarity, as a sign, and not as a likeness [Bild].
>
> The objective fact of similarity between what appears and what is meant, is accordingly irrelevant: it is not, however, irrelevant where something is presented by way of a likeness [Bildvorstellung]. This shows itself in the possibility of fulfillment: it was only the recollection of this possibility which allowed us to bring in 'objective' similarity in this context. The likeness-presentation [Die Bildvorstellung] plainly has the peculiarity that, when it achieves fulfillment, the object which appears before it as likeness ['Bild'] gets identified through similarity with the object *given* in the fulfilling act. Having held this to be the peculiarity of a presentation by way of likeness [Bildvorstellung], we have admitted that *fulfillment of like by like internally fixes the character of a synthesis of fulfillment as imaginative*. (LI,711)

Similarity is not sufficient to define an icon – a Husserlian 'Bild' – for also mere signs (*Zeichen*, which in Husserl do not possess any iconic content) may share similarity with their object. In order to function as a *Bild*, similarity needs to be invoked by a certain type of conscious act, imagination, which is defined by addressing its object 'durch ähnliches'. Thus, Peirce as well as Husserl agree in the need for a further determination of the Icon than mere similarity. In Peirce, it is the functioning of a phenomenon as a sign for a similar object; in Husserl it is the notoriously asymmetrical concept of intentionality incarnated in a certain subtype named *Bildbewusstsein* which introduces the asymmetry (more about Husserl's concept of *Bild* in Chap. 14). Thus, there is a seminal difference between exactly which kind of support similarity needs in order to become an icon. In Peirce, the sign character is carefully defined without any reference to consciousness; in the still pre-egological Husserl of the *Untersuchungen*, similarity is defined objectively, but not so imagination, which is seen as an intentional act with a certain kind of 'Erfüllungssynthesis', that is, defined by phenomenological consciousness (a not yet transcendental consciousness, it must be admitted). The Peircean solution introduces the asymmetry not by means of an intentionality-like concept, but through the pragmatic concept of 'use as a sign', elsewhere 'function as a sign'.[61] Implicitly, then, Peirce makes the concepts of 'use' and 'function' more broad than (human) consciousness, true to his naturalized semiotics. For instance, the well-known biological phenomena of mimicry would by his definition (but not so by Husserl's) be unproblematic instances of iconicity.[62] To us, the Peircean solution is the one to be preferred in order to avoid any dualism between man and nature or the like, and it seems quite likely that Husserl's idea may be reinterpreted in a Peircean frame (so as to make intentionality a concept depending on the more general concept of a (teleological) system). On the other hand, Husserl's rendering of the special character of the *Bild* makes apparent an important feature which is in Peirce (at best) more implicitly present: similarity between two phenomena and the use of one as a sign of the other are not sufficient if just added as two independent features. To make up a *Bild* – or an icon, we could say – a sign must signify *through* its similarity to its object. We shall return to this elsewhere to concentrate on that other defining feature of iconicity, in addition to its intentionality/use/function part: its *similarity*, whose objectivity both authors invariably stress.

It is probably not to say too much to state that similarity has had a bad press during the last century or so. Many currents in recent philosophy, psychology, linguistics, etc. have seen the objectivity of similarity as highly questionable and has consequently sought to eliminate the concept as a piece of common-sense ideology, of a simplistic folk theory of signification. The attack has come from two principal sides. One is what could be called the Nietzschean side in general, claiming that similarity is only a construct by a weak mind in order to control an overwhelmingly non-self-similar ontological Being. In Nietzsche himself, we find the prototype of this idea in his ontological concept of the world as *werdende*, as a current of constant change, in which it is of course impossible to isolate stabilities other than by means of artificially stiffening and controlling what fundamentally cannot be controlled. This idea has been prominent in various existentialisms, nihilisms, vitalisms, *Lebensphilosophien*, and the like throughout the 20th century and one of its latest incarnations is of course the famed 'philosophies of difference' which teach that difference is always presupposed by any identity or similarity. In this kind of thought, similarity is most often identified with identity as being part of the despised 'Identitätsphilosophie'; by this operation all the non-trivial difficulties in the concept of similarity are concealed by reducing it to the trivial idea that A = A.

In linguistics and phenomenology, two traditions closely related to the history of semiotics, the anti-psychologism necessary for the very birth of these two traditions as autonomous disciplines (in, for instance, Hjelmslev and Husserl) implied the refusal of any attempt to account for semantic content by means of psychological representations. This crucial idea has been central for the idea of semiotics as a discipline, supported by the equally strong anti-psychologism in Peirce's increasingly influential semiotics. The interesting and often overlooked fact is that this foundational and necessary antipsychologism has imbued semiotics with an anti-iconic tendency throughout most of the twentieth century, in so far as iconicity has very often been spontaneously identified with psychological imagery. The surrounding scientific climate, moreover, served to underline this tendency: in the philosophy of mathematics, in philosophy of science (logical positivism, for instance), in quantum theory, and many other scientific currents, the abolition of iconical intuition of the object became conceived of as a necessary prerequisite for thought to become scientific. In semiotics, this anti-iconicity gave rise to a thoroughgoing conventionalism. Strangely enough, this semiotic conventionalism went hand in hand with the 60s reinterpretation of Saussure's – psychological! – version of linguistic structuralism, so that anti-iconism's roots in anti-psychologism were forgotten and replaced with a conventionalism teaching that all signs were due to 'codes' having the character of general social conventions and instantiated in the single language user's psychology. The whole critical impetus hopefully connected to this structuralism was concentrated in its resistance to iconicity interpreted as a resistance to ideology (to 'imaginarity', to 'Western metaphysics', to 'effects of reality', etc.). Despite currents of dissent (Lévi-Strauss' insistence on motivation in signs or Lyotard's on iconism ('figure')), this particular brand of anti-iconism was a commonplace of most variants of 60s structuralism (Greimas, Barthes, Lacan) as well as so-called poststructuralism (Derrida, Deleuze, Foucault).

Another line of attack on the notion of similarity has come from almost the opposite side of philosophy, analytical philosophy. The idea of a purely symbolic calculus as the ultimate aim of science and philosophy has been thriving at least since Leibniz, but during the last century this idea has enjoyed greater prominence than ever before. The history of this idea is long and cannot be mapped in detail here, but some crucial points can be sketched out. A source for it is, of course, the influential success of Newtonian mechanics which gave rise to a lot of philosophical attempts at explaining the reasons for this success. One of the ideas was the algebraic form of Newton's laws. Taken together with Descartes' analytical geometry which made transformations between geometrical figurae and algebraic expressions possible, this indicated that a formulation in terms of algebraic calculi were the ultimate aim for a science to be successful. This – in itself, sound – idea was sharpened through a series of developments in the nineteenth century. First, the problem of formalizing imaginary and complex numbers gave rise to the construction of autonomous algebra in the first half of the century: the insight that algebraic structures may be investigated with no reference to what the algebraic symbols might be used to symbolize, yielding the famous results of group theory. Second, the development of non-Euclidean geometries by variations on the famous parallel-axiom in Euclid's system. Originally these geometries sprang from the attempts at showing that the axiom might be proved as a theorem on the base of the simpler Euclidean axioms; one way to show this would be to assume the contrary and derive a contradiction. This procedure did not, in fact, result in contradictions, but merely in new, consistent, formal systems which could now be interpreted to yield Lobachevski and Riemann geometries, respectively. The crucial point now was that in these new geometries, the symbols of the old Euclidean system were to be given entirely new interpretations. What was in Euclid a straight line, for instance, became in Riemann a great circle. Thus, the interpretation of algebraically expressed propositions tended to be marginalized, a tendency which developed in different versions and degrees in the formal logic of Frege, Peirce, Schröder, Russell, etc. around the turn of the century. The idea of a purely formal reading originated in Moritz Pasch's scepticism against intuition – and rose to metamatemathical fame in Hilbert's famous idea of the possibility of making mathematical proofs by a mere 'formale Redeweise' – a formal mode of reading – in which the symbols involved were treated as mere letters equipped with certain rules of transformations, bracketing interpretation during symbol manipulation. It is necessary to emphasize that this whole development is extraordinarily fertile and by no means contrary to the idea of similarity *in itself.* The vast research in the autonomous regularities of algebra, symbol systems, and formal logic is one of the richest developments in science during the last centuries. But somehow two extreme versions of this ingenious idea emerged. One was that formalization was purely logical in a very restricted use of the word, as meaning consisting in a purely conventional, symbolical language, equipped with a purely combinatorial syntax without any semantic implications. Another was that formalization is not only crucial to science, but formalization *is* science, so that the very interpretation of formal systems tended to be conceived of as a source of error to be avoided. In Hilbert, it was still the case that the 'formale Redeweise' was only to be maintained within proof theory – in the previous determination of definitions, of axioms, rules for the manipulation of symbols, etc., the interpretation of the terms was ineradicable, as well in the subsequent understanding of the conclusion proved.[63] But in many heirs to his formalist program, this crucial distinction was forgotten, and the ideology gradually spread that, in establishing a formal calculus, the very aim was to *exterminate* intuition, not merely to control it. In the humanities, this idea spread through logical variants on structuralism, cf. for instance Hjelmslev's formal rendering of Saussure, in which the mere algebraic dependencies between terms was all linguistics might hope to map. All in all, it is no wonder this overall tendency to privilege symbolic and algebraic calculi at the expense of interpretation and intuitive presentation became hostile to iconicity and its basis in similarity. Thus, we find in two major figures in analytical philosophy, Quine and Goodman, attempts at reducing the notion of similarity completely – just like we in psychological or logical structuralism, for instance Greimas or early Eco, find analogous ideas.

In recent years, a new 'morphological turn' or even 'iconic turn' as a part of the vast domain of cognitive science has changed the picture.[64] Here, continuous models not reducible to algebra are introduced alongside feature-preserving mappings of such models between (mental) domains – in cognitive semantics, cognitive linguistics, in the Peirce renaissance in semiotics, etc. But still the anti-similarity movement is strong, both in its Nietzschean and its analytical variants.[65]

GOODMAN

Let us examinate the famous arguments of Goodman (1976) which have often, in an American context, been conceived of as the definitive burial of similarity, his 'Seven Strictures of Similarity' in which the harsh verdict goes as follows: 'Similarity, ever ready to solve philosophical problems and overcome obstacles, is a pretender, an impostor, a quack.'(437)

The first of the seven arguments runs as follows: 'Similarity does not make the difference between representations and descriptions, distinguish any symbols as peculiarly 'iconic', or account for the grading of pictures as more or less realistic or naturalistic.' Here, Goodman's idea is that the conviction of resemblance as 'the necessary and sufficient condition for representation is so deeply engrained that the evident and conclusive arguments to the contrary are seldom considered' (ibid.). Now, this conviction is at least not shared by any of the two great phenomenologists Peirce and Husserl, as we have already seen: each of them makes similarity a necessary but not sufficient condition for representation – and only for certain aspects of representations, namely the iconic ones. Accordingly, Goodman's argument precisely presents examples of resemblance lacking the asymmetry of the sign function: 'Yet obviously one dime is not a picture of another' etc. He realizes that this only proves that resemblance is not sufficient to define representation, and then turns to the idea that resemblance *and* reference should be sufficient. His example here, curiously, has exactly the same character as Husserl's with the *A* above – it involves the *suppositio materialis*:

> Consider a page of print that begins with 'the final seven words on this page' and ends with the same seven words repeated. The first of these seven-word inscriptions surely refers to the second, and is as much like it as can be, yet is no more a picture of it than is any printing of a word a picture of another printing. (ibid.)

But Goodman draws completely different conclusions from the example than does Husserl. He hastily concludes while continuing to the next issue:

> Still, once pictures are somehow distinguished from other denotative symbols – and this must be by some other means than similarity – does not comparative naturalism or realism among pictures depend upon their degree of resemblance to what they represent? Not even this can be maintained. (438)

It is strange that an heir to the Frege tradition like Goodman does not see that it is the oblique reference ('When I say 'A' I am referring to A...', etc.) so crucial to the Fregean distinction between *Sinn und Bedeutung* which is invoked here – and which is, of course, a type/token relationship, not an iconic relation. His argument may be summed up as follows: neither similarity nor the sum of similarity and reference is sufficient to define a picture. In this, he is perfectly right. But his conclusion that icons do not involve similarity at all does not follow from this. In both Peirce and Husserl, the defining feature is similarity in a signifying and referring function, which is not the same as the mere aggregate of similarity and reference. The reference of the picture is *dependent* on similarity; in Husserl's vocabulary the reference of a picture is an 'unselbständige Inhalt' because it depends on the similarity. Of course this is not the case in all kinds of signs, for instance in the *suppositio materialis* which both authors use as an example and where the icon is merely *mentioned* (or, object of an act of *Nennen*) and thus does not *function* as a sign. In these cases, the reference takes place as a result of deictic – or, in Peirce's wording, indexical – sign use. Thus, the failure of Goodman is here not to recognize that there are several different ways of referring (iconically, indexically, symbolically, at least). This fault is then, in Goodman's hasty style, mixed up with another in the argument already quoted. Similarity should be independent of the degree of realism of a representation for reasons of cultural relativism: realism is a function of culturally specific systems and hence similarity is an effect of such systems and not the opposite way around. This crude argument overlooks that the fact that similarity is 'culture-dependent' does not make it a mere effect of cultural norms. Cultural norms *require* similarity in their description because culture consists in people acting similarly in some respects; one could say that one culture differs from another because it emphasizes other similarities. In that respect, the similarities perceived are of course 'culture-dependent', but this is merely because the very notion of culture involves systems of similarities. The very concept of a *norm* presupposes similarity to the extent that it requires that similar cases be judged similarly. So to say that 'similarity is relative, variable, culture-dependent', merely amounts to saying that any *particular* judgment of similarity is dependent on the classes of similarities envisaged, which is a mere truism, more apt, in fact, to dissolve the concept of culture than the concept of similarity.

Now, the 'second stricture': 'Similarity does not pick out inscriptions that are 'tokens of a common type' or replicas of each other.' Of course, the token-type problem involves the problem of nominalism versus realism, and the very terminology of token and type is a (probably unconscious) Peircean heritage. Here, Goodman invokes various variants on written types, such as for instance a B, a B where the middle horizontal line is not attached to the vertical bar, and an O - and,

Figure 8.

again hastily, concludes that the second B is topologically more like the O than the B proper, because the first two both possess one interior part only, the second two:

Goodman simply fails to see here that more than one topology exists; for instance a topology distinguishing singularities would not hesitate to identify the to Bs as having similar (but not identical) systems of singularities, different from the O. In fact, Goodman's example does little more than say that a green circle and a red circle are not similar, because one is green and the other is red. But both are still round. The problem is, of course, to focus upon the right level of comparison, and this is not always as easily done as in this deludingly simple example. In fact, typefaces constitute a very complicated case of similarity, as argued by Douglas Hofstadter (1985) and myself (1992) – but complexity is not an argument against similarity either.[66] Goodman's topological deliberation permits him to conclude that

I suspect that the best we can do is to say that all inscriptions that are *a*'s must be alike in being *a*'s. That has the solid ring of assured truth, but is hardly electrifying. Moreover, notice that to say that all *a*'s are alike in being *a*'s amounts simply to saying that all *a*'s are *a*'s. The words 'alike in being' add nothing: similarity becomes entirely superfluous. (439)

This conclusion displays an overall strategy found in similar versions in various anti-similaritists: 'similarity' can be reduced to identity. But this short-circuit overlooks the non-trivial fact that many propositions of the form 'All A's are A's' conceal complex cognitive skills used to detect objective similarities. 'All A's are A's' is not trivial, because in this proposition, the first 'A' refers to tokens (which may vary widely), the second 'A' to the type; and the question at stake is: which bundle of (different, to be sure) similarity-preserving transformations connects the various subtypes of A's with each other? Or to put it differently: how do we recognize a given token A as a type A when we have no access whatsoever to an extensional definition of A-ness, that is, to the 'set of all A's', the unproblematic existence of which (and access to which) Goodman is much too quick to presuppose. This old extensionalist trick of trying to define similarity by the set of individuals which is said to possess the quality in question invariably fails, because this set can never be defined without recourse to that very quality (did anybody ever investigate, for instance, the set of all red objects in the world, before becoming able to use the word 'red'?) – that is, it does not escape a *circulus vituosis*.

The 'third stricture' says: 'Similarity does not provide *the* grounds for accounting two occurrences performances of the same work, or repetitions of the same behavior or experiment.' The structure of the argument is the same as in the first stricture, as supposed by our italicization of '*the* grounds'. Of course similarity is not sufficient, it takes a supplementary indexical indication of which aspects of similarity are relevant. 'Repetitions of the same behavior, such as hitting a tennis ball against a barn door, may involve widely varying sequences of motion' (439). Actually, in this stricture Goodman does not draw any distinction between two issues involved: kind of similarity and indexical reference. When he sums up 'In each of these cases, the grouping of occurrences under a work or an experiment or an activity depends not upon a high degree of similarity but upon the possession of certain characteristics. In the case of performances of a Beethoven symphony, the score determines what those requisite characteristics are...', he overlooks the fact that the score in this example actually performs two tasks: it determines 'requisite characteristics' which we must still suppose can be instantiated in a host of variants which consequently will be similar with respect to a subset of those characteristics – and it codifies this set of characteristics by making them explicit and indexically attaching a stabilizing name ('Beethoven's xth') to them. Thus, all in all, this 'third stricture' is no argument against similarity at all; the fact that '... the principle of classification varies with our purposes and interests' does not imply that there is no similarity at stake in each one of such cases, and the similarity then will be relevant to a standard selected by the *purpose or interest* in question (notice these categories and their relatedness to Peirce's *sign function* or Husserl's intentional *Bildbewusstsein*).

The fourth stricture is more controversial: 'Similarity does not explain metaphor or metaphorical truth' (440). Here, we find Goodman's most famous denial of similarity: 'Anything is in some way like anything else'. This leads him to conclude, like in the second stricture, that similarity is the *result* of a process, be it similarity between tokens or the parts of a metaphor, rather than its prerequisite: 'In both cases a reversal in order of explanation might be appropriate: the fact that a term applies, literally or metaphorically, to certain objects may itself constitute rather than arise from a particular similarity among those objects' (440). There is an interesting tension here between the two parts of Goodman's argument. The first is actually a statement of a stunning extension: 'Anything is in some way like anything else'. Of course, Goodman intends by this statement to reduce similarity to insignificance because of its omnipresence, but if we take his words at face value, it is a statement of enormous a priori breadth. It is of course impossible empirically to undertake the task of actually comparing all known phenomena two and two; how then is it possible confidently to state this amazing theory? The idea is probably that if the *tertium comparationis* is chosen sufficiently general, some property will show up shared by both. Yet, we cannot presuppose that this property will in all cases be an empirical property (given one phenomenon, we can simply define another phenomenon by giving it another empirical property, for each property of the first one – a property chosen beyond the limits of any chosen similarity function). Consequently, the property shared must in this extreme case be of an a priori character, having no opposite. For instance: both are phenomena. Both have a certain form (the formless being also a form type). Etc. Given this deliberation, we can surely subscribe to Goodman's a priori law: anything is in some way like anything else. But then this is in flagrant contradiction with his consequence: that metaphorical use should constitute similarities rather than being constituted by them. If similarities between anything always already exist, then metaphor must be constituted by the selection of certain among these similarities at the expense of others. This discussion is still relevant today when for instance Lakoff's cognitive semantics claims that similarities are not preexisting but are created by metaphorical mappings from one domain onto another. But if Goodman's a priori law is really correct, this cannot be the case: the similarity chosen must in some sense of the word exist beforehand, as a potentiality. This would, of course, be Peirce's solution and it seems evident that it is pertinent for most everyday judgments of similarity. The fact that it has never before been asserted that this orange on the table before me is similar in shape to the moon (given a certain granularity of similarity classes), might cause sensible souls to see me as a genius for creating metaphors, but, modestly, it seems strange that this similarity should be something created by me. I merely discover (no great effort) this similarity by applying a certain *tertium comparationis* (a circle, give or take a certain rate of deformation). In rare cases, of course, it may take great pains to establish a new complicated *tertium comparationis* to see a similarity (Newton discovering the similarity between the movement of the apple and of the heavenly bodies, Eliot discovering the similarity between cruelty and the growth of April flowers) but this hardly implies that the similarity was not there before. Of course, artists will rage against this conclusion, for does it not imply that the work of art is not their creation, was it already in some sense a possibility before their effort? This is probably – if we for a moment allow ourselves the Nietzschean pleasure of getting personal – one of the reasons why it is so hard for many to give up the idea of similarities as something constructed: by doing so, one also gives up the self-flattering romantic-nihilist idea of the subject as artist, as genius, as creative *Übermensch*.

The fifth stricture deals with similarity in science: 'Similarity does not account for our predictive, or more generally, our inductive practice' (441). Goodman sets out by stating that the fact that the future will be like the past is often regarded as highly dubious – but contradicts this idea: '... while I am sure the future will be like the past, I am not sure in just what way it will be like the past. No matter what happens, the future will be in some way like the past.' This idea Goodman illustrates by drawing a curve plotting the relationship between two sets of variables. Now, there will exist a curve covering any possible extrapolation of the sets of data already given, he argues.[67] The idea seems to be that curves are our means for retroactively assuming a similarity, which was we were not able to see during the process, and of course it is mathematically correct that any amount of points in the plane (if we leave out the possibility of several points with same x value) can be connected with one line with increasing x-variable. But this argument proceeds as if science had never, in a large number of cases, established any *laws* delimiting this infinite set of possible curves to a small class of related curves. 'Along which, among countless lines of similarity, do our predictions run?', Goodman rhetorically asks, and of course this question can only be answered in each specific case. When letting go of a stone, I of course suppose it will behave similarly to the stone

I let go of yesterday, that is, according to the same law of gravity, – Peirce's favorite example – and hence will follow a parabolic trajectory. As a matter of fact, Goodman's fifth argument is highly anti-scientific, behaving as if no scientific laws had ever been established.

The sixth argument: 'Similarity between particulars does not suffice to define qualities'. This is not the fact, Goodman argues, because it does not follow from the fact that each two of several particulars are alike, that they are all alike. Objects may pairwise have a color in common without all of them having any color in common. Hence, 'Dyadic likeness between particulars will not serve to define those classes of particulars that have a common quality throughout.' This is correct: similarity is not transitive. Similarity cannot define quality, it is rather defined by it. Goodman's argument here assumes that it should be possible to judge two objects alike without in any way stating in what the similarity consists – which is, of course, not the case. Even in the cases in which it might be difficult to point out the precise similar feature, as between parent and child, it is possible to reason about it and gradually close in on it: 'it is something about the form of the eyes' etc. Even if there thus might be cases in which the tertium is not obvious, it must exist as a not-yet-fully-articulated tacit knowledge prerequisite to the similarity judgment.

The seventh and last stricture: 'Similarity cannot be equated with, or measured in terms of, possession of common characteristics.' Even if it was precisely what Goodman was about to do with the Beethoven œuvre in the third stricture, this is proposed as the more general, conclusive argument, underlying some of the earlier ones. The commonsense idea of similarity between two things – that they have at least one property in common – is once again contrasted with the idea than any two things have a property in common and that consequently this is insignificant. An attempt to rescue similarity by saying that two things having more properties in common are more alike than two things having less properties in common (443) is counterargued by the idea that 'any two things has exactly as many properties in common as any other two'. Goodman gets this idea from the extensional definition of property by class membership (a property simply being defined by a subset of elements), and each element in a given universe of elements is a member of exactly the same amount of universe subsets as any other. This set theory triviality is then supposed to be the definitive argument, for

> I have, indeed, been counting only first-order extensional properties.(...) The inevitable suggestion that we must consider intensional properties seems to me especially fruitless here, for identifying and distinguishing intensional properties is a notoriously slippery matter, and the idea of measuring similarity or anything else in terms of number of intensional properties need hardly be taken seriously. (444)

Now, it is not necessary to maintain that the idea of similarity needs a quantitative measure to be sustained, but it is comical that Goodman at this late point refuses to concern himself with intensional similarity because of its slippery nature – of course this is where the whole problem lies when extensionally defined properties, as he rightly claims, make the concept trivial, just like it makes the whole idea of properties trivial (because the property 'red' should then be defined by the set of all red objects, and this, again, is a set like all other sets and nothing in particular distinguishes it from any other set. Then, how do you zoom in on this set, knowing nothing about redness before you know the set with all its elements . . .). Of course, he adds, one could restrict oneself to counting the *important* properties – a solution he discards rightaway because importance is a 'volatile matter'. But of course similarity is relative to what in a given case is considered 'important' – and this of course lies in the Peircean and Husserlian framing of similarity in icons by 'function' and 'intention' respectively.

In a concluding remark, Goodman sums up his alleged results: similarity is much like motion: 'Clear enough when confined by context and circumstance in ordinary discourse, but hopelessly ambiguous when turned loose' (444). Similarity like movement is meaningless taken per se: 'Talking about motion is pointless if no frame of reference is established. (. . .) We have to say what a thing is to the left of, what it moves in relation to, and in what respects two things are similar.' We could not agree more. Yet, this is not sufficient for Goodman, because unlike motion, similarity cannot be saved by recognizing its relativity: adding a specification of the property in common, similarity is simply rendered superfluous (444). To this, Goodman once more adds the superfluous argument that circumstances alter similarities (of course they do), and concludes

> Relativity, even volatility, is not a fatal fault. Physics does not stop talking of motion merely because motion is not absolute. But similarity, as we have seen, is a much more slippery matter. As it occurs in philosophy, similarity tends under analysis either to vanish entirely or to require for its explanation just what it purports to explain. (445)

Now, one could argue that physics does not stop talking of matter either, even if it be slippery. And Goodman has not, in fact, proved that similarity evaporates under analysis. The idea is that the concept of common property makes it vanish. But a central point is that common property is not enough to define similarity: this property should be measured in relation to a certain granularity on the property scale in question. I might say that two objects are similar because they are both red (two traffic lights, for instance) even if it is not precisely the same red shade. But the property in common thus depends on a certain topology on the quality scale in question, in this case the color continuum. In some cases, this topology may divide the relevant continuum in very few fields, for instance dark and light, in others, it may require thousands of shades, in some cases the property classes overlap, in others they do not. This still makes it correct to assume that similarity is a mere shorthand for something else – but it is not a shorthand for something equally simple (for instance common property), because a given judgment of similarity implies a certain set of possible topologies on the property space (many different granularities may make the two red colors 'similar'). Thus 'similar' means 'invariant under certain transformations in quality space with a certain topology and granularity'. But this is not trivial. And this is why similarity is not eradicable: it is the phenomenological mode for such invariances to appear. In many cases the topology and transformations in the property space in question are evident – and in these cases the concept of similarity of course is trivial – but in other cases similarity is a sign to be interpreted; the precise character of the property space in question and its topology and transformations is a question to be investigated. Thus, similarity calls for further explanation.[68] In the easy cases, the explanation is of course, as are any trivial questions, trivial.

THE YOUNG ECO

In the history of semiotics in our century, the preference for a purely symbolic calculus at the expense of iconicity got a strong wind since the spread of formalism in the interwar period. In the American context, the logical semantics taking its point of departure in Carnap saw truth-conditions of expressions in purely formal systems as the decisive feature of scientific endeavors in semantics, later to terminate in Chomskianism. In the European context, the formalism of Hjelmslev, with its claim that all language description should proceed in a simple algebra, was a forerunner of what was to come in structuralism, both in linguistics and in the structurally inspired humanities more broadly. This tendency had its heyday in the sixties with French structuralism, generative grammar, and symbolic Artificial Intelligence, supported by its biological counterpart in DNA-based neo-Darwinism, when all of the world might seem to be one symbolic calculus – but the general tendency seems to be that it has been slowly receding since then, allowing for a comeback of an interest in iconicity. But still, the anti-iconic tendency remains strong, and in semiotics it is necessary to respond to the arguments of the extreme formalists in order not to fall prey to too naïve a concept of iconicity. Probably the most extreme anti-similarity movement in general semiotics is Greimas' theory, giving rise to the so-called Paris School. Even if the overall architecture of the theory includes a Merleau-Pontyan 'sémiotique naturelle', allegedly phenomenologically functioning prior to the intervention of language, still the concept of iconicity is relegated to a very superficial level of discourse. It is merely a secondary adornment of established meaning in order to render it similar to a given (and always ideological) conception of reality, that is, as promoting Barthesian 'effects of reality'.[69] In this view, icons are simply seen as a secondary and necessarily illusionary addition to basic conceptual ('semio-narrative') meaning already well-established, and all of it supposedly describable in a Hjelmslevian meta-language in which the denominations remain algebraic throughout and hence arbitrary. In contrast to this consequent and untenable position, Umberto Eco's famous attacks on iconicity in the 60s and 70s (as collected in Eco 1976) are at the same time much more multifaceted and much more unclear. It almost seems as if Eco is steadfastedly determined at 'getting rid of' iconic signs, as he does not hesitate to say, because he brings forth a whole series of widely varying arguments, often not even mutually consistent. His 'Critique of iconism' forms a part of his 'Theory of Sign Production' framed in a funny semi-Marxist language of production and labor, and even if it does not figure explicitly as an argument, this overall ontology is probably crucial to his critique of iconism: iconicity, or natural, motivated, analogical signs of any kind seem to come into being 'without work' so to speak. But in Marxism, there must be no free lunch, and consequently these apparently spontaneous signs must have their basis in the human production of codes and conventions revealed: they are bourgeois signs whose apparent freedom is built upon ideological suppression of underlying, more hardworking signs. It might seem like kicking a more than dead horse to counterargue claims like these nowadays, but it still has relevance because of the underlying idea that human subjectivity is responsible for all semiotic activity. This idea is not a privilege for Marxism, and one finds it in various disguises in existentialisms, (de)constructivisms and other heirs to the subjectivist strand in German Idealism on the one hand as well as in extreme formalists like Hjelmslev on the other[70] – and this is probably a very widespread reason for skepticism against iconicity that the acknowledgment of it invariably forms part of a Copernican revolution depriving the human subject of some of its privileges.

Within this overall frame, Eco's arguments are, as mentioned, various, to say the least. One argument goes against similarity as 'shared properties' by attacking Morris' simplified version of Peirce including the fairly reasonable statement that icons resemble their objects 'in some respects' so that 'Iconicity is thus a matter of degree' (Morris, quote from Eco, 192). Instead of understanding the strength of this claim, Eco sees it as a weakness because when stretched to onomatopoieas it includes 'completely conventional' expressions. Here Eco, in a Hjelmslevian craze, compares various linguistic expressions for a cock's crowing; English /cock-a-doodle-doo/, French /cocquerico/, and Italian /chicchiricchi/; the iconic relationship of these to the cock's crowing is 'very weak' (192). In fact, this positive statement is not very plausible; the onomatopoeias mentioned are all close to possible phonetic transcription of the actual sound, and they even display a large amount of similarity in between them: all of them are four-syllable words with the same prosodic structure with a long and stressed last syllable; in all of them the two first syllables begin with /k/ and finally the first and last syllable contains the same vowel (all this goes as well for Danish /kykeliky/, by the way). Eco undertakes this – very weak – analysis in order to counterargue 'shared properties', and his other example is not much more convincing. In this, he compares the photo of a glass of beer in an advertisement with an actual glass of beer. The fact that the photo contains neither glass nor beer nor vapor nor coldness is not a very convincing argument against shared properties when the icon in question is actually based on a *visual* rendering of the object. Eco does, at least, admit the existence of 'visual stimuli', but his explanation of their role in the sign process minimizes their importance to the extreme:

> I feel certain visual stimuli, colors, spacial relationships, incidences of light and I coordinate them into a given perceptual structure. The same thing [sic] happens when I look at an actual glass of beer; I connect together some stimuli coming from an as yet unstructured field and I produce a *perceptum* based on a previously acquired experience. (193)

He later questions even the notion of 'the "same" perceptual effect' and asks if it is not better to assume that 'previous learning' is responsible for my viewing two different perceptual results as the same. Even if it is the 'same' thing that happens in the two cases, the sameness is relegated from the sign and made the product of the I which in turn is made a product of previous learning, that is, culture. That is, an empirically easily verified similarity in form, color, texture, etc. is replaced with a completely unverifiable piece of 'previous learning'. Furthermore, this pseudo-explanation does not account for why 'the same thing' does not happen when I am faced with other 'yet unstructured' stimuli; if Eco's idea were correct I might see beer glasses all around me. Maybe he does. This analysis shows how extreme formalism and its idea of the forming of a 'yet unstructured field' (cf. the structuralist arch-idea of language's partition of an amorphous substance, standard in Saussure as well as Hjelmslev) can be coupled unproblematically with subjectivism and, in turn, culturalism. His further examples are similarly unconvincing: the iconic rendering of a horse by means of its contour is explained by a 'graphic convention' instead of a geometrical transformation of 3-D objective properties of form onto a 2-D surface, coupled with knowledge of the animal as being one concluded object, and the sweetness of saccharine and sugar is allegedly 'not a property of the two compounds, but the result of their interaction with our taste buds'. Still, recent research in the physiology of gustation shows that it is a molecular property which is responsible for this similarity in function, namely the presence on the periphery of both macromolecules of morphologically similar 'active sites' (even if this were not the case, the two tastes would still be phenomenologically similar, i.e. lie close to each other in taste space). The fact that various cultures prefer various tastes – which he adds as a culturalist argument – does not change these biochemical and phenomenological facts and cultural tastes still require similarity functions making it possible to recognize what is considered tasty in the given culture. All in all, his rejection of the shared property idea builds on a construction unluckily uniting subjectivism, culturalism, and extreme formalism.

A geometrical case is now made against the notion of similarity (which he surprisingly finds a more scientific notion than shared properties – presumably because similitude has a use in geometry defined by invariance in certain transformations[71]). Here, Eco's argument is even more coarse, to put it bluntly. A Lockean *tabula rasa* is invoked in the shape of a 'naive interlocutor' or 'non-trained informant' who is not able to perform the transformations required (and so unable to compare the Cheops pyramid with a small model of it, for instance). That transformations such as these should be completely conventional is now argued with the presentation of some simple graph theoretical figurae (Figure 9) which are topologically similar to the extent that they connect the same set of points with the same set of lines (but where the points do not inhabit the same places in the plane and the lines do not have the same shape).

These graphs are of course not similar in the sense that their line shapes are invariant in scalar transformation – and this is for Eco an argument that they are not 'spatially (and therefore geometrically) "similar"' (196). Eco probably feels the danger of trying to drive out one kind of similarity with another, for he instantaneously adds: 'This kind of *isomorphism* may be called a form of similarity but it would be very difficult to assert that it is a geometric similitude. To call such a relationship 'iconic' is a mere metaphor' (197). Now 'mere' metaphor may in many cases contain genuine

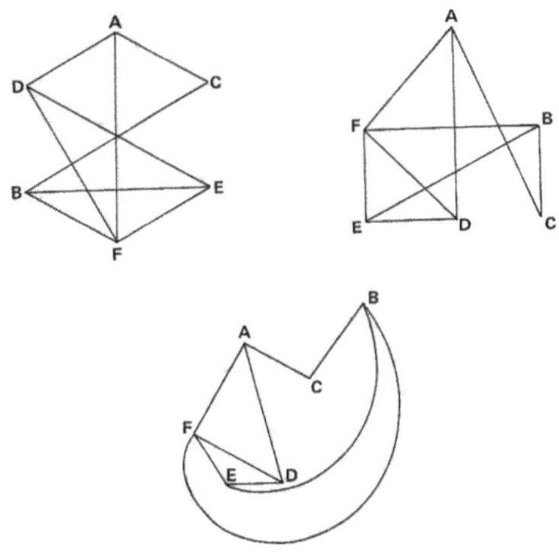

Figure 9.

insight, and such is the case here. Of course isomorphism is iconic; the question is why Eco refuses to admit it. A positive answer could be that he fears a concept of iconicity which is spontaneous and substantial, maintaining a similarity without a *tertium comparationis* – this would explain the strange fact that he considers it an argument against iconicity to reveal in each case the underlying principle of similarity. Of course, Eco's later work has shown a rationalist and well-placed fight precisely against unfounded similarity claims in various occult practices where no *tertium comparationis* and no correlative invariants under transformation are to be isolated.[72] But the fear of such 'similarities' should be no argument against well-founded similarities, and it seems that Eco in general mistakes the transformations guaranteeing the single types of similarities for being mere cultural conventions. Thus, in addition to the earlier arguments which pointed to large constructive competences on the part of the ego, this Lockean construction makes of the I a mere tabula rasa.

But more arguments follow. Eco confronts himself with a version of Peirce's iconicity and criticizes Peirce for not abandoning his reference to objects in his definition of it, because it makes iconism an 'umbrella-term that covers many different phenomena such as a mental image, a graph, a painting' (199). This argument is very strange; it is precisely the reference to similarity with the object which gives the unity of Peirce's definition and which allows it to include the various phenomena mentioned. There are two points in this: one is what we have already touched on and what Eco emphasizes over and over again: that a transformation between sign and object '... does not suggest the idea of natural correspondence; it is rather the consequence of rules and artifice' and hence remains a 'mere' convention. But rules can never be the *causes* of similarity, for they presuppose it in so far as their general formulation is of the type 'Do *the same* as what is prescribed here ...'. The same goes for his emphatic conclusion 'Similitude is *produced* and must be *learned*' – it would be impossible to teach anybody anything if one were not allowed to say 'Now, do *like I do* ...' and thereby presuppose similarity.[73] This argument, of course, is also valid against Goodman's similar claims. Another point is not stated explicitly, but it might seem as if Eco's implicit contrast to Peirce's alleged 'umbrella' term is a rather narrow idea of iconicity restricted to the domain of vision only. This of course brings him in conflict with the culturalism stated so far, because vision and the possession of eyes can hardly be interpreted as a cultural convention which must be learned. This comes to the fore in his next argument which is aimed against the symmetry of certain similarity phenomena which must consequently be excluded from the concept of iconicity: mirror images, doubles, tokens of the same type[74] and expressive signs. This argument is of course perfectly valid, cf. the central phenomenological distinction between symmetrical similarity and asymmetrical iconicity; yet it is no argument against the latter that it depends on the former from which it must be distinguished. But in this almost sound argument totally new criteria all of a sudden sneak in: 'Secondly the presumed 'iconism' that should govern the correspondence of a token to its type is not a *theorem* that semiotics could demonstrate; it is one of its *postulates* [...] The rules of this recognition are deeply rooted in the mechanisms of human perception and must be assumed as already given in any semiotic enquiry' (203). Here, semiotics is all of a sudden presented as a deductive doctrine with postulates, axioms, proofs, theorems etc., and token-type relations are wholesale reduced to physiology, that is, biology. But why should it not be important to guarantee the soundness of the postulates of one's science – which in this case are empirical matter from other sciences (cognitive psychology)? This sudden naturalist tendency, hastily forgetting the *tabula rasa* ideas, culminates in the refusal of iconicity in the so-called 'expressive' signs – Kandinskyan lines signifying emotions and the like: 'We may consider all these cases of empathy as mere *stimulations* that should be studied by the physiology of the nervous system.' (203) – as if they were not semiotic phenomena at all. A strange theory: seeing a picture of a beer as referring to a beer is a complicated cultural phenomenon, while seeing a drawn line as referring to a complicated psychological emotion is something perfectly natural and automatic. Just like some icons are thus 'mere' conventions for a completely blank Lockean culturalized subject, other icons are 'mere' neurological phenomena in a naturalized physiological subject brimming with innate semiotic competence. It is a wonder that these two subjects may thrive in one and the same head. The problem of cutting the cake and dividing the iconic signs amongst these two 'mere' categories is of course the numerous cases in which both are active and hard to distinguish. Eco elaborates on conventionality at length, making clear his contention that this is the semiotically interesting field (the natural 'signs' not being signs but something even simpler than Pavlovian conditioned reflexes), but finally embarks on still another type of argument by stating that measured on very general (we would call them formal ontological) terms, similarity becomes ubiquitous: '...on the level of very elementary formal phenomena such as high-low, right-left, or long-wide – everything resembles everything else' (212). Perfectly correct as this is, Eco takes it strangely enough as yet another argument for 'arbitrary underlying codifications' and is close to concluding with a complete conventionalism. Crude as it is, he finally offers a totally different classification between grammar-oriented and text-oriented procedures, in which iconicity must fall on the text-oriented side because of its lack of distinct elements and compositionality. Here, he suddenly admits that 'the drawing of a horse can be understood even by those who are not acquainted with visual conventions...' (214) because it is '*not further analyzable* either into signs or into *figurae*' (215). Now, the equivalent of an iconic sign is not a word or a phrase but a text (why not a phrase in many cases, actually, cf. traffic signs?) whose units 'are established – if at all – by the context' (216). It is difficult to see how this analysis at one and the same time can save iconic signs and reject their iconicity; the idea is probably that the context here is supposed to do the work which codes, conventions or physiology cannot do. But how does one recognize similar contexts if not by means of similarity (remember we are beyond conventionality now, so we must not posit codes for recognizing contexts...)

Eco's conclusion, under the witty heading of 'Getting rid of the "iconic signs" ', repeats the umbrella idea and adds a final stroke: it is not only the idea of iconicity, but also the idea of a *sign* which must be discarded in favor of the notion of *sign-function*, an in itself sound Hjelmslevian notion here diversified into various semi-Marxist modes of sign production. This idea should finally dissolve the 'iconic signs' because they are arguably produced by different procedures, interchangeably with other signs.

This final fallacy – to believe the product may be exhaustingly explained by the production – is as absurd as to claim that cars are not cars because some of them are produced by Germans, others by Japanese, but it permits us to conclude Eco's critique. His umbrella term thesis rests on this mode-of-production theory – and then the wildly differing critiques of iconicity might be understandable (to himself) as pertaining to different modes of production (even if they were not presented in this lingo). Conventionality, culture, and subjectivity constitute one source for apparent 'iconicity', the physiology of the nervous system is another, formal ontology is a third, and the context of signs constitutes a fourth. Even if a wedding of some of these positions into a culturally formed subject strangely and completely independent of its own natural, physiological bases can be constructed and is indeed typical for the period, then no construction can involve all of them: where should the formal ontological constraints intrude: in the subject? – this would attack culturalism – in the nervous system – this would attack the independence of nature and culture. Where would 'context' come in? – not in culture which is conventional through and through, not in the physiology of perception since no semiotics is possible here... Eco's iconic cleansing project is doomed to fail, not only because he attempts to dissolve a category phenomenologically well-defined, but also because his umbrella argument ends up umbrellaing his own implied ontology into a complete quagmire of a crude nature-culture dualism equipped with self-contradictory annexes. What can more generally be learnt from Eco's failures is that mode-of-production explanations in semiotics, be it of semi-Marxist brand as is the case here, be it of Nietzschean brand in various types of Lebensphilosophie and deconstructionism, or be it of neurophysiological or any other brand, can never stand alone but must be based on a prior phenomenological descriptive rendering of the field making coherent the objects, distinctions, and in general phenomena involved.

UMBERCEPTION AND ECONICITY – THE CONVERSION OF AN ICONOCLAST

Eco thus formed an anti-iconic front figure in the 'debate on iconism' taking place in semiotics in the 60s and 70s. In a recent book, *Kant and the Platypus*, however, Eco explicitly changes his basic conceptions on the issue. The main tenet of the book is the integration of a whole series of issues left out of consideration alongside with iconicity in Eco's early work: reference, cognition, truth, prelinguistic perception – rather serious and far-reaching questions, it must be admitted. In the present book's last chapter, Eco recounts an outline of his and his compatriots' position at the time, and I think it is fair to say that he to some extent misrepresents the scope and the radicality of the anti-iconism of the period, probably out of the same embarrassment that makes him characterize conversions from this semio-structuralism and to Peircean iconicity in some of his fellow-travellers as the result of confessions in a Stalinist show trial (341). This somewhat shrill description should probably be read as a testimony of the malaise inherent in the revision of one's own viewpoints, for it is, in fact, exactly the same conversion confession that is Eco's own purpose with this book: the admirable task of revising his 1976 position on iconicity. It is evident, furthermore, that a strong influence in this revision is what must be counted as one of the major developments in the semiotics of the turn of the millennium: the American 'cognitive semantics' tradition (involving Rosch, Lakoff, Johnson, Turner, Sweetser, Talmy, Fauconnier, etc.). Even if only rarely using the term 'semiotics', this tradition has strongly vitalized the development of a bouquet of core semiotics issues (general semantics and its relation to perception, to thought, to biology, etc.) by connecting them to the general cognitive science project. From an early point, Eco was one of the European contacts to this tradition, and he has done much to introduce it in Europe. Thus, *Kant and the Platypus* can also be read as his contribution to this development, involving a welcome revision of his own earlier viewpoints.

Even if this self-criticism gives rise to a whole series of really interesting ideas, the book is explicitly not intended as a systematic rejoinder to the (intended) systematical *Theory of Semiotics* of the book thus titled. The six essays in *Kant and the Platypus* come in seemingly arbitrary order and with sparse internal references, even if they circle around the same central set of issues. The first is a general (ontological, in fact) meditation on being as a positive issue – in contradistinction to semio-structuralism's preference for negative determinations and its resulting hesitations towards any ontological commitments – and it concludes with a discussion of structural semiotics' arch occupation with an amorphous continuum of being to be segmented by semiotic systems. The fact that this continuum possesses a 'grain', contains 'lines of resistance' as given before semiosis and to some extent governing it, is the anti-conventionalist point of the still rather weak ontology (references to Vattimo are weak but not missing, consequently) of this chapter. The second essay tackles the main problem head on: the explanation of the semantics of everyday empirical concepts (like 'dog', or, more conspicuously, 'platypus') cannot do without the (iconic) notion of 'schema' like in Kant's epistemology. The ideas from this chapter are taken further in the third essay where the Kantian impetus is reinterpreted in the light of actual cognitive science. Here, Eco constructs his own theory of the semantics of empirical concepts: they build on the prelinguistic perceptual generalization (in Peirce's terminology: perceptual judgment) giving rise to 'cognitive types' (CT) making recognition and identification of a phenomenon possible. This stability in perception seems to form the core of semantic content of the concept, the so-called 'nuclear content' (NC) involving a fuzzy spectrum of core knowledge attached to the CT. Finally, following Putnam's idea of a linguistic division of labor, various expert bodies of knowledge about the phenomenon in question may be elaborated with the NC as point of departure (and possibly contradicting it); they form elaborated corpuses of 'molar content' (MC). In the platypus example, the CT will be the perceptual type of the animal, making recognition of it possible; the NC will involve central pieces of knowledge about it (it has a beak, lays eggs, gives milk, etc.); and the MC will involve different sets of elaborated representations of it (the aboriginals' mythological ideas of the animal's role in cosmology, the scientific description and classification of it, etc.) – the latter possibly being subject to ongoing discussions and negotative reinterpretations. The fourth chapter recounts the polemic around the classification of the platypus raging in biology during most of the nineteenth century and draws some general conclusions as to the indispensability of both of Eco's well-known semantic description types of dictionary vs. encyclopedia in the forming of empirical concepts, permitting the revision and the integration into the new CT/NC/MC framework of these two terms. The fifth essay turns towards the question of reference and takes up Kripke's famous 'ontological' theory of reference: the idea of proper nouns as rigid designators devoid of any descriptive value whatsoever. Kripke's theory is relativized to become a regulative idea only in a more pragmatic theory of reference as subject to the same ongoing contractual negotiation between language users as content is subjected to. The final essay on iconism and hypoicons is somewhat disappointing in relation to the strong chapters in the middle of the book; we do not get the integration of the book's insights in a broader theory of iconicity, such as might be expected from its title. Still, the essay contains a interesting distinction between two forms of icons, *alpha*- and *beta*-, respectively, so that the *alpha* signs are perceived as signs irrespectively of any explicit intention of sign reading in the receiver (the dog seen as 'a dog', supposedly), while the *beta* signs presumes to be read as expressions of a sign function (the dog seen as a sign for a nearby kennel); this distinction apparently gives rise to the distinctions between primary and secondary icons equivalent to the distinction between 'perception surrogates' and more proper signs.

Let us begin by the reintroduction of iconicity in a central role (and it probably goes without saying that I perfectly agree with Eco on that central point). The question leading to the recognition of iconicity in Eco is: 'How do we assign names to things?' Simple versions of the scholastic (by essences) and the empiricist (by complexes of ideas) answers are both refuted in favor of the Kantian notion of schema. The schema is a *type* and hence no mere association of particular ideas (which would be impossible for the strong reason that empirical objects have an infinity of properties (Kant)), but it is on the other hand a result of construction and is thus no pregiven essence ready to be picked up. Eco runs through the Kantian doctrine of the reflective judgment in *Kritik der Urteilskraft* in order to underline the schema's constructed character: the reflective judgment seeks to constitute a general concept to subsume a particular phenomenon (and it is thus a precursor to Peirce's notion of abduction, cf. Chap. 16), and in so doing, it conceives of the thing *as if* it was a part of something general. In order to do so, furthermore, it conceives of the thing *as if* it was teleologically organized.[75] In any case, this teleological and generalizing feature of reflective judgment commits it to proceed by trial-and-error – that is, in a construction of a schema which will be able to subsume the phenomenon in question. In so far, the schema becomes the general type which permits the recognition of tokens of it. This is explicitly posed as a general prerequisite of signs: well before anything can stably stand for something else (the Stoic sign definition), a type must be able to stand for a token of itself. This sound theory is, it must be added, nothing new in semiotics; it is central in Peirce (the first trichotomy) but it also is inherent in e.g. Hjelmslev's version of 'semiostructuralism' – but still it is good to emphasize it as against the widespread and fateful superstition that signs begin with physical marks referring to something else. This primary iconicity is taken as a primitive: it *defines* similarity rather than the opposite way around. Primary iconicity thus satisfies Eco's old ambition of finding the 'lower threshold of semiotics' and is described in several ways: one is the psychological description as the adequate representation of a stimulus by a sensation. The prerequisite for this is a pure 'predisposition to correspond' (110). This last description corresponds to Peirce's idea that the icon is at the same time objective and vague, but there is an unnoticed tension between this objective and hence pre-subjective description of iconicity on the one hand and the just mentioned identification of iconicity with psychological and perceptual processes on the other. A mild conclusion here could be that this difference is not further elaborated in this book where the focus on the genesis of empirical concepts implies a natural bias towards psychology; a stronger conclusion could be that there is a problem in the fact that Eco's reintroduction of iconicity is so tightly intertwined with a reintroduction of psychological terms (which is strange, moreover, because Eco in general takes the cautious stance of seeing the mind as a black box which he, as a semiotician, abstains from peeping into). I shall return to this problem.

The schema thus satisfies the Husserlian requirement that there should be strong ties between linguistic and perceptual meaning. As is evident, this rendering of schema and iconicity lays a heavy burden on perception. Peirce's theory of perception – to which Eco returns over and over without tackling it head-on – is taken as the means to fill this gap: the relation between percept (a limit case with no autonomous mental existence in Peirce's account) and the consecutive perceptual judgment is quickly taken as a general description of the token-type relation in primary iconism.

This is taken up in the third essay where Eco outlines the most elaborated and original consequences of his iconic turn. The schema is an unsolved problem for truthfunctional semantics on the one hand just as it is for structural semiotics on the other, and the flora of schema-like concepts flowering in the cognitive semantics tradition these years (schema, prototype, stereotype, model, pattern, script, frame ...) testifies to the indispensability of schematic concepts. In so far, a schematic turn can be said to be underway, but, as Eco points out, without its proponents knowing much about the philosophical debts or depths in the concept. The main problem, according to Eco (here I only partially agree) – in continuation with the preceding chapter – is that Kant's schematism does not account for empirical concepts (the dog, the platypus, etc.). Accordingly, he takes up the development process of such concepts – Montezuma's and the Aztecs' gradually forming a sort of horse concept after the first meetings with the Spaniards and the polemic over the zoological classification of the platypus being his two main examples. In accordance with the central role accorded to perception, he supposes that the first stable structure formed is a 'cognitive type' (nicknamed CT) which is a perceptual schema, integrating a multimodal range of senses (including for instance the horse's general outline, neighing, smell, etc.). The CT, even if subject to ongoing trial-and-error negotiations, has the property of being a perceptual type, hence prelinguistic, and hence safe from all semiotic doubt, cultural relativism, etc. True, different cultures will form different CTs, but as a rule not *entirely* different – they will still be constrained by certain 'lines of resistance' as it is called in Eco's weak ontology. Its main function is to facilitate the recognition of yet-unseen tokens of its type. The important step here is that Eco succumbs to Peirce's insistence that the generality of the symbol is not the only generality in semiotics; it is preceded by – and conditioned by – typicality, the fact that phenomena tokens are organized in types before symbolicity and linguistic categorization further organize them. However, the CT is, due to its perceptual status, private, so how do we guarantee that a CT is in fact present? – only by the intersubjectively controllable detour of successful referring. If speakers pragmatically agree in referring to a phenomenon this must count as a proof of shared (or, in any case, sufficiently similar) CTs. This brings the next term in play, Eco's concept for the meaning (or better, the content) of empirical terms, the Nuclear Content (nickname: NC). It is defined as a set of public interpretants of a CT (and must be supposed to feed back onto the ongoing shaping of the CT) and becomes the possibility of the attaching of a substantive expression making the content communicable. The relation between CT and NC is conceived so that CT acts as disposition for the formation of a NC, while the presence of a NC, conversely, counts as proof of the existence of a CT. (There are problems in this claim: everyday substantives abound which have NC but hardly CTs – for instance 'furniture' – but this may probably be mended by a suitable theory of generalization and abstraction). The NC gives instructions for the identification of tokens of the type (its so to speak iconic side) as well as instructions for retrieval of tokens of the type (its so to speak indexical side). So it adds a considerable amount of knowledge to the perceptual CT of for instance a horse: where do we find horses, to what use are they typically put etc. The NC is still a common-sense-close notion, being subject to continuous negotiation and not necessarily internally consistent, being motivated by competing tacit folk theories, etc. The MC forms the third concept of this triad and constitutes some stable corpus of complex knowledge of the object (it contains more than is demanded for perceptual recognition) and is typically parceled out in different, sophisticated practical and theoretical discourses on the object, also subject to continuous negotiation in smaller subcultures but with stronger consistency constraints added. In this analysis, the very concept of 'concept' is left to unemployment, to the extent that everyday empirical concepts are CTs, expressed by NCs, while scientific and other special language concepts are defined by various MCs. It is evident that Eco's CT-NC complexes display close relationship to Eleanor Rosch's 'basic-level-categories': he is happy to state that they are interpretable even if primitives – and on the Kantian hand he maintains they act as general pictures acting as rules for the construction of specific pictures.

The more precise definition and description of CTs take up many efforts. CTs seem also to include qualia, basic acts (walking, jumping), basic opposition pairs (husband/wife), frames governing events, they must contain Gibsonian 'affordances' (what may be done with the phenomenon in question), they include face recognition, recognition of musical styles, fictive persons and artworks; we often use truncated CTs completely separated from NCs. One and the same CT may emphasize iconic, propositional and narrative meanings. CTs governing empirical concepts are juxtaposed 'cultural' concepts including abstract concepts, cultural relations, events, etc. CTs are compared to cognitive semantics concepts like stereotype and prototype and it is proposed that CTs have maximum extension and minimum intension, while the opposite is the case for prototypes.

These sketchy extrapolations of the CT concept display some weaknesses. First, CTs now seem to include a whole series of semantic phenomena traditionally seen as farther from perception: frames, opposition pairs, functional properties etc. The fact that phenomena like these in fact do form CTs leaves little doubt; but the idea that they spring almost automatically from perception seems a lot weaker in these more conceptual cases. The basic tie between CT and perception thus seems to be loosened without it being explained how CTs of these traditionally more 'conceptual' content types are formed. Moreover, the opposition between empirical CTs and cultural CTs seems very hard to uphold on the proposed basis which makes the latter farther removed from perception than the former: a very wide range of cultural objects (from Coca Cola bottles to saints) are recognized by means of simple perceptual schemata, while on the other hand it is an extremely doubtful claim that abstract concepts should in general be culturally specific through and through. As to the alleged opposition between CTs and prototypes, the idea that CTs should have minimum intension fits badly with the earlier claim that they include multimodal representations for object recognition (including smells, sounds, etc.).

While the basic layout of the CT/NC/MC theory thus seems a very promising idea, its elaboration suffers from serious lacks, supposedly because of their ties to (too) simple concepts of empirical phenomena, psychology, and culture inhibit their extrapolation. It is almost as if Peirce's much-quoted (also by Eco) remark – that if Kant had taken the consequences inherent in his draft of schematism, it would have overgrown his whole work – has a bearing on Eco's reflections as well.

The fourth essay recounts the platypus strife of the nineteenth century with the aim of reconciling the CT/NC/MC theory with Eco's earlier discussions of dictionary vs. encyclopaedia semantics, the former being characterized by hierarchical organization, linguistic determination, a limited inventory of semantic features, proceeding by definition (corresponding to structuralist analyses of meaning), the latter being characterized by an uncoordinated and indefinite mass of extralinguistic knowledge, proceeding by classification (corresponding to an empirical, everyday conception of meaning). In relation to the NC/MC distinction, a first glance might suggest that NC/MC correspond to Encyclopaedia/Dictionary, respectively (by the common feature unordered/ordered), but in the opposite direction points the fact that NCs are supposed to be simpler than MCs, and any precise correlation between the two concept pairs is given up. Eco after long deliberation places 'wild' categorization in NC while systematic categorization belongs to the MCs. Dictionary semantics is characterized – in both MC and NC – by limiting itself to register that a concept is located on a certain node in a classificatory tree, while encyclopaedic knowledge includes both locations of concepts in classifications as well as their further content. The conclusion by the platypus example is that an interaction between dictionary and encyclopaedia knowledge is necessary in the ongoing negotiation leading to from wild to less wild classification; it takes its point of departure in one and the same CT/NC which is impossible to doubt as a whole (the existence of such a strange creature as the platypus is beyond doubt, even if certain single properties (its laying eggs, its giving milk, etc.) may long be doubted), while the interpretation of this core in terms of MC is highly variable and subject to possible systematical doubt (is the platypus bird, mammal, reptile, etc. . . .) giving possibility for scientific progress.

A further possibility of development in this chapter is to elaborate on the distinction between CTs and NCs, prelinguistic categories and linguistic categories, respectively. Rita Nolan has in a strong paper proposed that the distinction between perceptual and conceptual categories may be drawn according to whether the category is contrasted to other categories.[76] This would point to the idea that the role of CTs – perceptual categories – is to facilitate identification (as a type), while the role of the NCs developed out of them is to facilitate classification. Identification and classification have much too often been identified in semiotics, but an extrapolation of Eco's proposal may lead to a theory where the latter is seen as a more complicated process presupposing the former.[77]

The fifth essay on reference is one of the book's best and takes the negotiation-semantics-idea from the field of icon and content to the field of index and reference. Kripke's well-known rigid-designator theory of proper names and their ontological reference is criticized with an Italian comical sketch as an example. Eco's argument is convincing: a completely naked reference without any descriptive content is only possible as a limit case, and rigid designation must be reinterpreted in a Kantian fashion as a regulative idea governing the ongoing research process rather than being the normal reference relation in ordinary language where pragmatic reference by negotiation is the rule: 'When people listen to acts of reference, they usually ask lots of questions'.

The reintroduction of primary iconism is nothing less than a semiotic necessity, and the discussion of Kant's schematism is highly relevant for the development of a semiotics between truthfunctional reduction of meaning on the one hand and various irrationalist claims of the ineffability of meaning on the other. The CT/NC/MC theory is a valuable outline for a theory of content between abstract concepts on the one hand and the multiplicity of perception on the other, and the negotiation theory of reference continues the valuable insight in Kripke reference theory in a pragmatic setting.[78]

THE SCHEMATIC TURN

Still, the lack of systematic disposition and conclusions in the book is highly regrettable. It contributes to the list of unanswered questions and problems in the position it puts forward. Let me name a few.

First of all, the delimitation of iconicity and schemata to empirical concepts with a close connection to the psychology of perception is very hard to understand. Of course, the understanding of empirical concepts is a spectacular problem, but if we should avoid the pitfalls of empiricism and psychologism making their re-entry in semiotics alongside the reintroduction of iconicity, it is crucial to see the problem of empirical concepts in a close relation to abstract concepts – just like it is the case already in Kant where schemata are seen to account for empirical concepts as well as a priori concepts. This idea in Kant is continued in Peirce's theory of diagrams – see next chapter – where we learn that empirical concepts are schematic only to the extent that they are built from pure diagrams invested with empirical constraints and references. This implies that Eco's account for schemata overlooks the crucial connection between observation and generality in schemata, to which we shall return in the next chapter (to some extent Peirce's version of the synthetic a priori). In diagrams, it is possible directly to observe generality, Peirce claims; Husserl had an analogous insight in his theory of categorial intuition in the sixth of the *Logische Untersuchungen* where he claims that in order for categories to be understood there must exist a kind of perception allowing us to grasp them (see Chap. 6). It is the observability of general, 'abstract' properties which is the crux of schemata. Here, all reference to human psychology is bracketed, and the corresponding extension of the icon category makes it cover anything from photos to algebra, from perceptions to graphs, in so far as these signs allow for the crucial operations to be performed, leading to more information about their object. As Eco fails to see this, his reintroduction of iconicity threatens to become the reintroduction of a culturalized variant of (psycho-)logical positivism (without his intention, no doubt) in semiotics. This becomes evident, when he (253) summarizes the contributions to the determination of the platypus in the following two currents: one body of perceptual observation sentences based on the intrinsic characteristic of the environment, and, on the other hand, one holist (in a Quinean sense), structural system of propositions which performs a 'cultural segmentation of the content' (corresponding to the strange subsumption of abstract meanings under cultural CTs). Protocol sentences, on the one hand, and on the other the means to organize them, be they logical or psychological of origin – the old couple so well-known from psychological or logical empiricism – with the one difference that the (psycho-) logical apparatus here is supposed to be culturally variable. It is well known that in all positivist ways of cutting the cake, what is deliberately left out is any reference to the synthetic a priori; no a priori is left but analytical tautologies. In Eco, this corresponds exactly to the underestimation of the general content of schemata, and it becomes very difficult to see which device should be able to mediate between the cultural concept systems on the one hand and the observation sentences on the other. Even if a huge step foward, Eco's newfound iconicity thus inherits some of the culturalist problems of his earlier position. The whole a priori set of concepts yielding the basic schemata of each of the single special sciences falls away – ironically in a period where analytically minded philosophers reintroduce the synthetic a priori (cf. the idea of 'fallabilistic apriorism', Chap. 8). Of course, the reason why the cultural reference is included here is in order to answer an important question of fundamental semiotics: how are cultural differences possible – but the answer to this question must not beforehand exclude the answer to an equally important question: how is scientific knowledge possibe? This disappearance of an autonomous account for abstract concepts in Eco's theory is indirectly apparent in the rather generous amount of information supposed to be present in the observation-close CTs. Here, it is for instance a pre-linguistic presupposition that a dog is an animal (in no scientific sense, it is true, but still supposed as a pre-cultural knowledge impossible to delete). But is it really appropriate to describe so complicated information as inherent in perception – even if admitting Peircean generality inherent in perception? – in any case it presupposes a very developed concept of perception. We can easily subscribe to Eco's observation that the idea that dogs are animals is culture- and language-independent – but is this fact not more satisfactorily described when we say that an a priori concept of animal is involved here – one that we might even be able to schematize (as something like a self-propelled metabolism looking for nutrition, cf. Chap. 9)?

The admittance of abstract relations as crucial to the construction of schemata would yield a whole series of further consequences: the variation of schemata, the experimentation with schemata, the fallibilistic reinterpretation of a priori schemata, the distinction between the icon and the psychological conditions for the processing of it. Let me conclude with arguing for this further list of steps to be taken in order to complete the schematic turn of semiotics.

There is a crucial variation procedure connected to schemata linking type and tokens, and this goes for empirical and a priori schemata alike. The general picture of the schema must be varied in order to yield subtypes and tokens of the typical schema. But not all schemata allow for the same formal kinds of variation nor the same scope of variation procedures. In dogs, e.g. the variation of the type must be rather extensive in order to cover races like Great Dane and Chihuahua, respectively, while the variation allowed in platypuses, presumably, is somewhat more restricted. In the concepts of animals, continuous variations (within a whole set of further *Bauplan* constraints) are relevant, while in the concept of chess games, discontinuous variations prevail. A huge issue is suggested by these remarks; suffice it to say that the types and scope of variation thus forms a part of the very concept of schema itself and the specification of the variation's type and scope forms an indispensable part of the content associated with each specific schema.

Eco touches upon this important idea in Gibson's concept of 'affordance', the fact that schematic content in some respects indicates 'what may be done' with the phenomenon in question. In addition to the variation linking type with token, this forms an even more extensive field of schematic research: which rule-bound manipulations may be undertaken on a schema in order to map which corresponding changes may take place in the object it depicts? Peirce's schema concept, that of a diagram, makes explicit this dimension of experimenting or manipulating in every diagram – in the animal schema it will be the manipulation of it to simulate animal behavior; in more abstract schemata – like logical or algebraical expressions – it will be the possibility of proving theorems or solving equations. This property is what makes schemata fit for *Gedankenexperimente* of every kind, cf. next chapter.

If we apply this experimenter's stance to abstract schemata we may get a crucial corollary as to the notion of a priori. In Kant, the notion covers knowledge which is valid without regard to empirical facts and hence pre-given for the transcendental subject. What diagrams make clear is that these two properties are not in any way synonymous. What is valid regardless of empirical facts is not for this reason self-evident for us (which the history of mathematics should suffice to convince us). This implies that a priori knowledge is exactly as fallibilistic as empirical knowledge and our access to it takes place via experimentation on schemata (see Chap. 8). This idea even fits very well into Eco's idea of a negotiation semantics which may easily be enlarged so as to encompass the evolution of abstract knowledge.

Having asserted the central place of abstract relations in schematism, we may propose a distinction between iconism as such (cf. Eco's 'disposition to correspond') which needs no human or other psychology to be realized, on the one hand, and the psychological means at human disposal for recognizing iconicity, on the other. The former, iconicity as such, may be studied without reference to the latter. In doing so, semiotics will remain faithful to its anti-psychological and phenomenological foundations. Equipped with these means, semiotics should be able to take further the *schematic turn* which Eco so fruitfully proposes, and to build a semiotics which is iconic and thereby enlightened rationalist in the best sense of the word.

This repudiation of more and less consequent anti-iconists takes us back to the Peircean definiton of the icon. Nowadays, everybody seems to know and love the Peircean tripartition of signs into Icons, Indices, and Symbols, referring to their objects by means of similarity, contiguity, and habit, respectively. Yet, there is more to the Peircean notion of Iconicity than meets the eye in this deceptively simple definition. Let us run through some of the central descriptions of Iconicity in Peirce. It must here, as mentioned above, be borne in mind that Peirce, being a realist, rarely proceeds by *defining* his notions. Icons are to Peirce real existent phenomena, and they may be *described* in a series of different ways; no simple definition will exhaust them. In fact, as it will be clear, this feature of his realism is closely connected to the very notion of iconicity itself.

Let us present a range of his different descriptions of icons and discuss the crucial implications involved therein.

SIMILARITY AND QUALITY

First of all, as is well-known, iconicity is dependent upon similarity: icons are 'signs whose significant virtue is due simply to its Quality' ('Minute Logic', 1902, 2.92).[79] This quality amounts to shared characters between sign and object:

> An Icon is a sign which refers to the object that it denotes merely by virtue of characters of its own, and which is possesses, just the same, whether any such Objects actually exists or not. It is true that unless there really is such an object, the Icon does not act as a sign; but this has nothing to do with its character as a sign. Anything whatever, be it quality, existent individual, or law, is an Icon of anything, in so far as it is like that thing and used as a sign of it. ('Syllabus', 1903, EPII, 291; 2.247)

Here, similarity is simply equal to shared qualities. As already mentioned, a complete Icon would stand in a relation of identity to its object except for its existence which in Peirce's Kantian tradition is of course no predicate, no quality. Hence, Peirce does not subscribe to Leibniz' principle of the Identity of Indiscernabilities: two objects may be perfectly alike except for their existence in time and space which is not counted as a quality but as a haecceity. An icon which shares only some qualities with its object has so to speak a general side in so far as it is blank with respect to the aspects not partaking in the Icon; these are bracketed by prescission in the Icon. Even if the very Icon in itself is not, as mentioned, general, one could say that any Hypoicon (Peirce's notion for any sign which primarily functions by means of iconicity) thus involves a germ of generality. This becomes the possibility of various more or less strict stylizations of Icons, rendering certain qualities important, others not so, often exaggerating the qualities taken to be important, whereby the Icon acquires a certain degree of typicality, being able to subsume tokens under it. Thus, an Icon may be a type, a Legisign, without any intervention of Symbols; in fact, this forms the very basis of Peirce's realism.

PREDICATES

Because Icons are the means of representing qualities, they generally constitute the predicative side of more complicated signs:

> The only way of directly communicating an idea is by means of an icon; and every indirect method of communicating an idea must depend for its establishment upon the use of an icon. Hence, every assertion must contain an icon or set of icons, or else must contain signs whose meaning is only explicable by icons. The idea which the set of icons (or the equivalent of a set of icons) contained in an assertion signifies may be termed the predicate of the assertion.
> ('That Categorical and Hypothetical Propositions are one in essence, with some connected matters,' c. 1895, 2.278)

Thus, the predicate in logic as well as ordinary language is essentially iconic. It is important to remember here Peirce's generalization of the predicate from the traditional subject-copula-predicate structure. Predicates exist with more than one subject slot; this is the basis for Peirce's logic of relatives and permits at the same time enlarging the scope of logic considerably and approaching it to ordinary language where several-slot-predicates prevail, for instance in all verbs with a valency larger than one. In his definition of these predicates by means of valency, that is, number of empty slots in which subjects or more generally indices may be inserted, Peirce is actually the founder of valency grammar in the tradition of Tesnière. So, for instance, the structure '_ gives _ to _' where the underlinings refer to slots, is a trivalent predicate. Thus, the word classes associated with predicates are not only adjectives, but verbs and common nouns; in short all descriptive features in language are predicative.

This entails the fact that the similarity charted in icons covers more complicated cases than does the ordinary use of the word. Thus,

> . . . where ordinary logic considers only a single, special kind of relation, that of similarity, – a relation, too, of a particularly featureless and insignificant kind, the logic of relatives imagines a relation in general to be placed. Consequently, in place of the *class*, which is composed of a number of individual objects or facts brought together by means of their relation of similarity, the logic of relatives considers the *system*, which is composed of objects brought together by any kind of relations whatsoever.
> ('Detached Ideas Continued', 1898, NEM IV, 339)

This allows for abstract similarity because one phenomenon may be similar to another in so far as both of them partake in the same relation, or more generally, in the same system – relations and systems being complicated predicates. In this respect, a father is similar not only to other fathers, but to sons; the father-son relation is similar to the mother-daughter relation, etc.

But not only more abstract features may thus act as the qualities invoked in an icon; these qualities may be of widely varying generality (just like Eco noticed in the passing):

> But instead of a single icon, or sign by resemblance of a familiar image or 'dream', evocable at will, there may be a complexus of such icons, forming a composite image of which the whole is not familiar. But though the whole is not familiar, yet not only are the parts familiar images, but there will also be a familiar image in its mode of composition. (. . .) The sort of idea which an icon embodies, if it be such that it can convey any positive information, being applicable to some things but not to others, is called

a *first intention*. The idea embodied by an icon, which cannot of itself convey any information, being applicable to everything or nothing, but which may, nevertheless, be useful in modifying other icons, is called a *second intention*. ('The Regenerated Logic', 1896, 3.433).

What Peirce distinguishes in these scholastic standard notions borrowed from Aquinas via Scotus, is, in fact, the difference between Husserlian formal and material ontology. Formal qualities like genus, species, dependencies, quantities, spatial and temporal extension, and so on are of course attributable to any phenomenon and do not as such, in themselves, convey any information in so far as they are always instantiated in and thus, like other Second Intentions, in the Husserlian manner dependent upon First Intentions, but they are nevertheless indispensable in the composition of first intentional descriptions. The fact that a certain phenomenon is composed of parts, has a form, belongs to a species, has an extension, has been mentioned in a sentence etc. does not convey the slightest information of it until it by means of first intentional icons is specified which parts in which composition, which species, which form, etc. Thus, here Peirce makes a hierarchy of icons which we could call material and formal, respectively, in which the latter are dependent on the former. One may note in passing that the distinctions in Peirce's semiotics are themselves built upon such Second Intentions; thus it is no wonder that every sign must possess some Iconic element. Furthermore, the very anatomy of the proposition becomes just like in Husserlian rational grammar (of the forth *Untersuchung*) a question of formal, synthetic a priori regularities.

ICONS IN ABDUCTION AND IN REASONING IN GENERAL

Among Peirce's forms of inference, similarity plays a certain role within abduction, his notion for a 'qualified guess' in which a particular fact gives rise to the formation of a hypothesis which would have the fact in question as a consequence. Many such different hypotheses are of course possible for a given fact, and this inference is not necessary, but merely possible, suggestive. Precisely for this reason, similarity plays a seminal role here: an

...originary Argument, or Abduction, is an argument which presents facts in its Premiss which *presents a similarity* to the fact stated in the conclusion but which could perfectly be true without the latter being so. ('Minute Logic', 1902, 2.96, our italics)

The hypothesis proposed is abducted by some sort of iconic relation to the fact to be explained. Thus, similarity is the very source of new ideas – which must subsequently be controlled deductively and inductively, to be sure (cf. Chap. 16). But iconicity does not only play this role in the contents of abductive inference, it plays an even more important role in the very form of logical inference in general:

Given a conventional or other general sign of an object, to deduce any other truth than that which it explicitly signifies, it is necessary, in all cases, to replace that sign by an icon. This capacity of revealing unexpected truth is precisely that wherein the utility of algebraic formulae consists, so that the iconic character is the prevailing one. ('That Categorical and Hypothetical Propositions are one in essence, with some connected matters,' c. 1895, 2.279)

The very *form* of inferences depends on it being an icon; thus for Peirce the syllogistic schema inherent in reasoning has an iconic character: 'Whenever one thing suggests another, both are together in the mind for an instant. [...] *every* proposition *like* the premiss, that is having an icon like it, *would* involve [...] a proposition related to it as the conclusion [...]' ('Short Logic,' c. 1893, EPII, 24; 2.444) Thus, first and foremost deduction is an icon: 'I suppose it would be the general opinion of logicians, as it certainly was long mine, that the Syllogism is a Symbol, because of its Generality.' (PAP, c. 1906, NEM IV, 317) – but instead it is an icon, because the icon is the only type of signs that *shows* (318):

The truth, however, appears to be that all deductive reasoning, even simple syllogism, involves an element of observation; namely deduction consists in constructing an icon or diagram the relation of whose parts shall present a complete analogy with those of the parts of the objects of reasoning, of experimenting upon this image in the imagination, and of observing the result so as to discover unnoticed and hidden relations among the parts. ('On the Algebra of Logic', 1885, W5, 164; 3.363)

In this light, it is no wonder that synthetic a priori truths exist – even if Peirce prefers notions like 'observable, universal truths' – the result of a deduction may contain more than what is immediately present in the premises, due to the iconic quality of the inference.

TO LEARN MORE

This leads us to what is probably the most decisive feature in icons at all: the fact that they are the only signs through the contemplation of which it is possible to learn more: 'For a great distinguishing property of the icon is that by the direct observation of it other truths concerning its object can be discovered than those which suffice to determine its construction.' ('That Categorical and Hypothetical Propositions are one in essence, with some connected matters', c. 1895, 2.279) But this epistemologically crucial property is nothing but an elaboration on the concept of similarity. It is not only the only type of sign involving a direct presentation of the qualities of its object, it is also a sign through the contemplation of which one can learn more than lies in the directions for its construction. If one imagines the limit case of a pure, icon-less index, then it would have a quality-deprived character of being-now, of mere insistence, about which we would never be able to learn anything except if it became possible to form some kind of icon of it. And if one imagines a purely symbolic sign, say for instance the variable x, we cannot learn anything about it except when it is placed in some iconical, that is, predicative, context or other. This implies that iconicity also covers what we for formalistic reasons have been used to seeing as 'purely symbolic formalisms', algebras, and symbolic calculi of various sorts. They are not deprived of iconicity; the very fact that we can learn more about their object from them is proof of their iconicity. Thus, the 'learn more' clause prevents Peirce's definition of similarity from being circular like most usual definitions are; similarity always involves the possibility of learning more than what is at present obvious. In this respect, icons play the central role in Peirce's evolutionary epistemology and his idea of science as a transpersonal endeavor of the infinite community of researchers: it is a steady growth in complicated predicates, in iconicity which characterizes science and guarantees its asymptotic movement towards ultimate truth, provided the pragmatistic maxims are followed. The diagrammatical implications of this operational icon definition are traced in Chap. 4.

ICONS, ALGEBRA, AND SYNTAX

As just mentioned, this vision of iconicity implies that it includes all kinds of algebraical systems (at least to the extent it is possible to manipulate them in order to gain further information): 'As for algebra, the very idea of the art is that it presents formulae which can be manipulated, and that by observing the effects of such manipulations we find properties not to be otherwise discerned.' ('On the Algebra of Logic', 1885, W5, 165; 3.363). Many years before his development of existential graphs, while he was still working on algebraical logic representations, Peirce thus realizes that algebra is even '*icons par excellence* [...] no application should be made of such an abstract statement without translating it into a sensible image' (ibid.). The sensible image in question here refers to the fact that we are able to synthesize an abstract statement into an expression, for instance an equation which we can thereafter manipulate according to certain rules which are adequate to *die Sachen selbst* in question. The manipulation of an equation or of an algebraic statement is, in this respect, perfectly equal to the manipulation of a picture or a text or any other icon in order to make it reveal some more similarities than immediately observable. This might sound like an offensive idea – that symbolic calculi should now count as prototypes for iconicity – but the arguments for it are very strong, and what is more, it provides a critical tool for distinguishing between fertile – that is, iconic – and less fertile formalization. A formalization in this optics is namely sterile to the extent that it does not permit any interesting possibilities of manipulation. Mere formalization without motivated syntactical, generative possibilities is in this view a blind alley. Of course, it may not always be told beforehand whether a certain formalization is fertile, and the relevant experimentation might be very mediate: to write down a fifth grade equation might seem hopeless in so far as we now know it has no canonical solution, but the very fact that it is formulated in the same language as solvable polynomia of lesser grade ultimately permitted the proof that it in fact has no solution – which is an even more impressive manipulation of it. But the mere substitution for some objects or object categories by letters or the like makes no manipulable icon, and this is why so many algebraic attempts in the humanities have proved sterile: they have merely exchanged some concepts with letters and have not furnished a motivated (that is, iconic), formal set of rules for their manipulation (cf. Chap. 4). An *iconic syntax* is invariably needed, and this is found already in *language*: 'That icons of the algebraic kind, though usually very simple ones, exist in all ordinary grammatical propositions is one of the philosophical truths

Boolean logic brings to light.' ('That Categorical and Hypothetical Propositions are one in essence, with some connected matters,' c. 1895, 2.280). We have already found icons in predicates, but the very syntax of language is in itself an algebra of partially iconic kind, because it is built in accordance with the crucial second intention distinctions in the object described: the distinctions between various kinds of predicates, various kinds of subjects, various kinds of copulas. In this respect, Peirce's idea of an iconic grammar may meet Husserl's *reine Grammatik* of the forth *Untersuchung*; both are formal, general calculi in partial accordance with aspects of the mereological structure of the object described. Thus, 'Every assertion is an assertion that two different signs have the same object.' ('Short Logic', 1895, EPII, 20; 2.437) – namely an iconic sign pertaining to a certain quality and an indexical sign pointing out the object in question in some frame of reference. These two in themselves, taken separately, assert nothing; it is only their being integrated in iconic syntax which makes an assertion: 'Icons and indices *assert nothing*. If an icon could be interpreted as a sentence, that sentence must be in a 'potential mood', that is, it would merely say 'Suppose a figure has three sides ..' etc. '(ibid. EPII, 16; 2.291). Here Peirce in fact discovers a logical category found around the same time by the Austrian logician Meinong who coined it *Annahme* – assumption – an assertion deprived of its assertive character and hence consisting of nothing but icons and indices (or, of symbols of them, to be precise) in Peirce's terminology, a kind of weaker copula.[80] In fact, his deliberation makes it possible to distinguish two kinds of *Annahmen*, assumptions, the merely iconic and the both iconic and indexical. 'Suppose three sides ..' would be an example of the first, while 'Suppose Socrates has three sides...' would be the other. A purely indexical assumption is probably impossible except as a limit case – or is at least empty: 'Suppose something is the case right now...' In the Brentanian tradition, to which Meinong also belongs, Husserl's *Bildbewusstsein* would be an example of the as if-character of assumption. This idea of an iconic grammar permits Peirce to develop yet another iconic parallel to this Austrian tradition, namely Stumpf's famous coinage of *Sachverhalt* as a notion for the signification (in other Austrians: the reference) of a proposition. In Peirce, we find a direct expression of this idea when he states that 'What we call a 'fact' is something having the structure of a proposition, but supposed to be an element of the very universe itself' ('New Elements' 1904, NEM IV, 239). Of course, this definition explicitly deals with what is supposed to be the case, that is, supposed reference and not signification, but it is entirely consistent with Peirce's intensional semantics to operate with 'possible facts', phenomena supposed to be able to be the case, which would then be entirely coextensive with *Sachverhalte*. Thus, in general it is the possible iconicity between the structure of grammar and the structure of facts that permits sentences to be understood:

> The arrangement of words in sentences, for instance, must serve as *Icons*, in order that the sentence may be understood. The chief need for the icon is in order to show the Forms of the synthesis of the elements of thought. For in precision of speech, Icons can represent nothing but Forms and Feelings. ('Prolegomena to an Apology for Pragmaticism', 1906, 4.544)

As a consequence, Peirce may of course reproach logicians which claim they do not use icons but merely symbols; they must invariably involve iconicity as the basis of their syntax of manipulation. The character of whole propositions in contrast to mere icons is that a proposition, or, in Peirce's coinage, a 'Dicisign [...] *conveys* information, in contradistinction to a sign [such as an icon] from which information may be derived' ('Syllabus', c. 1902, 2.309). We have already seen that a mere icon does not assert anything; still it is possible to derive information from it by experimentation, which so to speak makes it speak, in using it as material for a proposition. As opposed to this, a proposition already asserts something by attributing icon qualities to an object or event singled out by an index. Looking back on what is usually conceived as pure icons, they now differ on a scale between mere icons and dicisigns. A proposition, Peirce says elsewhere, is a sign making explicit the object to which it refers. A mere icon does nothing of the kind: a painting of a man is an icon of many possible men, but as soon as an index of some object is added, for instance 'Portrait of Mr. P.', the painting now becomes a proposition saying something like 'Mr. P. looks like this'. Of course the index needs not be in the title, but may reside in some other information conveyed by the painting itself.

All in all, the idea that algebraic, formal, and linguistic syntactic systems must involve iconicity in order to be able to convey information is a remarkable consequence of Peirce's concept of iconicity.

MATHEMATICS AND DIAGRAMS

As an implication of this fact, icons get a very special relationship to mathematics:

> The reasoning of mathematicians will be found to turn chiefly upon the use of likeness, which are the very hinges of the gates of their science. The utility of likenesses to mathematicians consists in their suggesting in a very precise way, new aspects of supposed states of things. ('The Art of Reasoning,' 1894, EPII, 6; 2.281)

Mathematics is merely hypothetical, all its statements are of the form if-then; given such-and-such axioms, such-and-such symbols, such-and-such rules of transformations, this and that will be a consequence. This makes it a science of icons, in so far as all this inference deals with iconic structures. Icons cannot deceive and make you believe in something illogical (albeit of course in something unreal), because they always portray something logically possible (you can not make an icon of the round square etc., but easily of the bald king of France). Hence, 'It will be observed that the icon is very perfect in respect to signification, bringing its interpreter face to face with the very character signified. For this reason, it is the mathematical sign *par excellence*. But in denotation it is wanting. It gives no assurance that any such object as it represents really exists.' ('New Elements', NEM IV 242–3). Mathematics is so to speak a mapping of the field of formal iconic possibilities, and as Peirce conceives of the field of qualities to be – at least potentially – one vast continuum, mathematics in fact ought to be able to detect the hypothetical relations of all qualities (including even, in principle, sense qualities). Mathematics does not assert anything about the world; it is more like a vast repertoire of possible complicated quality interrelations to be used in descriptions; for this reason every science depends on mathematics as a large stock of hypothetical relation structures – even if it does not derive its *actual* truths from there. It contains only perfect truths which Peirce in a play of words coins as imperfect – that is, not actual: 'A proposition is not a statement of perfectly pure mathematics until it is devoid of all definite meaning, and comes to this – that a property of a certain icon is pointed out and is declared to belong to anything like it, of which instances are given. The perfect truth cannot be stated, except in a sense that it confesses its imperfection. The pure mathematician deals exclusively with hypothesis.' ('Truth and Falsity and Error,' Baldwin's Dictionary, 1901, 5.567) The special kind of instrument used for deriving mathematical and other hypothetical truths is to Peirce the subset of icons named 'diagrams'. I shall not go into the special problems of them here, to which I shall return later, but let me run through the central features: 'Remember it is by icons only that we really reason, and abstract statements are valueless in reasoning except so far they aid us to construct diagrams.' ('The Logic of Quantity', 1893, 4.126) Diagrams are Peirce's heirs to Kant's schemata; they make possible the inferring of synthetic a priori propositions, or, in Peirce's terms, they make it possible to infer iconically about general matters. A diagram is an icon governed by a symbol such that it stands generally as a type for a whole set of token instantiations like it - just like the premises of a general syllogistic inference. This guarantees that the manipulation of the diagram holds for all these token cases involved. Certain rules for manipulation are implied by the structure of the icon itself, others are imposed by the symbol regulating it, and the outcome of this is that it is not possible to predict beforehand what the results of the transformations of it will be. This in turn becomes Peirce's explanation of the surprising fact that mathematics (as well as the special sciences using mathematics) is still able to discover new regularities: all truths are not given in and by the definition of a branch of mathematics, new truths may yet be discovered by formal experimentation observations. Thus, this opens all of a sudden the way to a 'Geschichte der reinen Vernunft': it is in fact possible to obtain a unforeseeable growth in synthetic a priori knowledge – and at the same time it introduces in a certain sense experiment and observation on the a priori level: the mapping of this field is due to the experiments of imagination in the 'reiner Anschauung', Kantianly spoken, or in the Husserlian *Bildbewusstsein* including the *kategoriale Anschauung*.

It is hardly surprising in a continuistic theory like Peirce's to find traces of diagrammatic features in most icons. Still it must be possible – in a further synthetic a priori research – to construct a typology of possible diagrams. Peirce himself gives us the idea of such an endeavor when, in a description of diagrammatic reasoning, he notes in the passing that

> Modern exact logic shows that every operation of deductive reasoning consists of four steps as follows: 1st, a diagram, or visual image, whether composed of lines, like a geometrical figure, or an array of signs, like an algebraical formula, or of a mixed picture, like a graph, is constructed, so as to embody in iconic form, the state of things asserted in the premise (there will be but one premise, after all that is known and is pertinent is collected into one copulative proposition). ('On Quantity', c. 1895, NEM IV, 275)

The following three steps consist in scrutinizing the diagram and attempting an experiment; in observing the results of this and trying to find a new relation between the parts of it 'not mentioned in the precept by which it was constructed...' (276); finally in repeating the experiment and inferring inductively that every diagram constructed the same way would yield the same result. But the crucial point in this context is the three subtypes of diagrams mentioned; Michael May and I have preliminarily proposed to call them *maps, algebra*, and *graphs*, respectively. I shall return to this idea later.

THE IMAGINARY MOMENT

Now, at a certain moment in the diagrammatical procedure just mentioned, a certain phase is reached, in which the icon performs its full impact, a moment of imagination, one might call it. It lies already in the very description of its character of pure quality: 'The role of an icon consists in its exhibiting the features of a state of things as if it were purely imaginary.' ('Logical Tracts, No. 2,' c. 1903, 4.448); here we find *in nuce* how Peirce's icon entails the idea of purely fictitious *Sachverhalte*. In an early quote, Peirce describes it in more detail as follows:

> Icons are so completely substitutions for their objects as hardly to be distinguished from them. Such are the diagrams of geometry. A diagram, indeed, so far as it has a general signification, is not a mere icon; but in the middle part of our reasonings we forget that abstractness in great measure and the diagram is for us the very thing. So in contemplating a painting, there is a moment when we lose the consciousness that it is not the thing, the distinction between the real and the copy disappears, and it is for the moment a pure dream – not any particular existence, and yet not general. At that moment we are contemplating an *icon*. ('On the Algebra of Logic', 1885; W5, 163; 3.362)

This moment of fiction is crucial to the possibility of thought, of imagination, or of contemplation of pictures to approach the object intended. In many cases, of course, this imaginary moment may be unrestricted as the notorious 'free play of imagination'; in others it is constrained by various, more or less severe, regulations pertaining to the object, for practical, æsthetical, scientific, or other purposes. In all cases, however, this moment of identification where the manipulation of the icon in a certain sense is a manipulation of the object itself, is crucial to the possibilities of solving the constraints and success of the experiment. One could counterargue here: it is precisely to avoid the wildways of intuition in this imaginative moment that we bind science to formal calculi – cf. Hilbert's formalist idea of bracketing intiution entirely while manipulating formal symbols. But here the Peircean will answer: formal calculation *also* requires this imaginative moment; it may not disclose for me a very large and sensuous picture of its object, but it still, thanks to the syntax employed, can never be completely deprived of an – even if very highly stylized – iconic link to its object. And, what is more, it permits us to interpret Husserl's version of the *adequatio rei ac intellectus*: the iconic ecstasy only lasts for a moment, and after the experiment it becomes possible to check if the iconic result is in fact pertinent for (some partial behavior of) the object it was supposed to depict. If not, another diagram may be invoked, a modification of the diagram present, maybe to construct a new subset for the object not entirely grasped... etc. The movement through this imaginary moment alternates continuously between on the one hand the icon-and-object-is-one supposition and on the other the split consideration of diagram and the more or less spontaneously conceived object, the *Widerstreit* of Husserlian *Bildbewusstsein* (see Chap. 14). The imaginary moment is described in psychologically-sounding terms in Peirce, but given his pragmatist conception of iconicity, psychology or the presence of consciousness form no necessary aspect of the imaginary moment. When iconicity is described in terms of manipulation and experiment possibilities, the condition of possibility for the imaginary moment lies in the fact that actions are undertaken on the icon which due to the structural similarity might as well be performed on the object. But the structural homology of actions does not necessarily require psychology.

The aesthetic possibilities in the hypothetical character of this imaginary moment are also suggested in Peirce. For even if all of mathematics is hypothetical, not all hypotheses are (explicitly, anyway) mathematical:

It cannot be said that all framing of hypothesis is mathematics. For that would not distinguish between the mathematician and the poet. But the mathematician is only interested in hypotheses for the forms of inference from them. As for the poet, although much of the interest of a romance lies in tracing out consequences, yet these consequences themselves are more interesting in point of view of the resulting situations than in the way they are deducible. Thus, the poetical interest of a mental creation is the creation itself, although as a part of this a mathematical interest may enter to a slight extent. Detective stories and the like have an unmistakeable mathematical element. ('On Quantity', c. 1895, NEM IV, 268)

Given this difference between mathematics and poetry, one could add that they share a similarity to the extent that both, in fact, are interested in the result of hypothetical reasoning, not the very steps leading there; this interest is rather the matter for the logician and the literary critic (all or some of whom may of course be incarnated in the same empirical person). As is evident, this rendering of the imaginary moment in icon contemplation has its counterpoint in the *Bildbewusstsein* of Husserl, whose eidetic variation to a large extent plays a role analogous to diagrammatic experimentation here, and it is no meagre consequence of Peirce's concept of iconicity that it permits mathematics, logic, and art to be understood as different practices all related to this imaginary moment of thought. Furthermore, as a rendering of *Bildbewusstsein* or the more diffuse 'imagination' or *Einbildung*, this is not conceived of as a separate and thereby ununderstandable faculty of mind ('fantasy'), but is seen as a specific phase in a well-described diagrammatic and more generally iconic process of interpretation.

In the last chapter, we shall return to a genre related to the detective story, namely that of spy novels.

THE ISSUE OF EVIDENCE

Still, a uncomfortable peculiarity remains. Is it not strange that icons so to speak 'hide' certain similarities whilst on the other hand they are supposed to be the very source of evidence? As we saw, the most crucial property in them is their ability to enclose hidden insights, making it possible for experiment to 'discover unnoticed and hidden relations among the parts' ('On the Algebra of Logic', 1885, W5, 164; 3.363). It is possible to give a recipe for constructing an icon without revealing all of its possible similarities. A most striking example of this is the simple process of digitalizing a picture. Seeing the line of 1s and 0s of a computerized image does not display the similarities inherent in the picture, even if the information in question in some sense is available with the right system of transformations, the right means of interpretation. Of course, in these wordings, a perceiver of some kind is implied, a subject (albeit not necessarily a human being) for whom certain similarities may be hidden. And even if we may not construct the icon digitally, there will in many cases be more information in it than it is possible to exhaust. In some sense, the information already lies in the digital sequence, but only if it can be processed, not only by some instance processing the icon onto a screen, but also by some instance able to, by some kind of *Bildbewusstsein*, to see the icon as an icon of some object. In the most simple of icons, these hidden similarities are almost non-existent. If I contemplate a patch of a certain shade of the color red, I contemplate an icon, but the only experiment I may venture here is varying the color to see when it changes or varying the form and size of the patch to see when it disappears. Not very promising, but as soon as we reach an icon with the (still not very high) complexity of a circle, it is ripe with hidden properties (just think of the issue of the exact relation between its diameter and periphery). In mathematics, an object as simple and apparently intuitive as the natural numbers are (after Gödel's incompleteness theorem) ripe with theorems (Goldbach's conjecture, solutions to Diophantian equations, etc.) which are hidden from immediate appreciation, and which may even in many cases be formally undecidable. In a painting are concealed many enigmas which the painter himself has never contemplated (and since the sheet of the painting is continuous, the number of possible similarities is infinite which means that he would never even be able to contemplate them all in finite time – so much for the biographical method) – what is, for instance, the precise distance between the main characters, which geometric figure do the persons describe if we draw lines between them, what is the sum of the height of all the trees depicted. These examples go to show that almost all of such 'hidden informations' are of course completely useless; the trick of the analyst is to make his experiment find somehow significant information among them. Take for instance a tour-de-force analysis by Erik Fischer of an altar piece by Eckersberg, where he drew the perspective lines of Judas' empty chair at the last supper, only to find that this chair was skew – probably as a symbol for the peculiarity of the missing person (see Chap. 13). Principally conceived, the reason for these 'hidden similarities' in the icon has several sources. One is that the icon does not have to be very complex, in order to be unable to be 'perceived in a glance'. This goes not only for very large paintings from which it is impossible to get at a sufficient distance to be able to synthesize them in one gaze (a feature deliberately used by the abstract expressionists as a means of calling forth awe in the viewer), but also sufficiently complicated predicate structures in a text, in a mathematical problem, in a building, etc. But even if a very competent observer (or hard work on the part of a less competent one) makes it possible to 'see it all in a glance', there's still all the possible relations between the parts to be worked out, relations between relations between the parts, etc. We need hardly stress that the amount of relations between elements grows exponentially with the number of elements, and that the higher order relations grow exponentially with order – amounts which in turn must be multiplied with the set of possible interpretation systems for these relations. Almost all of such higher order relations are meaningless, but in aesthetic cases a few of them often miraculously 'fit' together in order to produce a striking, provocative, pleasing, beautiful or otherwise noteworthy effect. Of course, this is why analysis is necessary, an analysis which in all these cases consequently cannot be anything but diagrammatical – including all the features mentioned above. Such kinds of 'hiddenness' are of course due to the 'finitude of man', so central to the critical tradition in epistemology. Yet, one might loosen this fact from mankind's special *Dasein* and its destiny by stating that *any* physically instantiated analytical apparatus whatsoever will have to be finite and hence possess some limit or other for the size and complexity of the iconic structures it may process. Human language generally cuts any *Sachverhalt* into phrase pieces involving no more than three or four actants (three being for Peirce the highest irreducible relation so it is an advantage for us we are not biologically limited to two)[81], but there seems to be no principal reason for not admitting computers able to work with iconic syntaxes with far higher valencies. Another more objective reason for iconic 'hiddenness' is that the icon in question may appear in various versions, not similarly difficult to interpret. The obvious example is the sculpture *Etant donnés* by Duchamp to be watched through a hole in a door, an artwork which may appear simple until you discover the hole – but it is well known that the solvability of a problem in general depends highly on how it is posed. Of course, this is also dependent upon the means available for the interpreter, but not only so. Here, one might propose a generalized use of Husserl's idea of *Abschattung*; the icon may so to speak display itself from several sides, and even if the same amount of information principally may be available from all versions, it might take much more computation to derive it from one icon than from another, analogous to Charles Bennett's concept of logical depth. An addition on page 3 saying 'It was the butler' of course requires less computation than reading all of the following 399 pages, collecting evidence all along. Another issue is the very structure of the icon, no matter how it is presented – cf. the supposed distinction between P and NP problems in computer science, requiring different computation time to be processed. Conceived of at a great distance, in some way it remains an enigma why it is that icons exist with this intriguing property of hiding similarities, luckily a prerequisite to art as well as science; if gods exist, then they probably see everything directly – not *per speculum in aenigmate* – and hence know neither icons nor indices nor symbols. In Peirce as in Kant, the plurality of faculties of the human mind is due to the fact that we do not, maybe unlike the gods, possess immediate 'intellectual intuition'.

The special force of the icon, however, is its connection to *possibility*:

Each Icon partakes of some more or less overt character of its Object. They, one and all, partake of the most overt character of all lies and deceptions – their Overtness. Yet they have more to do with the living truth than have either Symbols or Indices. The Icon does not stand unequivocally for this or that existing thing, as the Index does. Its Object may be a pure fiction, as to its existence. But there is one assurance that the Icon does afford in the highest degree. Namely, that which is displayed before the mind's gaze – must be *logically possible*. ('Prolegomena to an Apology of Pragmaticism', 1906, 4.531)

The icon has the undeniable quality of showing something possible – it cannot, unlike symbolic speculation, yield neither Husserlian *Unsinn* nor *Widersinn*, the first being only possible through the breakup of iconic syntax, the other only by constructing a contradictory predicate. We may talk and wonder about the xlypf, the which are has, the round square, the rational square root of two, or the married bachelor, etc., but no simple icon can display any of them. But when this principle applies to the icons at stake in logical reasoning, it has very strong consequences:

Now necessary reasoning makes its conclusion *evident*. What is this 'Evidence'? It consists in the fact that the truth of the conclusion is *perceived*, in all its generality, and in the generality the how and why of the truth is perceived. What sort of a Sign can communicate this Evidence? No index, surely, can it be; since it is by brute force that the Index thrusts its Object into the Field of Interpretation, the consciousness, as if disdaining gentle 'evidence.' No Symbol can do more than apply a 'rule of thumb' resting as it does entirely on Habit (including under this term natural disposition); and a Habit is no evidence. I suppose it would be the general opinion of logicians, as it certainly was long mine, that the Syllogism is a Symbol, because of its Generality. But there is an inaccurate analysis and confusion of thought at the bottom of that view; for so understood it would fail to furnish Evidence. It is true that ordinary Icons, – the only class of Signs that remains for necessary inference, – merely suggest the possibility of that which they represent, being percepts *minus* the insistency and percussivity of percepts. In themselves, they are mere Semes, predicating of nothing, not even so much as interrogatively. It is, therefore, a very extraordinary feature of Diagrams that they *show* [...] that a consequence does follow.... ('PAP', 1906. NEM IV, 318)

Evidence thus comprises the sensuous making overt a quality in the object as well as the evidence *that* some proposition follows from certain premises; in both cases icons are responsible. But this raises again the crucial question of how the same figure may be responsible for evidence and for hidden similarities. There are two reasons for this. The first lies in the process of iconic reasoning: evidence is what takes place *when* the experimentation reveals new aspects of the icon. Of course, the other parts of the predicate must – as icons – be evident themselves, but this evidence is backgrounded while the newly revealed quality is foregrounded and adds itself to the evidence already present. But this implies an important corollary to the phenomenological category of evidence: the fact that something is evident does not imply that it may not conceal further hidden aspects.[82] A problem may be perfectly clear to me without me knowing how to detect the solution. Or, the motif of a painting may be clear without the single strokes of paint being evident.

To use a Husserlian metaphor which is not perfectly apt: the sight of one side of the object does not tell us how the back side looks. The metaphor is deceptive, for in the problem case, the front side does in fact in some scrambled way contain the information necessary to reconstruct the back side. The other reason is that any icon fails to portray its object with final precision; this possibility is only asymptotically open for the community of researchers. This, of course, is Peirce's answer by means of continuity to Kant's *Ding-an-sich*, it may be reached, but only in an indefinite future. But this implies that given any present icon, it is always in principle possible to find an even better icon which will then yield more evidence than the former. Thus, as a consequence of both these reasons, iconic evidence is always relative.

But still, is it not a dubious idea to make icons the source of evidence? Husserl would say that evidence is constituted by the act in which the object is given in 'anschauliche Fülle' – intuitive richness – be it sensuous objects or logical truths, and that consequently evidence cannot be the product of any 'Zeichenbewusstsein' (*Ideen*); this would lead to an infinite regress, for in which *Bewusstsein* should the *Zeichen* now be constituted? – and so on. Ergo, he argues, evidence must lie outside of the *Bild-* and *Zeichenbewusstsein*. Is it not the case that the icon as evidence locks us up in a Piranesian prison of icons of icons of icons of icons ... from which we are never able to escape in order to ascertain true evidence? It seems to me that Peirce's theory of evidence provides an answer to this objection: evidence in the Husserlian *anschauliche-Fülle*-version actually has two sources: one is the Peircean index-consciousness of something being given with the force of haecceity, the evidence that something *exists*. The other is evidence that something *is as it is*, and this part is always dependent on icons with the relativity just mentioned: iconic evidence is dependent upon the continuous shift between icon-consciousness (to construct a Peirco-Husserlian bastard) and meltdown between the two sides of the iconic sign during the moment of imagination.[83] It is this very process which makes it possible to reflect upon the object and ascribe it still more qualities; it is this process that makes possible that the object is given (seen from one side,[84] it is true) in evidence once and for all. This is a strong fallibilist consequence of Peirce's theory of the icon: evidence is always potentially problematic – which is not a skeptical claim, for the contrary is true as well: problems are always potentially evident.

The strength of the Peircean concept of iconicity is thus that it permits the rational comparison of objects normally very far from each other: mathematics, logic, diagrams on the one hand, epistemology, phenomenology and psychology of perception on the other, and language, semiotics, and aesthetics on the third.

CHAPTER 4

MOVING PICTURES OF THOUGHT

Diagrams as Centerpiece of a Peircean Epistemology

Recent developments in semiotics, semantics, and linguistics tend to give concepts like 'schema,' 'frame', 'gestalt,' and the like a renaissance in the description of signification processes. The actual cognitive semantics tradition (Lakoff, Johnson, Talmy, Turner, Fauconnier, etc.), for instance, highlights the central role of schemata and their mappings between conceptual spaces in the description of many levels in linguistics. Another related development is the renewed interest in diagrammatic calculi in the computer science and AI communities, documented in e.g. the influential *Diagrammatic Reasoning* volume (Glasgow et al. 1995) – where the diagram category, however, is most often taken for its common sense value as an opposition to the symbol category; little effort is spent on determining the general status of the diagram as such.

This return of schematic iconicity in semiotics is probably the main event in semiotic scholarship during the recent decades, but it has not, until now, received a proper meta-theoretical treatment making clear the very concept of schema itself. This is a strange fact; in Peirce we find drafts for precisely such a theory in his general observations on the concept of *diagram*. While Peirce's systems for logic diagrams (his *alpha-*, *beta-*, and *gamma-*graphs implementing propositional logic, first-order-predicate logic, and various types of modal logic and speech act logic, respectively[85]) have received considerable attention in recent years because of their indication that iconic representations of logic are possible and even to some extent heuristically superior to symbolic logic systems, Peirce's general notion of diagram has passed much more unnoticed. This might be for editorial reasons – Peirce's central arguments concerning the general diagram category are not to be found in the *Collected Papers* – but still the diagram concept plays a central, not to say *the* central, role in the mature Peirce's semiotics. In particular 'PAP', a paper from 1906 (Robin 293, published in NEM IV), makes clear the crucial part played by the diagram and diagrammatic reasoning in Peirce. The present schema and diagram research would no doubt benefit from the knowledge of Peirce's general diagrammatic philosophy.

The aim in this chapter is twofold: firstly, to present and discuss Peirce's general diagram concept and its central role in his semiotics and in his philosophy as a whole, and, secondly, to argue for the significance, beyond Peirce philology, of this diagram concept for semiotics and epistemology of our day.

THE DIAGRAM AS ICON

The diagram is an icon. In the taxonomy of signs, thus, the diagram forms the second subcategory among the three types of hypoicons[86] – images, diagrams, and metaphors, respectively ('Syllabus' 1903, EPII 274; 2.277) – even if Peirce elsewhere notes that sharp distinctions among icons are not possible due to the inherent vagueness of the concept. Being an icon, the diagram is characterized by its similarity to its object – but while the image represents its object through simple qualities and the metaphor represents it through a similarity found in something else, the diagram represents it through a skeleton-like sketch of relations (mostly dyadic, apparently in an attempt at justifying the three icon subtypes triadically). Knowing the inclusive nature of Peirce's triads in general, it follows that non-degenerate diagrams will include images, while non-degenerate metaphors will contain diagrams (and images).[87] Still, this tripartition of icons is easy to overlook as yet another detail in the tree of ever trifurcating triads in Peirce's architectonic; it does not reveal the crucial role played by diagrams in Peirce's epistemology. To grasp this, a further investigation of the very definition of the icon is necessary.

THE NON-TRIVIAL ICON DEFINITION

The icon, of course, is defined as the sign referring to its object by virtue of similarity. Now, Peirce himself admitted the deliberate vagueness of this definition: an icon may refer to any object possessing the qualities in question – and as discussed in the previous chapter, a strong tradition in twentieth century philosophy has attacked such definitions for being so vague as to be completely meaningless. The dangers in the similarity concept are many: the trivializing of it to identity; the psychologizing of it to refer to subjective feelings or judgments of resemblance; the lack of criteria for judging two phenomena similar. These traditional drawbacks of similarity are overcome by Peirce's non-trivial because *operational* account of similarity. In 1895, it is stated as follows: 'For a great distinguishing property of the icon is that by the direct observation of it other truths concerning its object can be discovered than those which suffice to determine its construction' ('That Categorical and Hypothetical Propositions are one in essence, with some connected matters,' 2.279). This epistemologically crucial property of the icon is nothing but an operational elaboration on the concept of similarity. The icon is not only the only kind of sign involving a direct presentation of qualities pertaining to its object; it is also – and this amounts to the same – the only sign by the contemplation of which more can be learnt than lies in the directions for its construction. This definition separates the icon from any psychologism: it does not matter whether sign and object for a first (or second) glance seems or are experienced as similar; the decisive test for its iconicity rests in whether it is possible to manipulate the sign so that new information as to its object appears. This definition is non-trivial because it avoids the circularity threat in most definitions of similarity. At the same time, it connects the concept of icon intimately to the that of deduction. This is because in order to discover these initially unknown pieces of information about the object hidden in the icon, some deductive experiment on the icon must be performed. The prototypical icon deduction is the manipulation of a geometrical figure in order to observe a theorem – but the idea is quite general: an icon is characterized by containing implicit information which in order to appear must be made explicit by some more or less complicated procedure accompanied by observation. As early as 1885, Peirce writes ('On The Algebra of Logic'), discussing the syllogism, but with evident implications for the icon category as a whole, that '... all deductive reasoning, even simple syllogism, involves an element of observation; namely, deduction consists in constructing an icon or diagram the relations of whose parts shall present a complete analogy with those of the parts of the object of reasoning, of experimenting upon this image in the imagination, and of observing the result so as to discover unnoticed and hidden relations among the parts' (W5, 164; 3.363).[88] This property clearly distinguishes it from pure indices and symbols: If we imagine a pure, icon-less index (only possible as a limit case), then it would have a character completely deprived of any quality, a pure here-now of mere insistence, about which we would never be able to learn anything further, except exactly by some kind of icon of it. And if we imagine a purely symbolic sign (also a limit case), say e.g. the variable *x*, we could not learn anything about it except when placing it in some relation, syntax, system, context or other, that is, in some kind of iconical relationship. From this operational icon definition, connection lines run to a bundle of Peircean themes: the abductive guess as the suggestion of an icon as a general answer covering the particular question present; icons as providing the predicative, descriptive side of any signification process; the pragmatic maxim's conditional definition of concepts described by an icon showing which operations we could conceivably perform on an object subsumed under the concept; the scientific community's unlimited semiosis converging towards truth, that is, an ever more elaborate icon possessing still more operational possibilities. We shall touch upon some of these issues later during the discussion of the type of icon making all this possible: the diagram.

THE OPERATIONAL CRITERION AND THE EXTENSION OF THE ICON CATEGORY

It is a well-known fact that Peirce's icon definition sets it apart from spontaneous tendencies to privilege visual icons. It is a more controversial fact that the operational icon definition extends the icon category considerably, measured against the spontaneous everyday conception of resemblance. Peirce's logic graph systems as iconic calculi already indicates this change: they demonstrate that systems normally considered symbolic possess an ineradicable iconicity.[89] Using Peirce's sign concepts, namely, it is no longer possible to speak about iconicity and symbolicity as two concurrent modes of representation of the same content: if the same logical calculus may be represented in two ways, this indicates that the 'symbolic' representation did, in fact, already possess an iconic content: the possibility of experimentation on the calculus resulting in new insight grants – due to the operational icon criterion – that it is in fact an iconic calculus.[90] Thus, when the operational criterion is adopted, icons become everything that can be manipulated in order to reveal more information about its object, and algebra, syntax, formalizations of all kinds must be recognized as icons; in the 'That Categorical and Hypothetical Propositions…', Peirce adds that these types of signs are even icons par excellence due to their capacity for revealing unexpected truths: 'Given a conventional or other general sign of an object, to deduce any other truth than that which it explicitly signifies, it is necessary, in all cases, to replace that sign by an icon. This capacity of revealing unexpected truth is precisely that wherein the utility of algebraical formulae consists, so that the iconic character is the prevailing one' (1895, 2.279). This, in turn, implies that we, in the operational icon definition, find a useful criterion to distinguish fertile from less fertile formalization: the good formalization is one which permits manipulation in order to reveal new truths about its object; formalizations which only permit this to a small extent or not at all may be discarded.[91]

THE DIAGRAM'S STATUS IN ICONICITY

Given the operational icon criterion, we are now able to appreciate the central role played by diagrams in the icon category as such. As soon as an icon is contemplated as a whole consisting of interrelated parts whose relations are subject to experimental change, we are operating on a diagram. Thus, the inclusion of algebra, syntax, and the like in the icon category takes place thanks to their diagrammatic properties – but the same goes for your average landscape painting as soon as you stop considering its simple qualities, colors, forms etc. and move on to consider the relations between any of these parts and aspects. As soon as you judge, for instance, fore-, middle-, and background and estimate the distance between objects depicted in the pictorial scene, or as soon as you imagine yourself wandering along the path into the landscape, you are operating on the icon – but doing so in this way is possible only by treating it as a diagram. You may have no explicit awareness,[92] it is true, of the rules which permit you to follow the imaginary track (the laws of perspective permitting you to construct the scene, gravity keeping you on the ground etc.), but still they are presupposed due to the organization of your perception apparatus[93] and your tacit common sense knowledge. The principles *could* be made explicit, and this is what counts. Thus, it is hard to take a closer look at any icon without at least performing proto-diagrammatic experiments with it to reveal some of the implicit truths inherent therein. Thus, the use of a sign as a pure image is more like a limit case as when you enjoy the overall impression or *Stimmung* of a painting without going into any details – or when you use the icon as a trigger for an established code, such as the stylized woman figure on the door of a women's rest room. On the other hand, the appreciation of a metaphor may seem automatic, but recent metaphor research supports what lies implicit in Peirce's thought: that a diagrammatic analysis – be it conscious or not – precedes any metaphor consisting in the recognition of diagrammatic schemas in one phenomenon which may be used in understanding another. The metaphor of an 'ancestral tree' thus presupposes that the formal branching diagram is mapped from a tree onto family structure. Far from all metaphorical mappings are so easy, of course, but it seems reasonable to assume that the mapping of diagrammatic structure between conceptual spaces plays a central role in metaphor in general.[94] Thus, the diagrammatic way of interpreting an icon seems central as soon as any part of the internal mereological structure of the icon is taken into consideration. The diagram's skeleton-like, relational, and highly stylized picture of its object is at stake also when clothed in simple image qualities and hidden in the metaphors' import reference to other empirical phenomena.

Now, let us look closer at how Peirce dissects the single elements and phases in the diagrammatic interpretation process. As already mentioned, one essay stands out when it comes to detailed analysis of this process, namely one of the drafts for 'Prolegomena to an Apology for Pragmaticism' from 1906. The paper in question is Robin catalogue number 293 and is also known as 'PAP' from Peirce's own abbreviation. The next passage will present the central quote from that paper, describing the diagrammatic interpretation process *in extenso.*

To begin with, then, a Diagram is an Icon of a set of rationally related objects. By *rationally* related, I mean that there is between them, not merely one of those relations which we know by experience, but know not how to comprehend, but one of those relations which anybody who reasons at all must have an inward acquaintance with. This is not a sufficient definition, but just now I will go no further, except that I will say that the Diagram not only represents the related correlates, but also, and much more definitely represents the relations between them, as so many objects of the Icon. Now necessary reasoning makes its conclusion *evident.* What is this 'Evidence'? It consists in the fact that the truth of the conclusion is *perceived,* in all its generality, and in the generality of the how and the why of the truth is perceived. What sort of a Sign can communicate this Evidence? No index, surely, can it be; since it is by brute force that the Index thrusts its Object into the Field of Interpretation, the consciousness, as if disdaining gentle 'evidence'. No Symbol can do more than apply a 'rule of thumb' resting as it does entirely on Habit (including under this term natural disposition); and a Habit is no evidence. I suppose it would be the general opinion of logicians, as it certainly was long mine, that the Syllogism is a Symbol, because of its Generality. But there is an inaccurate analysis and confusion of thought at the bottom of that view; for so understood it would fail to furnish Evidence. It is true that ordinary Icons, – the only class of Signs that remains for necessary inference, – merely suggest the possibility of that which they represent, being percepts *minus* the insistency and percussivity of percepts. In themselves, they are mere Semes, predicating of nothing, not even so much as interrogatively. It is, therefore, a very extraordinary feature of Diagrams that they *show,* – as literally *show* as a Percept shows the Perceptual Judgment to be true, – that a consequence does follow, and more marvelous yet, that it *would* follow under all varieties of circumstances accompanying the premisses. It is not, however, the statical Diagram-icon that directly shows this; but the Diagram-icon having been constructed with an Intention, involving a Symbol of which it is the Interpretant (as Euclid, for example, first announces in general terms the proposition he intends to prove, and then proceeds to draw a diagram, usually a figure, to exhibit the antecedent condition thereof) which Intention, like every other, is General as to its Object, in the light of this Intention determines an Initial Symbolic Interpretant. Meantime, the Diagram remains in the field of perception and imagination; and so the Iconic Diagram and its Initial Symbolic Interpretant taken together constitute what we shall not too much wrench Kant's term in calling a *Schema,* which is on the one side an object capable of being observed while on the other side it is General. (Of course, I always use 'general' in the usual sense of general as to its object. If I wish to say that a sign is general as to its matter, I call it a Type, or Typical.) Now, let us see how the Diagram entrains its consequence. The Diagram sufficiently partakes of the percussivity of a Percept to determine, as its Dynamic, or Middle, Interpretant, a state [of] activity in the Interpreter, mingled with curiosity. As usual, this mixture leads to Experimentation.

It is the normal Logical effect; that is to say, it not only happens in the cortex of the human brain, but must plainly happen in every Quasi-Mind in which Signs of all kinds have a vitality of their own. Now, sometimes in one way, sometimes in another, we need not pause to enumerate the ways, certain modes of transformation of Diagrams of the system of diagrammatizaton used have become recognized as permissible. Very likely the recognition descends from some former Induction, remarkably strong owing to the cheapness of mere mental experimentation. some circumstance connected with the purpose which first prompted the construction of the diagram contributes to the determination of the permissible transformation that actually gets performed. The Schema *sees*, as we may say, that the transformate Diagram is substantially contained in the transformand Diagram, and in the significant features to it, regardless of the accidents, – as, for example, the Existential Graph that remains after a deletion from the Phemic Sheet is contained in the Graph originally there, and would do so whatever colored ink were employed. The transformate Diagram is the Eventual, or Rational, Interpretant of the transformand Diagram, at the same time being a new Diagram of which the Initial Interpretant, or signification, is the Symbolic statement, or statement in general terms, of the Conclusion. By this labyrinthine path, and no other, is it possible to attain to Evidence; and Evidence belongs to every Necessary Conclusion. (NEM IV, 316–19)

The remainder of this chapter tracks the implications of this passage, partly in terms of its relation to Peirce's thought, partly in terms of the actuality of its contents.

THE DIAGRAM AS AN ICON OF RATIONALLY RELATED OBJECTS

The diagram is a skeleton-like sketch of its object in terms of relations between its parts, but what makes it apt to reason with, to experiment on, respectively, is the fact that it is constructed from rational relations. In this requirement, Peirce explicitly continues a Kantian requirement of the foundations of science: the schematism. In Kant, the finitude of man entails that we have no access to 'intellectual intuition'; we can not – as may the gods – intuit the object in itself; we may only approach the object in a pincer movement with two flanks: concepts and intuitions, respectively. Concepts without intuitions are empty; intuitions without concepts are blind, as the well-known Kantian doctrine goes. The two may meet only in schemata, a priori as well as a posteriori, and the former constitute the condition of possibility for the famous synthetic a priori judgments.[95] Kant's central examples are mathematical: arithmetic is the schema rendering the concept of quantity intuitive, while the schema of the triangle is what permits an unlimited series of empirical triangles to be subsumed under the triangle concept. Peirce's demand that the relations in the diagram be rational is inherited from Kant's synthetic a priori judgment notion, just like his idea that rationality is tied to a generalized subject notion: rational relations are those known by 'anybody who reasons'. As is evident, Kant's 'transcendental subject' is pragmatized in this notion in Peirce, transcending any delimitation of reason to the human mind: the 'anybody' is operational and refers to anything which is able to undertake reasoning's formal procedures. In the same way, Kant's synthetic a priori notion is pragmatized in Peirce's account:

Kant declares that the question of his great work is 'How are synthetical judgments a priori possible?' By a priori he means universal; by synthetical, experiential (i.e., relating to experience, not necessarily derived wholly from experience). The true question for him should have been, 'How are universal propositions relating to experience to be justified?' But let me not be understood to speak with anything less than profound and almost unparalleled admiration for that wonderful achievement, that indispensible stepping-stone of philosophy. ('The Logic of Quantity', Chap. 17 of 'Grand Logic', 1893, 4.92)

Synthetic a priori is interpreted as experiential and universal, or, to put it another way, observational and general – thus Peirce's rationalism in demanding rational relations of the diagram is connected to his scholastic realism posing the existence of real universals. The relations which make up the diagram are observational and universal at one and the same time, and they constitute the condition of possibility for the diagram to exist as an icon (observationality) with respect to which it is possible to entertain generally valid experiments (universality). The extension of this concept of rational relations is only described negatively in Peirce's account; in a parallel version to the 'PAP' quotation above, he says:

But we do not make a diagram simply to represent the relation of killer to killed, though it would not be impossible to represent this relation in a Graph-Instance; and the reason why we do not is that there is little or nothing in that relation that is rationally comprehensible. It is known as a fact, and that is all. I believe I may venture to affirm that an intelligible relation, that is, a relation of thought, is created only by the act of representing it. I do not mean to say that if we should some day find out the metaphysical nature of the relation of killing, that intelligible relation would thereby be created. [...] No, for the intelligible relation has been signified, though not read by man, since the first killing was done, if not long before. (NEM IV, 316n)

Peirce's pragmatizing Kant enables him to escape the threatening subjectivism: rational relations are inherent in the universe and are not our inventions, but we must know (some of) them in order to think.[96] The relation of killer to killed, is not, however, given our present knowledge, one of those rational relations, even if we might later become able to produce a rational diagram of aspects of it. Yet, such a relation is, as Peirce says, a mere fact. On the other hand, rational relations are – even if inherent in the universe – not only facts. Their extension is rather that of mathematics as such, which can be seen from the fact that the rational relations are what make necessary reasoning in diagrams possible – at the same time as Peirce subscribes to his father's mathematics definition: Mathematics is the science that draws necessary conclusions – with Peirce's addendum that these conclusions are always hypothetical. This conforms to Kant's idea that the result of synthetic a priori judgments comprised mathematics as well as the sciences built on applied mathematics.[97] Thus, in constructing diagrams, we have all the possible relations in mathematics (which is inexhaustible, following Gödel's 1931 incompleteness theorem (Gödel 1986)) at our disposal. Moreover, the idea that we might later learn about the rational relations involved in killing entails a historical, fallibilist rendering of the a priori notion. Unlike the case in Kant, the a priori is thus removed from a privileged connection to the knowing subject and its transcendental faculties. Thus, Peirce rather anticipates a fallibilist notion of the a priori (cf. Chap. 8). In the alternative PAP version, Peirce continues: 'At any rate, a Diagram is clearly in every case a sign of an ordered Collection or Plural, – or, more accurately, of the ordered Plurality or Multitude, or of an Order in Plurality' (ibid.). We can say that the diagram is so to speak the redrawing of an icon in terms of a priori relations between its parts. In contrast to the wider term icon, defined by its relation to the object, the subcategory diagram is thus defined through its mode of rationally representing:

The Diagram represents a definite form of Relation. This Relation is usually one which actually exists, as in a map, or is intended to exist, as in a Plan. But this is so far from being essential to the Diagram as such, that if details are added to represent existential or experiential peculiarities, such additions are distinctly of an undiagrammatic nature. The pure Diagram is designed to represent and to render intelligible, the Form of Relation merely.' (ibid.)[98]

Thus, it is possible in a diagram to dissociate the pure diagram, built from rational relations, on the one hand, and its application: what the diagram may, in turn, be used to signify (via symbols) or refer to (via indices) on the other. Thus, the pure relational diagram forms a type.[99]

THE DIAGRAM AS TYPE

Taken separately from its signification and reference, a diagram is itself a type. Consisting of rational relations, it is no wonder that the diagram as such is an ideal entity which is, in turn, communicated through particular diagram tokens. The diagram in itself is not the graphic figures on the sheet before us or before our inner gaze, as we might spontaneously believe. The diagram-icon should not be perceived as a particular figure: already before ascribing to the diagram any content or reference whatsoever, there is a crucial process of abstraction (in Peirce's terminology, prescission, see Chap. 11) taking place, allowing the particular sinsign to be interpreted as instantiation of a type by bracketing all accidental features of the token at the profit of the type: 'One contemplates the Diagram, and one at once prescinds from the accidental characters that have no significance' (NEM IV, 317). When seeing a geometrical figure drawn on a blackboard, we immediately prescind from the stripe of chalk having any breadth, from the line's vacillating deviation from linearity, from the particular color of the drawing, and so on. This type-reading of a diagram token now depends on the set of rules, explicit or implicit, that is selected to govern its typicality. Thus, one and the same diagram token may be read as a type in widely differing ways according to the rules of interpretation used. A line may be interpreted in one diagram as a borderline, in another as a line of connection between two points, in yet another as a transport of some object between two locations. This may be banal, but nevertheless it is an important feature in the diagram's iconicity: the type only becomes apparent in light of the use of certain rules – long before the virtual application of the diagram on more specific meanings, not to talk about empirical reference. This implies that already the pure diagram is an icon governed by a rule, that is, by a symbol. For instance, the sinsign ∘ may be read as a token of the type *circle*, as a token of the type *circular disc* (including its interior), of the type *circular hole* (excluding its interior), of the type *conic section* (any other conic section, a point, an ellipse, a parabola, etc. would do as well as token), of the type *Jordan-curve* (a closed curve; here any other closed curve, e.g. a rectangle, would fulfill the purpose), of the type *hole in a 2-D surface* (a hole of any other shape would do as well), of the type *topological sphere in two dimensions*, of the type *closed and connected manifold*, etc., – each of these choices, in turn, yield different possibilities of which content the diagram type may be used to signify. In the language of Hjelmslevian semiotics, we could say that the diagram token is a unit of the expression substance referring to different types at the form-of-expression-level – all prerequisite to any reference to types in the content plane. Thus, the diagram type consists of two parts: a diagram token and a set of reading rules for the understanding of it as a type (which may, in many cases, be implicit); thus on the level of pure diagram types, the Kantian intuition-concept (talking Peircean: observation-universality) duality, is present in the very construction of the diagram as sign.

THE DIAGRAM AS THE INTERPRETANT OF A SYMBOL

In the next step, this diagram type only becomes a diagram *in actu* (recalling Peirce's basic dictum that signs are only signs *in actu*) when it becomes part of the inference process. To this end, the diagram type needs to be endowed with a symbolic signification – it must involve a 'Symbol of which it is an Interpretant'. Of course, it was only possible to construct the diagram type in the first place by precisely such a symbol (the reading rules just referred to), but these were all on the purely rational, pre-empirical level. The diagram being constructed as a type due to this symbol (the circle above e.g. taken to mean *totality* in a Neurathian cake-diagram) may now, in turn, act as the interpretant of another symbol (the population of Denmark, e.g.). The symbol in question refers to a general object while the diagram in question – being an iconic legisign, a type, – is in itself one. The condition of possibility for this connection is thus the generality of both terms; the diagram being a type and the symbol referring to it being general as to its object. This connection forms the defining semiotic link of the diagram. As the symbol refers to a general object while the iconic legisign, the diagram type, is in itself one, the possibility of the diagram lies in letting the latter constitute the signification of the former and hence refer to the same object. Of course, this is no merely arbitrary connection; what Peirce does not explicitly emphasize in this context (but elsewhere) is the fact that any symbol which is not a completely empty convention must always already refer to some icon (or, at least, it must make possible a process of inference leading to an icon), this icon being its initial interpretant before the symbol might be further elaborated in a diagram. The construction of the diagram, then, amounts to substituting for the initial interpretant of the symbol – the *Vorverständnis* of it, so to speak – a more precise and relationally elaborate icon.

This is a crucial point in order to understand the diagram's double determination – iconic and symbolic, perceptive and general – in Peirce. The diagram is an icon, but a special icon insofar as it is governed by a symbol, and in many cases doubly so, governed both by the type of rational relations used and the empirical phenomenon referred to (like the circle and the Danish population). But what does it imply to be governed by a symbol, to be the signification of a symbol? A symbol is defined by denoting a kind of thing, that is, an idea, not a particular thing ('Art of Reasoning',

1894, EPII, 9; 2.300); it does so by connecting a set of (possible) indices to an icon ('The Short Logic', 1895, EPII, 17; 2.295); it is a law, or regularity of the indefinite future ('Syllabus', 1903, EPII, 274; 2.293), and this implies that it is a rule which will determine its interpretant (ibid. 2.292). It is, simply, a sign which would lose its sign character if it had no interpretant (this in contradistinction to pure icons and indices, respectively). It is a sign referring to all possible entities acting according to some rule which is described by means of an icon: 'It is *applicable to whatever may be found to realize the idea connected with the word...*' ('The Art of Reasoning', 1894, EPII, 9; 2.298), and the habit or rule defining it links together icons: 'A *Symbol* is a sign which refers to the Object that it denotes by virtue of a law, usually an association of general ideas, which operates to cause the Symbol to be interpreted as referring to that Object' ('Syllabus', 1903, EPII, 292; 2.249). But the symbol does not determine the particulars which fall under it – except from precisely their falling under it. This is why it is necessarily general, and thus vague as to its extension. But it may also be vague as to its intension because being defined by a rule connecting icons: these need not be clearly defined, as is most often not the case in non-scientific concepts. Thus, the concept 'dog' is vague because it is not possible to determine beforehand all single creatures it may apply to now and in the indefinite future, but it is also vague for the reason that there is no sharp borderline between it and the concept of wolf. But still, it is defined by a rule-bound association of icons, constituting a general kind. Now, as is evident from these deliberations, any symbol in itself always already constitutes a proto-diagram, insofar as its predicative aspect is iconic. Peirce emphasizes this in 1904 in 'New Elements': 'A *diagram* is an *icon* or schematic image embodying the meaning of a general predicate; and from the observation of this *icon* we are supposed to construct a new general predicate' (NEM IV, 238). The rule in it needs not be explicit, as it is appropriately hinted at in the identification of rule with habit in Peirce. The diagram, then, can be seen as the making explicit (some of) the habits already inherent in a symbol.

Of course, it is important to keep in mind that the mode of existence of the symbol's object is here bracketed; it may refer to existent, future, past, imaginary, fantasy, or any other objects. The symbol 'unicorn' is no less a symbol because its object does not exist. It is perfectly possible to let a diagram make explicit the content of a symbol whose referent is fictitious merely. On the other hand, it is an important diagram property that it is beyond the reach of any diagrammatization to picture inconsistent symbols; this constitutes the very strength of diagrammatic formalization: every (correct, that is) diagram corresponds to a possibility.[100] For instance the grammatically correct symbol 'round square' which implies a rule connecting the two iconic qualities 'round' and 'square' reveals itself as inconsistent precisely when we try to construct a diagram to express these properties in one and the same figure.[101] The same goes for more complicated and less intuitive cases, for instance 'the rational square root of 2'; here a more complicated diagram manipulation is needed in order to grasp that symbol's inconsistency. Briefly, being an icon, the diagram cannot be inconsistent. It may display non-existent entities, but not logically inconsistent entities. Its object is necessarily *possible*[102] – in contrast to the object of a merely symbolic expression. This constitutes a motivation for diagrammatic reasoning: it can make explicit (parts of) the signification of a symbol and pragmatically weed out symbol inconsistencies.

Similarly, no distinction between more or less empirical symbols rules out explication by means of diagrams: both may give rise to diagrammatic explication. There are, of course, prototypical cases of pure and empirical diagrams, respectively, cf. for instance a diagram representing various parts of a population in a cake-diagram vs. a drawing of a circle as a diagram for the concept circle.[103] A pure diagram will be purely mathematical (for instance a map with no reference to its empirical interpretation but only referring to a 2-D surface with certain structures on it), while an empirical diagram will be the interpretant of some empirical symbol in the actual or some possible world (for instance, a topographical map of a country, fictive or not). This must not be confused[104] with the question of reference which is the issue whether the diagram is used in a proposition (a Peircean 'dicisign'), that is, applied to objects referred to by indices (for instance, a map of England). Thus, the empirical case covers two subcases: one where the diagram depicts relations of a material ontology with no factual content, the other when empirical facts are also represented in the diagram by means of indices.

Thus, the diagram may make explicit the consistent content of (parts of) both more and less general symbols – and these may in turn be used as predicates in propositions about indexically identifiable subjects (which also may be general, to be sure).

THE DIAGRAM AS A FORMAL MACHINE FOR GEDANKENEXPERIMENTE

Now, we reach the core point of Peircean diagrammatology: the diagram as vehicle for mental experiment and manipulation. The operational definition of the icon is intimately tied up with diagrammatic experimentation. Let us take a closer look at these connections. The central phase of the diagrammatic reasoning process, motivating the very construction of the diagram, is *deduction*: the demonstration of the fact that a certain version of the diagram necessarily follows from another. Thus, every deduction is diagrammatic of nature and the logic of diagrams is an extension of the traditional concept of deduction (tied to truth-preserving operations in symbolic logic) to cover a large range of phenomena not usually considered as deduction (that is, unless translated into the symbolic form of a formal language) – but describable as such in so far as they qualify as necessary movements of diagrammatic thought. A constructive geometrical proof is probably the arch example of a diagrammatic experiment – we have already in the Introduction discussed a simple and often quoted example – the diagrammatic version of Pythagoras.[105] The proofs given in classical Euclidean geometry may serve as core examples as they indeed were for Peirce who took them to be prototypical cases of diagrammatical reasoning.

Let us consider a couple of other examples. Diagrams play a huge role in both the research and the teachings of mathematics.[106] It is interesting that many proofs can be given in almost pure diagram form (with little or no accompanying text), such as the following proof of the angle sum of a star's vertices being 180 degrees (after Nelsen 1993):

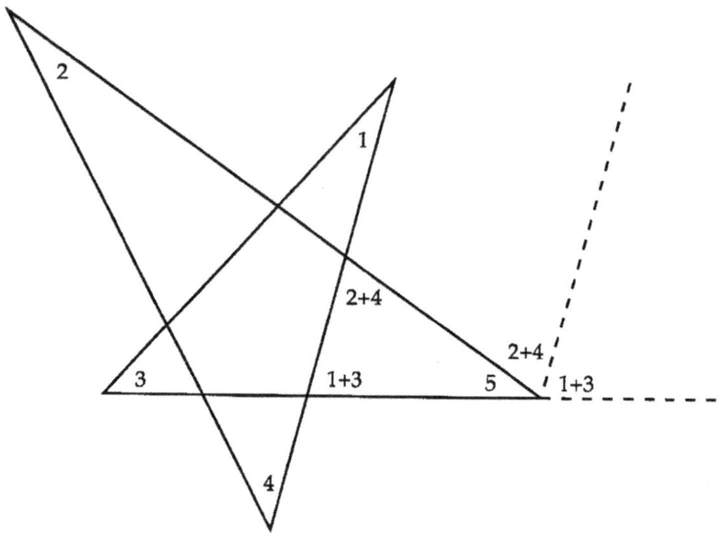

Figure 10.

Here, the angles are indicated by numbers. By seeing that angles facing the opposite way when a line crosses two parallel lines are equal, the sum of the angles in the small triangle to the right (which we know to be 180 degrees) can be seen to be equal to the sum of the star point angles. This, of course, involves the imaginary and experimental transformation and adding of angles – here strongly guided by the labeling of angles.

Another example can be taken from Peirce's own Alpha graphs – the proof of modus ponens in that system. The left side of the graph is a 'scroll' representing

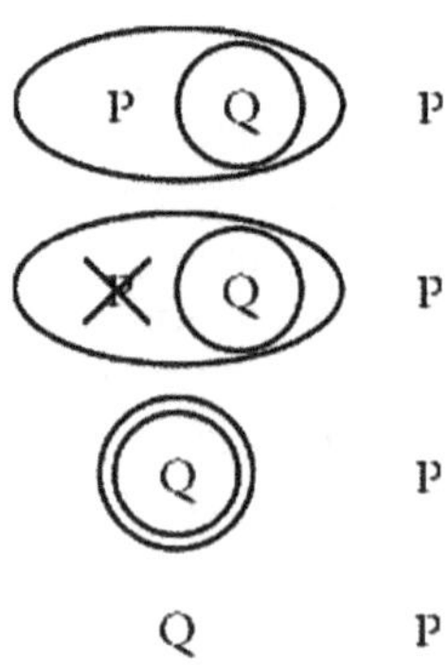

Figure 11.

the implication $P \Rightarrow Q$ – while the right side represents 'P is true'. The simple rules of Peirce's system now permit us to experiment with the diagram. The appearance of a graph in an outerly area permits the erasing of it from a more innerly area, so we are permitted to strike out the P in the implication, leaving us with Q in a double cut. But another rule – basically corresponding to double negation equaling assertion – permits us to erase the double cut, leaving Q alone on the basic sheet which equals 'Q is true'.

Perfectly simple – and somehow deludingly so – examples both of them, not implying the ingenious invention of new diagram manipulations or the introduction of new auxiliary objects into the diagram. The former is continuous, requiring the imaginary translation of geometrical objects on the surface; the latter discontinuous, involving the adding or erasing of whole discrete structures on the sheet. Thus, diagrams comprise both continuous and discontinuous variants – the latter traditionally experienced as more logical – but without any assumption of the latter serving as deep structure of the former. In both cases, diagrams facilitate, that you, thanks to the 'typical' idealization in the diagram, are able to work directly on the ideal state-of-affairs in question.[107]

Another prototypical example is the solution of an equation during a series of well-controlled steps according to the transformation syntax given by elementary arithmetic – mirrored by the solution of the same equation given its graphical representation thanks to analytical geometry.

These are experiments on pure diagrams – prior to indexical and (empirical) symbolical reference, but once an empirical diagram is constructed, the experiment follows the same formal procedure. A map permits you to find a route between two given localities (there is no unique solution, but any line connecting the two is necessarily one). An economic growth graph in a business magazine permits you to determine the actual tendency. These experiments are very simple, indeed, but the important thing is there is a continuum between such examples on the one hand and very difficult, even yet unsolved, problems in mathematics, on the other.[108] A very crucial observation here is that empirical diagrams continuously shade into ordinary icons. Take a photograph of a tree – it is an icon in so far as not previously explicit information may be gathered from it – say, e.g. the fact that the crown of the tree amounts to two thirds of its overall height. This fact was remarked nowhere earlier, neither by the photographer nor the camera nor the developer – and by noticing it you performed a small experiment of diagrammatic nature: you took the trunk of the tree and moved upward for your inner gaze in order to see it cover the height of the crown twice, doing a bit of spontaneous metric geometry, complete with the implicit use of axioms like the invariance of translation. Of course, this is an ordinary icon in so far as nobody constructed it with a diagrammatic intention. Nevertheless, you used it – *in actu* – that way. This continuum between diagrams proper (be it pure or empirical) and diagrammatic use of ordinary icons shows the centrality of the diagram for the icon category as such. It is with diagrammatic means that the operational use of the icon proceeds. Still, a distinction must be maintained between diagrams proper – that is, diagrams constructed with the explicit intention of experimentation and endowed with an explicit or precise syntax of transformation – on the one hand, and on the other, the more comprehensive class of diagrammatic unfolding of information from more 'innocent' icons. In any case, this defining feature of the diagram – its possibility of being rule-bound transformed in order to reveal new information – is what makes it the base of Gedankenexperimente, ranging from routine everyday what-if to scientific invention.

A famous example is the German chemist August Kekulé's discovery of the stereochemical arrangement of the Carbon atoms in Benzene (C_6H_6), forming a ring. The composition formula was well known, but it remained an enigma how six Carbon atoms, Carbon having a valence of four, might form a molecule with only six Hydrogen atoms. A normal Carbon chain would leave too many unsatiated connections in order for only six Hydrogen atoms to complete the molecule. According to his description in his 1890 25 years celebration speech, the scientist sat in 1863 daydreaming before the fire, exhausted by speculation. He then saw one of the flames assuming the figure of a snake which turned around and bit itself in the tail to form a ring-like structure which wiggled contemptuously before his gaze – and all of a sudden, Kekulé realized that the normally linear carbon chain in the Benzene case turned around to form a circle.[109] That discovery thus formed a spontaneous case of diagrammatical reasoning, realized in the shape of metaphors. The flame was taken as a metaphor of the snake which, in turn, was taken as a metaphor of the carbon chain – a structure of metaphors held together by the common diagram of a piece of line, able to bend. The spontaneous diagram experiment argued that the Carbon chain, just like a snake, was able to form a ring, and subsequent chemical analysis corroborated the idea, leading to a major breakthrough in organic chemistry.

THE DIAGRAMMATIC REASONING PROCESS

Before the crucial diagrammatic transformation is undertaken, however, preliminary steps take place in the overall picture of diagrammatic reasoning. The initial diagrammatic intention is in itself an interpretant of a Symbol (Peirce here refers to the Euclidean procedure of beginning with a statement of the general proposition to be proved, and then drawing a figure to illustrate the premiss of the conclusion). Thus, the reasoning process begins with the drawing of a diagram to exhibit the antecedent condition of its object, determining an 'Initial Symbolic Interpretant'. These two, taken together, now form the Peircean equivalent to the Kantian schema: the drawing constitutes its observable side; the initial interpretant its universal signification. Take an example: the drawing of a bridge construction equipped with the appropriate equations pertaining to its carrying ability. After this initial phase, Peirce in the long quotation above (PAP) considers the middle phase, albeit in a strange psychological tone alien to him: this initial schema determines 'a state of activity in the Interpreter, mingled with curiosity. As usual, this mixture leads to Experimentation.' Yet, he immediately admits that such a development must take place in any semiotic Quasi-Mind; we may discern the phenomenological core in the psychological shell: the central feature is the equipment of the initial diagram with transformation possibilities. Peirce here considers the sources for the transformation syntax: '... certain modes of transformation of Diagrams of the system of diagrammatization used have become recognized as permissible. Very likely the recognition descends from some former Induction, remarkably strong owing to the cheapness of mere mental experimentation.' One source for transformation rules thus comes from the diagram itself, and their deductive status untold, Peirce refers their recognition to 'some former induction' (say, the law of gravity involved in the equation system for the bridge's carrying ability has been corroborated by induction). This 'former induction' must, in fact, be taken to refer to at least two separate sources. First, what we introduced above as the symbol's pre-diagrammatic immediate interpretant; including the idea inherent in the symbol of certain developments being possible for its object, others not so. This signification is also iconic, even if not explicitly diagrammatic; it constitutes so to speak our common-sense *Vorurteil* as to the content of the Symbol which the diagram more rationally illustrates, in our example, the common-sense understanding of a bridge (implying that we gauge the effect of normal, mesoscopic vehicles, not planets or atoms, on it). But the vagueness here probably comes from the fact that the determination of the possibilities of experimentation on the diagram is twofold and has yet another source of transformation possibilities in addition to the vagueness inherent in the symbol's generality. Quite another comes, namely, from the very structure of the diagrammatic figure as a legisign (without any reference to which possible symbols it may be taken as an interpretant of): which purely formal possibilities does the law governing the sign allow for varying upon the diagram's arrows, amount of entities, forms, structure, etc.; in the bridge case the set of equations with variables taken separately, apart from their actual referent. The former has its source in the generality of the symbol's object, the latter in the generality of the diagrammatic sign itself – what Peirce calls its being a type. Peirce identifies one more source stemming from the diagram intention (the fact that we want to gauge our bridge's carrying ability) which makes us experiment in order to fulfill this intention (we may vary the weight carried in order to find the point where the bridge can carry no more, hopefully far above the average weight of expected vehicles).

After having performed the transformation, in any case, the transformate diagram now displays the result at the same time as it is evident that the transformate diagram was contained in the transformand diagram. The transformate diagram, the eventual, rational interpretant of the transformand diagram, has in itself the conclusion, expressed in symbolic terms, as its interpretant: the bridge may carry vehicles up to 100 tons. Thus, the steps in diagrammatic reasoning lead from an initial symbol through three consecutive phases of diagrams and to a final symbol. We may envisage the possibility that the diagram transformation chosen does not lead to the expected result so that a trial-and-error process undertakes a new experiment on the same diagram. Say, if the bridge is shown to carry only 100 g vehicles, a new experiment changing the size or the material (or the diagram construction) may give a better result. Michael May has presented this model for the core processes of diagrammatical reasoning (May 1999, 186):

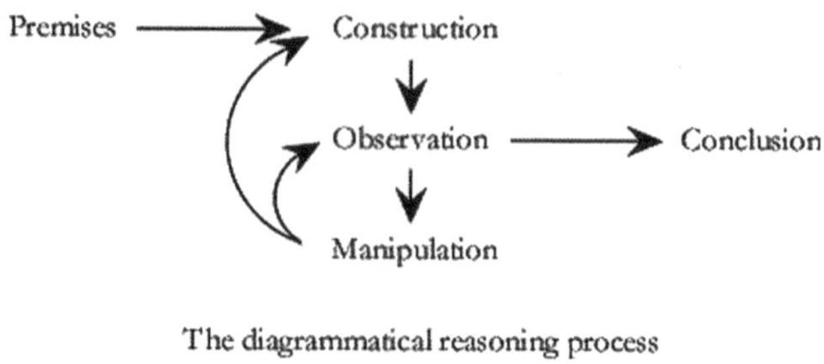

Figure 12. The diagrammatical reasoning process

Here, the short circuit leading from premiss over construction and observation to conclusion corresponds to corollarial reasoning, while the longer process involving one or several phases of manipulation and maybe even further construction corresponds to theorematical reasoning.

But there are a lot of possible prerequisites to be added to this ideal-type diagram transformation. First, the initial symbol already has its interpretant partly consisting of iconic material. (In our example: data about the bridge). Thus, the diagram may be said to be a rational analysis of parts of this pre-diagrammatic icon. But this entails the possibility of fallacies, if a diagram not conforming to the initial interpretant is chosen. Thus, experiments already in this phase may in some cases be expected; in cases less simple or less well-known than bridge building we might resort to a trial-and-error process here, experimenting with different formalisms in order to find those fitting best the intention. In that case, the deductive diagram transformation 'molecule' just described becomes a phase in Peirce's overall heuristics. An initial abduction makes a guess about how to formalize a given phenomenon, the deductive diagrammatic phase just described follows, and finally an inductive investigation concludes the picture, in which the diagrammatic result is compared to the actual empirical data: Does the diagram transformation actually, in some sense, correspond to an evolution in the phenomenon mapped in the diagram?

We can sum up the steps of the process as follows:

***a** Symbol (1)*

***b** . . . having a rule-bound, initial, pre-diagrammatic, immediate iconic interpretant*

***c** Initial interpretant diagram-icon*	***(a + c)** constituting the initial transformand diagram, the 'Schema'*

***d** Middle interpretant: the symbol-governed diagram-icon equipped with possibilities of transformation (with two sources, **a** as well as **c**)*

***e** Transformate diagram. Eventual, rational interpretant*

***f** Symbol (2) (Conclusion)*

***g** . . . having a post-diagrammatical interpretant differing from **b**. This interpretant being an interpretant of a as well, the diagrammatic reasoning has now enriched the total interpretant of the concept **a**.*

Thus, the process begins with 'some former induction' having given rise to the initial symbol's pre-diagrammatic interpretant, an inductive generalization sedimented as the meaning of the symbol. This meaning, of course, must be to some extent already structured,[110] and some of its rationally formalizable relations are now abductively selected, yielding a guess of which invariant properties may be sufficient to account for other central properties in the general object in question. Then, in the very construction process of the initial diagram, a constant feed-back comparison must take place between the general object as it is preliminarily and inductively grasped on the one hand, and on the other, the abductive guesses trying to establish against this background a more formalized diagram; in many cases this may take place almost automatically, due to the existence of well-established diagrams. It goes without saying that the fertility of the specific diagram chosen can only be fully measured with regard also to the deductive experimentation taking place later in the diagrammatic reasoning process which consequently also has a role in this constant feed-back trial-and-error process. But the overall picture of the initial phase of diagram construction is thus: the general knowledge contained iconically in the symbol – no matter if it be an empirical symbol or a mathematical one – is interpreted in relational terms so as to give (part of) the consistent general meaning an iconical illustration able to be manipulated, an illustration which is, on its side, also general. After few or many repeated transformations (subject to the three different sources of transformation syntax mentioned above), a transformate diagram is reached – its finality is of course only measured on how it accords with the initial intention. The evaluation of an interpretant for a candidate for final diagram status is in itself an abduction proposing a symbolic reading of that diagram, and this may, in turn, be inductively compared with empirical information present in the initial interpretant.

To sum up, the overall picture of the diagrammatic reasoning process is that it forms a formal deductive reasoning core, embedded, on each side, in the trial-and-error of abductive trials and inductive tests.

CARTOGRAPHY AS AN EXAMPLE

Maps are no doubt a good candidate to a diagram subcategory: rule-bound depictions of aspects of the shape of phenomena. The non-trivial icon definition is evident here: the construction of a map, be it based on triangulation from a set of selected measuring points in the landscape, or on the rational, stylized rendering of aerial photography, does not explicitly contain all information held in the map. What types of experiment which reveals this information may be performed with respect to a map? We may, for instance (1) find a route between two localities, (2) determine a distance or an area, (3) recognize landscape forms – and so on.

Of course, there is nothing very 'experimental' in a laboratory sense of the word in these transformations; nevertheless their status as diagram transformations are granted by their fulfilling the demand for revealing truths not stated in the construction of the diagram. Take for instance the distance between two cities. To measure the map distance with a ruler and figure out the approximate real distance from the map's scale is a typical map manipulation, depending on the fact that the map we imagine here is endowed with a metric topology. In this case, a middle interpretant will be the map with your route on it added; the transformate diagram will be the map with the ruler – and the final conclusion will be of the form 'The distance between New York and Pittsburgh is so-and-so-many miles', revealing a number nowhere present in cartographic triangulation nor aerial photography. The experiments possible of course depend on the type of map projection, some are area preserving but not distance preserving, some vice versa, some are distance preserving in some directions, not in other. Thus, different map types may be described simply with reference to which types of experiments they allow. Other maps do not even have a metric topology,[111] take for e.g. your typical subway map which does not keep invariant neither distance nor geometric form but which merely keeps invariant certain connexity properties: the connexity of the single subway lines – very often symbolized with one color for each – and the crossings and touchings of several subway lines, highlighted by a circle or other closed curve with another color indicating the weaker connection possibilities of changing line at certain stations. Here, it is easy to find one's localization and path relative to fixed points: the stations – but it is no longer possible to gauge metric properties (how far are we from the main station?) nor morphological properties (a curved track may be represented by a straight line, and vice versa, even in the same map), nor sub-area categorization.

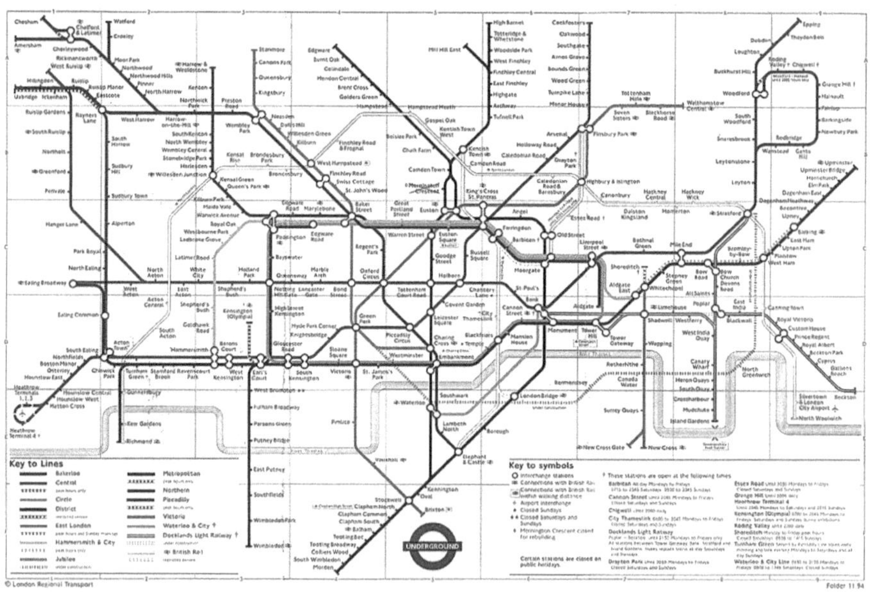

Figure 13.

In the London Underground Map case, all track curves have been stylized by the map admitting only four directions, north-south, east-west, northwest-southeast, and northeast-southwest, respectively. In this case, the typical diagram reasoning process will deduct a specific travel route from the combination of travel conditions (departure, goal) with map structure. Initial schema will be the colored spaghetti articulation interpreted as a diagram-icon by the symbol 'London Underground', the middle interpretant will be your present position and the end of your travel, and the transformate diagram will be the possible paths between the two tracked as continuous lines on the spaghetti figure, and, again, the conclusion will be symbolic statements of the type 'We gotta change at Piccadilly'; 'There's no way there without changing two times', 'It seems to be the shortest way to go via Victoria' (shortest here referring to transport time measured by means of number of stations passed, rather than to any metric property of the diagram). Even in as simple diagrams as road maps, we can appreciate the distinction between two experiment classes. One is the simple use of the diagram, following the transformation rules more or less explicitly given. Another is experimental in a stronger sense of the word: it experiments with the very layout of the diagram itself: the possibility of building a new subway line in order to mend London's traffic problems. These more ambitious experiments, in turn, may involve two dimensions: one is further information with respect to the object (or our ideas about it) making the extension of diagram possibilities desirable. Another is the change in the very formal apparatus of the diagram (these two may, of course, trigger each other), as for instance the development of Venn diagrams out of Euler diagrams by the addition of a rule (the shading of an area referring to an empty set), or the reinterpretation of Euclid's axioms in order to make non-Euclidean geometries. Already the first type of experiment is unlimited as soon as the diagram in question is continuous (like most maps), but in sufficiently complicated diagrams we must expect not to be able to account for the possibilities of interesting experiments beforehand (cf. the Gödel inexhaustibility of mathematics). On two different levels, these properties are what constitute the well-known 'depth' of icons and diagrams. This inexhaustibility is dryly remarked on by Peirce when he epigrammatically states about the content of a diagram simply that 'Everything is involved which can be evolved' ('Logic of Quantity,' from the 'Grand Logic,' 1893, 4.86).

THE GENERALITY OF DIAGRAMS

The strength of Peirce's diagrammatology lies in the extraordinary breadth covered by his diagram concept. In Jaakko Hintikka's Peirce interpretation, the audacity of Peirce's conception lies in his generalization of structures in geometrical reasoning to logic in general. To Hintikka, this generalization becomes evident in Peirce's distinction between 'corollarial' and 'theorematical' reasoning. Peirce himself called this 'My first real discovery about mathematical procedure' ('Parts of Carnegie Application', 1902, NEM IV, 49) – named of course after corollaries and theorems in geometrical proofs. 'Peirce's brilliant insight is that this geometrical distinction can be generalized to *all deductive reasoning*' (Hintikka 1983, 109). Corollarial reasoning refers to conclusions which may be read directly off the diagram, interpreted in the right generalized way – while theorematical reasoning, on the other hand, requires the introduction of auxiliary constructions not explicitly referred to in the premisses (in the initial diagram). In Euclid, such auxiliaries often have the character of the construction of further figures in the diagram – and in the geometrical tradition the distinction between corollarial conclusions and theorematic constructions is often spontaneously referred to as between logical and geometrical consequences (ibid.). In the standard conception of logic referring to formal languages, no such distinction is, of course, possible, all inferences being simply shorter or longer cases of symbol manipulation. In Hintikka's account, however, Peirce's distinction is not only due to superficial representation differences in geometry but refers to deep issues in logic. First, Hintikka emphasizes that theorematical reasoning necessarily implies the introduction of new variables in the inference process – often in the shape of a lemma in mathematical proofs. This points to the non-triviality of theorematical reasoning as opposed to corollarial reasoning. But then theorematical reasoning becomes a matter of degree – depending on how many new individuals are introduced in the construction, and thus yielding a 'rough measure of the nontriviality of an argument' (113). Thus we may imagine a continuum between completely trivial corollarial reasoning in one end and still more non-trivial theorematical reasoning in the other. Peirce's concept of diagram manipulation and diagram experiment thus refers to a whole range of theorematical reasonings of various degrees of non-triviality. Second, Hintikka emphasizes that Peirce's distinction solves the age-old problem he nicknames 'logical incontinence' (114): how can anybody possibly miss the knowledge of all logical consequences of one's premisses? Corollarial conclusions are indeed evident to everybody who is able, at all, to understand the diagram in question as a type. But theorematic conclusions require both the introduction of the right auxiliary entities and the right construction obtained from combining these entities with the initial diagram. Theorematic reasoning thus may require huge skills on the part of the reasoner so it is no wonder why such conclusions may be difficult to obtain. Third, Hintikka argues that the distinction throws a new light on Kant's analytic-synthetic distinction. It is unclear whether Kant's concept of analytic judgment covers corollarial deductions only, or both corollarial and theorematical deductions, simply because Kant did not have a corresponding distinction at his disposal (115). Hintikka argues that Kant's concept of analytic inferences covers corollarial reasoning only, so that theorematical inferences must be classified as synthetic – corresponding to Peirce's idea of diagrammatical reasoning as covering the synthetic a priori domain. We shall return to this connection between diagrams and the synthetic a priori in Chap. 8.

Hintikka's appreciation of the corollarial-theorematic distinction forms part of his overall charting of twenteith century logical thought in two main traditions, inspired by van Heijenoort's famous brief paper on logic as calculus versus logic as language. Hintikka's idea is that logic since the latter half of the nineteenth century has followed two different currents depending on basic ideas often not clear to the persons involved – currents involving huge parts of analytic and even continental philosophy. He nicknames them 'language as Lingua Universalis' and 'language as Calculus Ratiocinator' after Leibniz's famous dreams of a universal language and a reasoning machine, respectively.[112] The main current in logic falls within the former, comprising Frege, Russell, Wittgenstein, the younger Carnap, Quine. Here, logic is a formalization of (parts of) language, and it is universal in so far as it is impossible to address the world by any other means than this one language. This has a series of corollaries. Logic refers to the one existing world only (hence the hesitation to accept counterfactuals); semantics becomes ineffable and impossible to formalize because it must be discussed in that same language which is the object of the discussion; formal syntax becomes the only objective of logic; linguistic relativism threatens because there is no possible corrective to the given linguistic understanding of something; the relation between world and language is impossible to attack outside the one universal language (Wittgenstein: the limits of my language and the limits of my world coincide), and hence no truth definition may be given.[113] Surprisingly, what might at first glance seem to be a strongly realist position (logic as one, universal language) thus holds a series of antirealist or even relativist or skepticist consequences – becoming evident e.g. in Quine's linguistic holism and his ideas of the impossibility of translation. The alternative tradition has rather been an undercurrent running from Boole via Peirce, Schröder, Löwenheim to certain aspects of Hilbert, Gödel, the later Carnap – and, of course, to Hintikka himself. According to this tradition, logic is no universal language, but rather a calculus aimed at problem-solving in a given domain of discourse.[114] Thus, there is a possible plurality of logics, and this tradition has given rise to modal logic, epistemic logic, and model theory – very often, hence, Hintikka nicknames this tradition the model theory tradition because of its obsession with semantics and the issue of which formalisms fit best a given content. While this tradition might at a first glance seem antirealist with its emphasis on many competing representation systems, its corollaries are, quite on the contrary, realist. Semantics is possible, because one fragment of logic or language may be unproblematically put to use in the discussion of another such fragment; truth definitions may be given (even sometimes in the same language as it refers to – Hintikka's own 'independence-friendly logics' with free combination of quantifiers as a logical example) and correspondence between representations and world is not only possible but may be continuously refined by the use of many different iconic representation systems for different domains of reality, depending on their pragmatical purpose. Thus, possible-world-semantics grows out of this tradition, just like the insistence of model theory that the semantics of a logical expression must be charted by means of the variation of its possible references. Both traditions embrace formalization, albeit for quite different reasons, which has added to the difficulty in distinguishing them. The universalist tradition embraces formalization because of the ineffability of semantics – which leaves formal syntax as the only or central goal to achieve. The calculus tradition, on the contrary, embraces formalization because of the clarity and efficiency of formal representation and of calculation – and formalization here may range from very general systems able to subsume highly different semantic interpretations (Hilbert who, maybe surprisingly, holds a stock in both traditions) and to highly semantically motivated formalizations in the other end, so to speak encompassing a continuum covering far more possible worlds than universalism's one-world claim in one end and covering highly specific discourse domains within that world in the other end. Hintikka's daunting hypothesis no doubt throws a new light upon the history not only of analytical philosophy of the twentieth century. The work of Martin Kusch (1997) even points to the fact that also continental philosophy displays an exactly analogous split, Husserl falling on the language-as-calculus side while Heidegger falls on the universal-language side – thus placing Heidegger and Derrida along with Frege and Quine while Husserl sides with Peirce and Hintikka, quite different from standard history-of-ideas accounts. We shall return to this below. The anti-iconicity traditions discussed in Chap. 3 thus appear in a new light given Hintikka's distinction – making it easier to understand how iconicity could possibly be attacked from two otherwise completely opposed camps of philosophy.[115]

Hintikka's account for Peirce's special role in the calculus tradition ('The Place of C.S. Peirce in the history of logical theory', in Hintikka 1997) resumes Peirce's obsession with interpreted logic, as opposed to axiomatized systems; his interest in modal logic; his anticipation of Hintikka's game-theoretical semantics in his ideas of the dialogicity of logical inference, his lack of hesitation towards formalizing logic by means of logic. Hintikka takes care to note that the plurality of possible representations does, in fact, point towards realism rather than the opposite:

> In general, like all believers in logic as calculus, Peirce was not only ready to provide an interpretation for their calculi at the drop of a symbol. He could – or thought they could – discuss such changes systematically in an explicit language. The very freedom of choosing a formalism differently on different occasions was sometimes a consequence of their belief that it is the underlying representational realities that really mattered, not the formalism. (Hintikka 1997, 150)

One could rather say that exactly the possibility of having competing representations of one and the same state-of-affairs is what grants realism – not unlike Marvin Minsky's old idea that intelligence requires at least two different representations of the same thing. This idea does not, as Hintikka takes care to note, in any way prevent Peirce, just like Hilbert, from being interested in a purely mathematical definition of his logic notation – the important thing is the possibility of explicitly discussing its interpretation. This points to the issue which is most important in our general diagrammatological context: the iconicity of logic. To Hintikka, Peirce's notion of iconicity and its idea of the sign as a model of its object is 'not completely different' from the ideas of model theory. To Peirce – as opposed to the Fregean tradition's emphasis on thought at the expense of intuition – logic is, in itself, basically iconical. Hintikka here emphasizes Peirce's idea that 'The Icon does not stand unequivocally for this or that existing thing, as the Index does...' ('Prolegomena to an Apology of Pragmaticism', 1906, 4.531, quoted in Hintikka 1997, 153) – thus the icon necessitates interpretation and facilitates different uses (albeit, of course, not any possible interpretation or use). Hintikka sharply contrasts Peirce's iconicity of logic to the early Wittgenstein's picture theory of language:

the latter is not only semantically ineffable and may only be showed, not explicitly discussed, but it is also a static mapping of propositions onto states-of-affairs. Peirce's iconicity of logic, on the contrary, does not only depict states-of-affairs but also their iconical interrelation, making reasoning using icons possible: it thus covers the very 'iconic aspects of logical inference' (Hintikka 1983, 116) about which Wittgenstein necessarily remains silent. The very manipulations performed on a diagram do themselves iconically refer to possible experiments on the more or less ideal object referred to by that diagram – thus diagrams are not the accidental sums of iconical representations plus logical manipulation. They are icons depicting their object as well as the logical inferences performed on them.

Peirce's generalization of the geometrical concept of diagram to all of reasoning must be understood on this background. Diagrams are not Peirce's iconic equivalent to an all-embracing formal logic understood in the universal-language tradition. Diagrams is rather an umbrella notion for widely differing representation systems which may be picked, constructed and used for correspondingly differing domains and purposes – the diagram notion only insisting that all of them, despite all differences, possess an iconic basis and, in sufficiently complicated cases, make possible diagram experiments in theorematic reasoning. Hence the concept of diagram in Hintikka's calculus interpretation is perfectly aimed at understanding processes of logical inference clothed in widely different representation systems and in widely different domains. In the latter half of this book, we shall investigate three such domains: biosemiotics, pictures, and literature.

DIAGRAM TYPES

As in any branch of research, the possible establishment of an inventory of rational subtypes will constitute a major progress. Unfortunately, no simple diagram taxonomy seems to be at hand, at least not referring to pure diagrams – for the very simple reason that the category of pure diagrams is coextensive with mathematics as such. This implies that the question of pure diagram taxonomies is inevitably entangled in the large questions of the foundations of mathematics. Other taxonomies might refer to different diagram intentions, different diagram graphics, different diagram subjects, etc., but a comprehensive review of diagram taxonomies by Blackwell and Engelhardt (1998) reveals little agreement among scholars. Peirce, taxonomist of signs, never really attempts to develop a diagram taxonomy; the closest he gets might be the remark already quoted, made *en passant* in an early account for diagram experimentation in Robin (15), 'On Quantity' (c. 1895, in NEM IV, 275): '. . . a diagram, or visual image, whether composed of lines, like a geometrical figure, or an array of signs, like an algebraical formula, or of a mixed nature, like a graph . . .' so that we might envisage yet another trichotomy comprising maps, algebra, and graphs; that is, simple diagrams, construction precepts, and diagrams equipped with construction precepts, respectively? The construction of a rational taxonomy of diagrams will be a major future challenge for (not only) Peircean semiotics.[116]

THE IMAGINARY MOMENT IN DIAGRAMS: PEIRCE AND HILBERT

During the operational interpretation of an icon, a certain phase typically appears which at the same time exposes the icon's full range of possibilities and displays a central danger of iconic fallacy. As discussed in Chap. 3, this 'imaginary moment' involves momentarily suspending the distinction icon-object while operating on the former, as Peirce notes in already in the 80s:

> Icons are so completely substituted for their objects as hardly to be distinguished from them. Such are the diagrams of geometry. A diagram, indeed, so far as it has a general signification, is not a pure icon; but in the middle part of our reasonings we forget that abstractness in great measure, and the diagram is for us the very thing. So in contemplating a painting, there is a moment when we lose the consciousness that it is not the thing, the distinction of the real and the copy disappears, and it is for the moment a pure dream – not any particular existence, and yet not general. At that moment we are contemplating an *icon*. ('On the Algebra of Logic', 1885, W5, 163; 3.362)

This moment of fiction when we, operating on the icon, takes it for the object itself, is crucial for our operations: here, the constraints on our operations stemming from the icon's formal properties are identified with the constraints stemming from the object's properties and the constraints stemming from the question leading us to diagram experimenting (the three sources discussed above), and it feels like we are operating on the very object itself. This goes for all icons, from paintings where we leave our observer's position and momentarily insert our imaginary body on a stroll into the landscape and to equations where we cease manipulating only ink symbols on a sheet and tackle invariances in arithmetic directly. This 'imaginary moment', of course, is a description in psychological terms of a phase in a process that is not itself of psychological nature. But the important thing in our context is the virtual source of error inherent in this moment: properties stemming from our preformed folk understanding of the object in question may interfere, without our attention, in our experiments with the icon – with the result that we see things in a picture not really presented there, or we find regularities in a formalism which are not really implied by it. The latter was, of course, the case in Euclidian geometry where our everyday conceptions prejudiced us to assume the parallel axiom true – a fact which in the history of mathematics predisposed mathematicians to be on guard against intuition.[117]

Thus, there is a certain tension in this 'imaginary moment'. To the extent that the imaginary moment leads to the eventual interpretant, the conclusion seems to be directly 'read off' the diagram and so furnishing evidence. On the other hand, even if this fertile moment is the very source of evidence, it is precisely the seductive welding together of object and representation in this phase which constitutes the major source of error in diagrammatic reasoning and has long since been recognized as such. The whole formalist endeavor in the philosophy of mathematics and the emphasis upon symbolic calculi and mistrust of geometry since the late nineteenth century is based on attempts at getting rid of the dangers of seduction by intuition in this very moment. More precisely, this danger can be traced to the triple source of constraints on the possible experiments in this crucial phase of the reasoning process: they descend, as we saw above, from the initial pre-diagrammatic interpretant, from the diagram intention as well as from the internal regularities of the diagram-icon as iconic legisign. But the first two of these sources are of course ripe with common sense, with folk theories and virtually ideological preconceptions of the object and thus possibly with wishful thinking – and the imaginary moment may lure the reasoner into accepting these preconceptions and tacitly letting them govern the experiment so as not to discover crucial formal possibilities in the legisign or even to abandon internal legisign-constraints to the benefit of fallacious common sense assumptions in cases or aspects where the two are mutually exclusive. Hence the idea of formalism in mathematics; one could describe Hilbert's idea as that of getting rid of the imaginary moment precisely in the decisive part of the process leading from the diagram-icon via the middle to the eventual interpretant, bracketing the process from signification in these phases and then reinvesting it after having reached the transformed diagram, that is, the theorem. Of course, orthodox Hilbertians will be shocked to see the idea of purely formal proof theory (with intuition's role reduced to the level of meta-mathematical interpretation) transformed into iconic diagram manipulations: the Peircean process at first glance seems to be almost the opposite: one could leave out the possibly folk theoretic symbolic determination while manipulating the icon – and then reinvest the symbolic interpretation after having reached the theorem. But a closer analysis reveals the similarities: the diagram in Peirce is iconic indeed, but it is a formally controlled, 'rational' icon equipped with a syntax of transformations, while the Peircean symbols here are the possible source of error because of their immediate interpretants in the form of pre-diagrammatic ordinary icons, 'wild' icons, so to speak. The reason for confusion here comes from widely differing conceptions of 'symbol'.[118]

We can add that the well-known mainstream formalist idea from the full-fledged Hilbert doctrine of the 1920s that diagrams should be completely expelled from the proof to a role merely of heuristic support device (cf. Husserl below) was not always unanimous in Hilbert, as discussed in Michael Greaves' (2002) fine book on the somber destiny of diagrams in twentieth century geometry and logic. Hilbert's famous standard idea, as expressed in the early (1894) quote 'A theorem is only proved when the proof is *completed independent of the diagram*.' (72),[119] sometimes gave way to rather different ideas like the 1900 quote: '... arithmetical signs are written diagrams, and geometrical diagrams are drawn formulas' (74). Here, Hilbert in fact expresses a completely Peircean idea of equivalence between symbolic and diagrammatical expressions.

Correspondingly, formalist reading in Hilbert and rule-governed 'imaginary moment' in Peirce may thus be seen as parallel ideas of controlling a seductive phase of reasoning. Then, the isolation of the purely diagrammatic part of the process (Peirce) would be equivalent to the idea of keeping a pure mathematical reasoning apart form uncontrollable iconicity (Hilbert).

Moreover, Hilbert perfectly realized that a certain and inevitable minimum of *Anschauung* remains indispensable even in symbolic calculation, namely the basic ability to identify, count, and permute symbols on a string. In both cases, then, the crucial opposition ceases to be between symbolic and iconic and becomes rather the opposition between a controllable, rational intuition and a 'wild' pre-formal intuition. The crucial difference is rather, now, that the Peircean point of view will see the remaining controlled domain of rational intuition as a definitively iconic field, while the Hilbertian will often see it as purely symbolic, unfortunately, but unavoidably, to be exposed to a severely constrained finite intuition, corresponding to the simplest arithmetic, able to infallibly count strokes in a row (assuming the early digital idea that the finite symbol alphabet could be translated into a system of such strokes). Of course, Hilbert himself was no Hilbertian and he perfectly realized the unavoidable remnant of *Anschauung* in this 'formale Redeweise' (cf. Kreisel 1982).[120] Here, Peirce's technical research into iconic logic diagrams shows, as mentioned, that the task undertaken by 'symbolic' calculi may be equally well performed by apparently much more explicitly iconic systems.[121] The equivalence of Peirce's graphs to formal syntactical systems proves, of course, that the latter possess the same degree of iconicity as Peirce's – both may give rise to the extraction of the same amount of non-explicit information (as Hintikka also notes, Hintikka 1997, 154). Still, the problem that motivated Hilbert is still relevant for the Peircean account of diagrammatic reasoning: we cannot expect, even less can we demand, the imaginary moment to involve the whole process from initial interpretant to eventual interpretant. The very formal *raison d'être* of diagrammatic reasoning entails that purely diagrammatic constraints with no apparent interpretation, only motivated by the diagram as a rule-bound legisign, may take over in decisive phases of the argument and, just like in Hilbert, preclude 'wild' intuitions from intruding. Thus, the imaginary moment must be virtually split into two: an initial moment where diagram and symbol (1) are identified, and a final moment where transformate diagram and symbol (2) are identified, so as to keep a pure diagrammatic transformation phase in between them. In this case, the comparison between symbol (1) and (2) of course becomes crucial: in the empirical case, the question will be, does the symbol (2) give meaning as expressed in a proposition about symbol (1) – e.g.: has an object of type (1) ever empirically given rise to an object of type (2)? If not then the diagram may be invalid, or the observation insufficient. So the pragmatist trial-and-error feedback between initial and final symbols in the diagrammatic reasoning process must be the Peircean means of avoiding being caught up in the 'imaginary moment'.

DIAGRAM, CONTINUITY, CONCEPT, ABDUCTION, PRAGMATISM ...

The diagram has intimate connections to many central aspects of Peirce's doctrine. The prototypical diagram: a set of lines between points on a continuous sheet of paper, may serve to indicate the important relation between the diagram as epistemological device and the signification of Peirce's notion of the continuum for metaphysics. How do we immediately 'see' that the conclusion of a diagram experiment is valid for a whole class of cases referred to by the premisses? One source is, of course, the typicality of the diagram, but this typicality consists in the possibility of continuously deforming any token to the diagram type. Something analogous holds for the transformations. We see this by the fact that a continuum of possible realizations are built into the diagram. This may take place by different means: one is the continuity of the underlying sheet. By imaginatively performing the transformative change of angle size on the sheet we see that the tripartition of angles into acute, rectangular, obtuse is complete, because we can make the angle pass through all values between 0 and 180. The variable x is in the same way, so to speak, a hole in the sheet through which a whole continuity of instantiations may pass. Of course, discrete diagrams exist where this idea is less relevant (equations defined only with reference to natural numbers etc.) – but continuity, so Peirce's metaphysics, is more inclusive than discontinuity, so that we are only able to understand the latter against the background the former.

Continuity is also, as we saw in Chap. 2, the basis for Peirce's 'medieval' realism with regard to the existence of real universals which refer to natural habits and the continuity of their possible instantiations. But diagrams are intimately connected to symbols, as we have seen, in the diagrammatic reasoning process. Concepts are 'the living influence upon us of a diagram'[122] – this should be compared with Peirce's basic pragmatist meaning maxim, according to which the meaning of a concept is equal to its behavioral consequences in conceivable settings. This implies that signification of a symbol is defined conditionally: 'Something is x, if that thing behaves in such and such a way under such and such conditions' – 'Something is hard, if it is not scratched by a diamond'. But this maxim, developed on the basis of a conception of scientific experimenting, is formally equal to the idea of diagrammatic experiments: the signification of the concept is the diagram of the experiment. The aim of science is to try to make such conditional definitions as diagrammatic as possible. This is the diagrammatic component in Peirce's laconic enlightenment maxim, 'symbols grow': new symbols arise through diagrammatic experimentation.

DIAGRAM PERSPECTIVES

Peirce readily admits that his use of the word diagram employs it in '... a wider sense than is usual' (PAP, 1906, NEM IV, 315); precisely this is the great advantage of his diagram concept: a whole series of semiotic processes – the tropisms studied by biosemiotics,[123] the contemplation of pictures, metaphorical, analogical, and poetical reasoning, linguistic and narratological syntax, basic sensorymotor schemata, as well as mathematics proper – become understandable as different realizations of one and the same basic rational semiotic behavior, namely, diagram experimentation. Thus, it liberates semiotics from the static and narrow idea of the en- and decoding of signs, because the interesting part of semiotics lies elsewhere, in the epistemological dynamics of diagram interpretation – at the same time as it saves semiotics from the false and ungraspable 'dynamics' of irrationalist poststructuralisms, vitalisms, and constructivisms. Moreover, it constitutes a wholly actual attempt at making explicit René Thom's great intuition in philosophy of science, '... there is only science from the moment when you can embed the real within the virtual'.[124] You only understand a phenomenon in terms of a scenario mapping (some of) the possible ways of *changing* that phenomenon into related, virtual phenomena. Quite contrary to Quine, eager to expulse counterfactuals from science, this basic idea is what diagrams formalize: various counterfactual transformations of the phenomenon's real possibilities as the means of gaining insight into it.[125] The diagram as a central concept in epistemology thus unites a series of actual scientific and philosophical currents: cognitive semantics and linguistics, the resurfacing of diagrammatic reasoning in AI as well as, more generally, the renaissance of intuition, pragmatism, and scientific realism in the philosophy of science.

CHAPTER 5

EVERYTHING IS TRANSFORMED

Transformation in Semiotics

Rien ne se crée, rien ne se perd, tout se transforme.

Lavoisier

Transformation procedures are crucial to diagram understanding. Traditionally, however, the notion of 'transformation' has not played a central role in semiotics. If we check the main semiotic dictionaries and encyclopaedias, the article in Sebeok 1986 under that word mainly deals with Chomsky's transformation grammar, while the notion does not appear as the headline of a separate article but is only put to use *en passant* as an auxiliary notion in the description of certain theories.[126] In the otherwise more idiosyncratic dictionary of Greimas (1979) presenting mainly the concepts of Greimas' own theory, the notion surprisingly receives the broadest treatment with reference to its origin in European comparative linguistics and its mathematical use, primarily in the US. Greimas's subsequent description loosely hints at some general ways of using the term in semiotics, even if, as could be expected, its use in Greimassian semiotics (to which we shall return below) plays the main role.

Our hypothesis here is that these sins of omission tend to hide the decisive and central role played by transformations in semiotics in general. Our aim here is to outline its use in different semiotic fields and theories – to provide a brief overview over the possible roles and tasks which it has – or ought to have.

If we go to the history of semiotics,[127] the term 'transformation' is most openly central to exactly Chomsky's generative grammar with its idea that one and the same deep structure may be transformed to a set of different, cognate surface structures by means of a set of transformation rules. Transformation holds a prominent position, moreover, in Lévi-Strauss's brand of structuralism (even if here being subject to less explicit investigation) with its idea that a myth may be transformed into other, related myths by the transformation of structure, measured in relation to a reference myth, in principle arbitrarily selected. Lévi-Strauss himself refers – which is only taken up by rather few of his many disciples – to the Scottish biologist d'Arcy Thompson who early in the twentieth century tried to found a theoretical, morphological biology built on the notion of transformation. Inspired, among others, by both Lévi-Strauss and Chomsky, Greimas's semiotics takes up the concept, now under the notion of 'conversion', it being the procedure which takes its point of departure in simple deep semantics and develops that into realized meaning via a series of intermediary steps. A certain interest here is deserved by the narrative level where can be found the notion of 'narrative schema', involving the idea of a general regularity in the temporal process of texts, connected to concrete, empirical texts by transformation. In recent semiotics, transformation plays a main role in the systematical picture semiotics of Groupe μ, while cognitive semantics and linguistics as central procedures under different notions claim transformations ('mapping' and 'blending' of conceptual or mental spaces), just like transformation in the main inspiration of this current, the cognitive psychology of Eleanor Rosch, plays a main role as that procedure which connects the prototype of a semantic category with different, less typical instantiations of it. In the diachronous linguistics of the nineteenth century, a central transformation problem resides in the idea of sound laws as the nexus of linguistic development.

On the other hand, it is striking that no prominent concepts of transformation are to be found in most of European structuralism and structural linguistics, the Saussure-Brøndal-Hjelmslev-Eco lineage. Here, the exception is the theoretically not very explicit Jakobson (e.g. in his idea of a projection between the two main axes of language). In Peirce, such as he has become popularized as a taxonomist of signs, transformation might not seem central in a first glance, but a closer investigation finds transformations in the heart of his semiotics, namely in diagrammatical variation and the abstraction procedures facilitating diagrams. In Husserl's semiotics and phenomenology, the transformation concept explicitly plays a central role, first in the variation of profiles in the phenomenology of perception (where the continuous transformation of one *Abschattung* into another grants the unity of the object perceived), second in the idea of a cognate procedure on the ideal level in the concepts of *Wesenserschauung* where the eidetic variation finds the identity of an ideal object on the basis of the transformational possibilities of an empirical example. A related idea, more formalized, is to be found in Cassirer and his formalization of perception on the basis of group theory: an object perceived is defined by invariance in a group of geometrical transformations.

As might be evident in this list of semiotic concepts of transformation, there are certain recurrent sources present: mathematics, biology, linguistics. The former because here the formal treatment of various types of transformations (function, mapping, variation, etc.) may be found which may then be put to use in the empirical field of semiotics; the next because semiotics as an empirical science is often rooted in the more basic and comprehensive science covering those living beings having access to meaning; the latter because language is often taken to be a privileged and fundamental semiotic system. These three sources of inspiration are not parallel, now, mathematics is purely formal, general and a priori (without involving here the large infights of philosophy of mathematics, I just mean a priori as valid prior to empirical application) – while biology and linguistics are empirical sciences which to the same degree as semiotics (and other empirical sciences) must use mathematical formalisms, whether implicitly or explicitly, but both of them share some of the regional ontological prerequisites to semiotics as an empirical science.

As to the former, both Husserl and Peirce, qua practising mathematicians, had a direct interest in mathematics which inspired the variation type concepts of both – based on the mathematical concepts of function or mapping as the procedure taking one object into another.[128] An influential version of the concept of transformation is to be found in Felix Klein in his famous Erlangen-program which systematizes different geometries depending on which transformations leave their objects invariant, and more generally group theory as a formal concept for sets equipped with different procedures taking elements into other elements belonging to the set. In our days, a further development is category theory, generalizing the transformation concepts of different mathematical sub-disciplines to the notion of arrows between objects (as generalized functions) and adjungated functors operating between whole categories (thus providing of meta concept of transformation between whole subsets of mathematics: arithmetic, set theory, group theory, topology, etc.).

To the extent that biology is counted among the sciences of meaning (cf. 'biosemiotics', e.g. Sebeok 1992, see Chaps. 9–12) and the discussion of the ongoing use of meaning categories even down to biochemistry ('genetic information'), various transformation concepts of biology also inform semiotics, thus the metabolic cycle as basic for semiotic intention (Thom 1975, Uexküll 1982), the transport of information in genetics, epigenesis as transformation procedure in 'developmental' biology – but also the concept of mutation in the theory of evolution. Even if mutation often plays the role as an arbitrary basic concept which cannot be further explained (as an error in transformation), a rational explanation is sought in the actual complexity theory (The 'Santa Fe'-school, Kauffman, Goodwin, etc.) which reference back to d'Arcy Thompson and the 'rational morphologies' of the nineteenth century. Thus, there seems to be a connection between whether a biological theory involves a notion of transformation (in addition to causal concepts) and whether that theory recognizes a meaning concept in biology. In many actual semiotic theories, biological foundations often play the role of yet uncovered limit condition (the inneism of Chomsky, Lévi-Strauss's idea of the universals of human cognition in neurology, Greimas's reference to a 'natural world' prior to semantic investment, the body concept of phenomenology and, correlatively, the 'embodiment' notion of cognitive semantics) – the actual neurobiological research may be expected to achieve a large semiotic relevance in the more detailed investigation of these spontaneous ideas. In general, though, biology as a theoretically articulated science remains split into subdisciplines to such a degree that its explicit reflections on its formalisms used do not seem much more developed than semiotics.

Much more could be mentioned than this list of examples: transformations and the transformation of transformations pop up in many forms and in some sense it is strange that the discussion of the central status of the transformation concept in semiotics has not begun long since. Several reasons may be given for this striking omission. One is the general phenomenological rule that evident phenomena may be hard to make explicit. Another reason – more in the history of science – may be that European semiotics with its roots in linguistic structuralism has often, lead by Saussure's methodological distinctions between synchrony and diachrony as well as *langue* and *parole*, tended to see static structures as having ontological prominence over their transformations and thus has been interested primarily in 'codes' seen as stable relations between content and expression. The very concept of the *sign* may thus, paradoxically, have hindered the insight into the centrality of the concept of transformation. If the colocalization of expression and content are taken as central or essential, then the transformation of expressions, of contents, or holding between the two of them may be relegated to the periphery of semiotic investigation. The distinction between diachrony and synchrony in Saussure has often, in this tradition, been taken as an ontological barrier privileging synchrony over diachrony – and, furthermore, the place of synchronous description in the middle, between diachronous linguistic development on the one hand and linguistic use on the other has split this structuralism into two concepts of time without mutual contact and both ontologically underweight, with the often-noted implication that diachronous system change as the result of changes in use tends to become invisible. Correlatively, much criticism of this tradition – cf. post-structuralism – has referred to the transgression of those 'static structures', but then most often in an irrational vitalism referring to a deep, ungraspable movement which are not formally grasped as a transformation. Within linguistics, the focus upon use has given rise to the various pragmatical traditions sensitive to changes in linguistic usage, but then often in versions having little interest in the structural character of sign systems. These developments constitute, of course, part of the 'transcendental Jalta' diagnosed by Jean Petitot in twentieth century philosophy and science, between logical atomism and reductionism on the one hand and vitalist irrationalism on the other; in this tragic split, the notion of transformation seems a victim which is either reduced away on the one hand or mystified into a dynamic deep ontology beyond scientific reach on the other.

The field covered by semiotics is indeed vast, and there is no reason beforehand to assume that the transformations involved, not to talk about the theoretical tools needed to study those transformations, should be identical nor simple. For this reason we must approach the question in a so to speak botanical way and try to achieve an overview over the most important concepts of transformation used in the semiotic sciences as they are evolving. I do not claim the list of transformation ideas below to be exhaustive, but I hope to have picked a series of central and typical versions.[129]

LÉVI-STRAUSS: TRANSFORMATION AND STRUCTURE AS INTERDEPENDENT CONCEPTS

Claude Lévi-Strauss (1971, 73, 74, 88) is probably the scholar in semiotics, broadly taken, in whom the concept of transformation occurs most explicitly. Even if his formalizations most often take place ad hoc and are not tied to precise definitions nor consistent relations between the different formal subtypes,[130] transformation has a crucial place in his definition of structuralism. Lévi-Strauss is, of course, anthropologist and so to speak a practician comparing myths by transformation without first making those transformations explicit. The central idea lies in the assumption that it is the *narration* of a myth which provides its core – and this narrative structure can be understood only in comparison of the myth with other myths. This comparison is possible only by transformation: the myth is transformed into neighboring myths, cognate myths, etc., and only the location of the myth in this landscape of transformation decides its content. A myth, thus, consists of a group of variants. In this view, the concept of structure is solidary with the concept of transformation: it is meaningless to claim that a single myth 'has' a structure – a structure being a pattern only discernible in relations of variance and invariance by transformation between several myths – a structure only exists as realized in a multitude of related objects. Lévi-Strauss only rarely expresses this idea as directly as in the late interview volume (1988): 'But the notion of transformation is inherent in structural analysis. I would even say that all the fallacies, all the abuse made to or with the notion of structure come from the fact that their authors have not understood that it is impossible to conceive of the notion isolated from transformation. Structure cannot be reduced to system: a set of elements and the relations which unite them. To talk about structure, it is necessary that there are invariant relations between the elements and relations of several sets, so that you can pass from one set to the other by means of transformation.'[131] In this context, it is important to point out that the concept of transformation is open to both synchronous and diachronous investment – it refers to the structural relation between myths as well as to their possible origin and genetical family relations. Which myth is chosen, in the single case, as 'reference myth' is the choice of the investigator based on pragmatical considerations – an idea solidary with Lévi-Strauss' empirical work: the general deep structure of myth is an invariance of all possible transformations and thus not localized in any particular object. In a history of science context, it is interesting to note that Lévi-Strauss's concept of transformation does not at all stem from structural linguistics – which it is normal to assume to be his main theoretical source of inspiration outside of anthropology, just like the source of his preference for dual opposition structures in his friend Jakobson. His transformation idea, quite on the contrary, explicitly stems from d'Arcy Thompson to whom we shall return below.

It is interesting to remark that transformation in Lévi-Strauss seems to be prerequisite to the concepts of identity and difference: the identity of a myth is constituted as a group of variants (understood as variants due to transformation) – but at the same time one such myth relates to neighbouring myths due to specific transformations. Thus it is transformation which forms the basis of identity and difference on the level of myths – and there seems to be a continuum between the two because the delimitation of variants of 'one' myth and variants of another depends upon the generality of the transformation chosen. A further important consequence is that the concept of transformation refers to organized wholes (myths – more generally, gestalts, patterns, schemata, etc.) – structure resides in several, comparable organized wholes. Interesting in our context is also the fact that the concepts of transformation and structure are here related to a concept of abstraction of a non-inductive kind: 'Regarding comparative method, I have often said that it does not consist in comparing first and generalizing afterwards. Contrary to what is often believed, it is generalization which founds and makes possible comparison.'[132]

D'ARCY THOMPSON AND THE PRIMACY OF CONTINUOUS TRANSFORMATION

D'Arcy Wentworth Thompson is probably considerably less known than Lévi-Strauss; he was a Scottish 'naturalist' or theoretical biologist in the beginning of the twentieth century. His scepticism towards Darwinism notwithstanding, his general idea was to study biological form in a sort of phenomenological bracketing of the yet unknown – then even more than today – precise biochemical processes underlying them. The relative simplicity of macroscopic biological form allowed for a purely morphological science of biological appearance[133] which might, inter alia, investigate the question of *scale* of biological phenomena (originally Galileo's idea), and which prompted the construction of a doctrine of transformations ('On the Theory of Transformation, or the Comparison of Related Forms') where he was able to take one and the same bone form, organ form and in some cases the whole body shape in related species and demonstrate elementary, continuous Cartesian transformations holding between them. On the other hand, such transformation is possible only between related species being variations of one and the same type – they may never connect completely different *Baupläne*. D'Arcy Thompson thus is a partisan of natural morphology. Very illustrative – also in relation to the discussion of scale – is his'derivation' of the same bone structure in ox, sheep, and giraffe by a simple shortening of scale in one dimension and a logarithmic contraction in another (277):

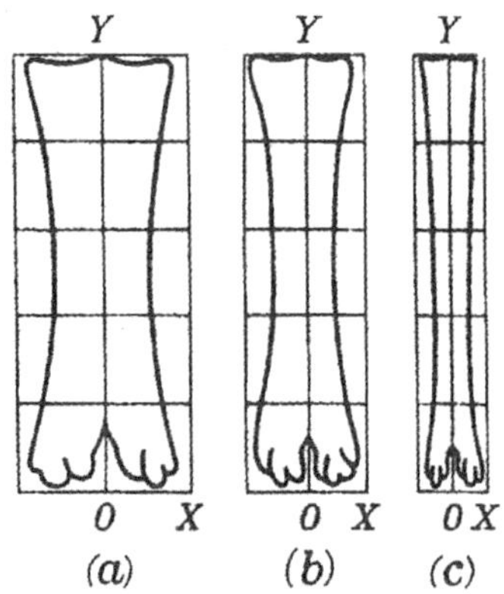

Figure 14. Structural similarity between foot bone structure in (a), ox (b), sheep, (c), and giraffe

Most famous – thanks to their illustrative evidence – is his morphololgies of different fish species where he introduces radial coordinates (ill. 149, 151). A corresponding treatment of skulls of horse, rhinoceros, and tapir does not yield the same degree of detail correspondence (they are more remotely related), and

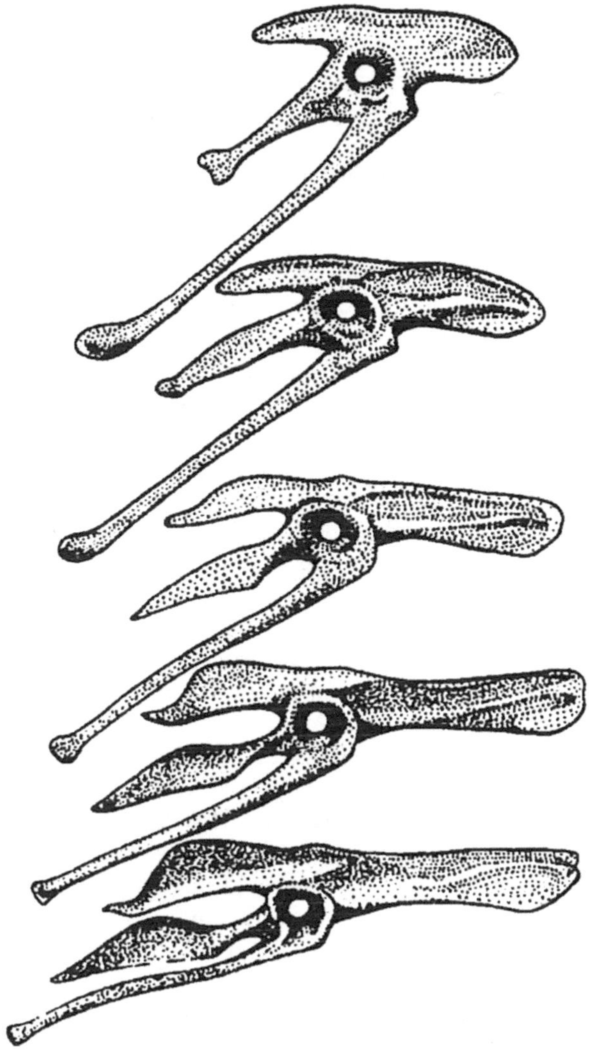

Figure 15. Pelvis from Archaeopteryx and Apatornis with three intermediate forms reconstructed by interpolation

yet – with the exception of certain details regarding eyes and dental positions – interpret the skull of a rabbit as a relatively simple bending (and diminishing) of the rhinoceros skull. There is thus an unspoken idea here that more general morphological similarities can be determined by the same type of transformation as the more specific ones, only with a increased tolerance towards detail deviance. With this basic idea, d'Arcy Thompson saw the possibility of theoretical reduction of the vast empirical biological complexity – at the same time this reduction permits the single species to maintain its specificity as the result of one specific transformation. By correspondence with the Danish draughtsman Gerhard Heilmann, he let produce a series of plates which tries to map intermediary forms between given morphological types, such as those (fig. 15) between the pelvis shapes in early birds (*Archaeopteryx* and *Apatornis* (309)).

From these immediately convincing diagrams, D'Arcy Thompson draws the quick conclusion that continuous transformation permits us to privilege the simplest transformation between two types and thus assume that they represent the real process of evolution connecting those types. Thus, he gives a rendering of the famous series of 'missing links' from the stipulated ancestral tree of the horse (the first such tree which seemed possible to sketch on the basis of paleontological findings) and compares it with the series of construed intermediary forms derived from a continuous transformation between the earliest known proto-horse *Eohippus* and to the *Equus* of our time. Several sources of error, however, are present her. Not only can we not assume that the empirical paleontological findings belong to one and the same line – several of them may, of course, belong to side-branches to the line leading to *Equus*. But, principally more grave, it is not in general the case that particular continuous transformations are unique, they are in general elements of a whole generic equivalence class of possible continuous transformations between the two forms (if not, the transformation would not be stable). In the horse example, this implies that there will be more than one possible route leading from the proto-horse to the horse, and we have no formal means for preferring any single one among them. These two things imply that we may not select one simple transformation and expect of evolution to have chosen exactly the same continuous series of transformations. Here, Thompson exaggerates the degree of information which can be extracted from morphological analogies. This does not, however, prevent his method from possessing a series of important properties: it compares morphology on the basis of the whole shape structure and not from singular, independent features (as did paleontologists in his time, e.g. in the comparison of human and ape skulls based on measuring size along a few axes). It constructs morphological classes on highly different levels of generality (cf. the rabbit example) determined by different transformations. It also provides an impression of the limits to morphological classification: transgressing the large animal groups in the animal kingdom, morphological transformation gives less meaning, because they pertain to wholly different *Baupläne*. It gives an indirect picture of the underlying complexity of transformations (the change of 'powers' forming the shape in question as he says with a deliberately vague notion). It introduces a phenomenological morphology category where structural classification becomes possible even if (potential) knowledge about the underlying causal dynamics is bracketed. Doing so, it provides a formalization of some of those 'inexact essences' of non-geometrical kind which Husserl referred to and which he – probably too hastily – condemned as inaccessible to axiomatic formalization.

A very important corollary to these continuous transformations is that they do not take an inventory of discrete elements as their basis (even if such elements may, of course, appear as their outcome) – which is why they allow that certain elements under certain transformations vanish completely. D'Arcy Thompson does not directly make this important implication explicit, but it lies implicitly in his observations regarding the small bone *interclavicus* in the *Ichtysaurus* which in the transformation to *Cryptocleidus* is 'minute and hidden' (305). In a continuous variation nothing prevents that a certain element can cease to be realized but which – knowing the transformation – may be said still to be *imaginarily* present (cf. next chapter) – an important fact because it demonstrates that the amounts of explicit large-scale elements in two shapes connected by transformation do not have to be identical – thus without this fact prohibiting the existence of such a transformation. Spontaneously experienced similarity (in the inventory of elements) hence does not count as a decisive criterion of the existence of a significant transformation between two phenomena. In discontinuous transformations, in contrast, a 1-1 mapping between elements would normally be expected and the lack of that possibility either as an indication of the non-relatedness of the two phenomena or as their difference must be explained by the addition of some explicit rule or reason.[134]

FELIX KLEIN AND THE ERLANGEN PROGRAM

Lévi-Strauss and D'Arcy Thompson have, each of them, sporadic references to the German mathematician Felix Klein and his work on the relation between transformation and invariance in mathematics. His famous, so-called Erlanger program from 1871, systematizes the bundle of different geometries of his time. The centuries of attempts at proving the parallel axiom of Euclidean geometry (given a line and point, only one line parallel with the given one may be drawn through the point) had produced the strategy of denying that axiom in the hope of deriving a self-contradiction as a proof of it. Instead, these denials of the axiom turned out to produce fully consistent, alternative geometries (Lambert, Saccheri, Taurinus, Gauss – formalized by Lobachevski, Bolyai, and Riemann) – namely the so-called hyperbolic and elliptical geometries, respectively (where infinitely many or none parallels might be drawn, respectively). At the same time, the so-called projective geometry was developed (Monge, Poncelet, Carnot) which, in contrast to the (non-)Euclidean geometries are not metric and introduced the principle of continuity (Poncelet) which allowed for the proof of general theorems for many different figures, even if some of the elements of the figures in certain cases became imaginary (Chasles) – cf. above how this principle plays an empirical role in d'Arcy Thompson. Cayley could now place metrical geometries as a subset of projective geometry. Klein's basic insight ('Vergleichende Betrachtungen über neuere geometrische Forschungen' (1872)) now was the fact that these different geometries might be systematized after which group of transformations each of them allowed for.[135] The aspects of objects each of them studied were interdependent with the transformations because the objects might be determined by invariance with respect to those transformations. In the Euclidean geometry, e.g. rotation, parallel displacement, mirroring in a line or in a point, (in short, 'stiff movements') e.g. are allowed transformations, because a figure remains the same when subjected to these procedures which keep shape, angles, size, etc. invariant. The group of projective transformations, e.g. is more general and gives rise to invariants as linearity, co-linearity, conic sections, harmonic sets of points (which, to be sure, also hold for the more specific Euclidean geometry). Klein's general classification had this character:

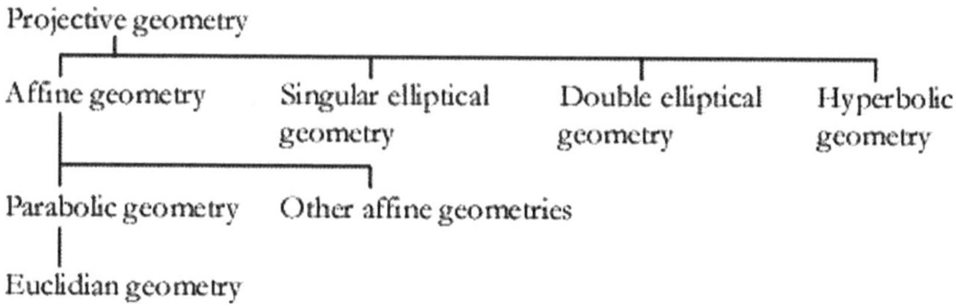

Figure 16.

even if he also determined even more general geometries, such as the algebraic geometry of the 'Cremona'-transformations, differential geometry leaving the second derivative invariant, and finally a geometry whose transformation group consisted only of different forms of inverse continuous mappings (so-called homeomorphisms): topology, the so to speak most general geometry where only invariants as connexity, compactness, open- and closedness, number of holes remain and

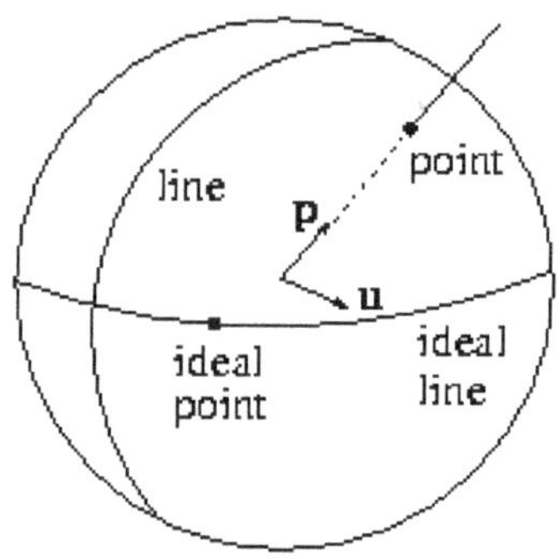

Figure 17.
Projective geometry may be considered as geometry of a sphere, projected from the center and onto a plane parallel to the equator plane and touching the sphere at the North Pole. Thus, all great circles on the sphere become lines in the plane – lines which end in ideal points at infinity of the plane – just like the projection of the equator becomes an ideal line at infinity. In projective geometry, several transformations are permitted which do not exist in Euclidean geometry, making, on the other hands, the correlated invariants fewer (Length, angle and ratio of lengths, e.g. cease to be invariant in projective geometry). This points to a general structure regarding transformations: the more identity preserving transformations allowed in a given diagram, the less invariants are preserved – and the more general those invariants which remain (such as topological invariants like connectedness and number of holes)

objects can be conceived independent of the space in which they are embedded. This classification has been further refined since Klein, and his endeavor is primarily relevant in our context because it provides the foundation for the idea that a vast and differing inventory of possible transformation groups (with each their corresponding inventory of invariant objects) can be made for the formalization of space. It is an open issue how many of these transformations have direct semiotic relevance but apart from Euclidean geometry there is a long tradition for pointing to the fact that various topologies due to their 'generalization ability' play a huge role in perception and language (cf. e.g. Leonard Talmy 2000; René Thom 1980).

The group of transformations characterizing each geometry may be formalized in group theory which opens up the possibility of a meta transformation where the optics may change by the changing of geometry and it becomes possible to signify one and the same object in different ways, because different invariances correspond to different properties – without this perspective shift dissolves the objectivity of the object in question. A figure with three straight edges is, of course, a triangle in Euclidean geometry while the same figure, topologically speaking, is a connected, 1-D manifold. Klein's duplicity of transformation and invariance thus allows, phenomenologically spoken, that the object's objectivity is maintained, while different acts pick out different aspects of it.

The concept of transformation in mathematics is explicitly tied to geometry as in Klein, even if the concept strictly taken is synonymous with the more general concepts of function and mapping. Transformations are thus also characterized by the different possibilities of describing a function. By a procedure (intensionally), by a graphical representation (intermediate between intensional and extensional), by correlated sets of points (extensionally). Extensional description is, of course, only possible to exhaust in finite, i.e. discrete, cases where the set of element pairs can be given explicitly. Intensional description by a procedure rule (e.g. $f(x) = \tan x$) in a certain sense comprises transformations of continua but suffers, on the other hand, from not being able to make explicit its extension (and thus not naturally state the field of validity of the function – e.g. that the given function has no value for $\pi/2$). As to graphical representations (which are continuous but suffer from imprecision and, in many cases, from being partial only), the main tendency in mathematics in the last century and a half has been to see them as heuristic tools only, which ought to be marginalized. Here, the actual interest in diagrammatic reasoning forms, of course, a counter-movement. Most transformations in semiotics take place without any conscious intention and explicit representation and it may for this reason be difficult to indicate preferred representation mode for them; maybe the mixed version could be expected to appear most often, even if the other two pure versions may also appear in different pragmatical contexts.

Let us from this preliminary triad of explicit transformationists in anthropology, biology, and mathematics look at a series of more implicit transformation concepts in mathematics (various aspects of qualitative dynamics) and biology (biosemiotics, complexity theory), cognitive semantics and linguistics (Lakoff, Turner), psychology (E. Rosch, S. Harnad), semiotics (Greimas, Groupe μ) as well as philosophy – Husserl, Peirce – in order to get an overview of the different semiotic role of transformations in these sciences.[136]

QUALITATIVE DYNAMICS

The most explicit influence from mathematics in semiotics is probably René Thom's controversial theory of catastrophes (1977, 1980), with philosophical and semiotic support from Jean Petitot (1985, 1992). Catastrophe theory is but one of several formalisms in the broad field of qualitative dynamics (comprising also chaos theory, complexity theory, self-organized criticality, etc.). In all these cases, the theories in question are in a certain sense phenomenological because the focus is different types of qualitative behavior of dynamic systems grasped on a purely formal level bracketing their causal determination on the deeper level. A widespread tool in these disciplines is phase space – a space defined by the variables governing the development of the system so that this development may be mapped as a trajectory through phase space, each point on the trajectory mapping one global state of the system. This space may be inhabited by different types of attractors (attracting trajectories), repellors (repelling them), attractor basins around attractors, and borders between such basins characterized by different types of topological saddles which may have a complicated topology.

Catastrophe theory has – cf. the general discussion of the Erlangen program – its basis in differential topology, that is, the branch of topology keeping various differential properties in a function invariant under transformation. It is, more specifically, the so-called Whitney topology whose invariants are points where the *n*th derivative of a function takes the value 0, graphically corresponding to minima, maxima, turning tangents, and, in higher dimensions, different complicated saddles. Catastrophe theory takes its point of departure in singularity theory whose object is the shift between types of such functions. It thus erects a distinction between an inner space – where the function varies – and an outer space of control variables charting the variation of that function including where it changes type – where, e.g. it goes from having one minimum to having two minima, via a singular case with turning tangent. The continuous variation of control parameters thus corresponds to a continuous variation within one subtype of the function, until it reaches a singular point where it discontinuously, 'catastrophically', changes subtype. The philosophy-of-science interpretation of this formalism now conceives the stable subtype of function as representing the stable state of a system, and the passage of the critical point as the sudden shift to a new stable state. The configuration of control parameters thus provides a sort of map of the shift between continuous development and discontinuous 'jump'. Thom's semiotic interpretation of this formalism entails that typical catastrophic trajectories of this kind may be interpreted as stable process types phenomenologically salient for perception and giving rise to basic verbal categories.

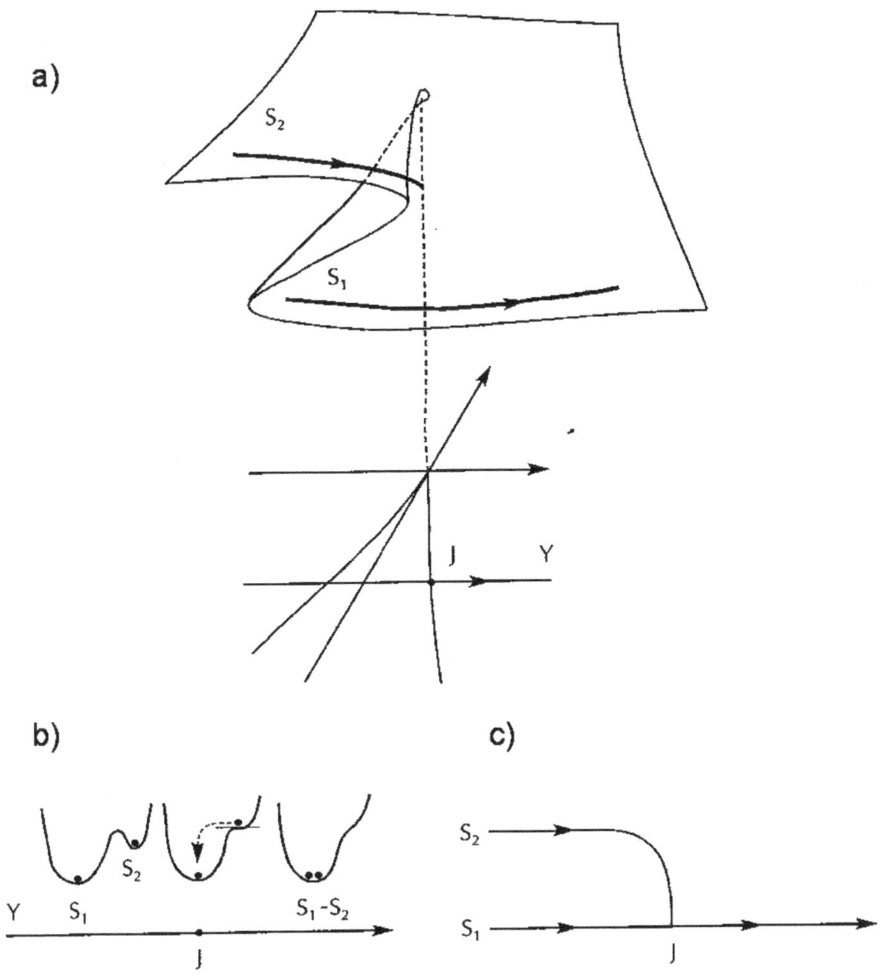

Figure 18.
One of the simpler catastrophes is the so-called *cusp* (a). It constitutes a meta-diagram, namely a diagram of the possible type-shifts of a simpler diagram (b), that of the equation $ax^4 + bx^2 + cx = 0$. The upper part of (a) shows the so-called fold, charting the manifold of solutions to the equation in the three dimensions *a, b* and *c*. By the projection of the fold on the *a,b*-plane, the pointed figure of the cusp (lower *a*) is obtained. The cusp now charts the type-shift of the function: Inside the cusp, the function has two minima, outside it only one minimum. Different paths through the cusp thus corresponds to different variations of the equation by the variation of the external variables *a* and *b*. One such typical path is the path indicated by the left-right arrow on all four diagrams which crosses the cusp from inside out, giving rise to a diagram of the further level (c) – depending on the interpretation of the minima as simultaneous states. Here, thus, we find diagram transformations on three different, nested levels.

The concept of transformation plays several roles in this formalism. The most spectacular one refers, of course, to the change in external control variables, determining a trajectory through phase space where the function controlled changes type. This transformation thus searches the possibility for a change of the subtypes of the function in question, that is, it plays the role of eidetic variation mapping how the function is 'unfolded' (the basic theorem of catastrophe theory refers to such unfolding of simple functions). Another transformation finds stable classes of such local trajectory pieces including such shifts – making possible the recognition of such types of shifts in different empirical phenomena. On the most empirical level, finally, one running of such a trajectory piece provides, in itself, a transformation of one state into another, whereby the two states are rationally interconnected.

We cannot here go further into the formalisms of catastrophe theory and their semiotic interpretation,[137] but suffice it to say that these three levels of transformations are interlinked, because the higher transformations take the lower as their objects, varying their conditions. Generally, it is possible to make a given transformation the object of a higher order transformation which by abstraction may investigate aspects of the lower one's type and conditions.[138] As an abstract formalism, the higher of these transformations may determine the lower one as invariant in a series of empirical cases (for Thom, e.g. the giving of a gift and in the structure of a sentence involving indirect object).

Complexity theory is a broader and more inclusive term covering the general study of the macro-behavior of composite systems, also using phase space representation. The theoretical biologist Stuart Kauffman (1993) thus argues that in a phase space of all possible genotypes, biological evolution must unfold in a rather small and specifically qualified sub-space characterized by many, closely located and stable states (corresponding to the possibility of a species to 'jump' to another and better genotype in the face of environmental change) – as opposed to phase space areas with few, very stable states (which will only be optimal in certain, very stable environments and thus fragile when exposed to change), and also opposed, on the other hand, to sub-spaces with a high plurality of only metastable states (here, the species will tend to merge into neighboring species and hence never stabilize). On the base of this argument, only a small subset of the set of virtual genotypes possesses 'evolvability' as this special combination between plasticity and stability. The overall argument thus goes that order in biology is not a pure product of evolution; the possibility of order must be present in certain types of organized matter before selection begins – conversely, selection requires already organized material on which to work. The identification of a species with a co-localized group of stable states in genome space thus provides a (local) invariance for the transformation taking a trajectory through space, and larger groups of neighboring stabilities – lineages – again provide invariants defined by various more or less general transformations (cf. d'Arcy Thompson). Species, in this view, are in a certain limited sense 'natural kinds' and thus naturally signifying entities. Kauffman's speculations over genotypical phase space have a crucial bearing on a transformation concept central to biology, namely mutation. On this basis far from all virtual mutations are really possible – even apart from their degree of environmental relevance. A mutation into a stable but remotely placed species in phase space will be impossible (evolution cannot cross the distance in phase space), just like a mutation in an area with many, unstable proto-species will not allow for any stabilization of species at all and will thus fall prey to arbitrary small environment variations.

Kauffman here takes a spontaneous and non-formalized transformation concept (mutation) and attempts a formalization by investigating its condition of possibility as movement between stable genomes in genotype phase space. A series of constraints turn out to determine type formation on a higher level (the three different types of local geography in phase space). If the trajectory of mutations must obey the possibility of walking between stable species, then the space of possibility of trajectories is highly limited. We shall return to Kauffman's semiotic ideas in Chap. 12.

Finally, self-organized criticality as developed by the late Per Bak (1997) belongs to the same type of theories. Criticality is here defined as that state of a complicated system where sudden developments in all sizes spontaneously occur. The prototype is boiling water where vapor bubbles of all sizes appear. Here, criticality is delimited to one, catastrophic point at 100 degrees centigrade. The prototype of *self-organized* criticality, in turn, is a sand pile rising from a continuous addition of sand from above. After some time, this pile reaches a maximum steepness after which any single new grain of sand may release avalanches on the side of the pile – the smaller the avalanche, the more frequently it occurs, and vice versa, size and frequency having a 1/f relation. This invariant property may be found in highly different systems sharing the property of such 1/f noise (widespread small, rarer larger events) and thus allows, parallel to catastrophe theory, a formal transformation of invariance between as different empirical phenomena as traffic jams, earthquakes, bourse cracks, extinction of species in an ecological system, Schumpeter 'creative destruction' in economics, etc.

These examples make evident a certain use of the concept of transformation which lay already in the Erlangen program, connecting it to the concept of form. Transformations identifies invariances: the fact that traffic jams, earthquakes, etc. all can be said to be self-organized critical systems is the same to say as that a related form may be found in all of them and thus may be mapped between them.

GROUPE μ – DOUGLAS HOFSTADTER

The Belgian group of semioticians Groupe μ in 1993 published what already stands as a new classic of pictorial semiotics, *Traité du signe visuel*. Here, they attack the linguistophile French semiotics of the 60s by involving psychological and cognitive research in the visual system, and one of the decisive concepts in their theory is exactly transformation. They introduce, on the characteristical third place between signifier and referent a meaning concept on the purely visual level named *type*, characterized somewhat diffusely by being a sum of visual paradigms (a head, in a paleo-Greimasian way, will then be composed of the marks superativity, roundness, closedness with the subtypes of eyes, ears, nose, and mouth . . .). The decisive step in our context, however, is that the verbal denomination of these features do not define them; they are types on the purely visual level (even if all of them, in the example given, have a denomination in natural language – far from all visual types necessarily have that). Such types are conceived as invariants in relation to a long series of possible iconical transformations, geometrical transformations (translations, rotations, displacements, congruences, projections, topological transformations, aggrandisations, diminishings), analytical transformations (various filters, continuisations, blurrings, differentiations, discretizations), optical transformations (changes of contrast or depth sensitivity, inversions), and kinetical transformations (integrations, anamorphoses). The decisive criterion here is that transformations stabilize the image as the type remaining invariant through transformations. It provides, in turn, for the possibility – cf. above – that these transformations are used inversely, resulting in a whole flora of *transformation rhetorics* whose effects can be investigated. Given, e.g. a head, it may be turned, filtered, discretizised, anamorphosed, etc. if it remains a type with a sufficient amount of redundant features to be recognized as such. Here, a crucial semiotic duplicity is evolved (which may also be found in Hofstadter): that possible transformations leave the type invariant (rather than define it), but, conversely, give it a series of variation possibilities to be used semiotically and aesthetically.

Douglas Hofstadter has analyzed related examples, e.g. in relation to designer typefaces (1986).[139] His question comes, so to speak, from the opposite side of Groupe μ's sender's point-of-view (seeing transformation as a rhetorical device), namely a cognitive receiver's point-of-view: how is it possible immediately to recognize a letterform, an A, e.g. given the enormous amount of very different and highly artificial designer typefaces we meet? Hofstadter runs through an argumentation for the fact that no computational algorithm is possible which may produce all possible As (thus, the A is yet another example of a Husserlian 'inexact essence'). The argument is an analogy to Gödel's theorem: a sufficiently complex formal system cannot be both consistent and complete. Analogously, no program to generate As can be both consistent (generating As only) and complete (generating all possible As). But the fact that such a transformation algorithm may not be constructed only gives transformation free reins to construct As in a trial-and-error process – thus Hofstadter sees this logical limitation as equivalent to a creative freedom explaining the fact that still new designer typefaces surprise the market year after year. The type A is defined by certain central characteristics, but they are not necessary and sufficient properties for the single As which may not, then, be defined beforehand. When the observer categorizes a strange designer A, he involves in a trial-and-error process as well (diacritically guided by the other letters of the same typeface, often appearing alongside it), whether it is possible to stably transform the apparent hieroglyph into a prototypical A. It may be expected that the single typeface is characterized by a specific set of such transformations. In this case, transformation defines identity: that is counted as an A which may be transformed to prototypical A. The transformative variation of prototypical As is thus also an ongoing investigation of the border areas of the type category. Here, transformation takes two roles, as in Groupe μ, it leaves types invariant and it opens up a stylistic field of variation possibility within that type. Hofstadter even extends the argumentation to artistic creation in general to be analyzed as 'theme with variations'. Analogies abound between Hofstadter and cognitive semantics, to which I shall return.

PARIS SCHOOL SEMIOTICS

The semiotics of Greimas also involves transformation concepts on several levels – the most explicit case being that of 'conversion' leading from simple semantic units (conceived of in Jakobsonian binary terms), over narrative dynamization of these units and to discursivation with iconization, enunciation and the whole set of surface sophistications of the text. Conversion is conceived as the transformation category which adds new, specifying properties to a pre-existing schema – and thus not as variable as the plastic continuous transformations in the two preceding examples.[140] Reference may be made to Petitot's attempt at a catastrophe theoretic modelization of the lower levels of this process, involving plastic transformation possibilities. Yet, the theory contains a less formalized, but analytically strong tool in the so-called 'narrative schema', an idealized textual event process schema based on Propp's more empirical formalization of Russian fairy-tales. Greimas' schema is constructed as a series of presuppositions. A later phase presupposes the earlier phases, not the opposite way around – this implies that any phase of it may deviate from the given norm. It provides, so to speak, a reference tale, and any particular empirical tale may be determined by how it modifies, that is, transforms this arch-fairy-tale. In this case as well, no necessary and sufficient properties may be listed, because any particular property may vanish during the transformation (but not, of course, all of them at the same time). Maybe the fairy-tale in question never produces a hero which is then only imaginarily present, just like in projective geometry. The generalization of Poncelet's principle of continuity thus relativizes the logical ideal of necessary and sufficient conditions, on a purely formal, pre-logical level. The prototypical tale has a hero, of course, but it is possible to transform it into a version with no hero.

This idea is connected to a general fact in the basic use of transformation defining invariants by their insensitivity to (a specific) transformation – this typification creates those entities which, in turn, may be involved in logical calculi. Transformation thus, also in Paris School semiotics refers to a basic geometry preceding logic, even if this is far from always explicitly admitted.[141]

COGNITIVE LINGUISTICS AND SEMANTICS

An actual semiotic current with central transformation concepts is the American cognitive linguistics and semantics (Lakoff, Johnson, Sweetser, Talmy, Turner, Fauconnier, etc., cf. Fauconnier and Turner 2002; Lakoff 1987; Talmy 2000; Turner

1996). Here, I shall only comment upon two related issues, Lakoff-Johnson's metaphor concept and Turner-Fauconnier's blending concept. Lakoff-Johnson's metaphor concept is motivated by the observation of the widespread occurrence of structural, conceptual metaphors in ordinary language: 'Love is a journey' – which then serve as a deep structure with an infinity of surface manifestations: 'Our relationship is not running well', 'We go to therapy in order to add som new fuel to our marriage', 'We cannot find our way together anymore', etc. The idea is that such metaphors map structure from one semantic domain upon another – with the aim of understanding parts of this second domain. This mapping, of course, is a transformation constitutive for the conceptual metaphor in question. Whether the structural similarity between the domains is now constituted by the metaphor or whether, conversely, it has a pre-existing similarity as its condition of possibility is a standing quarrel in metaphor theory, but in both cases the connection between the two fields presupposes a transformation which maps a more general schema (here the gestalt of source-path-goal) from one domain onto the other. Transformation of other bits of structure from other domains onto the same target domain are also possible, of course ('Love is an illness', 'Love is a revelation', etc.), but the decisive kernel remains transformation as a means of understanding. Another category of transformation here goes from the deep structure of the established metaphor and to its manifold of surface manifestations which may be found in a continuum from well-established 'dead' expressions in ordinary language and to more creative, maybe artistic ones. Linguistic usage here plays the same role as designer typography in Hofstadter's case: it investigates the possibilities of established metaphor by further transformation.

Turner's and Fauconnier's blending theory is, in fact, a further generalization of Lakoff-Johnson's metaphor model. They find many cases of coupling between semantic spaces which do not possess the orientation characteristic of metaphor where transformation indicates a semiosis so one domain becomes expression for the other as content. Examples include a philosopher today discussing with Kant and so blending his own arguments with Kant's imagined answers – without any of the dialogue partners being a metaphor of the other. Furthermore, they make explicit the general structure always at stake in transformation by giving it its own mental, 'generic' space, so that any blending has (at least) three typical inputs: those of the two more or less empirical ideas to be blended and that generic structure which facilitates the blend. In this analysis, widespread grammatical phenomena, e.g. compound nouns, are seen as results of semantic blending, characterized negatively by having no compositional explanation (there is no fire in a fire station while there are railways in a railway station – such combinations must hence be seen as blendings).[142] This definition opposed to simple compositionality is equivalent to Hofstadters non-computability claim, also in this case the result becomes a trial-and-error transformation in the lack of general procedure for the mapping of elements onto one another. The relation to Hofstadter's theory is becoming evident, furthermore, in the fact that parts of the two research teams now collaborate – and Hofstadter's work on analogy inferences is a good example of the meeting point of these traditions. Analogy questions are not in general computational; most often they have several possible, not equally brilliant answers. Hofstadter has created a trial-and-error computer program with random components able to solve simple formal analogy questions by running the same questions over and over and thus charting the different possible answers. Analogy questions of ordinary language are, of course, more difficult, because they require context knowledge. Who is Denmark's first lady? PM Anders Fogh Rasmussen's wife? But she does not participate in public life? The queen? But she has a formalized status. The best answer is probably Prince Henrik, the Queen's spouse – even if both of the other two are also possible answers and satisfies different amounts of the requirements of the analogy. The answer to such 'creative' questions thus requires blending – here between an American concept and a Danish context – but the decisive thing in this context is that the investigation of the different answers requires a transformation where an American structure is mapped onto a Danish empirical case which is impossible without certain smaller give-and-take-counter-transformations (in the case of Prince Henrik: he is not a woman) characteristical for blendings. Blendings are thus the formula for a calibrated compromise between different transformations which is why it may not be reduced to compositionality.

HUSSERL AND PEIRCE

After this hasty overview over a series of different and more or less formalized concepts of transformation we shall now go to a more principal level and view them in the context of the more explicit transformation concepts in Husserl and Peirce.

In Husserl, there is a transformation, of course, in his paradigmatic example in the *Ideen* (1980a [1913]): the perceptual synthesis of a series of profiles, shadings, or aspects of an object. These variations of the object are spontaneously synthesized in perception and are understood as aspects of one and the same object because we are able to transform continuously between them. Continuous transformability is thus what grants the objectivity of the object. An analogous procedure Husserl proposes for ideal entities. Already in the Prolegomena of the *Logische Untersuchungen* (1900–01/1975, 1984), the idea of ideal entities is introduced, characterized by the fact that one or more particular properties is substituted by a variable. A decisive phenomenological task is to distinguish between the mode of givenness of individual objects and that of abstract, ideal objects (as is always the case in language). The 6th Investigation investigates 'categorial intuition' as the act which directly presents e.g. grammatical aspects of language and other abstract entities – and Husserl takes care to note the important fact that in such ideal objects the distinction between perception and imagination vanishes, so that imagination is an equally valid way of access to such objects. In the *Erfarhrung und Urteil* (1939, 1985) this is taken further in the theory of *Ideenschau* or *Wesenserschauung* according to which we may grasp essences directly and thus understand an ideal content – by means of the method of eidetic variation.[143] This procedure is, of course, crucial for the determination of the phenomenological prerequisites to logic: in order for logical propositions to appear, involving ideal entities, we must know the way of access to such ideal entities.

We vary the idea by a transformation departing from an example, in principle arbitrarily chosen, incarnating the idea in question. Imagination varies this particular example over a continuum of virtual instantiations. This must presuppose, in order for the search to be exhaustive, that the variation procedure is continuous. Thus, eidetic variation presupposes infinity – the ideal category may be incarnated in a virtual infinity of different extensions. Husserl poses the question of what exactly forces or determines this free variation and prevents it from going anywhere, but he is unable to answer it exactly: there is no given limit (if it was at hand, no variation would be necessary), it is, rather (like in Hofstadter) the very transformative ability to return to the original example which decides how long variation can be drawn. In a certain sense you could say there is nothing further to be said about it if eidetic variation is simply a way for ideal objects to appear, then the investigation has reached its limits (cf. below). Another question refers to whether variation should be actually completed in imagination so that all the extension of possible variants of the idea is in fact covered. In most, if not all, cases this will be impossible, because of the infinity of possible variants (cf. Peirce's doctrine of continuity). If we imagine variation as a continuous change of some parameter in the object, then the temporal synthesis is what grants that all instantiations in the varied segment have been covered – no direct intuition of all those instantiations. We vary, e.g. the size of a triangle, the configuration of the angles, and see that we could in principle take this variation as far in any direction as we might wish. We must, in fact, make the very act of variation into the object of a founded higher-order act which synthesizes some pieces of accomplished variation and judges that it *could* be extrapolated to cover the whole of the idea. Here, of course, is a possible source of error in variation: this higher-order extrapolation may be wrong and overlook important areas of the extension of the idea. A formal equivalent to the *halting* problem of computer science is probably at stake here: it is not possible to devise a general procedure to determine whether a given idea has been searched to a sufficient degree to map all significant variants.

This problem is related to that of what Husserl calls exact versus inexact essences – which may be distinguished according to whether variation may predict all single concrete instantiations of the idea: in the concept of the triangle, variation is synthesized and it is realized that an extrapolated variation will be able to produce all possible triangles with all possible combinations of side and angle sizes. In inexact essences – e.g. empirical universals – this is not possible, which is why Husserl concludes that these remain 'vague morphologies' inaccessible to axiomatic treatment (*Ideen* §§71–75). We know, however (Gödel's incompleteness theorem, of which the *halting* problem forms a computational version) that axiomatic systems exist where not all concrete variations (theorems) may be formally predicted. To that extent, the distinction between exact and inexact essences is more blurred than Husserl realized. But the positive correlate to this incompleteness insight is that also essences which are not completely formally decidable permit more or less comprehensive formalization (e.g. already natural numbers). As the many different uses of the concept of transformations in semiotics also seem to hint at, it is thus not impossible to specify different formal properties in a variation mapping inexact essences, and even if there is a central point in the fact that they may not be axiomatized and rendered in necessary and sufficient properties, then a huge open field remains regarding the mathematical formalizability of these ideal objects (both a priori, tied to the material ontology of the essence in question, and empirically).[144]

The concept of eidetic variation has from time to time been wrongly taken as a strange, mystical ability – but it is important to maintain that it is a completely ordinary everyday process as a thought experiment isolating a type by transformation. Eidetic variation is so broad a concept that it probably founds the more specific semiotic transformation types we have here discussed. Even if ideal objects may be accessed through variation it is not their only mode of givenness: representation by a bound variable (an *x*, an empty slot, a place to be filled in, a variable property), be it explicit or not, is another representation of ideal objects – from literary works to equations. There is no real understanding without ideal types which are for that reason widespread all over semiotics (the typical sign as such, for one, is ideal).

As is evident, eidetic variation is closely connected to a non-inductive abstraction theory. This leads us to what is, in fact, Peirce's version of eidetic variation, also built upon a non-empirical theory of abstraction. Abstraction is, in Peirce, several different things (and just like in Husserl distinct from the issue of induction.).[145] First, it involves 'prescission' which is an act of focusing, which disregards certain irrelevant properties in order to focus instead upon other, relevant properties. In this procedure, the removal of properties corresponds to the introduction of a variable as the precondition of the isolation of a predicate (e.g. the isolation of the property 'round' from the other properties of an object). Another procedure is 'hypostatic abstraction' as the procedure which makes a new subject out of a predicate in order to facilitate further investigation (investigating, e.g. 'roundness' as such). These two transformations are the prerequisites for the construction of a diagram (in this example: a circle) making possible the schematization of the content of a concept and its ideal grasping (Peirce calls this the observation of universal propositions). By the variation of this diagram different transformations may investigate the extension of that idea, its limits, its relation to cognate diagrams (the point; ellipses and other conic sections; polygons), the impossibility of squaring it, circumscription by polygons, etc. – all results which appear with the necessity of ideality. The lawlike aspects of empirical investigation thus involve the incarnation of diagrams in the matter in question – thus Peirce's idea of the role of diagrams in knowledge thus provides his equivalent to Husserl's concept of eidetic variation. It even adds important details of the variation procedure, first its phases of focusing, hypostatization, and variation, second its distinction of different classes of variation. A decisive distinction to Peirce already mentioned is that between 'corollarial' and 'theorematic' diagram experiment. The former is directly read off the diagram while the latter requires the introduction of new entities in the diagram (cf. the helping lines and similar constructions in geometry). The latter requires an abductive 'jump' in variation[146] and thus makes evident why variation is not always an intuitively easy transformation (cf. the existence of unsolved and undecidable questions). As in Husserl where variation may distinguish ideas in still higher generality, leading to the issues of formal and material ontologies, transformation is also in Peirce recursive: a new variation may take the former as its subject in a new higher-order diagram.

Husserl's and Peirce's very general concepts of transformation, eidetic variation and diagrammatic experiment, respectively, closely connected to the mode of existence of ideal objects in general, make evident the very basic status of transformation. It can be argued that a series of ontological issues which are often taken to be primitives in semiotics (and elsewhere) are tied to transformation. Thus, e.g. the distinction between static and processual states-of-affairs which have given rise to so much vitalist discussion (e.g. in post-structuralism). Transformation precedes both 'static' and 'dynamic' issues – they refer to two different interpretations of a transformation as being a relation between two entities and as being something taking place in time – so to speak equivalent to the question of *Geltung* and *Genesis*.[147] Transformation does not choose any of the two sides but makes possible investigations of valid, a priori relations between concepts on the one hand and genetic investigations into the empirical change of objects on the other hand. Similar relations probably hold in the highly ideologized issue of identity/similarity/difference. An assumption of identity may probably be tied to any transformation – the invariance it allows for may be interpreted as the identity of an abstract object through change – an identity being so much more general, the larger the amount of varied properties are, and the less the core kept invariant. To this extent this 'semantical' notion of ideal identity (different from the numerical identity of empirical existence) is relative to the set of transformations used. This explains the trivial fact that any object is in some sense similar to any other object[148] – at the same time as explaining why this is not valid as a criticism of similarity as such: similarity is not arbitrary but relevant to the transformation chosen. Thus, transformation does not chose sides between harmonizing *Identitätsphilosophie* on the one hand and radicalizing philosophy of difference on the other: it is a means of description preceding both.[149]

This chapter has taken us from different specific uses of transformation concepts in semiotics and to their generalization in Husserl's and Peirce's theories for the ontology and epistemology of ideal objects and their transformations to which we shall return in more detail in the next chapter. We may here sum up the different pragmatical tasks for transformation: the emphasis may be placed on invariance (identity), on variance (the process), on the comparison between the two ends of transformation (similarity/difference), the comparison between different transformations of the same object, on repeated variations back and forth between two or more transformands (judgment/blending), on the transferral of further, more or less untransformed matter (metaphor), on a double grasping of the resulting object as the output of several different transformations.

As to semiotics especially, the following tasks may be listed:

- abstraction (cf. Peirce's prescission and hypostatic abstraction, respectively)

- categorization (categorical perception as the transformation of a continuum to a discrete system)
- to grant identity (of types, from phonemes and morphemes up to more comprehensive signifying structures)
  a) to understand a token as instantiation of a type
  b) to achieve overview over a type
  c) to constitute a type
- to synthesize a manifold
  a) in perception
  b) in categorical intuition
- to detect similarity (by comparative synthesis of the transformands)
- to map concepts (cf. Peirce: concepts as transformative conditionals, the meaning of a concept being coextensive to the transformative effects the object of the concept may be conceived to possess)
- to generate a variation of subtypes in a hierarchy of ideal entities (correlative to subsumption under a type), and connected to this:
- to constitute regional (material) ontologies
- to constitute and describe iconic signs as well as the iconical component in higher sign types (as correlation between types)
- to grasp the aspect of an object as the arche-form of the sign (selected by transformation) – cf. also the proposition ('this aspect is an aspect of this object /these objects')
- to reasoning as a transformation guided by the invariance of truth (and, more unfolded, by the invariance of possibility, truth, probability, corresponding to ab-, de-, and induction)
- to understand something in terms of something else (cf. Turner's 'literary mind': allegory as a basic procedure), and correlatively:
  a) analysis (of texts)
  b) integration of concepts (blending, analogy inference)
  c) recursivity (the transformation of transformations)
- answering questions/solving riddles (with the trial-and-error transformation involved) – cf. Hintikka's game-theoretical semantics

Many of the theories touched upon in this chapter are not traditionally classified as semiotics (mathematical, biological, philosophical theories). Why then keep that notion as a headline for the whole field of thought and meaning made possible by transformation processes? This issue comprises both empirical, natural phenomena (d'Arcy Thompson) and ontological and epistemological basic issues. The term semiotics only has a relevance for this vast field if it may be understood as based on meaning. This, in turn, is only meaningful if we take 'meaning' in an objective sense (cf. biosemiotics, Chaps. 9–12) comprising teleological processes in the empirical world without requiring any observing subject in anthropomorphous sense – and, if it comprises access types to the ontological zoo of different ideal objects on the other hand. Human thought and language are, of course, a privileged access to these ideal objects which must not lead to the idea that these objects are restricted to or created by the human mind, a fallacy which, depending on regional ontology chosen, will lead into biologism, historicism, psychologism, subjective idealism, or other reductive –isms – all of them springing from a genetic fallacy assuming that ideal objects and the transformations defining them should pertain to a specific and delimited ontological field. Transformative semiotics, in this use of the word, will cover much more than usually assumed – what is won, on the other hand, is the understanding of transformation as the basic process of semiotic intelligence: only a reifying forgetting of transformation (probably motivated in the efficiency and invisibility of most of the basic procedures of our cognitive apparatus) makes us overlook transformation being pervasive in our everyday perception, language, and reasoning.

CHAPTER 6

CATEGORIES, DIAGRAMS, SCHEMATA

The Cognitive Grasping of Ideal Objects in Husserl and Peirce

Diagram experimentation forms the prototype of the various transformation concepts encountered in different semiotic theories. Diagrams are, as we saw, types, or ideal objects – and they may be used to refer, in turn, to other general, ideal objects. A theoretical tradition with emphasis on the access to ideal objects is Husserlian phenomenology. Husserl's conceptions of abstraction and *kategoriale Anschauung* – categorial intuition – was first developed in his early work and played an important role in *Logische Untersuchungen*[150] and later in *Erfahrung und Urteil* and elsewhere. Here, the grasping of ideal objects pertains not only to mathematics and logic – even if they form an important case – rather, it forms a crucial parts of everyday cognition in so far as most cognitive acts are not simple and involve general elements in what Husserl calls 'sinnlich gemisschte', sensuously mixed form.

In this chapter, we shall outline the early Husserl's theory of access to ideal objects with the aim of comparing and integrating it with Peirce's diagram doctrines.

UNBELIEVABLE BUT TRUE

It is a strange fact that so little comparison between Husserl and Peirce has been undertaken.[151] Probably the historical reason is that the two philosophers both stand on the initial edge of the analytical/continental split and ended up as founding fathers for each their main currents in philosophy – phenomenology and pragmatism/semiotics, respectively. Seemingly rooted in each their specific tradition, the large bulk of common ideas and interests in their works has been ignored. Distorted parodies of the two – Husserl the transcendental solipsist, Peirce the pan-semiotician – have added to preclude a closer *Auseinandersetzung* between the two.

Correspondingly, it is a little known fact that the two of them *did* in fact know the other's work – albeit not, unfortunately, to a degree sufficient to reveal the crucial convergences to any of them. Husserl's interest in Peirce lies mainly in the beginning of his carreer; he refers to Peirce's early semiotic work at several occasions around the *Philosophie der Arithmetik* period, but there is no reference later than the nineties, so Husserl never came to know the mature Peirce's phenomenology and semiotics from around the turn of the century, and so he hardly influenced Husserl's later thinking. Conversely, Peirce several times refers to *Logical Investigations* which came out in the midst of Peirce's last fertile period of thought – but he only mentions it in a pejorative manner, classing it alongside other German logic studies as despicable examples of psychologism. This characterization evidently rules out that he may in fact have read very much of the book himself.

Husserl's Peirce references mainly surface in his review of the first volume of Schröder's *Algebra der Logik* (1890; a work much influenced by Peirce, to be sure) from 1891 (Hua XXI, 2–43) and his subsequent discussions with A. Voigt (1893, also in Hua XXI). Here, Husserl's main idea is to reject Schröder's view of logic as being extensional only, even if he also praises him for his work. He was subsequently reproached by A. Voigt (reprinted in Hua XXI) for overlooking the intensional logics of Frege and Peirce and replied to that comment. Even if Husserl mainly refers to Peirce as a precursor of Schröder, a couple of interesting comments are involved: Husserl frowns over Peirce's idea of assertions as presented by Schröder, it is 'unglaublich aber wahr' that he may think judgments are special cases of thought habits (20). Husserl probably here fails to grasp Peirce's wide and non-psychological idea of habits. On the other hand, Husserl praises Peirce's method from 'On the Algebra of Logic' (in Schröder's presentation) as one which '... impresses through its special originality, simplicity and elegance. It appears, especially after the simplification which *Schröder* gives it, as eminently useful, and for that reason it should be wished to receive a richer amount of related and calculated tasks.'[152] Mullin (1966) remarks that these recensions were in Peirce's library with editor's stamp – so if Peirce did not himself receive them from Husserl, he got them from Schröder or (less probable, though) bought them in an antique book store.

The early Husserl thus may, to some extent, be influenced by Peirce through Schröder. The degree of Peirce's indebtedness to Husserl is equally difficult to ascertain. The mature Peirce decides upon the term 'phenomenology' for a part of his study which is, in some respects, rather close to the early Husserl's use of the word as referring to the study of essences. It is the study of universal elements of all phenomena; it is a study explicitly anti-psychological (and ante-psychological) of nature; it is the most basic of all positive studies (by 'positive', Peirce does not refer to empirical 'Tatsachenwissenschaften' only, rather, 'positive' refers to sciences of the given as opposed to sciences of the merely possible), and logic is founded upon it; it is a study that includes all kinds of possible experiences, including dreams and abstract thought; and it is a study – almost an outline of a Peircean phenomenological reduction – bracketing whether the phenomena it studies exists or not. Of course, Peirce does not share the later Husserl's basing phenomenology on a study of conscious acts, but this difference seems more a difference of emphasis within a field than it is a foundational difference.

'Phenomenology' did not play any important role in the papers published in Peirce's lifetime, but it rose to prominence due to Hartshorne and Weiss' edition of CP where they picked it as a headline term – Hartshorne having studied by Husserl in Freiburg in the 1920s. That was no bad choice; Peirce's use of the term 'phenomenology' is indeed intense, but covers a rather short timespan:

> Phenomenology, whose business it is simply to draw up an inventory of appearances without going into any investigation of their truth. ('Minute Logic', 1902, 2.120)

Phenomenology ascertains and studies the kinds of elements universally present in the phenomenon; meaning by the phenomenon, whatever is present at any time to the mind in any way.
('Syllabus', 1903, EPII, 259; 1.185)
I will so far follow Hegel as to call this science Phenomenology although I will not restrict it to the observation and analysis of experience but extend it to describing all the features that are common to whatever is experienced or might conceivably be experienced or become an object of study in any way direct or indirect.('Harvard Lectures on Pragmatism', 1903, EPII, 143; 5.37)[153]
This science of Phenomenology is in my view the most primal of all the positive sciences. That is, it is not based, as to its principles, upon any other positive science. By a positive science I mean an inquiry which seeks for positive knowledge; that is, for such knowledge as may conveniently be expressed in a categorical proposition. (ibid. EPII, 144; 5.39)
... Be it understood, then, that what we have to do, as students of phenomenology, is simply to open our mental eyes and look well at the phenomenon and say what are the characteristics that are never wanting in it, whether that phenomenon be something that outward experience forces upon our attention, or whether it be the wildest of dreams, or whether it be the most abstract and general of the conclusions of science. (ibid. EPII, 147; 5.41)
The business of phenomenology is to draw up a catalogue of categories and prove its sufficiency and freedom from redundancies, to make out the characteristics of each category, and to show the relations of each to the others. (ibid. EPII, 148; 5.43)
There are three sciences according to me to which Logic ought to appeal for principles, because they do not depend upon Logic. They are Mathematics, Phenomenology, and Ethics. (Letter to John Dewey, June 1904, 8.242)
(...) what I have called phenomenology, that is, just the analysis of what kind of constituents there are in our thoughts and lives, (whether these be valid or invalid being quite aside from the question). It is a branch of philosophy I am most deeply interested in and which I have worked upon almost as much as I have upon logic. It has nothing to do with psychology.(letter to William James, Oct. 1904, 8.295)
Phenomenology has no right to appeal to logic, except to deductive logic. On the contrary, logic must be founded on phenomenology. (ibid. 8.297)

The word is also used in the passing in a letter to Lady Welby, Oct. 1904 (8.328), as well as in a 1906 commentary to a proposal by a Mr. Peterson to discuss philosophical terminology (*The Monist*, Vol. 16, 147–151 (1906), 5.610).

As is evident, the main period for Peirce's use of the term covers the years 1902–06 with the main emphasis on three years only, 1902–04. From around 1905, Peirce substitutes the terms 'phanerology', 'phanerochemy', 'phenoscopy', and especially 'phaneroscopy' for 'phenomenology'.[154] 'Phaneroscopy' is used in the 'Adirondack Lectures' (1905, 1.284), 'Logic viewed as Semeiotics, Introduction Number 2. Phaneroscopy' (1.285–87, dated by the editors of the CP 'c. 1904'); 'Phaneroscopy or the Natural History of Concepts' (c. 1905, 1.322); 'Phaneroscopy fan' (1906, 4.7); 'Phaneroscopy (phan)' (intended for the *Monist*, January 1907, 1.306); a letter to James (1909, EPII, 492ff; 8.303). Thus it seems that the substitution of 'phaneroscopy' for 'phenomenology' takes place rather systematically around 1904–05 – but due to Peirce's intensive use of the latter in the fertile 1902–04 period and Hartshorne and Weiss's subsequent use of it as a volume title in the *Collected Papers*, it has come to stand out as the standard notion for Peirce's investigations in the area.

As a matter of fact, Peirce's use of the word is not unrelated to Husserl's, as is evident from the characterization above claiming that phenomenology '... studies the kinds of elements universally present in the phenomenon; meaning by the phenomenon, whatever is present at any time to the mind in any way.' – given that 'the mind' in Peirce is not coextensive with the human mind, neither is it an empirical issue to be studied psychologically.

Thus, there is a striking temporal coincidence between Peirce's use of the term 'phenomenology' (1902–06) and his Husserl references. When Peirce discusses the origin of the concept, he refers to Hegel only, but this simultaneity points to the possibility of Peirce having borrowed the term from his – probably – cursory readings in LU.

As to Peirce's explicit references to Husserl, they are rather few – twice he is mentioned only as part of lists of German logicians ('Minute Logic', 1902; 'Review of John Dewey's Studies in Logical Theory', *The Nation* 79, Sept 15, 1904, 219–20; 8.188). Once, however, Peirce writes about '... the distinguished Husserl (Note: See, e.g. his *Logische Untersuchung*, Teil I, Kap. 3 (1900)) after underscored protestations that their discourse shall be of logic exclusively and not by any means of psychology (almost all logicians protest that on file), forthwith become intent upon those elements of the process of thinking which seem to be special to a mind like that of the human race, as we find it, to too great neglect of those elements which must belong, as much to any one as to any other mode of embodying the same thought.' – instead, Peirce's existential graphs are mentioned as an alternative, non-psychological rendering of logic (1906, 4.7). Peirce shares a widespread criticism of Husserl at the time – for falling prey, in the later Investigations, to the same psychologism which he himself attacks in the Prolegomena.

The missed encounter between the two thus seems to have a background in both of them misunderstanding the other's account of logic. Peirce emphatically saw logic as an ideal, normative science – Husserl likewise saw it as an ideal science, but equally emphatically not as a normative science. When Husserl rejected normative logic, however, it was simplistic accounts of logic as psychological 'Denkökonomie' he had in mind. Thus, it seems that this less decisive difference made both of them misconstrue the other's point of view as being psychological. Peirce directly attacks Husserl for psychologism, and Husserl's attack on Peirce's – unbelievable but true! – identification of judgments with habits seem to rely on a presupposition that habits are psychologically contingent facts. This minor difference between the two seemingly spreads fog over their basic agreement in antipsychologism to a degree such that none of them noticed it.

In any case, the closest similarity between the two's careers seems to hold between the mature Peirce and the early Husserl, that is, coincidentally, between their thought as it develops simultaneously around the turn of the century, that is the period when Husserl writes *Logical Investigations* while Peirce strives to mould the mature version of his philosophy including phenomenology and semiotics. Just to enumerate a series of virtual meeting points between the two (each of which could merit a separate study)

- their basis in an anti-psychological first-person perspective (even if very different outcomes eventually spring from this idea in the two of them)

- the giving up of the inner/outer dichotomy in favor of a functional relationship between subject and object (cf. Rudolf Bernet).
- the refusal of all Ding-an-sich-like notions by emphasizing the reality of the object as it is known (and, correlatively; the perfect knowledge of an object as a limit case to a series of knowing acts)
- the interest in making a taxonomy of different ways of approaching an object (in Husserl: the inventory of different intentional acts and their components: in Peirce: the taxonomies of signs and their components). There is even a certain structural similarity (albeit not identity) between Peirce's well-known icon-index-symbol triad and Husserl's imaginative-indexical-signitive intentions in the 1st and 6th LU.
- the foundation of logic upon phenomenology.
- the emphasis on autonomous forms in logic (Husserl: *nominale und propositionale Akte – Schlüsse*; Peirce: rhema-dicisign-argument) – and their ensuing a priori theories of linguistics and semiotics, respectively.
- the conception of complicated cognitive acts/signs as founded on a motivated complex of simpler acts/signs.
- the conception of meaning as being general (as a species of meaning acts (early Husserl anno LU) or a type (Peirce, at least in so far as regards sufficiently complicated and interesting signs))
- the anti-Cartesian refusal of absolute doubt in favor of doubt within a horizon
- the striving for a phenomenological solution to the problem of how mathematics and universal, objective knowledge in general may be shared by subjective acts of knowing.
- the discovery of (proto-)speech acts (Husserl: the notion of different act qualities and the frame question of the 6th Investigation; Peirce: the so-called 'Gamma graphs').
- the introduction of variables as empty 'slots' in logic and semantic expressions
- an Aristotelian realist tendency to claim the reality of species and our direct access to them (eidetic phenomenology and 'Wesenserschauung', Husserl; the generality of perceptual judgments and the 'pragmatic maxim' and diagram experiment as means of clarifying a concept in Peirce), along with the idea that the law of the excluded middle does not hold for ideal matters
- the positive redefinition of the synthetic a priori (as 'pure laws including material concepts' (Husserl, 3rd Investigation) or as 'universal propositions relating to experience' (Peirce))
- a criticism of empiricist and psychological theories of abstraction in favor of an 'ideierende Abstraktion' (Husserl) or Peirce's complex of abstraction types – and even the close connection between the conditions of possibility of abstraction and a mereology for the phenomenon (Husserl's 3rd Investigation; Peirce's definition of his three types of distinction).
- the attempt at giving a rational exposition of the loose Kantian idea of a schematic meeting between 'Sinnlichkeit' and 'Verstand' in epistemology (Husserl: 'kategoriale Anschauung', Peirce: diagrams).

The aim of this chapter is to compare Husserl's account of categorial intuition in the 6th LU (and its prerequisites elsewhere in that work) with Peirce's diagram and diagrammatical reasoning notions. Just like Peirce develops his notion of diagram-icons to understand the observation aspect of the access to ideal and universal objects, Husserl undertakes a daring extrapolation of the concept of intuition ('Anschauung') to involve the grasping of grammar and linguistic syntax, of essences, of states-of-affairs, etc. Thus, analogous roles are played by diagrams and categorial intuition[155] in Peirce and Husserl around 1900, and the immediate difficulties in admitting this probably stem from their different *prototypes* of these notions: that of grammar and linguistic *syncategorematica* in Husserl, that of geometrical construction proofs in Peirce. Our aim here is to introduce the cross-fertilization between those two concepts to be put to use in an actual understanding of the cognitive ability to grasp, understand, and manipulate ideal objects.

MEANING AND INTUITION IN THE LOGISCHE UNTERSUCHUNGEN

In Kant, of course, we find the idea that a crucial problem of epistemology is the possible mediation between what he considers two faculties of mind, *Verstand* and *Anschauung*, respectively. A simple integration between the two in an 'intellectual intuition' is deemed impossible and the belief in its possibility leads into the 'transcendental illusion'. The possible mediation between these two is only deemed possible in the construction of rule-bound schemata (the arch examples being arithmetics, geometry (the triangle) in the realm of pure schemata, and the concept of the dog in the realm of empirical schemata). Each in their way, Husserl and Peirce strive to clarify the more precise relationships behind the sketchy Kantian deliberations.

In the LU, this problem is highlighted in the second section of the 6. LU but with close ties to various other chapters throughout the book. The prerequisite to understanding the ideas given in the 6. LU is, of course, the structure of the intentional act as outlined in the 5. LU. Here, Husserl distinguishes between the act itself, its content and its object. The content is the act's meaning conceived of as species, and it, in turn, includes three dimensions: the quality of the act (its character of being propositional, imperative, wishing or whatever), the matter of the act (the way the object is presented), and the representative content of the act (the degree of fulfilment in which the object is presented: intuitive vs. signitive acts). In the intuitive acts – of which perception is the prototype –, the object is immediately given in 'anschauliche Fülle', but in other, more distant acts, the object is merely intended in a symbolic fashion, the so-called signitive acts, which prototypically comprise linguistic expressions (but not only, cf. below). The meaning of signitive acts has no intuitive character, but, on the other hand, they aim at being fulfilled by synthesis with decidedly intuitive acts (not unlike symbols requiring icons for making explicit meaning in Peirce). This more complicated, founded act forms the 'Erfüllungssynthese', that is, it fulfills an immanent striving present in the merely signitive acts. This very tension between signitive and intuitive acts forms Husserl's version of *Verstand* and *Sinnlichkeit*, respectively, and it is at stake in any linguistic and symbolical act in general as well in science more specifically. This fulfilling of an empty signitive act is thus a higher-order act, involving the *Leervorstellung* of the signitive act, its intuitive fulfilment, as well as the act founded upon the two which unites them.

A host of problems are involved here. For in sufficiently complicated acts, the signitive act contains lots of elements not immediately present in the fulfilling perception. Husserl's introductory example in the LU is sentential structure; all the *unselbständige* moments of the sentence apart from the material presented in the nominals subject and predicate: quantifiers, conjunctions, numerals etc. – the syncategorematica of the 4. LU. This leads to the idea of the *kategoriale Anschauung*, 'categorial intuition'. The basic claim of phenomenology: that the justification of any kind of knowledge ultimately derives from the possibility of grasping it in intuition, is extended to include these 'categorial' aspects of meaning. Thus the concept of intuition is generalized to encompass categories. The categorial content of the act also aims at its own fulfilling intuition, that is, the idea that the object of the logical and formal apparatus of the expression can in some sense be grasped intuitively as it is in itself, in perfect analogy with the perceptive fulfilment of the parts of the expression referring to sensible objects. Categorial intuition thus comes to bear an immense weight in Husserl's epistemology: it becomes responsible for the grasping of all that is not simply perceptual.[156] To be blunt, categorial intuition is what makes Husserl differ from a crude sensualism. Thus, it involves not only syncategorematica, but all kinds of ideal objects taken in a broad sense of the word: states-of-affairs, logic, mathematics, formal ontological categories, material ontological categories, all sorts of natural and cultural kinds, word meanings.[157] Thus the different essences grasped by eidetic variation already introduced in the LU are intended by categorial intuition which thus subsumes *Wesensschau*.[158] And thus the ideal objects grasped by categorial intuition include Ingarden's 'purely intentional objects' involving the objects of linguistic meanings in general and literary works in particular. In introducing the new distinction between signitive and categorial, Husserl clarifies aspects which were mixed up in the Kantian outline: on the one hand, we have merely signitive meaning aiming at its intuitive fulfilment. On the other hand, we have two different forms of intuition, sensuous and categorial, respectively – and the categorial intentions possess their own type of intuitive fulfilment, more or less remotely founded upon sensuous intuition, to be sure.

This gives a combinatiorial table as follows:

	signitive act	intutive act
sensuous object		
categorical object		

Figure 19.

There are two possible sides to the signitive act; one side which requires fulfilment in a categorial intuition, another side requiring fulfilment in a sensuous intuition – maybe both in one synthesizing act. Apart from very simple cases, most acts include categorial components (as soon as perceptual judgments are passed, for instance, or general properties of any kind are invoked, categorial meaning is implied). A pure intuition is only possible, it should be added, in the cases of certain types of objects (i.e. mental acts and certain species and universals[159]), but emphatically not in the cases of empirical objects, in which only partial fulfilments are possible due to their appearance in adumbrations.

The first axis (signitive-intuitive) is complicated by the fact that other types of approaches to the object are possible: intuitive acts comprise also imagination which presents its object through (partial) similarity, only without allowing immediate fulfilment. (Here, imagination should probably be taken to include imagery, fantasy, and memory alike, which were subsequently distinguished in Husserl's writings in the years after the LU (see Chap. 14)). Finally, the combinatorial table given above is complicated by the fact that signitive intentions typically involve categorial meaning;[160] only signitive intentions with no syntax nor generality (exclamation of simple word meanings) might constitute a limit case.

THE ISSUE OF MEANING

This ingenious construction leaves one question open: how is the fulfilment of a signitive act (a categorial form, respectively) performed? As Jocelyn Benoist (Benoist 1997, 136) has remarked, 'The paradox is that a signitive categorial form can only be satisfied by an intuition which is already itself categorized.'[161]

The problem is that no positive determination of the meaning part of the signitive intention, taken separately, is undertaken. On the one hand, signitive acts are distinct from mere indexical *Anzeichen*;[162] on the other hand, meaning is distinguished from all kinds of psychological imagery, representations, 'Vorstellungen'. Meaning, in the original 'species theory of meaning' of the LU, is seen as the species whose instantiations are the single mutually synonymous meaning acts, but this does not indicate how the meaning in the specific case can be characterized.

But without any further positive determination of the meaning concept, it remains difficult to describe precisely why a specific meaning finds its fulfilment in the exact set of intuitions it does. We know from the criticism of the empirical abstraction theories in the 2nd LU that the way we make an 'ideierende Abstraktion' is analogous to what Husserl later in *Experience and Judgment* calls eidetic variation; that is, we substitute for all non-essential parts of the phenomenon empty, algebraic variables, making it possible to focus upon the invariant species left. This variation procedure must not be understood, of course, as if the specific core left was itself untouched by the variation which investigates which range of variability the species left is able to assume.[163]

CATEGORIAL MEANINGS AND OBJECTS

The meaning respectively object side of the act are easily distinguished in the prototypical categorial case of logic. Logic categories include *subject, predicate, proposition*, etc., and the corresponding object categories include *object, property, state-of-affairs*, etc. In general, logical categories refer to object categories pertaining to what Husserl calls *formal ontology*,[164] the general science of objects without regard to their material qualities. This apparently simple duality between formal logic and formal ontology hides some complications. Both are species concepts, but meaning species and object species are not identical, even if the grasping of the second by means of the first plays a crucial epistemological role. Both, consequently, are species made explicit by the ideational abstraction described in the 2nd LU. At the same time, categorial meanings form a crucial part (that is, moment) of sufficiently complex, 'sinnlich gemisschte' empirical meaning species in general (cf. below), just as categorial objects form parts (that is, moments) of empirical object species. This forms the central link in Husserl's solution to the problem of epistemology: by means of the dependency calculus in terms of parts and wholes, complex objects characterized by specific sets of interrelated parts may be represented in signitive meanings characterized by analogous interrelation systems between their parts.[165] This implies the possibility for manipulating with empty, signitive meanings without constant reference to their intuitive fulfilments: the (partial) isomorphism of the manipulation rules guarantee the fulfilment possibility. This basic idea is analogous to Peirce's general diagram concept in which the crucial feature is that the diagram is a sign representing its object by a schematic figure connecting parts by means of rational relations, that is, precisely a mereological analysis of the object in terms of ideal relations graspable by abstraction. This analogy leads to the question of this chapter: what precisely does categorial meaning comprehend? It goes without saying that the basic logic categories form prototypical categoriality, but as categoriality is present wherever we rise from a purely sensous perception fulfilment, logic is not sufficient for describing categoriality.[166]

A whole series of problems is connected with this issue: the role played by categoriality in fulfilment of signitive intentions; the status of pictorial signs (and the categoriality inherent in them) in relation to the grasping of ideal objects; the principle of variation in the determination of species in general.

MEANING AS THE DETERMINATION OF A RANGE OF POSSIBLE INTUITIONS

Let us take a closer look at what meaning is supposed to *do*. In the beginning of the 6th LU it is laconically stated that 'The 'generality of the word' means, therefore, that the unified sense of one and the same word covers (or, in the case of a nonsense word, purports to cover) an *ideally delimited manifold of possible intuitions*, each of which could serve as the basis for an act of recognitive naming endowed with the same sense.' (6th LU, Vol. 2, 691–92/ Hua XIX 563, our italics) This is exemplified as so often before in the word 'red': 'To the word "red", e.g. corresponds the possibility of both knowing as, and calling "red", all red objects that might be given in possible intuitions. This possibility leads on, with an a priori guarantee, to the further possibility of becoming aware, through an *identifying synthesis* of all such naming recognitions, of a sameness of meaning of one with the other: this *A* is red, and that *A* is *the same*, i.e. *also* red: the two intuited singulars belong under the same "concept" ' (ibid.). The bound variation of the species meaning in question may singularize it in particular instantiations. Thus, it is the opposite operation of the variation undertaken in the abstraction process' isolation of the species in the first place.

This eidetic variation procedure can allegedly be applied from the most simple to the most complicated cases; in the *Prolegomena*'s conclusive and ambitious outline of a 'theory of theories' concentrating on theoretical form, we find the same idea at the level of whole theories: to substitute for its given parts undetermined variables to leave only the formal categorial structure of the theory behind (§67).[167] Furthermore, the variation can be extended to involve the formal structure itself: by the variation of basic factors in the theory, the conditions for the transposition of one theory into another may be made clear (§69). Correlatively, on the object side of the theory, the domain of knowledge corresponding to the purely formal theory will be the idea of pure mathematics in general, the *Mannigfaltigkeitslehre* (§70). If space, writes Husserl, is the categorial form of cosmos, studied by geometry, then this is only a part of a genus of categorially determined manifolds describing space in a generalized meaning of the word. Here, the categorial form of the theory is strictly correlated to its object side. To the formal logic of the former corresponds the formal ontology of the latter.

In the 2nd LU, Husserl returns to the question in the famous discussion of the general triangle in the British empiricists. He refutes Locke's claim that the non-existence of the general triangle should imply that it is only a mere invention of understanding, and he criticizes Hume's psychologistic and nominalist idea that a singular representation becomes general merely by means of the addition to it of a general name. On the other hand, Husserl is close to Berkeley at this point: the universal is a singular idea used to represent all other singular ideas of the same sort, provided that representation here is read as implying meaning rather than reference, as triggering rather than substituting. The single sign may not refer to an infinity in extension, but it means 'any triangle, no matter which one'.[168] Thus, the role of singular illustrations for universal concepts, tokens for types, should be, Husserl repeats over and over, taken as trigger ('Anhalt') rather than substitute ('Stellvertreten'). So it is a *means* of grasping the thought rather than a *substitute* for it, but it is not necessarily a less prestigious role to be *Anhalt* than *Stellvertreter*. For the role of the trigger seems to be an illustration – an illustration subsequently to be read in an eidetic fashion, and, in turn, to be varied eidetically in order to yield 'any triangle'.[169]

VARIATION AND ABSTRACTION

In the conclusion to the 2nd LU's eidetic abstraction theory, Husserl explains the extended meaning of abstraction as follows: 'Thus we directly apprehend the Specific Unity *Redness* on the basis of a singular intuition of something red. We look to its moment of red, but we perform a peculiar act, whose intention is directed at the 'Idea', the 'universal'. Abstraction in the sense of this act is wholly different from the mere attention to, or emphasis on, the moment of red; to indicate this difference we have repeatedly spoken of *ideational or generalizing abstraction*' (2nd LU, Vol. 1, 432/ Hua XIX, 226).

In this two-tier account for abstraction (emphasis – generalization) there is a surprising similarity to Peirce's abstraction theory, in which he puts great weight on distinguishing various types of abstraction having to do with the distribution of attention to selected aspects of the object on the one hand (involving three types of 'distinctions'), and the so-called 'hypostatic abstraction' on the other. The seminal attention focussing abstraction, which enables us to distinguish parts which can not act as distinct unities, is 'prescission'.[170] But this focussing mechanism, however important, does not in itself lead to higher degrees of abstraction. The property focussed upon must, in turn, be made subject to 'hypostatic abstraction' which makes of it a general noun as a subject for a new proposition with predicates to be determined. This two-step abstraction mechanism seems to make explicit what Husserl is more briefly outlining in the 2nd LU in the quote given with its distinction of the 'Hervorheben' and the 'generalisierender Abstraktion', the emphasizing and the ensuing generalizing abstraction.

To return to the issue of the 'illustrative' aspect of meaning, this abstraction account seems to clarify what we more exactly do when using a picture as 'Anhalt': first, we emphasize the moment of it in question ('Red', 'Triangle'), second, we generalize this moment of it by variation which is what, third, permits us to give it a specific, nominalized name ('Redness', 'Triangularity'). But once this has been achieved, there is, conversely, a way 'back'; by using variation we may now devise the 'ideally delimited manifold of possible intuitions', that is, we may, by variation, produce any particular triangle. The variation principle delimits the manifold. The question here is: what is the part played by categoriality in this relation of variation between signitive intention and intuitive fulfilment?

HUSSERL'S EXAMPLES IN THE LOGISCHE UNTERSUCHUNGEN

Husserl's general description of categorial intuition suffers from the same defects as does the analogous description of sensuous fulfilment – we do not know the precise road from signitive categorial intentions to fulfilled categorial intentions presenting categorial intuitions. But unlike the case in the perceptual counterpart, we do not even have a clear idea of what the relevant fulfilment looks like. Husserl's own primary examples point to formal logic and its use in linguistic syntax of empirical languages as for instance when he considers the example of conjunction ('and'). He writes, in a famous passage, that the act of conjunction is different from simple – non-categorial – perceptions of sensously given unitary sets, series, swarms etc., because it is a distinct act adding the contents of two former acts to mean the compound content 'A and B'. So this conjunction is a founded, categorial act requiring its own intuitive fulfilment.[171]

Some of the more complicated examples given in the course of the LU may throw some light upon this issue – e.g. the mathematical expression to be calculated; the map of England; the recognition of Goethe's handwriting; the model of the steam engine. Husserl himself does not make any categorial conclusion to this variety of examples, but taken together, they make it possible to outline what we may conceive categorial intuition to involve. Not all of these examples are given in the relation to the 6th LU's chapters on categorial intuition, but still they involve different aspects of it.

The first example concerns the mathematical expression $(5^3)^4$ and concerns mediate fulfilment. It is not the case that the meaning of a complicated expression is of the same kind as a simple word meaning. On the contrary, the complicated expression facilitates '... the possibility of *fulfilment-chains built member upon member out of signitive intentions*. We clarify the concept $(5^3)^4$ by having recourse to the definitory presentation: Number which arises when one forms the product $5^3 \cdot 5^3 \cdot 5^3 \cdot 5^3$.' (6th LU, Vol. 2, 723/Hua XIX, 601) In the same manner, this expression takes us back to simpler definitions, and every step in this operation is an act of fulfilment, prescribed by the signitive representation: 'A remarkable property of the cases just discussed, and of the class of significative presentations which they illustrate, lies in the fact that in them the *content* of the presentations – or, more clearly their 'matter' – *dictates a determinate order of fulfilment a priori*.' (6th LU, Vol. 2, 724/ Hua XIX, 602) What can be learnt from this example is that certain expressions allow their contents to be constructed by an ordered, stepwise operation with increasing fulfilment. If we generalize this to other mathematical expressions we can add that it is far from always the case that the procedure to be undertaken is unanimous nor clear. An equation may be solved in different ways, in different variables; maybe it may not be solved at all; maybe it is not even known whether it may be solved (Goldbach's conjecture). In short, in expressions like these, a (in some respects) rule-bound but otherwise (in other respects) free operation can be performed in order to seek fulfilment, but only in some cases is it clear that stepwise fulfilment is able to reach its goal.

This may be compared to an example given a couple of pages earlier on an intuitive series of fulfilment: 'Another example of an intuitive fulfilment-series is the transition from a rough drawing to a more exact pencil-sketch, then from the latter to the completed picture, and from this to the living finish of the painting, all of which present the same, visibly the same, object.' (6th LU, Vol. 2, 721/ Hua XIX, 599). This fulfilment series has a slightly different character from the mathematical case – also apart from involving imaginary rather than signitive intentions. In the painting series, the earlier stages may be left behind, once the latter more fulfilled ones are reached – not so in the mathematical example where it is important to remember the expression of which the number 244,140,625 is in fact the result. Here, conversely, $(5^3)^4$ might as well serve as a fulfilment of this number, in an intention pointing the opposite way, given the signitive intention (much more difficult, to be sure) to resolve it into its prime factors.

A peculiar case concerns what Husserl calls 'signitive intentions outside the meaning function' – referring to acts of classification without the relevant word being invoked. The recognition of an object as a Roman roadstone or of a tool as a drilling machine, for example – but separated from the uttering of the corresponding word. The classification of a phenomenon as belonging to a species, a token to a type, thus seems to be the pure function of the signitive intention.[172] 'Objects are, strictly speaking, only 'known', as they are given in their actual intuitive foundation, but, since the unity of our intention ranges further, objects appear to be known as what they are for this total intention. *The character of knowing is accordingly somewhat broadened.* Thus we recognize (know) a person as an adjutant of the Kaiser, a handwriting as Goethe's, a mathematical expression as the Cardanian formula, and so on. Here our recognition can of course not apply itself to what is given in perception, at best it permits possible application to intuitive sequences, which need not themselves be actualized at all.' (6th LU, Vol. 2, 716/ Hua XIX, 593). With the signifier of the expression placed in brackets, this paragraph in fact presents the signitive intention *in nuce*: it concerns the pure species and the problem of how, given a concrete perception, this perception is classified as instantiating the species: a piece of handwriting identified as Goethe's. This, in fact, is to our day still a problem hard to solve: how is it possible to identify a style of writing? From a general point of view, this question may be of the same kind as those about the emperor's servant and Cardan's theorem, but from a more detailed cognitive science point of view, there is a huge difference. The simple version of variation is certainly not possible in the Goethe's handwriting example: there are no simple parts which may be replaced with algebraic variables. Rather, the species are grasped through the variation of the whole with certain stylistic features kept constant: the variation of the type of ink and the type of paper is easily performed, but more difficult is the variation of the written expression with the style kept invariant. We can not seriously assume a variation which *de facto* covers all possible texts in the world, rewritten using Goethe's hand writing, rather we implicitly grasp the idea of such a variation and judge it possible in principle. Goethe's style is grasped as a set of certain, typical, stable variation aberrations as compared to a normal zero handwriting.[173] This is an adaptation to intuition processes which need not be actualized themselves: we need not see for our inner gaze other examples of Goethe's handwriting in order to recognize an example of it – this variation is presumably undertaken without being explicit in consciousness.

From this example, two things can be inferred: that the main problem resides not in the (mostly arbitrary) relation between word expression and meaning, but in the relation between meaning and its fulfilment; that the variation process involved in classification may vary the content continually while keeping general moments invariant which characterize the whole of the object and which are hard, maybe impossible, to make explicit as such.

A further example is the map of England, a prototypical diagram example. Husserl mentions it as an example of an indirect representative serving as partly fulfilling intuition: '... as when the use of a geographical name calls up the imaginative presentation of a map, which blends with the meaning-intention of this name...' (6th LU, Vol. 2, 727/ Hua XIX, 606). When the fulfilling of the name 'England' is performed by a map (instead of the object itself), it sure is an indirect object. 'The analogy of what appears and what is meant, which may be present here, does not lead to a straightforward presentation by way of an image, but to a sign-presentation resting upon the latter. The outline of England as drawn on a map, may indeed represent the shape of the land itself, but the pictorial image of the map which comes up when England is mentioned, does not *mean* England itself in pictorial fashion, not even mediately, as the country pictured on the map. It means England after the manner of a mere sign, through external relations of association, which have tied all our knowledge of land and people to the map-picture' (ibid.). The map referring to England is seen as a complicated expression with several levels; the iconic qualities of the map in relation to England is superposed by the use of it as a sign referring to England as the object, including the associative connections to our diverse knowledge of that country. So the map is no mere picture even if built on iconic qualities. It must be considered as a diagram which implies two things: a similarity between map and object plus, in turn, the use of the map as a sign for the object, including association with qualities not directly mapped therein.

What can be learnt from this example is that the map has a double foundation, composed of a moment of similarity on the one hand and a signitive intention on the other.

The last of Husserl's scattered examples concerns the most typical diagram example in the LU: the steam engine model. The example occurs in the context of the chapter of the 6th LU conclusive to the exposition of sensuous and categorial intuitions. The chapter introduces the crucial distinction between the categorial synthesis of simple perceptions on the one hand (e.g. particular states-of-affairs) – and general intuitions with general objects on the other, giving rise to *synthetical* and *abstractive* categorial intuitions, respectively. In the former, the founding acts' objects are included in the founded acts, not so in the latter – but both are categorial acts. Accordingly, we may distinguish at least three types of involvement of categorial intuition: one is present in the categorial moments of simple perception judgments, e.g. of concrete states-of-affairs. Another is the pure grasping of categorial structures in specie, in logic and mathematics. And yet another is the use of categorial means to grasp general empirical objects.

A crucial observation here is the following: 'Talk of 'perception' presupposes the possibility of correspondent imagination: a distinction between them, we held, is part of the natural sense of our ordinary talk about 'intuition'. But it is just this distinction that we cannot here draw. This seems to stem from the fact that abstractive acts do not differ in consonance with the character of the straightforward intuitions which underlie them; they are quite unaffected by the assertive or non-assertive character of such underlying acts, or by their perceptual or imaginative character.' (6. LU §52, Vol. 2 800/ Hua XIX, 691) The fact that the distinction between imagination and perception becomes irrelevant in the case of categorial intuition is very important: it implies that when talking about categories, an imaginative fulfilment is as good as any. This brief statement is ripe with consequences. The function of imagination as access to ideal structures is implied, just as the role of thought experiments in science and thought in general. This implies, moreover, that merely imaginative representatives of categorial structures may be used as completely fulfilling signs for them. As to the categorial structure of an object, an image of a special, general kind (or, as Peirce calls it, a diagram) permits us to directly grasp the very category in specie. This includes a general categorial 'reading' of a particular example, cf. the discussion of Locke's triangle above. An individual object can not serve as an analogy of itself, Husserl writes, but 'It is quite different in, e.g. the case where mathematical analysis has given us an indirectly conceived Idea of a certain class of curves of the third order, though we have never *seen* any curve of this sort. In such a case an intuitive figure, e.g. of a familiar third-order curve, perhaps actually drawn, perhaps merely pictured, may very well serve as an intuitive image, an analogon, of the universal we are intending: our *consciousness of the universal is here intuitive, but analogically intuitive, in its use of an individual intuition.*' (6th LU §52, Vol. 2, 801/ Hua XIX, 692, our italics). This interesting claim is what is examplified in the steam engine example: 'And does not an ordinary rough drawing function analogically in comparison with an ideal figure, thereby helping to condition the *imaginative character of the universal presentation?* This is how we contemplate the Idea of a steam-engine, basing ourselves on a model of a steam-engine, in which case there can naturally be no talk of an adequate abstraction or conception. In such cases we are not concerned with significations, but with universal representations by way of analogy, with universal imaginations, in short' (ibid.; the German version has: 'In solchen Fällen haben wir es mit keinen blossen Signifikationen zu tun', so the English quote should rather talk about '*mere* significations').

This characterization of the model of the steam engine[174] thus unites iconicity and generality, Peirce's two major characterizations of the diagram. Unlike the merely signitive word 'steam engine', the model implies a general imagination of the idea of such a machine – and the act of imagination is in abstract, categorial cases a complete fulfilment. But implicitly, it also displays the third major feature of diagrams: the possibility for experimenting. A model of a steam engine only reveals the idea of the working of this apparatus when conceived of in a temporal, operational, and experimental fashion. The model gives rise to a thought experiment, letting water be heated, steam to be produced and suddenly cooled with the characteristic working process of the machine as a result. Mobile parts of the object possible to manipulate physically may add to the efficiency of such thought experiments. This feature of the diagram is only implicitly present as a necessary feature in Husserl's steam engine example, but in the mathematical example above it was made explicit in the idea of a stepwise, operational *Erfüllungsreihe*.

To sum up: categorial intuition and its use in the direct 'erschauen' of meaning as species constitutes a strikingly close (even if much less explicit) parallel in the LU to the mature Peirce's diagrammatic epistemology. Peirce's distinctions between pure and applied diagrams find a counterpart here in the distinction between pure categorial intuition (in which categories are grasped in abstraction from the acts they spontaneously appear in) and categories put to use in the grasping of empirical species (as the steam engine) or of empirical state-of-affairs. Furthermore, Peirce's extrapolation of logic from formal inference schemata to cover a much wider range of signs finds its (implicit, that is) counterpart in the (few, but) widespread examples in Husserl's text making it clear that not only logic, but also geometry and the whole 'Mannigfaltigkeitslehre' of the *Prolegomena* form the content of pure categorial intuitions, possibly to be put to use, in turn, in applied – 'sinnlich vermisschte' – categorial intuitions.

Finally, it must be the more or less perfect grasping of categorially formed species that allows for the mysterious route leading from signitive intentions to intuitive fulfillings. How should the passage from the word 'steam engine' to the perceptive fulfilling of it be possible if not via the intermediary (maybe only parts or aspects of) a general, imaginative model of it?[175]

THE AMBIGUOUS STATUS OF PICTURES IN THE LOGISCHE UNTERSUCHUNGEN

Taken as a whole the LU remain ambiguous as to the role of pictures. It seems as if two tendencies are *verschmeltzt* in the early Husserl: one is the phenomenological turn against psychologism; the other is the formal turn against imagination to the benefit of formal calculi, and I believe there is a tendency in Husserl to identify or confuse the two. This can be seen in his repeated arguments against 'phantasies' in the question of semantics – all at the same time as the steam engine example admits the crucial role of imagination in categorial intuition fulfilment. But this identification is misplaced. There are two strands in this argument. One is the anti-psychological argument: semantics is not psychology and meaning must be conceived of as an ideal, phenomenological species and should not be taken to rely on more or less contingent, individual fantasy pictures only. But this does not entail that semantics is formal understood as devoid of intuition, rather it necessitates a concept of phenomenologically pure, eidetic pictures, the 'allgemeine Imaginationen' that Husserl points to at the end of the central chapter in the 6. LU – that is, diagrams.

The ambiguous attitude towards pictures is mirrored in a similarly ambiguous attitude towards space. Husserl writes for instance (3rd LU, Vol. 2, 455/ Hua XIX, 256) in connection with the redefinition of analytic/synthetic concepts that ideal objects comprise two types, the essences to which '... correspond the concepts or propositions which have content, which we sharply distinguish from purely formal concepts and propositions, which lack all 'matter' or 'content'. To the latter belong the categories of formal logic and the formal ontological categories mentioned in the last chapter of the *Prolegomena*, which are essentially related to these, as well as to all syntactical formations they engender. Concepts like Something, One, Object, Quality, Relation, Association, Plurality, Number, Order, Ordinal Number, Whole, Part, Magnitude etc., have a basically different character from concepts like House, Tree, Color, Tone, Space, Sensation, Feeling, etc., which for their part express genuine content', the two categories of concepts giving rise to formal and material ontologies as analytical and synthetic disciplines a priori, respectively. Here, surprisingly, space is classified alongside the other material species belonging to different material ontologies. This classification of it apparently runs counter to what is said in the conclusion of the *Prolegomena* where we find the idea that the correlate to a purely formally conceived theory is a field of experience in general, this field in turn to be studied by Husserl's general conception of mathematics, the 'Mannigfaltigkeitslehre'. But here, this study *includes* space, placing it on the purely formal level, far from the 'sachhaltige' rendering of it in the quote above: '...if we mean by 'space' the categorial form of world-space, and, correlatively, by 'geometry' the categorial theoretic form of geometry in the ordinary sense, then space falls under a genus, which we can bound by laws, of pure, categorially determinate manifolds, in regard to which it is natural to speak of 'space' in a yet more extended sense' (*Prolegomena*, §70, 242/ Hua XVIII, 252). Thus, the specific space concepts pertaining to each material ontology are but species of a formal genus of space belonging to formal ontology. But this general space category implies that space is also among the categories finding fulfilment in pure categorial intuition. This allows for the Husserlian counterpart to Peirce's pure diagrams (requiring space) with no reference to any actual existence.

This allows us to return to the issue of the possible role of pictures in the fulfilment of signitive intentions. As a geometer, Husserl agrees completely with the formalist tendencies of his time:

'It is a well-known fact that no geometrical concept whatsoever can be adequately illustrated. We imagine or draw a stroke, and speak or think of a straight line.' (1st LU, Vol. 1, 302/ Hua XIX, 70). The picture drawn is no representative of the geometrical object – cf. Locke's triangle – but is a mere 'Anhalt', a trigger for a more precise fulfilment, just like Peirce's diagram token is not in itself a representation but merely a precondition for the diagram type to be grasped. Now, given the possibility of a stepwise fulfilment with an increasing degree of fulfilment, this role of 'Anhalt' may be given a more detailed description: the picture is read in an eidetic manner, governed by the signitive intention present (for instance, the picture of a triangle accompanied by the word 'triangle' – as opposed to the same picture accompanied by the word 'manifold', 'polygon', 'Jordan-curve', etc., emphasizing other moments in it). This eidetic reading of the concrete picture is a higher-level categorial act, founded on the signitive and pictorial acts alike, and it makes possible the eidetic imagination of the general picture. The concrete drawing is not general, but the categoriality of the signitive intention present prompts such a reading. Husserl himself does not consider further this interplay between pictorial and signitive intentions leading to eidetic imagination, but Peirce's diagram concept does just that. It emphasizes the diagram's double determination: it is an icon in so far as it is a (skeletal) picture of its object, but it is governed by a symbol permitting the emphasizing of the relevant aspects of the picture intended. In so far, the interplay between symbol and icon, signitive and pictorial intentions, prompts eidetic abstraction permitting to imagine the pure species. This species may now, in turn, be used to map the relation structure of widely differing objects (triangle trade, erotic triangles, triangulation in navigation, etc.). The diagram category thus makes evident that the mereological dependency calculi of the 3rd LU kind are necessary but not sufficient for formal ontology. Mereology needs supplementation by other branches of mathematics; geometry, topology, and category theory are prominent candidates, but only ongoing investigation will show which formal disciplines will be needed adequately to map the categorial properties of diagrams and the corresponding categorial objects they depict.

HUSSERLIAN CATEGORIAL INTUITIONS AND PEIRCEAN DIAGRAMS

To sum up, in the relation between signitive and intuitive acts, categorial intuition plays the role of:

- permitting the synthesis of contents into all kinds of nominal objects and states-of-affairs
- permitting eidetic variation to be a crucial device for grasping species, that is, meaning. Once the species is constituted, the variation procedure may work in the other direction furnishing the species with possible particular instantiations
- permitting the rule-bound, stepwise fulfilment of certain signitive acts (the mathematical and the sketch examples)
- permitting the adequate grasping of structures in formal ontology, this implying the necessity (as Jean Petitot points out) of geometrising the basic structures in formal ontology[176]
- grasping the content of complicated empirical species (the steam machine example) by permitting rule-bound operations involving its parts in specific configurations.

All these points make categorial intuitions play roles analogous to those played by the diagram in the mature Peirce's theory of knowledge. Here, diagrams are similar to their objects in two crucial aspects: they form relational, mereological analyses of their objects, and – as we saw in Chap. 4 – they are subject to Peirce's operational criterion for iconicity: one phenomenon is an icon for another if and only if experiments or manipulations on the former may reveal new insight into the latter ('new' in the sense that the information in question is not explicity expressed in the icon). This implies that diagrams are the vehicles for all deductive reasoning – such reasoning simply being defined as manipulations on diagrams.

This procedural aspect of the diagram's iconicity is not explicitly thematized in Husserl's account for categorial intuitions, but it is, as shown, present in his examples. The crucial variation procedure itself is nothing but an operation on a diagram; the steam engine model permits to imagine the working of the machine in an operational procedure; the rule-bound transformation of the mathematical expression is another operation on a diagram.

In Peirce, this operational criterion for iconicity is tied to the continuity metaphysics: to perform an operation on the diagram implies the continuity between the single diagram instances which the operation connects – which facilitates the corresponding continuous unity of the depicted object in space and time.[177] From a Husserlian point of view, this central property in diagrams connects to several important issues. One is the idea clearly brought to the fore in *Ding und Raum*, that the very prerequisite for the unity of a logical entity is the continuum of *Abschattungen* of an object which makes their schematic synthesis possible. This forms the very basis of the founding of logic on phenomenology in Husserl, and it implies, as Jean Petitot remarks, that the categories of object and of logic both presuppose continuity. This casts a Husserlian light on the operational icon definition in Peirce's diagrams: it is because the object *itself* is defined by a range of continuous operations that a formalized icon may depict it by repeating (parts of) these operations. Furthermore, it connects time-consciousness intimately to diagrams – and to categorial intuition: it is only through the synthesis of temporal experience with the fulfilment series involved in diagram manipulation that the corresponding insight into its object becomes possible.

DIAGRAMS AS WHOLES WITH SENSUOUS MOMENT OF UNITY?

Having thus argued for a 'diagrammatizing' reading of categorial intuition and the LU, we may now let Husserl's conception throw a refining light back onto Peirce's ideas. For why is it that diagrams are so apt at capturing ideal objects? As Elmar Holenstein argues,[178] Husserl's arguments in §60 ff of the 6. LU places him at a delicate intermediate position with respect to the different schools of Gestalt Theory. Husserl claims, of course, that categorial acts are founded upon sensuous acts – but, on the other hand, he does not claim the same for acts intending sensuos Gestalts which are grasped immediately. In the first claim, Husserl agrees with Meinong's and the Graz school's 'theory of production'; in the second claim, though, Husserl is on a par with the Berlin School led by Koffka, claiming that the Graz view reintroduced sense data not pertinent in experience and invoking the direct perception of Gestalts instead. If Husserl is correct on this point, this may throw some light upon the efficacity of diagrams: they permit to grasp categorial contents *by the representation of them in sensuous Gestalts,* provided with signitive, categorial reading instructions. True, as we have seen, the sensuous Gestalt is by no means sufficient in itself (Locke's triangle) and it needs to be supplemented by general rules for its eidetic reading, for its variation, as well as for the experimenting upon it. The diagram (or, at least, the simplest significant part of it) must be graspable *in one glance* in order to represent the relevant species or type. There must be a minimum of spontaneous Gestalt grasping for the mind to be able in any way to construct more complicated Gestalts on the one hand or to abstract features from the Gestalt in order to represent categorial properties.[179] This would give a further Husserlian support to Peirce's claim that all necessary reasoning proceeds by diagrams.

Peircean diagrams or Husserlian categorial intuitions – both point to the necessity of the direct intuitive access to ideal objects as a prerequisite to a phenomenologically conceived realism. Categories and diagrams give intuitive access to idealities and, in turn, make possible the recognition of empirical objects instantiating analogically formed properties. This points to the necessity of further investigation a de-mentalized notion of icons, of *allgemeine Imaginationen*, in all degrees of generality, and of their role in categorial fulfilment.

CHAPTER 7

MEREOLOGY

Parts and Wholes in Phenomenology and Semiotics

In Peircean diagrams as well as Husserlian categorial intuitions, part-whole relations thus play a foundational role. Strange as it may seem, though, mereology – the theory of parts and wholes[180] – has only rarely caught the explicit attention of semiotics.

Semiotics, as a study taking signifying phenomena in general as its object, is faced with the issue of the signification of wholes in relation to the signification of its parts as a completely everyday phenomenon, and the recognition of distinct levels or layers of signification is also a well-known idea in most parts of semiotics. Taking the prototypical case of a text as an example, semiotics is faced not only with the traditional linguistic question of the organization of phonemes and morphemes into words, and in turn, words into sentences by means of syntax, but also, in turn, the successive integration of sentences into more extensive wholes of transphrastic discourse, periods, scenes, scripts, narrations, genres, systems of ideas, etc. As is the case in the sciences more generally, this mereological problem gives rise to two typical approaches; one, reductionist, takes a compositional attitude to the signification of wholes which is consequently seen as some kind of sum of its elements, so that an algorithmic syntax of some sort is supposed to make it possible to derive the whole's signification directly from the knowledge of the signification of its parts. The second, holist, stance takes the signification of the higher levels as irreducible, relying on their own phenomenological motivation, and, correlatively, the parts as being an analytical result of a partitioning of the whole, expanding the possibilities of the whole and making its signification in the single case more precise.[181]

In this chapter we shall briefly discuss the mereological implications in four major trends of semiotics with different degrees of connection to the Brentanian tradition in philosophy from which modern mereology originate, namely Husserl, Jakobson, Hjelmslev, and Peirce. Husserl, of course, is a Brentanian, and in his famous third Logical Investigation, he outlines a theory of parts and wholes as a part of formal ontology. The second, Jakobson, has direct connections to this tradition, primarily via Husserl; the third, Hjelmslev, displays striking similarities with the tradition without any direct influence admitted; the fourth, Peirce, working simultaneously with Brentano, has no relation at all to the tradition but still structural similarities are found.

HUSSERL

It is strange to so small extent Husserl's work is recognized or even known in the semiotic world. As hinted at in the previous chapter, most of his work either explicitly deals with or at least touches upon issues central to semiotics. In his early work, around the period of the *Philosophie der Aritmetik*, he even uses the word himself,[182] and later, in his first chef-d'oeuvre *Logische Untersuchungen* (1900–01), he investigates a whole series of central semiotic questions. *Logische Untersuchungen* ought to count as a classic of semiotics. Its long *Prolegomena* contains a detailed attack on psychologism in logic and semiotics – in so far it constitutes a major contribution to the fundamental anti-psychologism of general semiotics alongside Peirce's strongly related position. The first investigation draws a distinction between two sign types, *Ausdrücke* and *Anzeichen*, respectively: signs conferring meaning to an object vs. signs merely indicating an object. The second investigation is a critique of empiricist abstraction theories attempting to make induction the source of abstract knowledge – and it points instead to a phenomenological change in conception modus as responsible for abstraction's grasp of ideal objects. The third investigation, to which we shall return below, constructs the foundations for a formal ontology of wholes and parts which forms a basis for all phenomenological and semiotic investigations in so far as it makes possible to distinguish proper parts from non-proper parts, the last including what is usually called properties. The fourth investigation takes the mereology of the third as the foundation of a pure, that is, a priori grammar, mapping dependency relations between linguistic entities; nouns and sentences are taken as independent primitives upon which other linguistic entities are dependent. The fifth investigation is the first sketch of Husserl's intention theory, also constructed with the part-whole toolkit of the 3rd LU, distinguishing between the quality, the matter, the representative content, and the object of an act, respectively. The quality is in our days' terms a speech act category; it refers to the act's character of being propositional, imperative, wishing, etc. The matter of the act is the way its object is presented in the act, and the representative content, finally, is the degree of fulfilment with which the object is presented (perception representing the highest degree of fulfilment, and linguistic representations ('signitive acts') and imaginations ('imaginative acts') like phantasies, pictures, dreams, memories, etc. being act types with different lower degrees of fulfilment). All these aspects of the act are presented as moments, that is, 'unechte Teile', in the terminology of the third investigation. Finally, the sixth investigation takes up epistemology on the basis outlined in the former investigations; as we have just seen, the central problem of categorial intuition (how categories, among them linguistic categories, possess their own type of intuitive fulfilment) is discussed.

As is evident, most of the issues discussed in the *Logische Untersuchungen* lie at the heart of semiotics, and the mereology of the third investigation forms a crucial piece of formal ontology for the description of all these subjects. The main idea is that all objects may be described in terms of parts and wholes, and that two types of parts may be distinguished. Proper parts versus non-proper parts or parts versus moments, respectively. Parts – 'echte Teile', or 'Stücke', or concrete parts – are parts which may be separated from the whole they constitute, while moments – 'unechte Teile', or aspects, or abstract parts – are parts which may not be so separated. This sparse definition may be extended to relative autonomy and dependency, respectively, so that one object is relatively dependent on another if that content may only exist in connection to the other or parts of it. This idea makes possible the crucial structure of three possible dependency relations between parts:[183]

If we consider any pair of parts of a whole, the following possibilities obtain:
1. There is a relation of foundedness between both parts.
2. There is no such relation. In case 1, the foundedness can be:
   (a) reciprocal
   (b) one-sided ... (LI, 466)[184]

Husserl's mereological investigations include further points of interest – so as for instance the difference between wholes requiring a moment of unity and wholes not requiring it. Smith 1994 (236) summarizes Husserl's ideas in a small taxonomy. Between wholes, Husserl distinguishes those which do not require additional objects to exist (such as nut and bolt), opposed to those that require additional unifying objects such as nails or glue. The latter category yields two subtypes, depending on whether the unifying object is a concrete part or an abstract moment. In the latter case, the moment of unity will correspond to von Ehrenfels's 'Gestaltqualitäten'.

This distinction is related to different versions of Gestalt theory, cf. Barry Smith's recapitulation of the Austrian Gestalt school's 'production theory' requiring such a moment of unity in addition to the parts, as opposed to the Berlin Gestalt school's claim that no parts of a gestalt are genuine and all parts are moments only accessible by abstraction (Smith 1994, Chap. 8). Both schools' theories have their advantages and flip sides. The Graz school has the advantage of distinguishing between a part and the role played by that part in the gestalt in question, while the Berlin school tends to blur this distinction and thus opt for holism. The Graz school, on the other hand, tends to come close to Helmholtz's old idea of Gestalt-like phenomena being the result of 'unbewusste Schlüsse' – unconscious inferences – so as to make them an additional feature added to sense data by the intellect – while on the other hand the Berlin school does not follow this subjectivist idea: its holism has the merit of integrating both subjective and objective determinants as responsible for the Gestalt, and so the Berlin school will find Gestalts not only in the physiology of the gestalting subject, but also in the purely objective, even physical surroundings. Smith's conclusion is not unanimous, but it seems as if the two schools correspond to different gestalt possibilities on a continuous scale rather than being mutually exclusive, so that both very subjective and very objective gestalts as well as a large range of intermediate types are possible. We can not go further into this vast discussion here, but a further clarification of types and subtypes of Gestalts and their relation to their parts will no doubt enrich the semiotic discussion of mereology. Husserl's mereology further forms the basis for his reinterpretation of the a priori, see the next chapter.

Finally, Husserl's fourth *Untersuchung* should be considered. Here, he outlines a pure, a priori grammar using the mereological tools of the third Investigation – an idea, as a matter of fact, closely related to Hjelmslev's idea (cf. below). In contradistinction to Hjelmslev's empiricist idea of using mereology as a descriptive metalanguage for linguistics, Husserl's idea is to base the mereological description of language on certain ontological presuppositions, namely the privileging of the noun and sentence, respectively, as independent entities (after the Scholastic distinction between categorematica and syncategorematica, respectively; the former possessing an autonomous signification). Furthermore, Husserl defines the important distinction between *Widersinn* and *Unsinn*, respectively – logical and grammatically nonsense, respectively, where the latter depends on irreconcilable syncategorematica being combined, while the former is grammatically correct while semantically contradictory. Husserl's sketches of a pure grammar received more interest in logic than in linguistics: they became important for Ajdukiewicz, Lesniewski and the development of categorial grammar. But even so, a volume like the brillant *Rational Grammar* by Jean-Louis Gardies not only outlines the *Wirkungsgeschichte* of the fourth Investigation, it also gives a detailed account of its possible implementation in linguistics.

JAKOBSON

The fact that Jakobson's version of structural linguistics involves strong influences from Husserl's *Logische Untersuchungen* is clearly stated at several occasions in Jakobson's oeuvre, but it has only received general recognition after the work of Elmar Holenstein who, in a period of relative phenomenological oblivion during the fifties-sixties-seventies, never ceased to underline the crucial lines of connection between structuralism and phenomenology (Holenstein 1975, 1976). He even traced three or four variants of phenomenology taking each their characteristical departure in the LU, and thus setting the premisses of much twentieth century thought, as follows:[185]

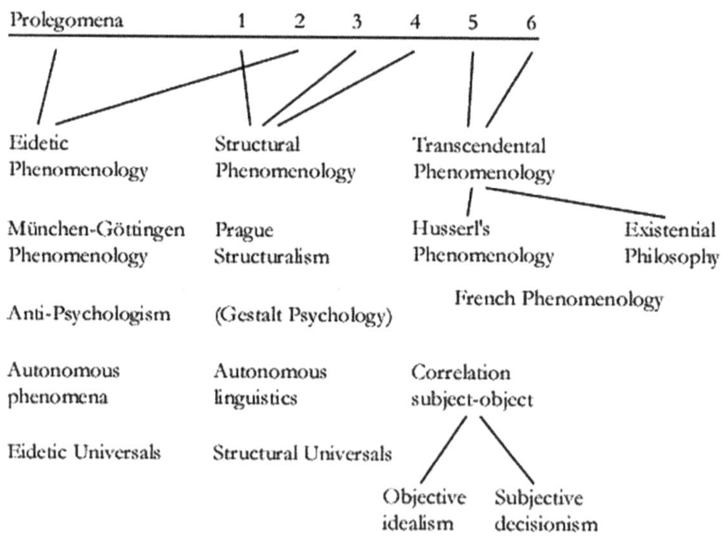

Figure 20.

In this respect, the first, third, and fourth investigations, on the sign, mereology, and pure grammar, respectively, become a founding part of structural semiotics with its emphasis on the possibility of unfolding a set of a priori foundations for the study of semiotic phenomena.

As early as a young member of the Moscow Linguistic Circle, Jakobson was acquainted with the *Logische Untersuchungen* through the Russian Husserl disciple Gustav Spet;[186] the Prague Circle which he joined in the twenties was influenced by Husserl through its founder Vilém Mathesius,[187] and to the end of his life, Jakobson did not cease to emphasize the central role of Husserl in the development of semiotics[188] and structural linguistics.[189] As a main figure in Prague Structuralism, Jakobson placed a great emphasis on the 3rd and 4th LU especially, and several times he underlined his view of linguistics as a science investigating a hierarchy of wholes and parts,[190] and he envisioned language as a whole as a 'pattern of relations'. In one of his major accomplishments as a linguist, the definition of the phoneme, he used Husserlian concepts not only to underline the anti-psychological character of the phoneme,[191] but also to describe the phoneme as composed of inseparable aspects. Jakobson never made one over-all theory of language, convinced as he was that linguistics must be made up of pieces taken from widely differing sources, ranging from anthropology to mathematics, but the mereological issue is also to be found in his most well-known contribution to the formal research of language, his notion of the 'marked' versus 'unmarked' units of language. His main idea here is that language at many levels makes use of a paradigmatic opposition between parts which are defined by asymmetric dependency.[192] Markedness is defined as follows: 'One of the essential properties of the phonological correlation consists in the fact that both members of a correlation pair do not enjoy equal rights: one member possesses the mark in question, the other one not; the former is called *marked*, the second *unmarked*...'[193] The opposition between these two is contradictory, in so far the unmarked term does not imply the absence of the feature implied by the marked term, it only implies the absence of any reference to that feature, be it positive or negative. Later, a correlated idea is presented in the theory of *zero-signs*, referring to the 'opposition de quelque chose avec rien', with a Saussure quote.[194] The marked term is dependent on the unmarked, not vice versa. This asymmetry implies that the unmarked term may play the role as the more general term of which the marked term forms a part. A semantic example will serve:

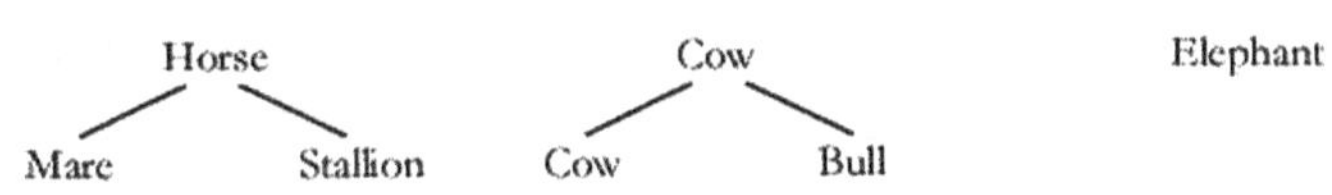

Figure 21.

In the two extremes, horses and elephant, symmetry prevails; there is a mutual dependency between 'mare' and 'stallion' and a one-sided dependency of both on 'horse', so here markedness is used to erect co-ordinate subclasses of a class. In the elephant case, English does not admit special terms for the sexes (except for a duplication of the cow-bull distinction).

With respect to cows, however, 'bull' is the marked term, while 'cow' is unmarked. This implies two different meanings of 'cow', general and specific, respectively. This is to say that in neutral contexts, the unmarked term is used; when you for instance see a field with cows and bulls on it, you can indicate them all by pointing to them and stating: 'See the cows', while if one of your children points to a bull and adds: 'See the big cow', you will answer: 'That's no cow, it is a bull.' So the term 'bull' is unilaterally dependent on the term 'cow'. The unmarked term, so Jakobson, has a zero-meaning (in this case, with respect to gender) in contrast to the marked term, but it is characteristical that the semantics of the unmarked term now oscillates between referring to the marked feature being absent on the one hand or referring to the absence of any marked feature on the other. (Cf. the specific and the general use of the word). 'Cow' consequently oscillates between entertaining a one-sided and a mutual dependency with 'bull'. This feature is, of course, not only found in at many levels in language structure, but also for pragmatic reasons in use, when you want to single out some (small) marked subset of a set: 'All linguists are stupid, except for cognitive linguists':

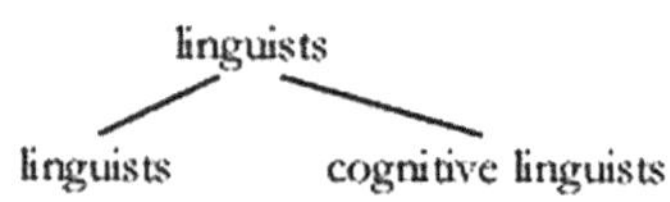

Figure 22.

In short, this distinction may be invoked when you want to express that something is part of a larger whole, but yet an atypical part.

Thus, the linguistic distinction between marked and unmarked corresponds to a cognitive and phenomenological relation pertaining to prototypicality. If you take as a basis a prototypical case, then the appearence of a non-typical case will possess an ambiguity: it is, on the one hand, part of the category, but, on the other hand, it differs from the prototypical case at the center of the category. Hence, there is a motivation to distinguish this case from the prototypical case, so that a seemingly symmetrical opposition is constructed. Yet, the prototypical case's categorizing power still extends to the marginal case, so that will still be subsumed under its main category. Thus, the marked category is at one and the same time in opposition to the unmarked category and constitutes a subtype of it – the core characteristic of the marked/unmarked relationship. This corresponds to the fact that in semantics, the tendency is that the case considered most normal, widespread, prototypical, stereotypical (or any other typicality measure) case is referred to by the unmarked term, while the less typical case is referred to by the marked term, the marked term's expression typically being longer, more complicated, and more rarely used than the unmarked term.[195] Thus, the marked/unmarked distinction finds its foundation in a phenomenological mereology.

HJELMSLEV

A much more ambitious and reflective theory is Louis Hjelmslev's glossematics (partly conceived in cooperation with Hans-Jørgen Uldall). It is probably not very well known that this theory is founded almost unanimously on a mereology. In oppositon to Jakobson, however, this theory's relation to the central European phenomenological mereology is much less clear. Glossematics takes as its point of departure the necessity of basing the cultural sciences taken as a whole on a relation as unanimous as the concept of quantity in natural sciences. This relation is taken to be quality, measured in dependencies. The dependency of one phenomenon on another is taken to be the very basis of the theory in Uldall's *Outline of Glossematics*. In his magistral introduction to the glossematic project, *Prolegomena to a Theory of Language* (Danish *Omkring sprogteoriens grundlæggelse*, 1943),[196] Hjelmslev takes the central object of the theory of language to be the sign, which he analyses as follows:

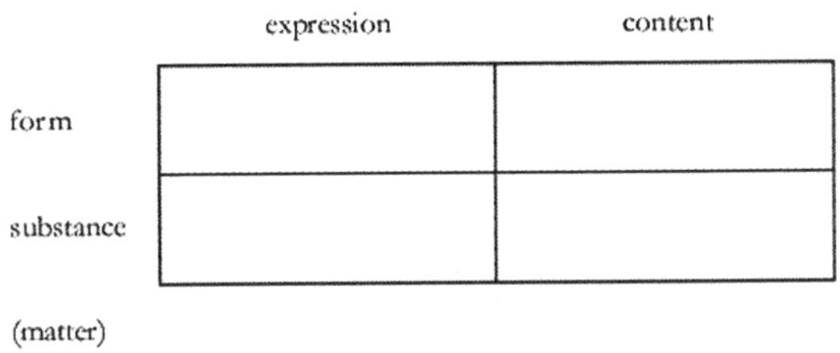

Figure 23.

The idea, now, is that the central object of the study of language as a system is the two boxes of form. Form of expression and form of content, respectively, are the two areas which may be grasped by glossematics, and even the sign, correlating units from those two domains, does not belong to the linguistic system, but to the use only. This implies that not only the matter of expression and content, respectively, that is, phonetic matter and the phenomenological world referred to is left out of scope, but also expression and content in so far as they are substances shaped by linguistic form. As the central object left, the respective systems of form of expression and form of content are now made the objects of linguistic analysis. This is pursued by beginning with the discourse as an undivided whole, and analysis now is supposed to partition this object into invariant parts, named functives, registering the internal function relationship between them. Having exhausted this description at a given level, analysis goes on to repeat the procedure as to the internal structure of the elements, and the procedure is supposed to go on until a bottom of 'figurae' are reached on each of the two domains. Thus, Hjelmslev adheres to what Langacker calls the building block metaphor. The first partitioning is supposed to give the two functives expression and content, thereafter follows (e.g.) periods, sentences, morphemes, etc.

Here, the idea of a glossematic algebra of dependencies finds its place. Between two functives on a given level, three so-called functions may be discerned, defined by types of dependency. Three possible dependencies may hold between two functives: dependency, interdependence, and correlation (which is the absence of dependency). Dependency occurs when one part requires another for its presence (but not vice versa), interdependence occurs when two parts mutually require each other and consequently only appears together, while constellation occurs when the occurrence of two parts is free, and both, one, or none of them is equally possible. Hjelmslev now distinguishes between dependencies in the domains of linguistic linearization and system, defined by both-and relations and either-or relations, respectively, which in his vast terminological system are christened selection, solidarity, combination, and specification, complementarity, autonomy, respectively (37). We can illustrate Hjelmslev's idea with an example from the syntactic field. Selection, one-sided dependency, is at stake, for instance, in the relation between main clause and relative clause (a relative clause may not occur without a main clause, while the opposite is not the case). Solidarity, two-sided dependency, occurs for instance at the sentence level between noun phrase and verb phrase, and combination, zero dependency, is found e.g. between two main clauses.

It is striking that Hjelmslev here as the basis for his theory of language takes three mereological types of dependencies very well known in the Brentanist tradition. We find them in Brentano, for instance, and at a prominent place in the 3rd LU where we have already seen the identical distinction between 'gegenseitige', 'einseitige', and no relation, respectively (264–65, cf. above).

There is not, however, any mention in Hjelmslev as to where he is inspired to his triad of dependencies which he merely 'predicts' for purely formal reasons. While his co-founder of the Copenhagen circle and enemy Viggo Brøndal refers to Husserl, just like their common disciple Paul Diderichsen does decades later, there is no mention of any phenomenological inspiration in the *Prolegomena.*[197] At several occasions, Diderichsen remarks upon the complete similarity between the dependence calculi of the *Prolegomena* and *Logical Investigations*, but no further explanation is given.[198] The reasons for this are hard to guess, but three possibilities (at least) are at hand. One is, of course, that Hjelmslev simply came upon the idea of a mereological grammar independently; another is that the absence of references is due to the very radical and autonomy-claiming linguistics he is about to found. Unlike his companion Brøndal, much more Jakobsonian in spirit in his reference to the philosophical tradition and to a multiplicity of sources for his version of structuralism, Hjelmslev wants to free himself from any metaphysics, inspired as he is by logical positivism, especially in Carnap's version. Maybe he saw too much metaphysical heritage in references to the phenomenological tradition? A third possibility is influence via an intermediate (so as for instance Anton Marty;[199] both Jakobson and Brøndal seem unlikely in that role) or from a common source of inspiration (Brentano).

If we go into the history of glossematics in more detail, an even more complicated relation to mereological considerations shows up. In addition to the dependency calculus of the *Prolegomena*, Hjelmslev has a further concern with mereological issues in his calculus of so-called 'concept zones' (on the content side, approximately corresponding to 'semantic domains') and their parts. This idea appears as early as 1933, in the context of the semantic motivation of grammatic categories and in direct discussion with Jakobson's markedness concept (Hjelmslev 1985, 35ff). Jakobson's binarism of course implies that paradigms with three terms must be analyzed as degenerate versions of four-term systems obtained by the combination of two two-term systems. Arguing against binarism, Hjelmslev proposes – probably with inspiration from Brøndal – a tripartition instead of a bipartition of the zone of a given conceptual substance, yielding two opposed parts with a neutral domain between them. (To see which use Hjelmslev makes of these ideas, let us mention his analysis of grammatical numerus which is seen as founded on the concept zone of discrete versus compact, including a neutral zone between them. Hjelmslev 1972, 94f). Interestingly, Hjelmslev calls this calculus 'sublogical'; it is inspired by Lévy-Bruhl's idea of 'participation' in primitive thought where opposed terms may share content. Thus, formal logic is supposed to be but one possible derivate from this sublogical basis, an idea not unrelated to Husserl's idea of a phenomenological foundation of logic (Husserl 1985).

Such a three-part zone now may be occupied by different terms, defined by placing each their emphasis on differing combinations of the three parts of the zone, this emphasis indicated by a slash in the relevant part(s) of the concept zone:

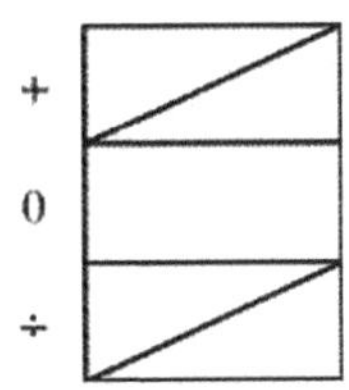

Figure 24.

So a term of a paradigm is now seen as a specific combination of the three parts of the paradigm's concept zone. Thus, Jakobson's markedness-unmarkedness distinction corresponds to two of these possibilities only, namely the term occupying only one of the opposed terms (markedness), together with the term indistinctly occupying the whole concept zone (unmarkedness). Thus, unmarkedness is no longer the mere absence of the marked term, unmarkedness is reconstructed as the vague presence of the whole of the concept zone. But the distribution of the emphasis/non-emphasis over the three-part concept zone yields seven possibilities (with the exclusion of the case without any emphasis at all) instead of two. Later in the development of glossematics, this calculus is further complicated by the introduction of two emphasis degrees (already in 1934, cf. Hjelmslev 1972), and in the elaborated technical presentation of glossematics in the 'Resumé' of the early forties (Hjelmslev 1975), a new set of seven possible emphasis patterns over three-part concept zones is established. These different emphasis terms now combine pairwise to give twelve possible different polar opposites within one and the same concept zone (Hjelmslev 1975, 42), just like systems with from two up to six internally opposed terms appear as a possibility of different emphases of a concept zone (ibid. 31–32). It must be said, though, that the implicit constraints preventing free combination to give an even larger set of possibilities are never made explicit, and neither of two recent and very thorough reconstructions of Hjelmslev (Gregersen 1992; Rasmussen 1992) make clear the nature and role of these constraints. In our context, it must be added, though, that the concept zone calculus in its 'bound' variant is argued to give exactly the three dependency types as a corollary (Hjelmslev 1975, 60). Thus, it might be said that just like in Husserl, an explicitly mereological calculus (the partitioning of the concept zone it into three parts and their possible combinations) results in a dependency calculus – even if the route of derivation is much more labyrinthine in Hjelmslev's case and necessitates further research surpassing the scope of this chapter.[200] As to the influence question, there seems to be a thin thread leading from the third investigation via Jakobson's markedness concept to Hjelmslev's complicated three-value markedness calculi and further on to his three dependency types – but without any explanation as to how exactly the same dependency calculus appears at each end of the thread.

Anyway, the radical purism in Hjelmslev's dependency calculus as well as in his concept zone calculus also have other consequences which is my main reason to bring him into the discussion in this context of actual mereological thought. Hjelmslev's purism namely displays some dangers in a too consequent mereological approach. To see this, take Hjelmslev's mereological treatment of linguistic tradition. The whole inventory of morphology and syntax, of distinctions between syllables, words, flexions, sentences, hypotaxis, parataxis, etc. must be given up completely in favor of a purely mereological description. We are not supposed to distinguish preposition and government, e.g. in any other way than by knowing that one selects the other, the whole complex of the two again being independent of the clause as a whole on sentence level. The same goes for semantics where the shortcomings of the theory were most easily felt; the consequent mereological approach prohibited any phenomenological semantics in so far the elements of meaning were allowed to receive purely arbitrary denominations only.[201] Being functives, they were to be referred to by algebraical letters, and their semantic content was supposed to be read off of their mereological dependencies only.[202] In semantics, the theory restricted itself to banalities such as that the meaning of 'bull' was dependent on the meanings of 'ox' and 'male', respectively.

What is to be learned from the partial failure of Hjelmslev's grand mereological project? The set of restrictions the theory admits deliberately cuts it off from possible insights, first in the letting out of sight language's reference to any context ('matter'), and second in its dogmatic decision that any relation between expression and content is arbitrary merely. Both these ideas have been excellently attacked by the cognitive semantics tradition.[203] The main implication in this context, however, to be drawn concerns the consequences of the idea of a purely mereological dependency calculus used as a descriptive metalanguage. Such a calculus so to speak conceives its object from outside, sees it constructed by discrete building-blocks holding a highly restricted set of dependencies between them. To describe language and other semiotic phenomena (which is, implicitly, the ambition of glossematics, the matter of expression being of secondary importance) it is necessary not to delimit oneself beforehand to one selected calculus (even if it is a fertile one) of description. Moreover, Hjelmslev's use of it repeats some of the Berlin School's less lucky consequences without gaining its advantages: Hjelmslev ceases to use traditional linguistic terminology, so that for instance morphology versus syntax should be mereologically reinterpreted. But doing so, the Graz School advantage of being able to distinguish a part (a word's morphology) from the role played (in sentence syntax) in a gestalt, is lost.

Another drawback is the extremely discontinuous character of the calculus, given by the definition of the dependency calculus to hold between well-defined units of a lower level (which is not a necessary implication by dependency calculi). All continuous phenomena in signification are a priori bracketed by the choice of so restricted a metalanguage (on the linguistic expression side intonation, prosodic features, gestures; on the content side the whole question of continuous schemata and their (continuous) eidetic variation in semantic description.)

Finally, a drawback is a fact which Barry Smith has often referred to: mereology's explicit and admitted weakness. In Lesniewski's version, this was even picked as a special privilege of the theory, making it independent of ontological assumptions. The flip side of this is that mereology's weakness makes it unfit to describe most empirical wholes without further formal equipment. In linguistics, this further equipment is most often tacitly presupposed – in for instance the idea of co-existence of terms in a sentence. But dependence relations do not in any way imply the existence of parts in the same place. A whole consisting of the tone A, my left shoe and the contour of England is perfectly admissible. Contiguous wholes thus require at least additional topologies for their description, making it possible to distinguish connected and non-connected wholes – and they may require metrics, spatio-temporal embedding and much more in order to map further properties of interest; this goes for objects in general as well as for objects of semiotics specifically. Mereology and its dependency calculi do remain a very important formal part of semiotics, but we have no reason to assume that they exhaust the formalisms necessary, just like a considerable work in formal ontology will be required to yield a more refined taxonomy of gestalt types.

In Husserl, mereology forms a crucial part of formal ontology, relevant for any ontological domain whatsoever (but not necessarily with any claim for being exhaustive for formal ontology), and it subsequently plays a central role in his pure grammar. In Hjelmslev, exactly the same calculus is supposed to be the formalism relevant to describe all possible formal properties in semiotic systems as a presuppositionless metalanguage deprived of any of the ontology-like ambitions of phenomenology. Yet, Husserl's refusal that 'vague morphologies' may be formalized may approach him to Hjelmslev, and their strong reliance on dependency calculi may probably have its reason in their reliance on the former's weak, methodological reliance upon logic, and the latter's strong methodological reliance upon language, respectively, both being discontinuous systems. Jakobson, on the other hand, makes a less theoretical use of mereology, but his theory points towards a pragmatical grounding of mereology in prelinguistic, phenomenological perception of wholes with atypical parts – generalizing an experience of discovery: the sudden appearance of a new, strange phenomenon within the bounds of the supposedly well-known.

AS A WHOLE

Consequently, a much more liberal stance must be taken with respect to which forms may count as significant. It comes as no surprise when we claim that an outline may be found in Peirce's diagram concept. But it is noteworthy that the diagram is connected to mereology in two ways.

The diagram is a stylized picture of its object – and this stylization involves two kinds of abstractions. One is the so-called 'prescission', the second of Peirce's three distinction types, dissociation, prescission, and discrimination, respectively. These separation types, in fact, form the equivalent in Peirce to the dependency calculi we isolated in the other three semioticians (see Chap. 10). Dissociation separates independent objects; prescission separates objects of which one may be supposed to exist separately; discrimination separates objects which may only be represented separately. The precise relation between these distinction types and the dependency calculi of Husserl and Hjelmslev has not been established yet, but the following is a first attempt: dissociation is the distinguishing ability corresponding to constellation in Hjelmslev and independence in Husserl, while prescission seperates a founding part from a founded part and discrimination vice versa (so that interdependent parts may only be discriminated, while unilaterally dependent parts may be prescinded (the independent part) or discriminated (the dependent part)).

When making a diagram, we must prescind it from the particular token drawn on a piece of paper so as to grasp it as an ideal object. Furthermore, the activity we may picture with a diagram by manipulating it may itself be made the object of a higher-order diagram. This is 'hypostatic abstraction' in Peirce's system, to be sharply distinguished from the distinction types: it makes a noun out of a predicate and thus makes it possible further to investigate the properties of this predicate. Prescission thus is a focusing mechanism, leading to the predicative isolation of still more general properties of an object – while abstraction is an objectifying mechanism, making an object of thought out of a predicate, or, as Peirce puts it sloganlike, it makes a thing out of a thought.

Now, the second mereological feature of the diagram is that it – be it pure or applied – forms in itself a whole consisting of interrelated parts. Murphey, in his 1961 classic survey of Peirce's development, writes on Peirce's continuum concept: 'Indeed, the very idea of the continuum as that every part of which is similar to the whole so obliterates the distinctions between class and member, individual and general, concrete and abstract that it is impossible to make the notion precise without running into contradiction' (343). Maybe this was exactly Peirce's intention. The vanishing of the element/subset distinction was a defining feature of Lesniewski's mereology, and it is correct that Peirce's continuism repeats it. For to what does a diagram refer – a continuous field of cases which may not be sorted in elements and subsets – it s herein its generality resides.

The character of the interrelations in the diagram makes certain experiments possible, and these experiments are now interpreted as holding also for the object depicted:

Deduction is that mode of reasoning which examines the state of things asserted in the premisses, forms a diagram of that state of things, perceives in the parts of that diagram relations not explicitly mentioned in the premises, satisfies itself by mental experiments upon the diagram that these relations would always subsist, or at least would do so in a certain proportion of cases, and concludes their necessary, or probable, truth. For example, let the premiss be that there are four marked points upon a line which has neither extremity nor furcation. Then, by means of a diagram,

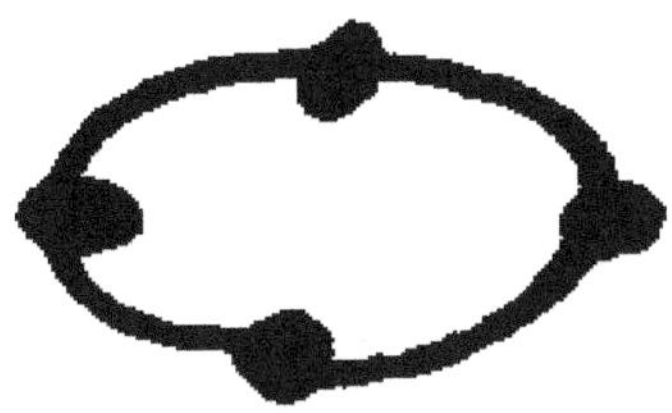

Figure 25.

we may conclude that there are two pairs of points such that in passing along the line in any way from one to the other point of either pair, one point of the second pair will be passed an odd number of times and the other point an even (or zero) number of times. This is deduction. (Untitled manuscript, c. 1896, 1.66)[204]

An important diagrammatical operation is the working together of prescission of abstraction, that is, a diagram property is selected as an object which may itself be subject to rule-governed manipulation (cf. Chap. 11). Abstraction permits the diagram to be recursive and embed one diagram with its whole set of procedures as an object in a more abstract diagram making it possible to investigate the first one. To stay in the map example, we can for instance generalize from the single map and abstract the subject of 'mapness', leading to the question of possible projections giving rise to maps with various properties. In Husserlian slang, this makes possible to investigate still more abstract moments as if they were *Stücke*. In a diagram, thus, mere aspects of the object may appear objectified as autonomous parts. This is, as a matter of fact, a crucial means in diagrammatically making implicit information explicit.

The strength of the diagram category is that it – in contrast to the dependency calculi of the early Husserl and of glossematics – displays an open variety of continuous and discontinuous mereological devices used in the construction of meaning. In any case, namely, a diagram analyses its object into a collection of interrelated parts, the relations of which may be specified in many terms in addition to dependency: connectedness, boundedness, quantity, locality, form, metric relations . . . In short all relations giving possibly rise to necessary reasoning[205] which is a set the extension of which we may not be in a position to determine beforehand. Husserlian mereology may, however, contribute to an aspect of diagrams only marginally hinted at in Peirce: their connection to the issue of the synthetic a priori. Husserl's doctrine of dependency and foundation in the 3rd LU was, in fact, intended as defining the analytic and synthetic a priori anew – as the disciplines of formal and regional ontologies respectively. Both were to be charted by mereological calculi, the former in a pure, the latter in a material version. In the next chapter, we shall attempt to extend Husserl's brilliant insight to the continuous mereology representations in Peirce: the diagrams.

CHAPTER 8

DIAGRAMMATICAL REASONING AND THE SYNTHETIC A PRIORI

It would be a great mistake to suppose that ideal experimentation can be performed without danger of error...

Peirce, 'The Logic of Relatives', *The Monist,* Vol. 7 (1897), 3.528

In the broad tradition of Austrian thought to which Husserl belongs, a conception of the a priori has developed which differs in many respects from the Kantian notion of a priori which has been the target of so much criticism in the twentieth century.[206] Kant's famous idea was that there was a third category alongside mere analytical knowledge ('All bachelors are unmarried') where the predicate was contained in the subject and which was thus a priori true, on the one hand, and synthetic a posteriori knowledge (all kinds of empirical knowledge of states of the world) which depended upon the contingent behavior of the world. Between the two were the 'synthetic a priori judgments' which differed from the latter in being independent of empirical knowledge, and which differed from the former in being not reducible to tautologies. Kant's main examples were arithmetic, geometry, and physics – and the leading question of his whole critical philosophy became, as is well known, 'How is synthetic a priori knowledge possible?' Kant's answer depended as we touched upon in Chaps. 3–4 upon his idea of 'schematism' which may make concept and intuition meet – and thus his account for synthetic a priori judgments was tied to his overall idea that science is made possible by the application of categories and forms of intuition by the subject upon the world which does not, in itself, possess or support such forms. Synthetic a priori knowledge hence was characterized by being infallible – because tied to our own conceptualizations – and subjective – because being the result of the imposition on the world of means of knowledge production inherent in the subject.

The Austrian idea of the a priori is wholly different and far more Aristotelian in spirit. In economics, a tradition running from Carl Menger through Ludwig von Mises to Joseph Schumpeter, Friedrich von Hayek and Murray Rothbard incarnate Austrian apriorism; in philosophy a closely related set of ideas springs from several of Brentano's pupils – Carl Stumpf, Husserl, Adolf Reinach – to be taken up nowadays by the renaissance of that tradition championed inter alia by the 'Manchester gang of three', Kevin Mulligan, Peter Simons, and most explicitly, Barry

Smith. To put it briefly, this idea of the a priori moves its foundation from subject to object. A priori regularities hold, not because of the imposition on the world of some forms or faculties in the subject, but because of recurrent, stable structures of reality itself. And, for the same reason, the a priori does not, as in Kant, refer to anybody's 'judgment' but rather to the object of such judgments: propositions, and, in turn, to those states-of-affairs referred to by propositions. This has the consequence, now, that the a priori covers a much wider range of domains than Kant's notion, centered upon mathematics. Mathematics continues, to be sure, to occupy a central place in the Austrian a priori, but a priori foundations lie, for an Austrian point of view, at the bottom of every single special science by virtue of the fact that every science requires its own stable, interrelated network of basic concepts – providing what Husserl called 'regional ontologies' and Carl Stumpf nicknamed 'Vorwissenschaften' (Stumpf 1907) – proto-sciences. Austrian a priori, thus, comes in packets, where different parts and aspects of the object of a science are tied together in a relational whole which forms the prerequisite to – rather than the result of – empirical research in the area. As Barry Smith remarks, the knowledge of such a priori structures is most often implicit or tacit (cf. the Austrian Michael Polanyi's notion of 'tacit knowledge') – being made explicit, a priori structures often – not always – seem disappointingly trivial. Focal examples discussed in this tradition include claims like 'nothing can be red and green all over'. This observation explains why such a widespread phenomenon can at the same time escape attention by scepticists of all sorts, from positivists claiming nothing exists between analytical and empirical knowledge and to historicists and culturalists claiming the basis of all knowledge varies potentially infinitely in time and space. Finally, as Murray Rothbard and after him Barry Smith have strongly argued, the Austrian notion of a priori is not, unlike the Kantian notion, infallible. As a priori structures are objective, we have the same distance to them as we have to contingent, empirical knowledge – and we may make huge and long-lasting mistakes in grasping those structures, as has in fact been the case in many cases in that most central a priori domain of mathematics. Thus, Austrian a priori may give rise to 'fallibilistic apriorism'[207] where the a priori field is vastly extended in comparison to the Kantian version, but where our access to it is subjected to constraints parallel to those on empirical knowledge. Even if often tacitly presupposed in both common sense reasoning and scientific investigations, it demands a huge effort to make explicit and coherent our knowledge of the a priori – and, what is more, such work may end up contradicting and revising parts of our more or less implicit assumptions. In what follows we shall intermittently use Wojciech Zelaniec's abbreviation of the synthetic a priori: 'synap'.

In the Brentano tradition, the central articulation of the Austrian conception of the synthetic a priori takes place in Husserl, more precisely in the 3rd LU. The mereological calculus presented here forms the very basis of Husserl's account of synap structures – because such structures have the character of consisting of relations of necessity between parts and wholes. As we discussed in the preceding chapter, Husserl's idea was to chart such relations using three different relations of dependency. Real instantiations of such relations simply form the flesh of his idea of the a priori. The mereological distinction between abstract and concrete (dependent/independent, inauthentic/authentic) parts forms a nucleus of formal ontology holding for every conceivable object, alongside with subject/quality, individual/species, genus/species, relation/collection, entity, number, series, size, etc. The very existence of a dependent part – 'Moment' – implies a law of essence presupposing the existence of certain pure species of which it forms a part and on which it is thus dependent (§7). It is important, though, not to conflate this distinction with the distinction between exact and vague phenomena – which is orthogonal to the distinction between genuine parts and moments (§9). Any dependent part presupposes an ideal law describing its relations of dependency to other specified parts and wholes – and vice versa: if any part participates in an ideal law, it is dependent (§10). Such laws are motivated by the specificity of the domain in question and are expressed in material concepts – to distinguish them from formal ontological concepts as the above, referring to the materially empty idea of an Object as such (§11). This distinction between formal and material is, in Husserl's doctrine, the basis for the genuine distinction between the analytical a priori and the synthetic a priori. Here, a crucial difference to the Kantian tradition appears: Kant identified the distinction between a priori and empirical with form and matter, respectively, making form the a priori contribution of the subject. Here, matter concerns the 'sachhaltige' concepts – while the formal is identified with the general concepts holding for all possible objects. This does not imply, however, that the material/synap regularities can not be *formalized*, that is, expressed in formal languages or other generalized forms of representation. Husserl attempts to outline a series of pure mereological laws in the formal ontology of parts and wholes (§14) which he foresees should be further developed in mathematical research – and it is important to see that this formal ontological endeavor runs in parallel with what he expects to take place in the different material domains where these laws will be further specified in material concepts and their interrelations. Another way of expressing the dependency of a part on another or on a whole is to say it is *founded* on it or that it is 'completable' ('ergänzungsbedürftig') by it. In the domain of color, for instance, color and extension are in a mutual relation of foundation. The founding part may, in turn, be founded in still another part, etc. Thus, Husserl's reformulation of the a synthetic priori can also be resumed in the laconic statement that 'Alles wahrhaft einigende ist Fundierung.' (§22) – all really uniting is founding – relevant for formal ontology when no material concepts are involved, relevant for materiel or regional ontologies when material concepts are involved.

Husserl's theory of the synthetic a priori plays a crucial role in his overall development: the results of the 3rd LU are put to use immediately thereafter in the fourth to sixth investigations. Thus, in the fourth investigation, a synthetic a priori theory of grammar is outlined,[208] in the fifth the first steps in Husserl's theory of intentionality are taken, seeing the different determinations of intentionality as founded aspects in the sense given and thus constituting an a priori account of conscious acts, and in the sixth investigation epistemological issues are treated on that basis, as we saw in Chap. 6. Thus, Husserl's whole phenomenology is in a certain sense a synthetic a priori theory of the region of pure consciousness and its objects, and the different idealist or realist interpretations of it can, in fact, be rephrased as discussions of which parts of intentionality are founded on which other parts: is the object of intention an independent (realism) or a dependent (idealism) part of the act?

Later, in the *Ideen* III, Husserl outlines a taxonomy of different regions of reality: material thing, body, and soul, respectively, but it is not necessarily the implication that each synthetic a priori complex of these three large domains could be immediately constructed 'from above'. Rather, the material ontologies or regional ontologies seem to have to be built 'from below' so to speak, from salient object complexes treated in the different domains and in close connection with the empirical scholarly disciplines investigating them. Thus, some of the most successful results of these ideas among Husserl's immediate pupils cover much more restricted domains. Prime examples are Adolf Reinach's legal phenomenology of social acts which has gained a reputation as the first thoroughly developed speech act theory, with 'social acts' as occurring in court as its central object,[209] and Roman Ingarden's theory of literature built on a phenomenological theory of language (see Chap. 17) – both of them selecting nested parts of language as their domain for synthetic a priori mappings of regional ontology.

The basic idea in Reinach, however, remains that legal speech acts are constrained by the synthetic a priori structure of social actions – so that the edifice of social institutions is possible only because they conform to such a priori regularities. As Barry Smith wittily summarizes: 'Thus, there could be no culture or society in which the validity of contracts is in general made conditional on the parties' undergoing painful surgery, or on the finding of a proof for some hitherto unproved conjecture in mathematics' (Smith 1992, 6). Such facts are to a large extent immediately graspable by people with only a common sense understanding of legal matters. Such regional ontological regularities, moreover, generally come in packets. Take the institution of *promising*. It involves at least two parties, one giving the promise and another taking it, it involves an obligation, it involves a time scale within which the promise is to be fulfilled, it involves a type of action being the object of the promise, it involves the possibility of a sanction if the promise is not kept.[210] This whole complex forms a sort of synthetic a priori molecule which can not be disrupted in different cultural or historical circumstances – even if it may, of course, be instantiated and shaped in such circumstances in a high variety of ways and different connections to other facts and social acts. Furthermore, the concepts involved in promising: responsible parties, trust, obligation, action, time limit, sanction, etc. refer further on to other basic concepts in the synthetic a priori networks of social acts. Thus, the Austrian conception of the synthetic a priori is resistant to all kinds of psychologisms, historicisms and culturalisms and the unavoidable skepticism connected to them – but it is also, on the other hand, resistant to a positivist interpretation in terms of logic and fact alone. Such regularities involve – again Smith – 'the ineliminable presence of materially specific concepts' (7) in their formulation. So they may not be expressed by purely logical concepts and relations only. Rather, they possess the character of constituting whole networks of a priorially connected material (and formal) concepts. Even in a domain as far from the natural sciences as that between law, social science and linguistics which occupies Reinach, such laws can not be mere conventions only. Smith argues against Searle's idea that such institutions are built from human conventions, by pointing to an infinite regress: '*the very practices involved in formulating and adopting conventional rules presuppose universal categories of the given sorts*'(9) – the very possibility of *making* a convention in the first place is the result of a certain type of contractual speech act which cannot, then, itself be conventional.

Another, closely related tradition is Austrian economics where the a priori stance has dominated a tradition running from Carl Menger to Murray Rothbard.[211] The idea here is that the basic structures of economy is entrenched in the more basic structures of human action as such. Nowhere is this more evident than in the work of Ludwig von Mises – his so-called 'praxeology' is built from the idea that all basic laws of economy can be derived aprioricallly from one premiss: the concept of human action. Of course, this concept is no simple concept, containing within itself purpose, product, time, scarcity of resources, etc. – so it would be more fair to say that economics lies as the implication of the basic *schema* of human action as such.

Even if the Austrian economists' conception of the a priori is decidedly objectivist and anti-subjectivist, it is important to remark their insistence on subjectivity *within* their ontological domain. Thus, the Austrian economics tradition is famous exactly for their emphasis on the role of subjectivity in economy. From Carl Menger onwards, they protest against the mainstream economical assumption that the economic agent in the market is fully rational, knows his own preferences in detail, has constant preferences over time, has access to all prices for a given commodity at a given moment, etc. Thus, von Mises' famous criticism of socialist planned economy is built on this idea: the system of ever-changing prices in the market constitutes a dispersed knowledge about the conditions of resource allocation which is a priori impossible for any single agent – let alone, any central planner's office – to possess. Thus, their conception of the objective a priori laws of the economic domain perhaps surprisingly had the implication that they warned against a too objectivist conception of economy not taking into account the limits of economic rationality stemming from the general limitations of the capacities of real subjects. Their ensuing liberalism is thus built on a priori conclusions about the relative unpredictability of economics founded on the role played by subjective intentionality. For the same reason, Hayek ended up with a distinction between simple and complex processes, respectively, cutting across all empirical disciplines, where only the former permit precise, predictive, quantitative calculi based on mathemathical modeling while the latter permit only recognition of patterns (which may also be mathematically modeled, to be sure, but without quantitative predictability). It is of paramount importance, though, to distinguish this emphasis on the ineradicable role of subjectivity in certain regional domains from Kantian-like ideas about the foundational role of subjectivity in the construction of knowledge as such. The Austrians are as much subjectivists in the former respect as they are objectivists in the latter. In the history of economics, the Austrians occupy a middle position, being against historicism on the one hand as well as against positivism on the other. Against the former, they insist that a priori structures of economy transgress history which does not possess the power to form institutions at random but only as constrained by a priori structures. And against the latter, they insist that the mere accumulation of empirical data subject to induction will never in itself give rise to the formation of theoretical insights. Structures of intelligible concepts are in all cases necessary for any understanding of empirical regularities – in so far, the Austrian a priori approach is tantamount to a non-skepticist version of the doctrine of 'theory-ladenness' of observations.

A late descendant of the Austrian tradition after its emigration to the Anglo-Saxon world (von Mises, Hayek, and Schumpeter were such emigrés) was the anarcho-liberal economist Murray Rothbard, and it is the inspiration from him which allows Barry Smith to articulate the principles underlying the Austrians as 'fallibilistic apriorism'. Rothbard characterizes in a brief paper from 1957 what he calls 'Extreme Apriorism' as follows: there are two basic differences between the positivists' model science of physics on the one hand, and sciences dealing with human actions on the other: the former permits experimental verification of consequences of hypotheses, which the latter do not (or, only to a limited degree, we may add); the former admits of no possibility of testing the *premisses* of hypotheses (like: what *is* gravity?), while the latter permits a rational investigation of the premisses of hypotheses (like: what *is* human action?). This state of affairs makes it possible for economics to derive its basic laws with absolute – a priori – certainty: in addition to the fundamental axiom – the existence of human action – only two empirical postulates are needed: '(1) the most fundamental variety of resources, both natural and human. From this follows directly the division of labor, the market, etc.; (2) less important, that *leisure is a consumer good*' (316). On this basis, it may e.g. be inferred, 'that every firm aims always at maximizing its *psychic* profit' (317). Rothbard draws forth this example so as to counterargue traditional economists who will claim that the following proposition could be added as a corollary: 'that every firm aims always at maximizing its money profit'. This can not be inferred and is, according to Rothbard, an economical prejudice – the manager may, e.g. prefer for nepotistic reasons to employ his stupid brother even if that decreases the firm's financial profit possibilities. This is an example of how the Austrians refute the basic premiss of absolute rationality in terms of maximal profit seeking. Given this basis, other immediate implications are: 'the means-ends relationship, the time-structure of production, time-preference, the law of diminishing marginal utility, the law of optimum returns, etc' (318). Rothbard quotes Mises for seeing the fundamental Axiom as a 'Law of Thought' – while he himself sees this as a much too Kantian way of expressing it, he prefers instead the simple Aristotelian/Thomist idea of a 'Law of Reality' (ibid.). Rothbard furthermore insists that this doctrine is not inherently political – in order to attain the Austrians' average liberalist political orientation, the preference for certain types of ends must be added to the a priori theory (such as the preference for life over death, abundance over poverty, etc.). This also displays the radicality of the Austrian approach: nothing is assumed about the content of human ends – this is why they will never subscribe to theories about Man as economically rational agent or Man as necessarily economical egotist. All different ends meet and compete on the market – including both desire for profit in one end and idealist, utopian, or altruist goals in the other. The principal interest, in our context, in these features of economical theory is the high degree of awareness of the difference between the – extreme – synthetic a priori theory developed, on the one hand, and its incarnation in concrete empirical cases and their limiting conditions on the other.

Barry Smith, in 'In Defense of Extreme (Fallibilistic) Apriorism' (1996b) directly plays on Rothbard's title in his articulation of his position. The article takes the form of a skiing slalom where a question is posed and a position rejected at every turn of the trajectory, thus gradually narrowing the field, only to end in the conclusion that synthetic a priori knowledge exists – as the final summary of the paper goes:[212]

Do the empirical theories with the help of which we seek to approximate a good or true picture of reality function without any non-empirical presuppositions?

Yes: Extreme empiricists

No: *Are the propositions which express these pre-empirical assumptions in every case analytic (tautological, lacking in content)?*

Yes: Logical positivists

No: *Do we have an infallible knowledge of all the synthetic pre-empirical propositions which are presupposed by the various sciences in the different phases of their development?*

Yes: Extreme Cartesians

No: *Could these assumptions, which are presupposed by the empirical sciences, be arbitrary?*

Yes: Feyerabend

No: *The propositions in question must therefore be characterized by a certain plausibility. Is this plausibility always a contextual affair?*

Yes: Hermeneutic relativists

No: *There is therefore something like an intrinsic plausibility. Are the intrinsically plausible pre-empirical synthetic propositions which play an indispensable role in the sciences given only individually, so that we have only a few isolated examples thereof between which no systematic relations would obtain?*

Yes: Ad hoc Kantianism

No: *Is it really true, as the Kantians assert, that the intrinsically plausible or intelligible pre-empirical synthetic propositions here at issue are read into or imposed upon the world by us?*

Yes: systematic Kantianism

No: *Might the intrinsically plausible pre-empirical synthetic propositions all be false?*

Yes: Epistemology Nihilism

No: *Certain pre-empirical synthetic intrinsically plausible propositions thus require ontological correlates which are their truth-makers: there are intelligible structures in the world, which we could also call 'a priori structures'.* (189–90)

Of course, *some* of our conceptions regarding such structures may, at any moment, be false – herein fallibilism. But just like the case in the Peirce-Popperian version of fallibilism with respect to empirical knowledge, *not all* such conceptions can be wrong at the same time, and we have fair reason to believe that the amount of correct synthetic a priori knowledge in fact increases during the history of science. In that sense, fallibilistic aprorism is not without analogies to certain brands of Neo-Kantianism attempting to historicize the a priori, seeing it as dependent on our best scientific theories at any moment – so as e.g. Cassirer's position. All depends, of course, on the nature of this historical dependency – if the claim goes that validity of a priori positions is *per se* determined by history, then it amounts to nothing more than historicizing the transcendental subject imposing its changing forms on a formless unknowable reality, 'hermeneutic relativism' will be the result, and nothing is gained. If, on the other hand, the claim is that history permits science and philosophy to grasp ever more structural aspects of reality, then the position is close to the one discussed here. Cassirer's position is especially ambiguous as to this issue; in some expressions, he seems very close to the realist position, and in any case, some of his results may easily be reinterpreted in an Austrian framework, cf. for instance his famous 1945 paper on the history of the emergence of structuralism from Goethe via Maxwell and Gestalt Theory to Jakobson which may easily be read as a charting of the successive discovery of a structural a priori.

The American philosopher Michael Friedman in fact makes a case for the ineradicability of the a priori along such Neo-Kantian lines in his *Dynamics of Reason*. Let us compare his argument to the Austrian tradition. In his *Parting of the Ways,* Friedman compared Carnap, Cassirer, and Heidegger as three different roads taken in modern philosophy from a common Neo-Kantian point of departure – based on the empirical fact that all three of them were present at the Davoser meeting in 1929, so to speak the last summit meeting before twentieth century philosophy split in differing traditions. Comparing the analytical and continental excesses of Carnap and Heidegger, respectively, Friedman ends up proposing to continue the middle road of Cassirer, then made impossible by World War 2 and the split tradition in philosophy. This is, in fact, what he himself attempts in *Dynamics of Reason* – not to reconstrue Cassirer's own position, but on the basis of a discussion of an intertwined history of philosophy and of science through the nineteenth and twenteeith century to construe a position Cassirerian in spirit. His point of departure is the analytical tradition from Carnap to Kuhn and the different conceptions of 'stratification of knowledge' at stake here. Carnap's well-known distinction between L-rules and P-rules in the epistemology of physics – logical rules versus physical rules – is seen as a forerunner of Kuhn whose project was enthusiastically greeted precisely by Carnap. The physical rules are internal and synthetic as opposed to the logical rules being external and analytical, and Carnap's idea is that when problems may not be solved internally, physically, it may be time for a revision of the L-rules (which, in the spirit of logicism, also include mathematics) (32). They thus constitute *a relativized and constitutive a priori* (40–41) and are thus comparable to Kuhn's notion of a scientific paradigm. Friedman now argues for two indispensable addenda to this picture. One is an intermediate layer between the supposedly purely empirical and logical layers in Carnap's brand of positivism.[213] Guided by history of science, Friedman points to a necessary even if overlooked level of 'coordination principles' guiding the application of logic and mathematics to empirical data,[214] principles as for instance the equivalence between accelerated movement and movement in a gravitation field in general relativity, or the principle of absolute light velocity in special relativity. This layer of coordination principles is presumably part of the 'relative a priori'; it may, evidently, change during the history of science. In an almost Peircean wording, Friedman emphasizes that these coordination principles contain 'real possibilities' (84), and he even stresses the basic idea that they must possess 'counterparts in reality' (87). Here, a decidedly Neo-Kantian framework like Friedman's seem to come very close to an Austrian idea of the a priori – from a completely different point of departure. Having thus supplemented the Carnap-Kuhn conception with a realist addendum, Friedman adds another important idea: when comparing different possible paradigms in a Kuhnian revolutionary situation, the scientist is not only faced with a crude pragmatism ordering him to chose the (unnegotiable) paradigm which explains most. He also has to take advice from what Friedman calls 'philosophical meta-paradigms' (44). Also in this idea, Friedman is informed by history of science – pointing to the fact that major and well-described breakthroughs in physics were most often extremely well-informed in the philosophical debate of the time, so e.g. Newton and his background in the metaphysical discussions between Descartes, Gassendi, Huyghens, Spinoza, and Leibniz; or Einstein being well-versed in the discussions on the foundations of geometry in Helmholtz and Poincaré (44–45). Philosophy does not supply direct *answers*, but it supplies possible 'meta-frameworks' and thus forms a source for 'suggestions and guidance'. In this reformulation of the Carnap-Kuhn two-tier model into a four-tier model, Friedman by the same token relativizes the radicality of paradigm shifts. For such transformations must, as Cassirer insists (65) 'leave a determinate stock of principles unaffected'. When the two-tier model with its direct theoretical determination of empirical understanding is given up, we may imagine different kinds of paradigm shifts with changing amounts of continuity comprising selections of empirical regularities, certain coordination principles, or mathematical models. New coordination principles could evolve continuously out of the earlier ones, and their transformation could be guided by a changing philosophical meta-framework. Friedman consequently proposes that the a priori realm (now consisting of synthetic 'coordination principles' and (more or less) analytical logico-mathematical structures) ceases to be constitutive in the trans-historical meaning of the word (but remains constitutive in the indispensability use of the word) – giving way to a regulative use of reason. Such a conception of the constitutive a priori, hence, rejects its unrevisability and its certainty – in short it becomes fallibilist, to use Smith's vocabulary. Thus it is interesting to take note of a certain convergence between leading followers of the Austrian paradigm and of sophisticated defenders of the Neo-Kantian paradigm – both of them committed to the role played by a priori assumptions in actual history of science.[215]

But which kinds of propositions does the synthetic a priori area contain, and what is the basic property shared by such propositions? The Polish philosopher Wojciech Zelaniec has undertaken a so to speak empirical piece of botanization among simple examples of 'synap' sentences, as he calls them, in order to explain their seeming obviousness. Thus, he does not seek any foundational definition of the synthetic a priori nor of the analytic, and his investigation does not depend on that – rather it is a contrastive investigation comparing synap sentences with other sentences with related properties: tautologies, sentences which may be transformed into tautologies, highly plausible empirical sentences, seemingly obvious sentences allowing for imagination of counterexamples, seemingly obvious sentences allowing for the conception – not the imagination – of counterexamples, etc. He begins by listing a series of central synap examples taken from philosophical literature which is worth repeating here:

1. every color is extended (Kant, Berkeley, Hume, Husserl, Stumpf)
2. for every two events, if one of them is later than the other, the other is not later than the first one (Pap)
3. if something is beautiful and real, then it is good (Roth)
4. everything red is colored (Chisholm)
5. every three tones are ordered linearly with regard to their pitch (Roth, Husserl, Stumpf)
6. the pleasant is preferable to the unpleasant (Scheler)
7. no surface, if it is red all over, is at the same time green all over (Schlick, Wittgenstein, Russell, Ayer, Pap, (Aristotle))
8. everything that is square has a shape (Chisholm)
9. only good actions can be the object of a duty (Scheler)
10. man acts (Hoppe, von Mises, (Aristotle))
11. if any tone-quality is eliminated, a tone-intensity will also be eliminated (Husserl)
12. every promise gives rise to – mutually correlated – claim and obligation (Reinach, Lipps)
13. pink is more like red than black (Austin, similar examples in Locke, Hume, Reinach, Hering)
14. every judgment comprises a presentation within itself (Stegmüller, (Brentano))

(Zelaniec has extensive references to the authors merely indicated here, 19–21; those put in brackets do not directly quote it as a synap example but have a related idea)

As is evident, Zelaniec's examples differ from Friedman's examples which much more have the status of basic ontological hypotheses in the research process. Zelaniec's synap sentences deal with many different domains but they share a basic common-sense quality – and Zelaniec's idea is precisely to investigate the reason in 'normal people' for assuming such sentences to be synap. Among 'normal people', he excepts people believing only tautologies are obvious as well as people with special information of the domain in question – and then he undertakes a comparison with a set of other seemingly obvious sentence types, as mentioned. His overall result (44) is that synap sentences are seen as obvious because they do not admit counterexamples: '... it is the obviousness of sentences whose counterinstance cannot be conceived.' The sentence type closest to this is sentences whose counterinstance cannot be *imagined*, yet can be conceived (sentences dealing with hectagonal coins and curved space are the examples given here).[216] What Zelaniec claims to have achieved is of course not in any way *defining* the synthetic a priori – merely to establish its existence by means of the listing of prototypical examples, and describe the reason for the conception of them as being obvious. Yet, there is not necessarily any tension between Zelaniec and those accounts of synap sentences which focus more upon their status in the sciences. As philosophical examples, the sentences picked are intended as being prototypical and easy-to-grasp illustrations of the synap category, and there is no reason not to assume that much more complicated and difficult-to-decide examples such as Friedman's 'coordination principles' may be found farther from the center of the category, so to speak. What is interesting in our context, however, is Zelaniec's central result of 'counterinstance cannot be conceived'. In this is implied that when trying to make sense of a synap sentence, one is urged to try to find a counterexample, but fails in this endeavor, and, furthermore, sees that one *must* so fail. In the easy cases, this insight is immediate, but we can easily imagine cases when that insight requires centuries of thought and experiment to be realized.

This brings us to the so to speak 'epistemological' issue connected to the synthetic a priori. By which means is our access to such structures made possible? Smith says little about it – except for pointing to such knowledge often being tacit or implicit and hence requiring great efforts to be made explicit. But which efforts? Husserl, as is well known, has a surprisingly simple answer which we have already discussed, namely his well-known doctrine of 'Wesenserschauung' or 'eidetic variation'. Already in the *Logische Untersuchungen* there are traces of this doctrine in the idea of the Prolegomena that concepts in theories can be replaced with variables, thus leaving the general formal skeleton of the theory behind which may then be varied in order to obtain a 'theory of theories'. In the first chapter of the *Ideen*, he returns to the issue in the discussion of 'sciences of essence' (Wesenswissenschaften), including both formal (logic, mathematics, etc.) and material ontologies. He places great emphasis on the claim that such grasping of essence is indeed a form of intuition ('Anschauung') analogous to empirical perception (14), so as to distinguish it from induction:

> The Eidos, the *pure essence*, may exemplify itself intuitively in givennesses in experience such as perception, memory, etc., but just as well *also in pure fantasy givennesses*. Thus we may grasp an essence *originarily* by taking our point of departure in corresponding intuitions, *but just as well also in intuitions not experienced, not involving being, but rather 'merely imagined' intuitions.*
>
> If we in free fantasy create some spatial figures, melodies, social processes, etc., or if we pretend to have experience acts, of pleasure or unpleasure, will, etc., so can we also there, through 'ideation', originarily intuit many pure essences and maybe even in an adequate way: be it the essences of spatial figure, of melody, of social process, etc. *in general*, be it the relevant special *type* of figure, melody, etc.[217]

This process of 'ideation' takes place in fantasy, and it takes place '... auf Grund einer freien Fiktion und Variation von einem solchem Dinge', as he says a little later about the physical object as example.

This grasping of essences thus takes place by what in *Erfahrung und Urteil* is called 'eidetic variation' – we vary in fantasy outline the object investigated until we find the ontological limits it cannot possibly transgress without ceasing to be the kind of object it is. When he describes the process in *Erfahrung und Urteil*, more detail is added. The process, also called 'ideation', is supposed to take place in three main steps:

1. The productive activity which consists in running through the multiplicity of variations.
2. The unitary linking in continuous coincidence.
3. The active identification which brings out the congruent over against the differences.
   (Husserl 1973a, 347)[218]

The process is supposed to differ from that of forming merely empirical generals, which are constructed by a method close to that often described as induction: a series of repeated, similar experienced objects give rise to an expectation of further real possible instantiations.[219] Thus, the extension of such empirical concepts is open, but their unity is purely accidental. Otherwise with 'pure generals' ('reinen Allgemeinheiten') which are reached by the three steps above. In that process, the point of departure is one conceived object – 'eine Erfahrungsgegebenheit' – which is then subjected to variation, not in real examples, but in a controlled process of imagination, of fantasy:

It is based on the modification of an experienced or imagined objectivity, turning it into an arbitrary example which, at the same time, receives the character of a guiding 'model', a point of departure for the production of an infinitely open multiplicity of variants. It is based, therefore, on a *variation*. In other words, for its modification in pure imagination, we let ourselves be guided by the fact taken as a model. (ibid. 340)[220]

This process of variation gives rise to the grasping of the unity of the concept in what remains *invariant* in the variation process as the necessary and general form without which the type of object would be unthinkable. In this 'Ideenschau', the general essence is intuited directly. This idea obviously entails problems: the variation process is open and endless – so we should not expect the variation process to cover all possible instantiations given in their individuality – a sort of problem akin to finitism in mathematics: we cannot expect the mind actually to perform an infinity of operations. Husserl of course realizes this problem: this '...does not imply that an *actual* continuation to infinity is required, an actual production of all the variants' (ibid. 342).[221] Rather, the very variation process itself is subject to an act of imagination: 'On the contrary, what matters is that the variation as a process of the formation of variants should itself have a *structure of arbitrariness*, that the process should be accomplished in the consciousness of an arbitrary development of variants' (ibid. 342).[222] The termination of the imaginary variation process is imagined only, in a sort of second-order imagination: 'This remarkable and truly important consciousness of 'and so on, at my pleasure' belongs essentially to every multiplicity of variations. Only in this way is given what we call an 'infinitely open' multiplicity; obviously, it is the same whether we proceed according to a long process, producing or drawing arbitrarily on anything suitable...' (ibid. 342).[223] This second-order imagination makes an object out of the primary imaginary variation, synthesizing it in imagining it finished. I think this is what Hintikka sees when he states that the '... *Wesensschau* accomplishes two different things. It separates the essence of an empirical object from its hyle and it makes this essence an object of intuition' (2003, 183). Separation and object-making are analogous to Peircean prescission and hypostatic abstraction, respectively (cf. Chap. 11).

It is not exactly clear which criteria Husserl imagines when he talks about the formation of variations until a suitable ('Passende') point where the unity is evident. Maybe we should see here a germ of fallibilism in Husserl's description: when does one know the variation has been undertaken to a sufficient degree? It is easy to imagine cases where a certain variation seems to display smooth growth, for instance, but where further variation, if not broken off prematurely, would give rise to discontinuous surprises (cf. geometrical singularities in the grasping of a manifold). This does not, evidently, constitute any kind of obstacle to the process – but it seems to account for the fallibility of the process in cases only to be discovered later. In fact, it constitutes a sort of equivalent to the 'halting problem' in computer science: you'll never know, in each single case, when a sufficient amount of semantic phase space has been covered by your variation process, and there is no shortcut permitting to decide it beforehand; the most effective search algorithm in semantic space is the very variation process itself.

Now, Husserl imagines that the whole of this manifold gives rise, at a higher level, to the '*eigentliche Erschauung des Allgemeinen als Eidos*' (413) – the true seeing of the universal as eidos – because the single images are made to cover each other so that difference in details are seen as metamorphosis of one and the same Eidos. Husserl's favorite metaphor of covering ('Deckung') can here be compared to Peirce's idea of such a process as a 'composite photograph', a concrete metaphor with closely related implications.[224] The result of this covering process is the emergence of something being itself ('ein Selbiges') which is taken to appear in a 'passive preconstitution', part of Husserl's idea of a 'passive synthesis' as part of the ante-predicative experience. This unity is a 'unity in contradiction', it is no individual but a fluctuating unity of constantly changing, mutually exclusive individuals – but it forms the basis for the intuition of the essence being the constant in the flux. Only then, this 'Selbiges' is taken as object for an active, intutive grasping (414), but it is important that the open manifold of variants is kept in mind, even if the trajectory through it is realized only passively. Thus, the very idea of difference presupposes the unity of the idea, because *'nichts in Widerstreit treten kann, was nichts Gemeinsames* hat.' (418) – nothing may enter into opposition which does not have anything in common.

A very crucial point in the 'Wesenserschauung' is that it must take place in fantasy in order to reach the completion necessary for the grasping of an essence. Contemplation of ever so many single cases will never be sufficient (so to speak Husserl's equivalent to Popper's claim that there is no logic of induction or Peirce's idea that induction is bound to remain probable only) – hence Husserl's constant criticism of inductive theories of abstraction (already in the 2. LU). In the world of fantasy, all regularities of a real, particular experience can be repeated, in 'quasi-mode', and the single cases contemplated must be treated as mere possibilities among others in this '... puren Phantasiewelt, einer Welt *absolut reiner Möglichkeit.*' (424) – this pure fantasy world, a world of absolutely pure possibility. The reason for this primary status of fantasy in variation is only hinted at in Husserl: it is the *continuity* of the flow of variants in fantasy which is impossible in the mere comparison between concrete cases.[225] A corollary to this – exactly like in Peirce – is that the extension of generals covers a continuity of pure possibilities ('reine Möglichkeiten', 429) which do not possess individual specificity (ibid.), so that differences between fantasy worlds can not be measured in either identity nor not-identity.

Just as in the LU, this procedure is taken to cover a priori truths of essence, characterized by their a priori necessity – and the procedure is deemed relevant for any domain of objects as such, from mathematics and the natural sciences and to 'jederlei Gegenstandssphäre' (427). The relevance of this procedure in Husserl is thus enormous: 'Von *jeder* konkreten Wirklichkeit und jedem an ihr wirklich erfahrenen und erfahrbaren Einzelzuge steht der Weg in das Reich idealer oder reiner Möglichkeit und damit in das des apriorisches Denkens offen' (428): from any concrete reality the road into the ideal realm of pure possibilities lies open – and thus into a priori thought. Thus, the eidetic variation as a procedure is quintessential for all science, but it must also be supposed to occur ever so often – albeit less explicit and controlled– in everyday life.

Thus, an implication of eidetic variation is that the grasping of pure concepts necessitates the workings of fantasy at a large scale and in many cognitive processes where we might not usually expect it. This has crucial corollaries for the place of fiction – and hence of literature – in a Husserlian perspective, to which we shall return in more detail in Chaps. 14 and 17.

In Husserl's combined description of the formal and regional ontologies by means of foundation relations on the one hand, and his epistemology of such domains by means of eidetic variation on the other, a strange tension stands out: the ontological relations of necessity in themselves are taken to be *discrete* – they consist of dependency relations between aspects – but the epistemological access to them is *continuous*, it requires the flowing fantasy variations of the ontology investigated. Of course, the continuous search might seem relevant for the detection of discontinuities, much like we fumble our way with the hand in the dark continuously along the wall in order to discover the discontinuity of the light switch. Still, there is a strange mismatch here which we also remarked in the preceding chapter: why not admit continuous dependency relations (of the type: the larger a given spatial extension is, the more color does it have – or any other continuous function between variables)? Husserl himself lists mathematical physics among the a priori regularities (427) without noting that most of physical laws are continuous regularities connecting different aspects of physical processes – thus, in fact, being prototypical instances of such continuous dependency relations.

This 'Austrian' doctrine of the synthetic a priori and our access to it has obvious corollaries in Peirce. My basic idea in this chapter is, as you will guess, that Peirce's concept of diagrammatical reasoning substantially enriches our understanding of the epistemological access to synthetic a priori structures of ontology.

Here lies, though, a major terminological problem to be sorted out. Even if not central in the many skeptical developments of twentieth century philosophy, the vocabulary used by the Austrian tradition is not far from that generally understood by philosophers. Not so with Peirce – for whom we hope to have to some extent prepared the ground in the first part of this book. Peirce is no friend to the a priori, he writes, for instance,

> But, it will be said, you forget the laws which are known to us a priori, the axioms of geometry, the principles of logic, the maxims of causality, and the like. Those are absolutely certain, without exception and exact. To this I reply that it seems to me there is the most positive historic proof that innate truths are particularly uncertain and mixed up with error, and therefore a fortiori not without exception. This historical proof is, of course, not infallible; but it is very strong. Therefore, I ask how do you know that a priori truth is certain, exceptionless, and exact? You cannot know it by reasoning. For that would be subject to uncertainty and inexactitude. Then, it must amount to this that you know it a priori; that is, you take a priori judgments at their own valuation, without criticism or credentials. That is barring the gate of inquiry. (Untitled ms. 1897, 1.144)

Blocking the way of inquiry, as we know, is the worst thing research can do. So Peirce's criticism of the a priori turns on the idea that it should be possible to reach any certain knowledge by means of uncritical inward reflection only. So, it is a variant of the Kantian a priori he attacks. Still, as we have seen in the central PAP discussion of diagrams from 1906, he adopts Kant's notion of 'schema' and sees diagrams as his own contribution to the explanation of access to general objects by observation. The same point is expressed as early as 1893:

> Kant declares that the question of his great work is 'How are synthetical judgments a priori possible?' By a priori he means universal; by synthetical, experiential (i.e., relating to experience, not necessarily derived wholly from experience). The true question for him should have been, 'How are universal propositions relating to experience to be justified?' But let me not be understood to speak with anything less than profound and almost unparalleled admiration for that wonderful achievement, that indispensible stepping-stone of philosophy. (Logic of Quantity, 1893, 4.92)

The problem, according to Peirce, can be rephrased as the question why universal propositions based on experience are possible – covering, of course, Husserl's empirical as well as pure generals. The direct observation of real generals is possible in Peirce who sticks to the Aristotelian maxim that 'Nihil est in intellectu quod non prius fuerit in sensu' ('Harvard Lectures on Pragmatism', 1903, EPII, 126; 5.181). This is interpreted, however, in a not quite Aristotelian and definitely not empiricist fashion because Peirce claims experience, perception, to be the source of such universality, present already in our perceptual judgments with their basis in percepts which are not necessarily directly accessible.[226] This provides the phenomenological foundation of diagrammatical reasoning in Peirce's theory of perception: by the observation of diagrams, universal regularities can be made *directly* observable, given a certain typifying interpretation of those diagrams. But this displays a striking degree of structural analogy to Husserl's doctrine of eidetic variation which is precisely the idea of founding general concepts in perception – although perception

('Anschauung') in a extended sense of the word. In Peirce, there is exactly the same identification between relations of necessity and diagrammatical reasoning as there is in Husserl between necessity and the synthetic a priori structures revealed by ideation. But Peirce's conception of diagrammatical reasoning is considerably wider than is Husserl's eidetic variation – at least taken at face value from his descriptions of the procedure. In Husserl, eidetic variation accounts for the perceptual genesis of pure concepts, succeeded by the search for dependency relations between such concepts – but in Peirce, diagrammatical reasoning accounts both for the genesis of such concepts and the possible relations between them – and for the experimental use of diagrams to discover new, possibly undetected such relations. Peirce's diagrams accounts for eidetic variation – both at the level of the single diagram-concept and on the level of the reasoning performed in it (recall Peirce's continuity insistence that diagram reasoning should make sure the result holds in all similar cases) – but also in the special form of variation seeking completely new – implicit – information by means of diagram manipulation. Jaakko Hintikka has rightly pointed to Peirce's self-declared 'most important discovery' as being the distinction between corollarial and theorematical reasoning, the former easily reading a conclusion off a given diagram, and the latter introducing some entirely new factor in the diagram so as to facilitate a new and unforeseen conclusion.[227] But still, in Peirce the diagram plays the same role in providing the phenomenological foundation of concepts – it is, for instance, possible in Peirce to attempt to make the content of a symbol explicit (or, in any case, *more* explicit) by using the symbol as a recipe to draw a diagram.

Where is the borderline between analytic and synthetic in Peirce, then? It goes without saying that empirical generals – the real possibilities – must belong to the synthetic domain, but where is the precise limit to the analytic? Of course, in Peirce a distinction is possible between pure and applied diagrams – but diagrams' necessary realization in space-time (which they NB share with the variation procedure in Husserl) makes it difficult to imagine any *completely* pure diagram. Take the example of a simple map giving the coastal profile of an island. This map may be seen as a geometrical curve with no further interpretation, in some sense as 'pure' a diagram as they come (here 'pure' to be taken in another sense than in Husserl, rather like 'formal' in Husserl). Still it can be used as a vehicle for diagram experiments – we may establish the curvature of parts of the curve or the area of its interior. A next step could be to see the drawing as a geographical map, that is, investing it with basic contents from geographical regional ontology, so for instance the concepts of sea, coast and inland. These concepts entail certain constraints on the drawing on the map – now all coastlines must form closed curves, so-called Jordan-curves.[228] Geographical ontology does not permit any coast suddenly to stop – unlike, of course, a geometrical line which may virtually end at any chosen point. This ontological interpretation now makes new diagram experiments possible: how much new land may be gained by the drying out of a certain shallow bay, for instance. A next step may add political ontology to the map – maybe the island is divided by several states so that a manifold of borders can be superposed on it – this, in turn, giving rise to new diagram experiments, maybe of a military sort: which borderlines ought to be changed, seen from an possible power's point of view? Still, all these experiments may still be 'pure' in a sense: they are material, they are 'sachhaltige', but they are still general in the sense that we have not yet interpreted the map in terms of references to *this* world, we have only invested it with synthetic a priori concepts stemming from different ontologies. Now we invest the island with the names of Hispaniola and the countries Haiti and the Dominican Republic – only now the diagram map receives an empirical concept structure (but still being general, because the map covers the island in all of the relevant period with constant coastline and borders).

Peirce's identification between diagrams and necessity relations mirrored Husserl's identification between a priori and necessity relations, we said. But there is one more term in Peirce's doctrine: pure diagrams and necessity are, in his doctrine, coextensional also with *mathematics*. But mathematics as such Peirce, as against Kant, tends to categorize as analytic. In an intriguing paper, Stephen Levy (1997) has charted Peirce's discussions of the status of mathematics with respect to the analytic/synthetic distinction (and the relation between mathematics and logic in Peirce where many different countercurrents goes against his 'official' view that logic is dependent on mathematics, not vice versa). Levy quotes Peirce's 'official' view that mathematics is analytic: 'An analytical proposition is a definition or a proposition deducible from definitions ... Mathematical propositions ... are in truth only analytical' (A reply to Carus in *The Monist* (1893), 6.595, quoted in Levy 1997, 194). This is as clear as it could possibly be, one thinks, but later Peirce has second thoughts. Levy quotes a refutation of Kant's idea that $7+5=12$ is synthetic – Peirce argues that this equation follows from the definitions of arithmetic via a series of purely logical steps (The Carnegie Application, 1902, NEM IV 58–59). Here, however, Peirce finishes his argument by claiming: 'There is, in short, no theorematic reasoning required to prove from the definitions that $7+5=12$.' Levy rightly concludes that by 1902, mathematics is thus partitioned in an analytic and a synthetic part following the division between corollarial and theorematical reasoning.[229] After this idea, corollarial mathematics is analytical while theorematical mathematics – along with the investigation of real possibilities in the special sciences – belongs to the synthetical domain. The closer establishment of this distinction remains, of course, an issue for future research. As Levy also argues, this distinction coincides with the appearance of abduction in mathematics. As he rightly claims, Peirce only rarely discusses this almost omnipresent issue (as many different necessity implications follow from any axiom or theorem, which route should we pick in our proof? – this question may be determined by hypothesis – abduction – only). Levy (86) discusses one of the few times Peirce relativizes his definition of mathematics by necessary reasoning in explicitly introducing hypothesis in mathematics:

> ... it is an error to make mathematics consist exclusively in the tracing out of necessary consequences. For the framing of the hypothesis of the two-way spread of imaginary quantity, and the hypothesis of Riemann surfaces were certainly mathematical achievements.
>
> Mathematics is, therefore, the study of the substance of hypotheses, or mental creations, with a view to the drawing of necessary conclusions. ('On Quantity', c. 1895, NEM IV, 268)

Hence, the hypothesis-making, abductive side of mathematics must fall within the domain of the synthetic, conforming to Peirce's overall tendency to let the analytic cover necessity and deduction, while the synthetic covers probability and ab- and induction. In a late quote, Levy finds a deliberation in which Peirce compares theorematic reasoning and abduction: 'theoremic reasoning ... is very plainly allied to Retroduction, from which it only differs, as far as I now see, in being indisputable.' (MS CSP 754, c. 1907, quoted from Levy). The question remains open, however, how this intricate web of ab- and deduction is to be parsed: are all conclusions reached by deduction on the basis of a theorematic abduction, synthetic, or is it only that hypothesis itself? Peirce seems to remain silent on the issue – but it is evident that we will here argue for the primacy of the former interpretation. In that case, diagrams and abductive diagram experiment become a royal road to the investigation of the synthetic a priori.

As to the synthetic a priori in the special sciences, it is a corollary to Peirce's architectonics of sciences that all sciences study mathematical structures, albeit incarnated in further ontological and empirical clothing relevant for the science in question. Necessary structures in all sciences are diagrammatical, precisely because they may – potentially, at least – be studied by means of mathematics, but they place upon their subject further material constraints of metaphysical character, depending on the science in question.[230] The regional ontology provides further constraints, guiding which parts of mathematics are relevant in the science in question. In this respect, Peirce's doctrine holding mathematics as first science is at odds with Husserl's idea. Even if the procedure of eidetic variation is generalized from mathematics, Husserl vacillates as to the possibility of mathematics to study the *outcome* of such variations. We have already remarked the differences between the first and second editions of *Logische Untersuchungen* on this point: in the first issue, it was seen as a major phenomenological task to describe vague forms in exact mathematical language – in the second issue, this task was given up in favor of the idea of 'vague essences' in experience which are assumed impossible to map mathematically. Why Husserl gave up that central idea is hard to say – it has not necessarily any inner connection to the well-known 'transcendental' turn he began to initiate in between the two issues of the LU. Jean Petitot has, as we discussed earlier, insisted on the idea that any ontology – or objectivity, as he as a Neo-Kantian prefers to say – must be formalized with the use of mathematics relevant for the essences of that ontology. Doing so, he makes, in fact, a Peircean move, requiring suitable diagrammatical structure to cover the synthetic a priori structure in question.

Again and again, a certain problem in a position connecting an Austrian a priori ontology with epistemological access comes up: *where is the precise borderline between the synthetic a priori laws of a given domain on the one hand, and the contingent, empirical data and tendencies recorded in that domain, on the other*? Here, Zelaniec's common sense treatment of the a priori and Friedman's embedding of the a priori in scientific development may, surprisingly, support each other. There is no exhaustive *definition* of what synthetic a priori regularities are – we can *describe* them and their role in reasoning, philosophy and science. Some of them are tacit, almost trivial regularities which we all depend on already in everyday reasoning (Zelaniec),[231] others have the form of necessary, non-trivial hypotheses called for in the ongoing development of science informed by both metaphysical discussions, empirical findings, and mathematical models (Friedman) – but in both cases we can not beforehand make explicit the exact borderline between synthetic a priori regularities and mere empirical tendencies found in the object studied. Rather, this distinction forms one of the basic *issues* in the ongoing development of science and philosophy. It is simply a crucial part of the interplay between common-sense basic assumptions, the different special sciences, mathematics, and philosophy to approach and refine the charting of that difference. To say it is an empirical issue would make no sense – but it remains, pace Barry Smith's 'fallibilistic apriorism', a question fallible both for empirical and a priori reasons and a question which may not be solved by pure armchair philosophy – but only by a philosophy involving itself in the ongoing development of the special sciences. Such investigation must be continued in specific regional ontologies. This is why we pass to the next three parts of this book where we search for diagrammatical and synthetic a priori regularities in three areas: those of biology, pictures, and literature.

PART II

BIOSEMIOTICS, PICTURES, LITERATURE

CHAPTER 9

BIOSEMIOTICS AS MATERIAL AND FORMAL ONTOLOGY

> Therefore, if botany and zoology must perforce rest upon metaphysics, by all means let this metaphysics be recognized as an explicit branch of those sciences, and be treated in a thoroughgoing and scientific manner.
>
> Peirce 'On Science and Natural Classes', 1902 (EPII, 116).

The notion of biosemiotics is no doubt so intriguing because it unites the foundational problems of no less than two sciences within its name. The first is, of course, biology. The concept of biosemiotics is coined by Ju. S. Stepanov[232] in 1971, and the idea is that biology must take serious the amount of semiotic terminology used in it, seemingly impossible to reduce. Biology, even in its most recent Neo-neo-Darwinian brands, seems not to be able to get rid of a whole vocabulary taken from linguistics, semiotics, informatics, philosophy, logic, and related sciences of meaning. Thus, concepts more or less explicitly borrowed from these domains now fill the gap which was earlier filled by vitalist dreams of yet undiscovered physical forces and the like. All currents in biology – from vitalism proper with its exaggerated ontological commitments to strange, yet unknown forces, and to neo-Darwinist reductionism – each in their manner bear witness to the predicament so brilliantly predicted by the often overlooked masterpiece of eighteenth century theoretical biology, Kant's 'Kritik der teleologischen Urteilskraft'.[233] Here, Kant predicts that no 'Newton of the grass straw' will ever appear, because biology cannot be reduced to mechanistic physics. Yet, on the other hand, no substantialist vitalism is deemed possible; the only means left to interpret biological phenomena is to understand them in analogy with the teleological principles also used in the appreciation of goal-oriented activities. The semiotic metaphors abundantly used by the biology of our day have often been noticed:[234] 'genetic information', 'messenger-RNA', 'DNA code' etc. pointing to the semiotic role of certain entities in a goal-oriented process and thus bearing witness to the indispensability of the teleological or intentional concepts even at the most tiny orders of magnitude in biology. Now, the introduction of biosemiotics is an attempt to take this mass import of metaphors seriously without on the other hand falling prey to vitalist excesses. Such recurring expressions must be a sign that biology is basically a semiotic science and that it – pace Kant – will never get rid of its semiotic vocabulary. Thus, the basic idea of biosemiotics is to establish the sign as a primitive concept in biology, and something like a Bohrian complementarity seems to appear: no matter how thorough a purely biochemical description of biological processes might become, it seems it will still be lacking the intentional understanding conveyed by the semiotic concepts or metaphors – and, probably, vice versa, semiotics needs the biochemical underpinning before becoming real biosemiotics. The biosemiotic program remains, it must be admitted, still a mere program; until now it has not made significant contributions to the development of biology and there is even a considerable confusion as to the precise meaning of semiotic concepts when transferred to biology as well as to the amount of semiotic assumptions being carried with them.

The other science shaken in its foundations by biosemiotics is, no surprise, semiotics. In the substantial part of semiotics inspired by linguistics, the assumption is often made (tacitly, that is) that semiotics is a human science; the ability to produce, communicate and understand signs is in this tradition spontaneously seen as a human privilege. Thus, the claim of biosemiotics to be a candidate for a real science also threatens this type of foundation for semiotics. The biosemiotic project invariably involves the idea that a substantial part of the semiotic vocabulary may be meaningfully exported from semiotics proper to biological phenomena. Thus, the very idea of a biosemiotics tends to dissolve the borderline between natural and human sciences, to attack it from so to speak both sides. But if this attack should really prove successful, it will invariably entail deepgoing consequenses for the sciences on both sides of this borderline: biosemiotics as a science is not possible as a simple expansion of well-established semiotics onto a new field, on the contrary, it necessarily implies a foundational crisis for semiotics and a thorough discussion of the basic semiotic assumptions in a biological light.

It is well known that in semiotics as a whole, there is a very wide disagreement as to which domains do fall under the scope of semiotics, ranging from very restricted doctrines, admitting only conscious, completed acts of communications by human agents as semiotic phenomena, and to ultra-Peircean versions of pan-semiotics implying that any physical process in the universe falls under the reign of semiotics. It is less well-known that a subspecies of this vagueness – so fatal to semiotics' claims for being recognized as a unified science – characterizes the attempts at founding a biosemiotics. One school simply identifies biology and semiotics on the basis of the strong intuition that the physical world (or, more properly, the simple physical world, apart from more complex phenomena as biology) in itself does not contain sign processes. This idea – first advocated by Th. Sebeok (e.g. 1991) and supported by many, for instance A.U. Igambardiev (1992), Floyd Merrell (1992, 1997), the actual Copenhagen biosemiotics school (Hoffmeyer, Emmeche) – is counterargued by various arguments, not only by a version of Peirceanism (cf. Peirce's claim from his absolute idealism period that matter is just inert mind and thus part of semiosis proper), but also by the idea that a continuity must prevail between simpler and more complex physics (e.g. John Deely 1992, 63). Thus, the foundational question raised by biosemiotics in biology by the same token becomes a foundational question for semiotics: how far does semiotics extend?

This and the following three chapters investigate these issues and thus form an integration of my contributions to the Copenhagen School of biosemiotics since Jesper Hoffmeyer's pioneering effort took its beginnings around 1990.[235]

In this chapter, I shall try to trace the possible consequenses for both disciplines, semiotics and biology. Let us begin by investigating the basic antinomies of biology as laid out by Kant in his already mentioned basic doctrine on the possibility of theoretical biology, 'Kritik der teleologischen Urteilskraft'.

THE BASIC ANTINOMY OF BIOLOGY IN KANT

This treatise takes as its object the realm of physics left out of Kant's critical demarcations of scientific, that is, mathematical and mechanistic, physics in the *Kritik der reinen Vernunft*. Here, the main idea was that scientifically understandable Nature was defined by *Gesetzmässigkeit*, by lawfulness. In his *Metaphysische Anfangsgründe der Naturwissenschaft*, this idea was taken further in the following claim: 'I claim, however, that there is only as much *proper* science to be found in any special doctrine on nature as there is *mathematics* therein ...', and further '... a pure doctrine on nature about certain things in nature (doctrine on bodies and doctrine on minds) is only possible by means of mathematics ...'.[236] The basic idea is thus to identify Nature's lawfulness with its ability to be studied by means of mathematical schemata uniting understanding and intuition. The central schema, to Kant, was numbers, so apt to be used in the understanding of mechanically caused movement. But already here, Kant is very well aware of a whole series of aspects of spontaneuosly experienced Nature in the 'letzten Grenze der Dinge' is left out of sight by the concentration on matter in movement, and he calls for these further realms of Nature to be studied by a continuation of the Copernican turn, by the mind's further study of the utmost limits of itself. This is what he undertakes in the 'Kritik der teleologischen Urteilskraft' where the very judgments we use in seeing Nature as endowed with purposes are investigated. Why do we spontaneously see 'Naturzwecke', natural purposes, in Nature? Purposiveness is wholly different from necessity, crucial to Kant's definition of Nature. There is no reason in the general concept of Nature (as lawful) to assume that nature's objects may serve each other as purposes. Nevertheless, we do not stop assuming just that. But what we do when we ascribe purposes to Nature is using the faculties of mind in another way than in science, much closer to the way we use them in the appreciation of beauty and art, the object of the first part of the book immediately before the treatment of teleological judgment. This judgment is characterized by a central distinction, already widely argued in this first part of the book: the difference between determinative and reflective judgments, respectively. While the judgment used scientifically to decide whether a specific case follows a certain rule is *bestimmende*, that is, it results in explanation by means of a derivation from a principle, and thus constitutes the objectivity of the object in question – the judgment which is *reflektierende* lacks all these features. It does not proceed by means of explanation, but by mere analogy; it is not constitutive, but merely regulative; it does not prove anything but merely judges, and it has no principle of reason to rest its head upon but the very act of judging itself. These ideas are now elaborated throughout the critic of teleological judgment.

In the section 'Analytik der teleologischen Urteilskraft', Kant gradually approaches the question: first is treated the merely formal *Zweckmässigkeit*: We may ascribe purposes to geometry in so far as it is useful to us, just like rivers carrying fertile soils with them for trees to grow in may be ascribed purposes; these are, however, merely contingent purposes, dependent on an external telos. The crucial point is the existence of objects which are only possible as such in so far as defined by purposes: 'That its form is not possible after mere natural laws, that is, such things which may not be known by us through understanding applied to objects of the senses; on the contrary that even the empirical knowledge about them, regarding their cause and effect, presupposes concepts of reason.'[237] The idea here is that in order to conceive of objects which may not be explained with reference to understanding and its (in this case, mechanical) concepts only, these must be grasped by the non-empirical ideas of reason itself. If causes are perceived as being interlinked in chains, then such contingencies are to be thought of only as small causal circles on the chain, that is, as things being their own cause. Hence Kant's definition of the Idea of a natural purpose: '... an object exists as natural purpose, when it is cause and effect of itself.'[238] This can be *thought* as an idea without contradiction, Kant maintains, but not *conceived* (begriffen). This circularity (the small causal circles) is a very important feature in Kant's tentative schematization of purposiveness. Another way of coining this Idea is now: 'Dinge als Naturzwecke sind organisierte Wesen' (op. cit. 235) – things as natural purposes are organized beings. This entails that naturally purposeful objects must possess a certain spatio-temporal construction: the parts of such a thing must be possible only through their relation to the whole – and, conversely, the parts must actively connect themselves to this whole. Thus, the corresponding idea can be summed up as the Idea of the Whole which is necessary to pass judgment on any empirical organism, and it is very interesting to note that Kant sums up the determination of any part of a Whole by all other parts in the phrase that a Naturzweck is possible only as a '... organisiertes und sich selbst organisierendes Wesen' (op. cit. 237) – as an organized and self-organizing being. This is probably the very birth certificate of the metaphysics of self-organization. It is important to keep in mind that Kant does not feel any vitalist temptation at supposing any organizing power or any autonomy on the part of the whole which may come into being only by this process of self-organization between its parts. When Kant talks about the 'bildende Kraft' (so beloved and hypostatized by later romanticism and vitalism) in the formation of the Whole, it is thus nothing outside of this self-organization of its parts. It is, he maintains, analogous to the arts – but without the artist. This implies an interesting consequence: even if we conceive of these organized beings by analogy to our own soul, these beings cannot be organized by means of any soul (if they were, an artist was added or the organized matter would already be there for the artist to collect). And it is thus not explainable – *erklärlich* – through the analogy with our art, Kant adds in an interesting remark, because we ourselves – including our own purposes – in this respect belong to nature (we are so to speak an empirical part of the problem, not of the solution!).

This leads to Kant's final definition: an organized being is that in which all alternatingly is ends and means. This idea is extremely important as a formalization of the idea of teleology: the *Naturzwecke* do not imply that there exists given, stable ends for nature to pursue, on the contrary, they are locally defined by causal cycles, in which every part interchangeably assumes the role of ends and means. Thus, there is no absolute end in this construal of nature's teleology; it analyzes teleology formally at the same time as it relativizes it with respect to substance. This definition, Kant affirms, lacks a constitutive concept of understanding – like the concept of movement for mechanics – and can thus not be mathematized and reach the status of a principle but must remain a mere 'maxim' (op. cit. 237).[239] Kant takes care to note that this maxim needs not be restricted to the beings – animals – which we spontaneously tend to judge as purposeful: 'this concept now necessarily leads to the idea of the totality of nature as a system after purposive rules, an idea under which all mechanisms of nature must be subsumed after principles of reason.'[240] The idea of natural purposes thus entails that there might exist a 'plan' in nature rendering processes which we have all reasons to disgust purposeful for us. In this vision, teleology might embrace causality – and even aesthetics: 'Also natural beauty, that is, its harmony with the free play of our epistemological faculties in the experience and judgment of its appearance can be seen in the way of objective purposivity of nature in its totality as system, in which man is a member.'[241] As is evident, these speculations entail that it is not implied in this maxim which size these *Naturzwecke* may possess: theories like Lovelock's 'Gaia'-hypothesis belong to the same domain in so far it is not metaphysically decidable beforehand which size the self-organized system may possess. But no matter the size of organisms, an important consequence of Kant's doctrine is that their teleology is so to speak 'secularized' in two ways: (1) it is *formal*, and (2) it is *local*. It is formal because self-organization does not ascribe any special, substantial goal for organisms to pursue – other than the sustainment of self-organization. Thus teleology is merely a formal property in certain types of systems. This is why teleology is also *local* – it is to be found in certain systems when the causal chain form loops, as Kant metaphorically describes the cycles involved in self-organization – it is no overarching goal governing organisms from the outside. Teleology is a local, bottom-up, process only.

This Idea of *Naturzwecke* is now subjected to the Kantian discipline of dialectics which reveals an antinomy of judgment. The *bestimmende* judgment has no principle of itself, but acts according to a principle given from outside, guided by which it subsumes objects under concepts. In this way, it presupposes a schematism which exposes a concept in intuition (that is, makes it possible to list which beings in time and space fall under the concept). On the other hand, the *reflektierende* judgment must, lacking appropriate concepts, serve as its own principle (what is meant by this seemingly strange idea is that judgment is in itself a purposeful use of the faculties of mind, so that reflective judgment amounts to finding processes in nature similar to its own purposeful judging procedure). These two opposed types of judgment, then, form the antinomy of judgment: (1) all material things and their forms must be judged according to mechanical laws (2) certain material beings can not be judged by mechanical laws only (op. cit. 250). These two maxims obviously do not contain any contradiction, only if they are converted into assertions of constitution: *all material things can (can not, respectively) be constituted by mechanical laws.*

Of course, it is possible that in Nature *an sich*, these two relations between things do possess one and the same root, but we are not gifted to tell due to our finite mind, so we stand split between determination and reflection. And any idea of assuming the absolute antinomy between them derives only from our tendency to mistake the reflective judgment for a determination. Thus, Kant does not in any way doubt the existence of organized beings, what is at stake is the possibility of dealing with them scientifically in terms of mechanics. Even if they exist as a given thing in experience, natural purposes can not receive any concept. This implies that biology is evident in so far as the existence of organisms cannot be doubted. On the other hand, biology will never rise to the heights of science, its attempts at doing so are beforehand delimited, all scientific explanations of organisms being bound to be mechanical. Following this line of argument, it corresponds very well to present-day reductionism in biology, trying to take all problems of phenotypical characters, organization, morphogenesis, behavior, ecology, etc. back to the biochemistry of genetics. But the other side of the argument is that no matter how successful this reduction may prove, it will never be able to reduce or replace the teleological point of view necessary in order to understand the organism as such in the first place: no Newton of the straw of grass is ever going to appear.

Evidently, there is something deeply unsatisfactory in this conclusion which is why most biologists have hesitated at adopting it and cling to either full-blown reductionism or (today, the very few of them) to some brand of vitalism, subjecting themselves to the dangers of 'transcendental illusion' and allowing for some Goethe-like intuitive idea without any schematization (some kind of *élan vital, entelechy, Lebenskraft*, or the like). It seems obvious that the author of the 'Kritik der teleologischen Urteilskraft' has felt the same uneasiness. In the strange heap of posthumous writings, the famous *Opus postumum*, he returns over and over again to the question; never, however, to allow for a substantialst vitalism which is precluded by the critical consciousness about the dangers of the transcendental illusion. Instead, he tries to soften up the question by philosophical means. Here, he tries to establish an *Übergang* from metaphysics to physics as he awkwardly puts it, that is, from the metaphysical constraints on mechanical physics and to physics in its empirical totality, including the organized beings of biology. Pure mechanics leaves physics as a whole unorganized, and this organization is sought established by means of 'mediating concepts'. Among them is the 'bildende Kraft' which is not conceived of in a vitalist substantialist manner, but rather a notion referring to the means by which matter manages to self-organize. It thus comprehends not only biological organization, but macrophysic solid matter physics as well (the 'texture of fibers, laminae and blocks, which is formed by crystallizing minerals' (Kant 1993, 35)). Here, he adds an important argument to the critic of judgment: 'Because man is conscious of himself as a self-moving machine, without being able to further understand such a possibility, he can, and is entitled to, introduce a priori organic-moving forces of bodies into the classification of bodies in general' (op. cit. 66) – and thus to distinguish mere mechanical bodies from self-propelled organic bodies. This argument is akin, in fact, to present-day cognitive semantics: it is because we ourselves are natural bodies that we are able spontaneously to understand the biological aspect of physics. It is probably here we shall find the explanation of the idea that reflective judgment takes itself as its principle: this is to say that the very teleological orientation of judging makes it possible for judgments to find judgment-like processes in nature – still without being able to further understand it.

To sum up Kant's masterful achievements in these early sketches of a theoretical biology: biology is trapped between, one the one hand, a mechanist physics, being able to explain things only in causal terms mathematically formalized in the differential calculus, and, on the other hand, the spontaneous recognition by the judging body of other judging bodies in the physical world. The scientific entry to this experience is forever banned, but it may be possible to integrate it into philosophy by new 'mediate concepts'.

The relevance of Kant's musings for biosemiotics ought to be evident. The very reason for the coming into being of biosemiotics is the apparent indispensability of semiotic, teleological concepts in the would-be mechanist biology of our day. It immediately implies that biosemiotics is not, and can never be, a science in Kant's mechanist, causal use of the word. Instead, biosemiotics must be seen as a refined attempt at formulating the 'mediate concepts' of the *Opus Postumum*. In fact, Kant already anticipates biosemiotics in so far as he, in the idea of the reflective judgment's taking itself as principle, articulates the idea that we – because we are judging bodies – that is, biosemiotic beings – are able to see other biosemiotic beings and formulate the Idea of self-organized *Naturzwecke* to understand them.[242] The teaching of Kant's theoretical biology for biosemiotics will then be: biosemiotics can never free itself from this formal, teleological root; its mediate concepts of sign, meaning, information, communication etc. will invariably express aspects of a teleological kernel – and thus stand in opposition to the attempts in Darwinism and Neo-Darwinism to expel all teleological concepts from biology.[243]

Various Neo-Kantians have sought to elaborate this third way between vitalism and reductionism. Ernst Cassirer who elaborated a brand of Kantian semiotics in his chef-d'œuvre *Philosophie der symbolischen Formen*, also left a – albeit little noticed – sketch of a philosophy of biology in a chapter in the last of the four volumes of his impressive *Das Erkenntnisproblem*. Even if this account is now more than half a century old and lies well before Crick and Watson's DNA breakthrough, it is still surprisingly fresh. Cassirer concludes his historiography of biological thought with the *Vitalismusstreit* and its consequenses in the beginning of the century. As a Kantian, he opts for a third way and sees both the Darwinists' reductionist claims and vitalisms, like Hans Driesch's reference to a *Seele* or a *Psychoid* in the living being, as aberrant developments. Instead, he makes a synthesis of the leading biologists of his time, especially his colleague in Hamburg, von Uexküll, Ungerer, and von Bertalanffy.[244] Their 'Organizismus', as he coins it, does not aim at neither mechanist reduction nor at the acceptance of vitalist forces, but towards the recognition of a 'nicht-stoffliche Ordnung', (209) a non-material order, in addition to the causal organization of matter. Thus, in this analysis, the holist character of Kant's analysis of the *Naturzweck* is pulled into the foreground. The notion of 'Zweck', so Cassirer argues, tends to make us believe the parts in question possess an intention they do not; the concept of 'Ordnung', 'Form', 'System', 'Struktur' better satisfies Kant's analysis: 'They substitute for the concept of purpose the concept of order and system, and they characterize life by attributing to it the property of a system.'[245] This step, we could add, already lies as a possibility in Kant's ingenious analysis of the concept of ends in terms of causal circles, necessarily possessing a morphological organization. In these passages, Cassirer in fact paves the way for an epistemology of a structuralist biology in so far as he sees the consequence of Uexküll's thought as the fact that '... es ein selbständige *Formproblem* gibt, für das die Biologie eigentümliche Begriffe und Denkmittel auszubilden hat.' (209) – there is an autonomous *form problem* for which biology must build special concepts and means of thought. What is even more interesting is his fertile correction of Kant: when physics now long since has supplied Newton's mechanics with other modes of thought, the narrow constraints of a Newtonian epistemology are no longer valid for physics where holist concept like that of 'fields' now prevail; then why shouldn't they do as well for biology? Cassirer thus points in the direction of the notion of 'Form' as the Kantian 'mediate concept' which will enlighten biological thought.

If we now measure the claims of biosemiotics against these epistemologists of biology, it is easy to see that it stages itself as a further investigator of these third-position 'mediate concepts' of the Kantian tradition. A crucial question for biosemiotics, then, will be: what does the semiotic vocabulary add to the problem of form, of order, of *Naturzwecke*? What is won with the import of concepts like 'sign', 'codes', 'semiosis', 'information', 'messenger', 'semiosphere', and the like into the description[246] of biological form? One should here keep in mind René Thom's cautious remark: '... when we speak of 'information', we should use the word 'form' ' (Thom 1975, 127).

It must immediately be added that the word 'form' here plays several roles. First, one central role is the reference to the many different aspects of internal architectonic organization of biological beings, ranging from the structure of the cell and to multicellular organisms. Second, it refers to the global, stable structure – the organism – which results as the sum of the architectural building-stones just mentioned, kept stable by metabolism and the organism's register of specific actions to avoid states too far from metabolic stability. This fact opens two ways of inquiry: first, what does the very concept of life as organized beings imply semiotically, that is, does an a priori analysis of the concept of life in itself entail semiotic phenomena? second, the question of organized form is more general than living beings and hence connects biology to the cross-disciplinary question of emergence, the coming into being of 'new properties' in sufficiently complex systems.

CATEGORICAL PERCEPTION: BACTERIAL CHEMOTAXIS AS SEMIOTIC CASE

Let us take a concrete example as introduction to the investigation of semiotic a priori constraints on life. Since the classic paper by H.C. Berg, 'A Physicist Looks at Bacterial Chemotaxis' it has been known that as simple an organism as E. Coli is able to process semiotic information from its environment and act accordingly: in the presence of a sugar gradient, it is able to swim 'upstream' in the gradient and so approach a locality with higher nutrition concentration.[247] How are the details in this process? The bacterium undertakes this goal-oriented behavior due to a specific coupling between sensors – a primitive perception – and action, organized in what Berg calls 'biased random walk' (see Figs. 26–27).[248] The unicellular organism has

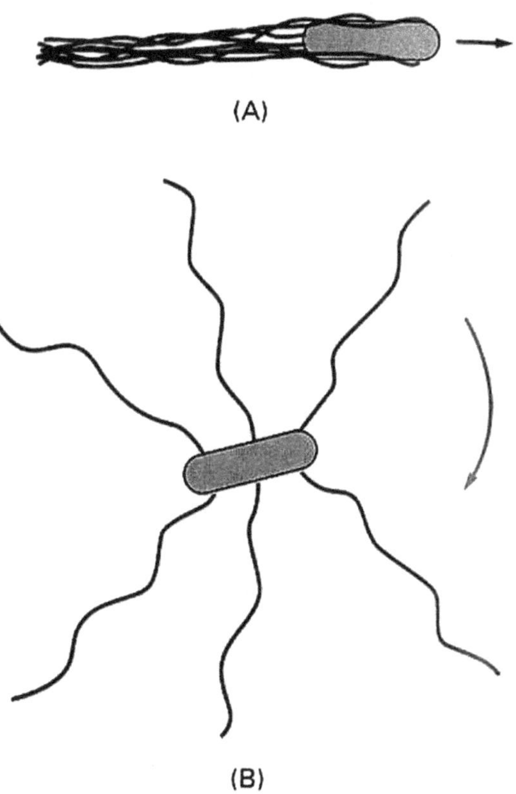

Figure 26. A) Swimming B) Tumbling. *Source:* This and the next pages: Copyright 1994 from *Molecular Biology of the Cell*, by Alberts et al. Reproduced by permission of Garland Science/Taylor & Francis LLC

six thin filaments protruding from its cell membrane, and in the absence of any stimuli, it simply wanders randomly around by changing between two characteristical movement patterns. One is performed by rotating the flagella counterclockwise. In that case, they form a bundle which pushes the cell forward along a curved path, a 'run' of random duration with a mean length of 1 second. These runs interchange with 'tumbles' with a mean value of 0.1 second where the flagella shift to clockwise rotation, making them work independently and hence moving the cell erratically around with small net displacement. The biased random walk now consists in the fact than in the presence of a chemical attractant, the runs happening to carry the cell closer to the attractant are extended, while runs in other directions are not. The sensation of the chemical attractant is performed temporally rather than spatially, because the cell moves too rapidly for concentration comparisons between its two ends to be possible. A chemical repellant in the environment gives rise to an analogous behavioral structure – now the biased random walk takes the cell away from the repellant. The bias saturates very quickly – within 0.2 seconds – which is what prevents the cell from continuing in a 'false' direction, because a higher concentration of attractant will now be needed to repeat the bias. The reception system has three parts, one detecting repellants such as leucin, the other detecting sugars, the third oxygen and oxygen-like substances. The biochemical machinery for signal transduction and for flagella activation are rather well-understood, and the whole mechanism has long since become standard textbook material, cf. the illustrations reprinted here.[249] These well-established empirical results, however, seem to await their proper semiotic interpretation.

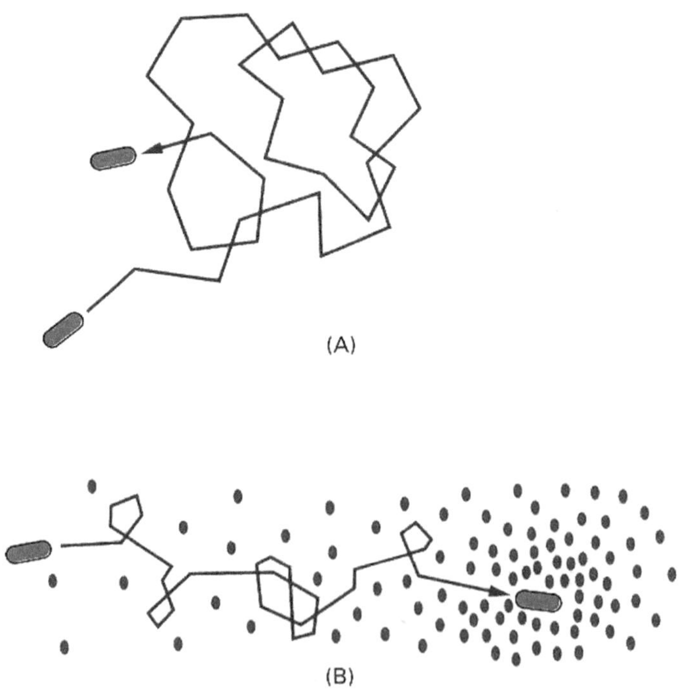

Figure 27. A) Random Walk B) Biased Random Walk.

As always, it is interesting to note the ubiquity of spontaneous semiotic vocabulary: 'information-processing pathways', 'measurements of external concentration', 'sensory information', 'the cell cannot solve this problem', 'The cell sets the mean run length at 1 second to provide ample head room for response', 'When watching such tracks evolve on a computer screen, one gets the impression of a bloodhound following a scent', etc. (all examples from Berg, 4–6). Our contention is that this terminology cannot be mere verbiage but reflects core properties of the process investigated. What are those properties, then, in semiotic terms? The cell's behavior forms a primitive, if full-fledged example of von Uexküll's functional circle (see Chap. 10) connecting specific perception signs and action signs. Functional circle behavior is thus no privilege for animals equipped with central nervous systems (CNS). Both types of signs involve categorization. First, the sensory receptors of the bacterium evidently are organized after categorization of certain biologically significant chemicals, while most chemicals that remain insignificant for the cell's metabolism and survival are ignored. The self-preservation of metabolism and cell structure is hence the ultimate regulator which is supported by the perception-action cycles described. The categorization inherent in the very structure of the sensors is mirrored in the categorization of act types. Three act types are outlined: a null-action, composed of random running and tumbling, and two mirroring biased variants triggered by attractants and repellants, respectively. Moreover, a negative feed-back loop governed by quick satiation grants that the window of concentration shifts to which the cell is able to react appropriately is large – it so to speak calibrates the sensory system so that it does not remain blinded by one perception and does not keep moving the cell forward on in one selected direction. This adaptation of the system grants that it works in a large scale of different attractor/repellor concentrations. These simple signals at stake in the cell's functional circle display an important property: at simple biological levels, the distinction between signs and perception vanish – that distinction is supposedly only relevant for higher CNS-based animals. Here, the signals are based on *categorical perception* – a perception which immediately categorizes the entity perceived and thus remains blind to internal differences within the category.

One further peculiarity deserves to be mentioned: the mechanism by which the cell identifies sugar, is partly identical to what goes on in human taste buds. Sensation of sugar gradients must, of course, differ from the consumption of it – while the latter, of course, destroys the sugar molecule, the former merely reads an 'active site' on the outside of the macromolecule. This has, however, huge semiotic implications, for this entails that E. Coli – exactly like us – may be fooled by artificial sweeteners bearing the same 'active site' on their outer perimeter, even if being completely different chemicals (this is, of course, the secret behind such sweeteners, they are not sugars and hence do not enter the digestion process carrying the energy of carbohydrates). This implies that E. coli may be *fooled*. We recall Umberto Eco's definition of semiotics as the study of all devices which may be used to lie. Bacteria may not lie, but a simpler process than lying (which presupposes two agents and the ability of being fooled) is, in fact, being fooled (presupposing, in turn, only one agent and an ambiguous environment). E. coli has the ability to categorize a series of sugars – but, by the same token, the

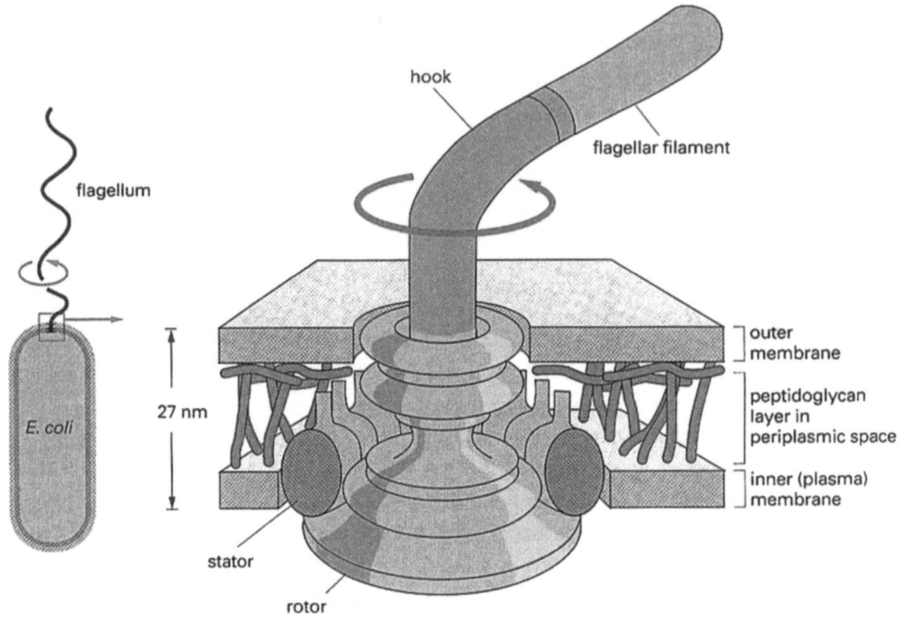

Figure 28. Flagellum rotor

ability to categorize a series of irrelevant substances along with them. This is semiotically extremely important – semiotics seems to begin with this duplicity: On the one hand, the ability to recognize and categorize an object by a surface property only (due to the weak van der Waal-bonds and hydrogen bonds to the 'active site', in contrast

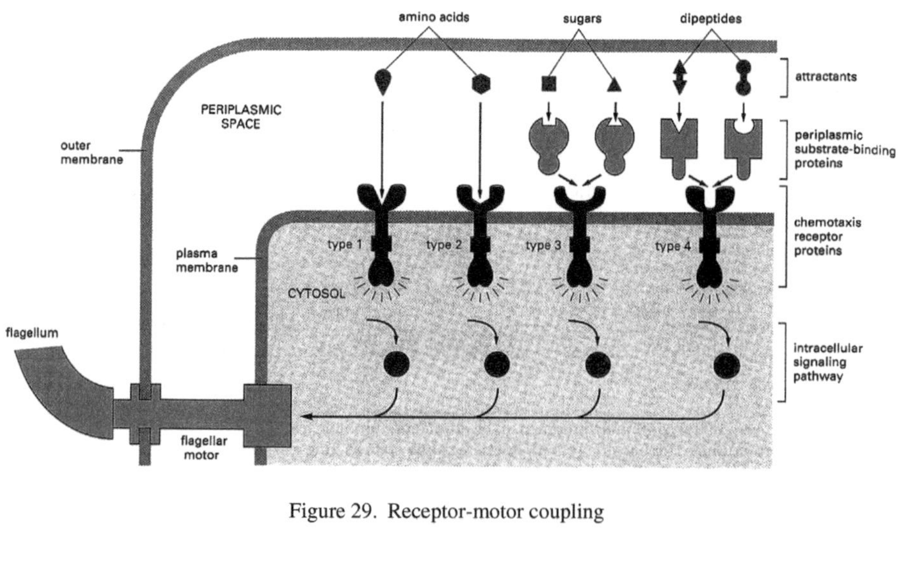

Figure 29. Receptor-motor coupling

to the strong covalent bonds holding the molecule together) facilitates perception economy and quick action adaptability. On the other hand, the economy involved in judging objects from their surface only has an unavoidable flip side: it involves the possibility of mistake, of being fooled by allowing impostors in your categorization. So in the perception-action circle of a bacterium, we find some of the minimum requirements for semiotic processes: the self-regulatory stability of a metabolism involving categorized signal and action involvement with the surroundings. An a priori semiotic hypothesis is that these properties characterize semiotic processes as such and that the types of processes described here function as building-blocks in more sophisticated and complex semiotic phenomena, from intercellular communication in multicellular organisms and to complicated perception and communication in higher animals.[250] A long-range goal of biosemiotics will be to analyze all such types of sign processes in biology and discern the network of mediate concepts of semiotics necessary to characterize them, from the simplest to the more complex cases. In a certain sense, these constraints on life would constitute the material ontology of biology.

SEMIOTIC A PRIORI CONSTRAINTS ON LIFE?

It has often been noted (e.g. de Loof & Broeck 1995) that no concise definition of the concept of life exists.[251] Most attempts at definitions end up as lists of properties required. An extreme opponent to this view, however, is the 'strong' school in the 'Artificial Life'-community, parallel to so-called 'strong AI'. The idea here is that computer modelings of life may themselves be considered living because they instantiate certain formal characteristics of living beings: they reproduce, fight, develop and even form symbiosis- and ecological niche-like compromises between them (cf. Langton 1997). This trend is right in leaving the 'carbon-centrism' of empirical biology behind and seeing life as a theoretical and purely formal question in its own right, but, on the other hand, an argument analogous to the AI case seems valid: merely formal calculi do not result in intelligence nor life like we usually understand them. Computer-based AI lacks qualia, genuine insight and the discussion of its limitations due to Gödel's theorem is an ongoing and yet unsolved strife in AI studies (the same goes for the discussion of life, cf. Rosen 1989, 1990[252]); computer-based AL lacks all the same properties (and in so far life comprises intelligent beings, AL must obey the same constraints) as well as the central biological property of metabolism. This calls for an a priori analysis of the concept of life and related concepts[253] in order to determine the 'mediate' concepts required for the study of the living. Taking the Austrian interpretation of the a priori, these mediating concepts will, simply, constitute the conceptual network of regional ontology of biology. Recurrent features in the various list definitions of life include the following:

Metabolism. Life is characterized by (at least in decisive phases of its being) the necessity to organize and sustain itself by means of a metabolism, during which low-entropy energy is consumed while the same energy is excreted in a higher entropic form. This rise in entropy makes possible a lowering of entropy in a smaller part of the energy used (so that, of course, the total entropy change is governed by the 2nd law of thermodynamics) which may consequently be used for the construction of the organism and its ordered patterns of behavior. This implies that the organism somehow must be located in an environment permitting a relatively steady supply of highly organized (low entropy) energy in some form. Some marginal kinds of life (virus in its non-active phase) may display lack of metabolism, but we would not call virus living if not for its possibility for an active, self-replicating phase when coupled to the metabolism of more complicated host organisms. The same goes for simple animals (e.g. some shrimp species) able to stay completely metabolism-less during long, cold seasons: it must display metabolism in other phases of its existence, otherwise we would not call it living at all. Metabolism involves a negative feed-back coupling making stable its circular process, preventing it from going off in any of two directions – the speeding up which would destroy the organism viz. the slowing down which would starve it to death.

Metabolism as a crucial feature in the concept of life is ripe with proto-semiotic categories. Already Kant's definition of the *Naturzwecke* as defined by organisms in which every part is interchangeably means and ends, implies the cyclical organization of metabolism. **Life** and **death** are defined by the continuation versus the definitive ending of metabolism. A priori categories of **food** and **poision** must pertain to any metabolist being: food will be energy-bearing elements of the environments necessary to develop and sustain metabolism; poison will be elements of it which hinders metabolism to an extent such that the metabolism or parts of it are damaged. **Illness** will be an a priori notion for conditions implying a reduction in the stable possibilities of metabolism. **Excrements**, on the other hand, will be an a priori category for high-entropy energy in some form (faeces, urin, heat radiation, etc.), leaving the organism. These simple semantic consequences of metabolism yield, of course, an enormous amount of semiotic primitives when combined. The very concept of **intention** has its foundation here: the metabolism's cyclical structure implies that it includes a self-preserving 'intention' in so far it is constituted by a structure of cyclical attractors for the metabolic system. A later part of the cycle may be seen as intended (in a teleological, not a psychological, sense of the word) by a former part of it, and the same goes for the external means (food, for instance) necessary to obtain the later phase of the cycle. The very concept of **narration** finds it bases here – the possible dramas unfolding in cases of **crises** of lack of food, threats of poisoning, pollution by own excrements, etc., and the more or less happy **solution** of the crises forms the basis of narrative drama.

Boundaries. Apart again from the limit case of virus, all life as we know it is spatially defined by membranes, conchs, shells, skins, shields, armor, layers, etc. which delimit and shield the internal finely tuned homeostatic system of metabolism from the uncontrollable energy fluxes of the environment. This, of course, entails that the traffic across these boundaries are subjected to control so that a priori categories of **mouth** and **anus**, respectively, is a consequence, each of them being selective and equipped with some means for selection. Santa Fe theoretical biologist Stuart Kauffman – whose work we shall later return to – argues that the inside of a cell must necessarily be subcritical with respect to the formation of new chemical compounds, in opposition to the biosphere as a whole which must be supercritical. The boundary of the organism introduces a topological distinction between spatial regions and thus implies semiotic basic categories like **proto-subject**, **inner/outer, here/there**.

Perception-Action. As a sophisticated means of fulfilling the needs of metabolism is the possibility of perception and ensuing action. As soon as the metabolism acquires means of (ever so slightly) changing its relations to the environment, that is, it displays a variety of behavioral options facing different environment aspects, this will imply proto-perception (not necessarily with any psychological implications, of course). This use of the notion of 'perception' is a generalization and does not necessarily include the presence of consciousness, qualia, map-like representation of the surroundings etc. It goes without saying that organisms in environments with a plethora of relevant energy would not need to develop organs of perception of any kind in order to locate food, but typically, the contingent changes in food supply makes some kind of perception add to the metabolism's fitness. It is well-known that plants may orient leaves towards the sun, just like E. coli will swim upstream in the gradient of nutrition. These phenomena are, of course, ripe with consequences of seemingly semiotic kind: the **feed-back** between the organism and its environment is tempting to interpret in semiotic **dialogue** notions: the organism **reads** some **sign** in the environment and **reacts** accordingly. Other species in the environment may now register the organism's change of behavior and in turn 'decode' it as a 'sign' to the extent that it has any consequence for their own metabolism.

This interdependence of the organism with its environment received its first semiotic interpretation in von Uexküll's 'theory of meaning', but the interdependence of perception and action in Peirce's pragmatism forms an important predecessor. We shall return to the details of Uexküll's classic of biosemiotics in the next chapter; as we saw in the E. coli example, the definition of an organism's environment, *Umwelt*, on the basis of the 'functional circle' connecting the organism with its environment via two classes of signs, perceptions ('Merkmale') and actions ('Wirkmale'). The *Umwelt*, then, is the parts and aspects of surrounding world in so far it enters into cyclical interaction with the organism. The functional circle may be seen as an extension of the cyclical structure of the metabolism; it consists of those metabolic cycles which cannot be completed except for their active integration of some outer segment into the cycle. Thus, metabolism and functional circle form the bases of so-called 'endosemiotics' and 'exosemiotics', respectively. In so far one accepts the semiotic vocabulary in the description of these cycles, it must be added that the signs in question do not yet obey the distinction between simple signs and propositions. The organism's recording of the presence of an edible part of the surroundings might be translated as 'food' or 'this is food', interchangeably. An important aspect here is that the engagement of metabolism's perceptual parts in an *Umwelt* makes necessary the reduction in complexity of the possible 'Merkmale'. The organism excludes potential 'Merkmale' not relevant for the metabolism, and among the perceived 'Merkmale', it arranges a certain **granularity**.[254] The simpler the organism, the simpler and more 'corny' will this granularity probably be, as a tendency. As discussed above, E. coli is unable to discriminate edible glucose from irrelevant saccharine – the two substances are categorized together due to a morphological similarity between the 'active sites' on the macromolecules' outside as perceived by the bacteria. In general, granularity implies that a host of different, more and less cognate, 'Merkmale' will give rise to the same type of chemical reaction in the metabolism; they will possess the same immediate signification for the organism and hence involve a primitive version of **categorization** or **categorical perception**.[255] The very notion of a sign seems to imply this much: a distinction between an immediate signification and a mediate signification – proto-forms of **meaning** versus **reference**, *Sinn* versus *Bedeutung*. The fact that sugar and saccharine tastes similarly to man or bacterium rests on their immediate resemblance due to the active site on the molecule (a particular atomic configuration on the periphery of these molecules) – but the nutrition value differs highly, that is, the 'same' signification ('sweet' – the word is used here, of course, with no necessary assumption of the existence of sweetness qualia in the process) refers to different objects ('food'/'non-food'). This possibility for the metabolism's perception to be **fooled** or **surprised** thus seems to be basic for the category of sign. Seen from the point of view of efficiency, the metabolism must do with in an important sense of the word *small* signs (cf. the active sites' specific 3-D configuration as opposed to the whole of the macromolecule) in order to be categorical, general and hence efficient, the flip side of this being of course the possibility of **mistaking**. Seen from the point of view of form, the substance of the macromolecule as a whole is represented through (a small subset of its) form: the 'active site'. This **'smallness' of the sign in comparison to its signification** is very important for the necessity of the sign concept at all – if the sign was not 'smaller' in this way, we would have no reason to use it as referring to a category (This is what Bateson (1979) refers to as 'collateral energy', his third criterion for what he called 'mental process'.). But what does this 'smallness' consist in? One level is energy. The sugar molecule in the E. coli case is discriminated by means of the weak polarity on the outside of the macromolecule due to hydrogen bonds; energies far smaller than the energies released by the ensuing consumption of the macromolecule involving the breaking down of the ordinary, strong chemical covalent bonds between its parts. If it was not for this decisive energy level difference, we would not speak of signs.[256] Another level is size and dimensionality; the 'active site' on the macromolecule's outside is less in size and dimensionality than the macromolecule as a whole. This implies a natural instantiation of the Aristotelian distinction between **form** (active site) and **matter** (molecule), the same form being able to subsume (and hence signify) several different matters.[257]

The first of these differences permits the sign to save energy on behalf of the organism: at a very low energy cost, it may perceive the possibility of obtaining energy to fulfill a phase of the metabolic cycle. This economy (measured on the background of the metabolism's demands) is crucial to the primitive sign. The second of these differences permits the sign to erect categorical and hence general classes. In so far, any sign is underdetermined[258] which, as we noted, makes it at the same time efficient and fallible. These speculations immediately entail certain abilities on the side of the sign reader: it must be able to amplify the low-energy sign perception in order to let it result in (higher energy) action; it must possess stockpiled energy in order to do so; and it must be able to restore the perceptual organ quickly and with low energy cost in order to function in later metabolical cycles (the eye must not be destroyed by the light; the cell membrane not by the active site).[259] Between sign and reader must persist a weak, stable interaction of resonance type, while the reader must possess a high energy, comparatively unstable system able to amplify the form perceived in the sign and take one among a whole range of different action measures.

These deliberations show that even at this primitive level, the notion of virtuality or potentiality – Peircean real possibility – is crucial to the sign. The sign – the active site – releases a specific behavior in the organism (f. inst. swimming up-gradient) which will result in one outcome among several possible outcomes (staying, fleeing). On a higher level, the reading of other active sites may result in other sets of specific behaviors. Thus, the meaning of one active site for the organism is embedded in a virtuality of other active sites leading to other specific meanings. The meaning of the primitive sign, consequently, is only possible as one choice among a (maybe open-ended) set of alternative choices. Here, the linguistic idea of **paradigm** finds a primitive proto-version.[260] As soon as signs become more complicated than at the molecular level, more possibilities within the interpretation of one sign come into being, and the immediate meaning of a sign will constitute a **schema** able to flexibly incorporate a range of different but cognate empirical instantiations of it.[261] As soon as animals with CNS become able to distinguish species fellows, predators, preys, from perceived visual profiles etc., we must expect that such simple schemata complexify into action schemata involving several different slots to be filled in, forming simple **diagrams** for characteristical action sequences.

A simple organism – defined by a simple metabolism with few systematic stable variations – will typically possess an extremely simple *Umwelt* with few categories and coarse granularity while higher discrimination ability will require more complexity on the part of the organism and its behaviors. Similarly, the more complex the organism, the more complex the schemata it must embody, and, in turn, the more action possibilities the schemata must possess. This further separates meaning and reference, in so far as the schemata leading to a series of different possibilities of action must now to some extent contain representations of those actions with specific sign 'slots' connecting the parts of each action series. Thus, a sign may 'mean' a range of different schematic action possibilities, each equipped with further releasing signs to determine the choice between them. (The perception of a predator may 'mean' escape, freezing, attack ...). We shall return to the more precise status of the 'Umwelt' below.

In multicellular organisms with central nervous system, the integration of a whole set of cells in complicated, schematic perception-action cycles becomes possible, and it seems likely that it is only at this stage that a detailed inner representation of the environment – be it conscious or not – becomes available. That is, only at this level perception, as we usually understand it, involving such mappings of the surroundings, becomes possible. This implies a crucial consequence: that signs are much simpler and much more biologically widespread than full perception – this in contradistinction to the phenomenological idea of signs as derivative entities only, as compared to the primacy of perception. In higher animals, it is true, special, complex signs, like in animal communication or human language, may appear which take sets of possible perceptions as their objects – signs which for that reason presuppose full perception. Signs on this primitive level transgress the distinction between sign and perception in higher animals. They might just as well be called primitive perceptions, but their simplicity makes it more apt to call them signs. Perception involving representation in the shape of environmental maps probably only appear in animals equipped with CNS – and only in such mappings, a distinction between primary perceptions and secondary signs, signs in a more narrow sense, becomes possible. Here, signs in the narrow sense will refer to information-bearing entities foregrounded on the background of perceptual environmental maps. But perception, in turn, is possible only as a composite of simpler, biosemiotic signs which are hence primary both in complexity terms and in evolutionary terms.[262]

Inner architecture. The fact that life necessarily is articulated in 'organized' beings is already implied in Kant's definitions. In biology, the conception of the cell as a 'protein bag' has long since given place to a charting of its complicated inner architecture with various organels, the kernel including the genome, mitochondria, chloroplast etc. Already the boundary constitutes a sophisticated minimal architecture, and the flux of energy running through metabolism from outside to inside and back to outside necessitates a further minimum of internal organization, in so far the energy flow must be separated in its two characteristic high and low entropic components, respectively. The very process of this separation has in itself low entropic character and requires a spatial organization in contrast to the idea of a disorganized inside; such a structure would be unable to perform the low entropic teleological proto-action connected with life. This inner organization immediately implies the partwise articulation of metabolism tasks giving rise to organels (in the cell) and organs (in multicelllular organisms). Life is thus necessarily **modular**, seemingly more modular, the higher the complexity of the tasks involved is, implying the pertinence of specific types of **mereo-topology** for the study of life, making it possible to discern modular parts defined by inner boundaries etc. Metabolism, of course, is only controllable by means of the genome for this very reason: a specific gene must 'code for' a specific type of reaction in a specific modular part of metabolism. Evolution makes possible the further differentiation, articulation and modularization of metabolism's cycle which is then ripe with potential new categorizations and discriminations relevant for the single 'organs' and their partial tasks. This modularity of life entails the possibility of the **modularity of signification**.

Intelligence. In its primitive form, intelligence is no doubt already implied by concepts like the just mentioned, metabolism, boundary, perception, inner architecture. In this view, the concept of intelligence must be completely discriminated from the concept of consciousness, this being probably a property in a much smaller class of biological beings equipped with central nervous system and highly developed cortex. In any place, it is impossible yet to tell whether primitive animals might possess some diffuse and marginal proto-consciousness, as has often been suggested.[263] It is probably an empirical question whether this be the case, and the question must be abandoned presently and left for possible future research. It is still a subject of much discussion whether high intelligence requires consciousness or whether all intelligent tasks could in principle be fulfilled by non-conscious apparatuses (cf. David Chalmers' much-discussed book); in any case it seems reasonable to use the concept of intelligence in connection with metabolisms' way of coping with their *Umwelten*. The biosemiotic notions of sign, meaning etc. must for this reason be cleaned for assumptions of consciousness and must be defined only with reference to function and form. Still, the evident skills displayed by biological species at coping with their *Umwelt* makes the notion of a higher or lower degree of **intelligence** an a priori concept of biosemiotics.

Reproduction. In light of our Kantian exegesis, we remarked that metabolism is one crucial empirical process instantiating Kant's cyclical alternatingly-means-and-ends definition of *Naturzwecke*. Now reproduction will be another. But it is important here to underline that it ontologically depends on the first. We can imagine metabolic life without the ability to reproduce, but not reproducing life with no metabolism. Reproduction is hardly a necessary feature in the definition of life; in principle, an eternally living cell in an optimal environment supplying all chemicals needed could be imagined, a cell which consequently would not have to perform mitosis or any other reproduction in order to survive.[264] However, the empirical unstability of nutrition flow and the competition from other individuals and species make the basic Darwinist principles generic: the pressure of selection entailing that reproduction pays and ensures growing fitness (and of course also that gendered reproduction and the following larger pool of selection pay even better). The secondary status of reproduction, however, entails that the Neo-Darwinian attempts at letting the principles of reproduction (scarcity of nutrition, offspring variation, competition, survival of the fittest etc.) count as the basic principles of biology, from which all other features should in principle be drawn, do not hold. Metabolism, organization, inner architecture are logically primary to the Darwinian concepts which consequently cannot define them. As Stuart Kauffman has pointed out, a certain amount of 'order for free' giving rise to stable, self-reproducing systems is the condition of possibility for selection to do its work: there must exist possible biological entities for selection to be able, in turn, to select between them and force them to evolve. Even if the Darwinist principles of reproduction, variation, selection, evolution are thus ontologically secondary to the principles of metabolism, it probably remains true that sufficiently complicated forms of life are impossible without evolution. Pure metabolic life with no reproduction would most probably remain extremely primitive. The metabolic stability of complex life forms is not *created* by evolution (just like neo-Darwinist metaphors taking genes to be effective causes of most if not all aspects of biology often imply), but must be supposed to exist beforehand as domains in the overall space of all possible species, like Stuart Kauffman argues. The possibility of reaching these optima in the space of species nevertheless depends on evolution as the most efficient natural searching algorithm in the space of possible species. Even if ontologically secondary, then, the principles of evolution seem necessary for the historical coming into being of complex organisms. Thus, the Darwinist scenario gives rise to further primitive a priori concepts: **competition, relative fitness, partner selection** (in gendered reproduction), and, in connection with the inner-architecture, boundary and metabolism properties, it yields categories like **co-working societies of cells, symbiosis, multicellular organisms** etc.

This series of interrrelated a priori concepts for biology is intended to satisfy the role of 'mediate concepts' from the alternatingly-means-and-ends analysis of *Zwecke* in Kant. It should come as no surprise that the majority of them are well-known biological concepts, even if maybe not technically so. The important idea is that they form an interconnected cluster of material-ontological concepts which organize biological thought as such rather than being mere human conventions. The use of semiotic vocabulary in biology thus seems to form a crucial part of a whole conceptual network of theoretical biology. This implies, on the other hand, that all sorts of semiotic terminology developed about human signifying processes may not effortlessly be exported into biology. Rather, a certain caution is required – biological signs probably involve a large range of complexity levels below that of human semiotics. On the other hand, caution should also be taken in not taking biosemiotics as indication that all of nature all of a sudden possesses semiotic aspects. One reason why it might be difficult to decide whether there are Peircean symbols and even life in inorganic nature is that Peirce is not himself unambiguous on the issue. Many authors supporting versions of the pan-semiotic ideas of symbols everywhere like to quote this famous Peirce idea:

> The October remarks [i.e. those in the above paper] made the proper distinction between the two kinds of indeterminacy, viz.: indefiniteness and generality, of which the former consists in the sign's not sufficiently expressing itself to allow of an indubitable determinate interpretation, while the [latter] turns over to the interpreter the right to complete the determination as he please. It seems a strange thing, when one comes to ponder over it, that a sign should leave its interpreter to supply a part of its meaning; but the explanation of the phenomenon lies in the fact that the entire universe – not merely the universe of existents, but all that wider universe, embracing the universe of existents as a part, the universe which we are all accustomed to refer to as 'the truth' – that all this universe is perfused with signs, if it is not composed exclusively of signs. (Note from 'The Basis of Pragmaticism', 1906, 5.448)

Most often, only the last, audacious claim is quoted – so John Deely in the very conclusion of his recent 1,000-page semiotic history of philosophy (Deely 2001, 742). Peirce's idea here seems to be that indeterminacy is generic in sign use because indeterminacy, following Peirce's ontology, is generic in the universe as such. Strictly spoken, it does not follow from the idea that indeterminacy exists in the universe plus the idea that indeterminacy exists in signs that all universal indeterminacy has a sign character. But the idea thus seems to be that the objective existence of generality (like in any physical law as for instance gravitation) gives the cosmos a sign-like structure, because individuals embodying generals function as signs for those universals.[265] And as we saw in the discussion of 'real possibilities', Peirce often extends this claim to embrace also the existence of symbols, teleology, and life in inorganic physics. But this runs counter to other, more modest ideas in Peirce. Take for instance the claim that

> In short, the problem of how genuine triadic relationships first arose in the world is a better, because more definite, formulation of the problem of how life first came about; and no explanation has ever been offered except that of pure chance, which we must suspect to be no explanation, owing to the suspicion that pure chance may itself be a vital phenomenon. In that case, life in the physiological sense would be due to life in the metaphysical sense. ('Some Amazing Mazes, Fourth Curiosity', 1909, 6.322)

Here, an idea closer to Thomas Sebeok's claim is produced: the growth of genuine triadic relationships seems to be coextensive with biology. Again, indeterminacy, now in the guise of 'pure chance' plays the decisive role, but in order to mend the obvious gap between biology and all of the universe, a metaphysical distinction is made, that of biology in a metaphysical viz. a physiological sense. Peirce's idea here seems to be that while biology proper should be taken in a restricted, physiological sense, referring to the evolution of existing organisms, biology in a wider, metaphysical sense should be taken as referring to the conditions of possibility for life inherent in the universe as such. But the trivial fact the such 'vital' conditions must in some sense exist in prebiological physics is not the same as to claim that the cosmos as such is alive or that the cosmos as such consists of signs or symbols. In Peirce's doctrine, the problem is connected to the issue of the status of thirdness. In Peircean phenomenology (at a more general level than semiotics), thirdness is defined as habit, generality, tendency, etc. – but at the same time it is often claimed that thirdness and genuine triadic relations are in themselves semiotic and possess a sign, or even a symbol structure. If that be the case, and if thirdness is generic in the universe, then all of the universe will be semiotic and even symbolic. There is ample evidence that Peirce (sometimes) believed this. As already mentioned, I do not belong to those who see a great metaphysical point in this idea. I see few interesting semiotic aspects in the mere fact of events being governed by laws or tendencies, so I prefer the restriction of the whole terminology of semiotics to biology proper, biology in Peirce's narrow, 'physiological' sense of the word. I have nothing against, to be sure, the idea that physical processes may be described in semiotic vocabulary, but I just do not see that vocabulary adds anything to our knowledge of such processes. Thus, they seem to constitute a sort of semiotic zero-case where semiotic terminology may be added or not. To say that the crater on the moon is an index of the meteor responsible for it does not add to our normal understanding of the physical process having taken place. Rather like you could describe colorlessness as a zeroeth degree of coloring or your could describe the absence of mind as a zeroth degree mind – without assuming that this implies that colorlessness is a special sort of color or that matter is a special sort of mind. Quite the opposite of the physics cases, however, seems to be the case in biology where more or less explicit semiotic vocabulary seems positively indispensable: even the most die-hard reductionists invariably use such terminology and seemingly are unable to avoid it. I would not hesitate to support Peirce's argument that biological nature must, of course, possess conditions of possibility in pre-biological physics, but I think it is to go unnecessarily much too far to claim that these conditions should, in themselves, in any non-trivial way make all of physics alive or vital or semiotic or symbolical in any acute sense of the word. This leaves open, in Peircean terminology, the task of accounting in more detail for the intricate relations between thirdness, biology, and semiotics, to be sure. This issue, however, may not be solved by mere Peirce philology. Rather, it requires much further research, ranging from empirical biophysics and -chemistry and to theoretical biosemiotics.

The primacy of the notion of metabolism in these deliberations (which we have underlined by beginning with it) supports Kant's embryonic idea of the cyclic structure as basic to the 'mediate concept' needed for living beings.[266] Following further Kant's lead, the above deliberations should – despite their clothing in earthly-biological specific terminology – be pertinent for any form for life, not only the terrestrial variants. This leads us to the question of their more precise status.

MATERIAL AND FORMAL ONTOLOGY FOR BIOLOGY

Our hasty sketch of an a priori analysis of the concept of life immediately entails an indefinite host of seemingly a priori concepts (in bold during the preceding paragraphs) – concepts which we recognize in unproblematical everyday use not only in biology, but in a whole range of other sciences where they most often – but not always – count as metaphors. Without further notice we may talk about the metabolism of an economy, the society as an organism, the narration of a text, and, of course, the signs' life in a society. What is the status of these concepts? Let us begin the answer to this question via a detour around the concept of 'emergence'. Emergence refers, in the epistemological debate in theory of science in the recent decades, to the idea of 'new properties' showing up in systems of sufficient complexity.[267] Thus, it seems intuitively suitable to speak of the emergence of life and hence biology out of physics, the emergence of minds and societies, hence psychology and sociology, out of biology, but also e.g. the emergence of multi-cellular biology out of cellular biology, etc. The existence of the concept is, of course, due to the conceptual knot presented by the theory of evolution: if nature, on this scale, is basically continuous – as the theory of evolution maintains (*natura non facit saltus*), then what is to be done with the phenomenologically striking 'domains of being' that nature displays? They must, consequently, have evolved as a sort of discontinuous cuts into continuity, and these cuts are then eo ipso unexplainable within each of these 'domains of being' which the single sciences occupy. 'Emergence' is a philosophy-of-science denomination for this problem. It implies several consequences for its strange in-between position between epistemology and ontology. The level-jump of emergence is for a first glance of epistemological character, because it is given by cross-domain unexplainability (an epistemological property) in the domain-defined disciplines. On the other hand, the phenomenological 'domains of being' seem so strikingly evident that it is tempting to give them an ontological status and hence not to reduce the problem to mere epistemology implying that we in a number of generations might reach an all-encompassing, unproblematic, and continuous unity science. From this point of view, then, the distinction between single sciences of our day would appear left behind like a curious Middle-age cul-de-sac, just like we today see the distinction between the sublunar and supralunar worlds in Aristotelism. The distinctions seem much too evident ever to be subject to such a reduction, and the place of emergence between epistemology and ontology is (yet?) undecidable. Maybe, even, is it in principle undecidable, because every observation of emergence implies unexplainability and hence ontological undecidability – while only an emergence-explained-away can have its ontological status revealed, that is, as a non-existing case!

If emergence shall not, consequently, be only a vague and negative concept for the not-yet-understood, then it must receive more positive determinations. The very idea that one concept may be used to subsume all cases of 'appearance of new properties' does rest on this supposition: that there is, somehow, something in common between the emergence of biology out of chemistry on the one hand and the emergence of psychology out of biology on the other – *in addition to* the mere 'appearance of new properties'. The philosophical reason for accepting a concept of emergence must hence rest on an (most often not explicit, that is) idea of *formal ontology* – formal to the extent that the various level-jumps so named do not have the same material base in common. The concept of formal ontology refers to Husserl's famous distinctions. On the one side, 'formal ontology' is the science of abstract objects and their properties and interrelations, just like formal logic is the science of abstract concepts and propositions, their properties and interrelations. Prominent candidates to formal ontology in Husserl (cf. for instance the famous 3rd *Logische Untersuchung*) is mereology, the science of wholes and parts, the distinction between genuine parts and moments, dependency, the category of object etc. – formal properties assumed to be relevant for any field of knowledge whatsoever. On the other side, formal ontology is contrasted to what Husserl interchangeably calls 'regional' and 'material' ontologies. They are plural and constitute, in fact, his way of dealing with the different 'domains of being'. To him, the regional ontologies are three: 'materielles Ding, Leib, Seele' (cf. *Ideen*, Vol. III), crudely corresponding to physics, biology, psychology/sociology, respectively. What, now, constitutes the very relation *between* these regional ontologies? Husserl does not, to my knowledge, pose this question: this very distinction between material ontologies, is it material or formal? Of course the distinction has a material side (cf. the relation physics/chemistry and biology, where the matter of the latter presupposes and in some sense contains the former). But the distinction between material ontologies must, if the concept of emergence has any philosophical interest whatsoever, have a formal side. If it does not have a formal aspect – that is, an aspect that can be described without reference to the material character of the particular level-jump – then the concept of emergence would degenerate into a purely nominalist denomination of a bundle of problems having nothing ontologically in common. It follows that 'formal ontology' applied to this distinction has another meaning than ontology in the discussion of epistemology and ontology above. There, the question was whether the level-jump did exist as anything but a lack of satisfactory scientific explanation. Here, it is supposed it does indeed exist, and that it possesses both material (specific) and formal (general) properties. But as the concept of emergence is characterized precisely by being unable to 'fill in' the hole in the continuity of explanation, then the formal ontological version of it must necessarily be no exhaustive explanation, but a necessarily not-exhaustive, general description comprising all cases of emergence (if not, they would not be conceiveable as instantiations of the same category). Thus, it seems that either emergence is part of formal ontology, or it does not exist.

Which formal and general properties are now central to emergence? If physics in its present-day version is assumed as basic ontology, in the sense that all phenomena are based on interactions between microphysical elements, then emergence seems to be the coordination of a larger amount of particles into a pattern not itself physical, a pattern which is now the bearer of the 'new properties'. This implies that the objects on the emergent level necessarily have a larger physical extension than the elementary level of physics, as well as the fact that these objects must be characterized by some kind of coordinated, collective behavior taking place between the physical entities of the lower level. Here, the concept of *form* becomes central to understand the pattern formation of collective behavior. But if this model – generalized from the prominent level-jump physics-biology – is valid, then it sets certain restrictions on which other phenomena may obtain status as emergence. The jump from cellular to multicellular biology seems to satisfy these requirements, in so far we here have a case of pattern-forming behaviors of entities defined on the lower level, and multicellular organisms are then the entities pertaining to the upper level. This consequence now meets the hindrance that we usually understand biology as being one, from virus to primates – but this intuition might be a consequence of our own being multicellular beings. If we were ourselves thinking bacteria, then the jump from cellular to multicellular biology might assume the character of a violent and unexplainable jump . . . This formal ontological point of view implies another consequence for the jump biology-psychology. If the emergent level is characterized by being composed by collectively acting parts from a lower level, then the *object* of psychology can not be the psyche; then its object must be the brain (or even: organisms equipped with brains), brains which, in turn, as a new and unique *property* have the psyche. The idea is not that the object of psychology should be the brain in a merely physiological use of the word (if so, it would not be a case of emergence), but that the object of psychology – as any higher order science, must be a complex, coordinated whole consisting of entities from a lower level. In the other end, nothing prevents us from assuming that formal ontological concept of emergence may already be indispensable in physics, cf. the emergence of macrophysical objects (crystallography, hydrodynamics, solid matter physics, meteorology, etc.) out of interactive patterns of microphysics. Here, it seems, formal ontological concepts related to emergence are needed in addition to the basic formal ontology at stake in microphysics and elsewhere: concepts such as self-organization, complexity, level, etc.

BIOSEMIOTICS AS MATERIAL AND FORMAL ONTOLOGY

These deliberations now take us to the last main point of this chapter. If we accept the regional a priori character of the basic concepts of biosemiotics, and if we accept the formal ontological character of the concept of emergence, then the road is paved for the conception of biosemiotics as a special branch of formal ontology, as pertaining to a specific subclass of emergent interactive pattern-forming systems. Valid for the material region of biology, the basic biosemiotic ontological categories also hold for higher emergent levels of psychology, sociology, etc. (even if these will add, of course, their own further categories). Material ontologies are *nested*, and they inherit the material categories of the lower levels as formal constraints on their own level. Barry Smith has taken up the study of formal ontology in the wake of the Austrian tradition from Brentano and Husserl, and in some recent papers he suggests to introduce the concept of 'niche' as a fundamental concept in formal ontology. As predecessors, he refers to Aristotle (his definition of place, 'khora'), von Uexküll (the *Umwelt*), to the biologist Evelyn Hutchinson's modern definition of the 'ecological niche' (as a 'volume in an abstract space determined by a range of physical parameters pertaining to food, climate, predators, parasites, and so on', Smith and Varzi 1998b, 201), to the psychologist Roger Barker ('physical-behavioral units'), and to the founder of the 'ecological' theory of vision J.J. Gibson. Smith remarks the strange fact that despite all human action is defined by the niche it belongs to, very little attention has been paid to the subject, and almost no attention to the subject as an ontological issue. In his own development of formal ontology, he takes his departure in the Husserlian idea of mereology (which, in contrast to set theory, does not assume knowledge on any basic elementary level and thus is phenomenologically 'bottomless', cf. Chap. 7). Because of its high generality, mereology still lacks notions to grasp connected objects (it can not distinguish between, say, a composite red object like a red car, and the mereological sum of all red objects), and some sort of *topology* must be added to the formal ontological repertoire, making it possible to formalize the notion of bounded, connected objects. Furthermore, formal ontology must be able to assign its objects places in a space, giving rise to *locality* as a third formal ontological basic concept. Now, the concept of *niche* (or, physical-behavioral unit, setting, environment) must be added in order to understand the organized space-time cell which an agent and its specific actions unfolds around it. In Barker's version, the 'unit' is a bounded space-time region in which a human activity takes place and unites its parts into a whole (a chess game, a football match, a real estate sale, an education process, etc.). 'Environment, person and bodily interior are thus combined together topologically within a nesting arrangement.' (Smith, 12) – thus agent and environment is connected in a non-random pattern. Gibson's 'ecological niche' permits Smith and Varzi (1998b) to generalize this idea to any object which is tuned into its environment by a specific set of perceptions and actions.

Now, if 'niche' is a sufficiently general formal ontological concept as to cover the bowling bag containing a number of bowling balls (Smith and Varzi (1998b), 22), and Barker's 'physical-behavioral unit' is a special subcase with a human agent –

then biosemiotics seems to be the formal ontology for those niches containing a living agent equipped with metabolism and *Umwelt*. So, we would get a relation as follows:

Physical-behavioral units $\subset$ *metabolic niches* $\subset$ *niches in general*

In our a priori analysis of the concept of life, metabolism with its ensuing networks concepts was a central idea, and the whole biosemiotic vocabulary seems, in fact, to be constructed in order to suit into the metabolic cycles of an organism. The genes are codes in so far they direct the details of the metabolic growth of an organism; the messenger RNA brings a message relevant for the direction of the metabolism, just like animal perception is relevant for the continuation of metabolism. In short, it is the cyclical organization of metabolism which makes it meaningful to speak of 'intention' (whether conscious or not), because the directedness of intention, be it inside the organism or directed outwards into the niche is governed by the cyclical attractor of metabolism. René Thom once remarked that extrapolation of the concept of formal cause in biology made it possible to let it subsume the effective and the teleological cause, respectively, in so far they are only two different viewpoints on one and the same metabolic process.[268] Thus the biosemiotic vocabulary centered, like Kant predicted and Cassirer further argued, around the concept of intentionality, of telos, formally interpreted as cyclic pattern or order, gives meaning in relation to the notion of the cyclical flow of metabolism.

But is this not a bunch of anthropomorphisms, the sceptic will say. Is this not only a glorious illusion projecting semiotic concepts from the human sphere onto our innocent surroundings? In so far we ourselves engage in biological niches (and even extend and develop them into sophisticated psychological, sociological, etc. niches) we of course know the semiotic vocabulary from phenomena in our own everyday life – but this does not imply it is a subjectivist vocabulary; it simply pertains to objects which possess the same ontological structure as our niches. Of course, one must be careful that the biosemiotic vocabulary is kept objective: there is no need to assume, for instance, that our idea of an intuitive content, including qualia, consciousness, etc., will be pertinent for lower-order biological niches. When speaking about 'signs' at these levels, a qualia-less, consciousness-free sign concept must be assumed before, in each case, to investigate whether anything makes probable the presence of qualia. Biosemiotics as a formal ontology must be completely emptied for any presuppositions about consciousness, this being (maybe only until further research, it should be kept in mind), a purely empirical question (even if possible empirical facts veri- or falsifying the assumption of consciousness in other beings are of course hard to come by).

But the decisive hint which makes us think biosemiotics is in fact the relevant formal ontology for higher niches is precisely the fact that its notions seem to be transcategorical. We may use the vocabulary not only about organisms and human beings, but also about those meta-organisms at higher levels made out of patterns in organism and human behavior – families, groups, firms, countries, cultures, etc. – in economics, sociology, political science, psychology, historiography. One animal society may signal to another, one anthill compete with another, one species with another; one state may bring message to another, compete with another, wage war on another, enter into symbiosis with another. Biosemiotics as a material ontology for biology, in fact, has formal bearings for all complex systems and thus become part of a formal ontology for emergence and the behavior of complex systems in general.[269]

This finally raises the issue of the relation between formal and material ontologies, respectively. Semiotics, on the one hand, seems to be transcategorical and its categories applicable on all complex systems with a metabolism; on the other hand, it seems to have a privileged relation to biology as its most simple instantiation.[270] Thus, the seemingly inevitable semiotic vocabulary in biology has no equivalent in physics (that is, in non-biological physics). Hence, even if being a piece of formal ontology, biosemiotics does not seem to possess the same generality as for instance mereology. But this points to the fact that material ontologies must also be described by formal means. We have no direct, extra-formal access to these ontologies; hence they will rather be characterized by the amount of formal tools appearing necessary for their description. Thus, biology and higher complex systems can not be described adequately without a set of Kantian 'mediate concepts' of biosemiotic kin while physics (present-day physics, that is) seems to be doing fine without it. Simple, that is, pre-biological physics seems to be a semiotic zero-case where the introduction of the formal ontological biosemiotic vocabulary does not lead to new insights: the crater on the moon as an index makes no harm, indeed, but does not add anything to the normal physical understanding of the process. On the other hand, sufficiently complex physical systems call for the vocabulary of 'mediate concepts' and are indeed, despite hard attempts, impossible to describe without it.

What is, on the other hand, gained by the recognition of semiotic vocabulary in biology? It is striking that biosemiotics has hardly, until now, yielded any new empirical biological insights. The recognition of semiotics as formal ontology rather has the role of explaining the necessity of the omnipresence of semiotic terminology in biology as it already is pursued and thus liberates biology from vain attempts at reducing it. Furthermore, it encourages biology to develop its spontaneous semiotic vocabulary into a common, agreed-upon, ontological conceptual system – just like the case in theoretical physics where the material physical ontology of powers, matter, fields, mechanical causes and effects, etc. is under constant critical development and are never taken for dispensable metaphors only. It consists, then, rather of a material ontological framework necessary for the empirical findings of biology to be understood precisely *as* facts of biology – and, in addition, serves as a formal ontological tool for higher domains.

CHAPTER 10

A NATURAL SYMPHONY?

Von Uexküll's Bedeutungslehre and its Actuality

The Baltic-German biologist Jakon von Uexküll, one of the founders of ethology, enjoys a renaissance these years, both in the discussions of biosemiotics and of 'embodiment' in philosophy and cognitive science. Thus, von Uexküll's basic diagram of the 'functional circle', characterizing every animal species, is central to the current biosemiotic attempts at understanding the foundations of biology. But von Uexküll's fertile idea is not unanimous – several different and even mutually contradictory ideas compete in his conception of it. This gives rise to certain dangers in biosemiotics which must be tackled in an analysis of what is involved in the ontology of 'functional circles'. For an overall historical view, von Uexküll's theoretical biology is a main contribution to the 'developmental' or 'epigenetic' trend in the biology[271] of the recent centuries – and like so many of the contributions to this undercurrent of biological thought it is not without a certain involvement with vitalism. Since the establishment of the so-called 'Neo-Darwinist' synthesis and the discovery of DNA, this tradition has been marginalized in the biology departments concentrating upon biochemical empirical research and a Neo-Darwinist philosophy privileging the genotype as the core object of biology. Consequently, the developmental or morphogenetic point-of-view with its interest in biological form and phenotypes has, as a tendency, been relegated to characters on the periphery of the biological institutions, e.g. ethologists, philosophers, semioticians, catastrophe theorists, complexity theorists, etc. From this exile, morphogeneticists keep claiming that biochemical reductionism will never be able to explain the phenomenon of the living being in its entirety – and is consequently, from Neo-Darwinist side met with routine accusations for vitalism, the harshest four-letter word in contemporary biology.[272] My contention is that there remains indeed valuable thought in this tradition, but, on the other hand, the vitalism accusations are often correct. What needs to be undertaken, then, is a task of criticism: the insights of the epigenetic tradition must be purged for vitalism in order to fit into our actual level of knowledge where *élan vital*, unknown organic force fields, and the like are definitely out of question.

Jakob von Uexküll is a perfect example of the ambiguity of this tradition – his work contains many acute critical corrections to our days' Neo-Darwinian orthodoxy in biology – at the same time as being marred with an irrationalist vitalism which is, at best, to consider aged. Its insistence on founding a theory of life *sui generis* – that is, not a theory with a physicalist reductionist perspective – is sound, in so far as it points to central issues nowadays left to ecologists, ethologists, and artificial life philosophers presumably dealing with more superficial layers of biology than biochemistry. Its central concept of 'Umwelt'[273] as the stable world of possible influences and reactions forming the environment of an organism is an important correction to a Darwinism taken to the extreme, taking genes and their mutations as its only ontological entities endowed with causal efficiency. Yet, on the other hand, von Uexküll's ardent anti-Darwinism and more general anti-evolutionism makes his theory a bizarre brand of creationism: the biological world is created by a 'composer' – inferred from a throroughgoing musical metaphor in von Uexküll's thought, in turn inherited from von Baer[274] – and even if von Uexküll must admit that the span of biological life must be counted in millions of years and whole kingdoms of animals, he does, in fact, little more than stretch the seven days of creation into a longer timespan for a composer god's work to unfold. These latter features of his thought, unfortunately, make it a hard job to evoke a serious scientific interest in his endeavor.

To some extent, the picture is parallel in semiotics. Being one of the founders of biosemiotics *avant la lettre*, von Uexküll is, on the other hand, a semiotician whose conception of *Bedeutung*, of meaning, is to a large extent caught up in an subjectivist and skepticist ontology and epistemology which has little attraction for a modern scientific semiotics. Now that von Uexküll is once more being rediscovered – largely thanks to Thomas Sebeok's efforts[275] – there is no reason that semiotics should inherit what has hindered his influence in biology. In this chapter, I shall try to distinguish von Uexküll's fertile contribution to semiotics from its clothing in irrationalist vitalism – being fully aware that in so doing, we might deprive him of some of his attraction for more *schwärmerisch* or constructivist admirers as well.

THE SEMIOTICS OF THE FUNCTIONAL CIRCLE

The backbone of von Uexküll's theory of meaning, of course, is his 'functional circle' constituting and delimiting the *Umwelt* of the organism in question. The circle closes the two processes of perception and action, *Merken und Wirken*, into one continuous movement so that one becomes the presupposition of the other, and, consequently, the one becomes the *telos* of the other, in a never-ending teleological circuit with no static goal.[276]

Perception is undertaken with respect to subsequent action, and action permits survival and further perception. In doing so, von Uexküll performs, in fact, a formal 'secularization' of the so contested notion of teleology. The telos does not consist in any goal outside the functional circle, but is simply defined by it.[277] The organism is able to perceive a limited range of stimuli from the surrounding world only, and furthermore, is able to respond to those stimuli in a correlated series of specific actions, both to be conceived of in semiotic terms: both form signs to be read and performed by the organism. These stimuli and responses make up its functional

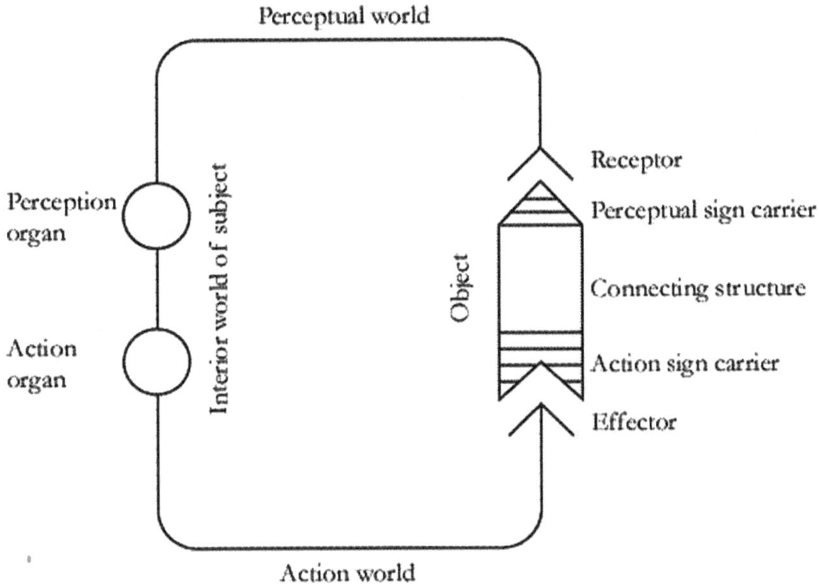

Figure 30.

circle which, in turn, constitutes its *Umwelt* as the repertoire of relevant perceptions and actions.[278] Thus, there are very large differences between the possible *Umwelten* of different species. To take one of von Uexküll's favorite examples, the tick,[279] waiting patiently, maybe for years, on a tree for a mammal to come by, is deaf, blind except for a primitive luminosity sensitivity and has a very limited sense of smell, tuned to detect butylic acid as a sign of mammal sweat. This *Merkmal*, if present, now releases a *Wirken*: The animal drops, and if hitting the mammal, its temperature sense guides it through its furs to the warmest – and thinnest – spot on the skin where the next *Wirken* takes place when it sinks its jaws into it and sucks blood in order to become able to procreate by parthenogenesis. Guided by light, the young ticks crawl up on a straw or a leaf to repeat the cycle. Correspondingly, the *Umwelt* of the tick is very poor, constituted by light, temperature and the smell of butylic acid and containing mammals only – mammals, to be sure, which have very little (but not nothing) in common with our mammal concept in so far as they consist only of sweat, heat and blood, so to speak. On the action side, the *Umwelt* contains little more than waiting, falling, crawling, and sucking. Higher animals, of course, hold much more complicated *Umwelten*, but the common and defining feature of all of them is the fact that these *Umwelten* are constructions of signs – *Merkzeichen* and *Wirkzeichen*, respectively: perception signs and action signs. These signs as read and enacted by the organism give rise to the objects of its *Umwelt*, any object being composed of *Merkmale* and *Wirkmale*, respectively: perception properties and action properties. Thus, an equivalent name for the functional circle simply is: meaning circle.

NATURALIZED KANTIANISM

This, in itself groundbreaking, theory of a large variety of phenomenal worlds in different animal species is now interpreted in a strange ontology. The basic framework is inherited from Kant, and von Uexküll thus forms part of a tradition that could be called 'neo-Kantian semiotics'.[280] Kant may, of course, be interpreted in several ways, and von Uexküll's Kantianism is not without problems. Taking his point of departure in a subjectivist reading of Kant, von Uexküll states that

> 'All reality is subjective appearance'. This has to be a great, fundamental understanding also of biology [...] The task of biology is to widen the results of Kant's research in two directions: 1. to take into account the role of our body, especially of our sense-organs and our central nervous system and 2. to explore the relations of other subjects (the animals) to the objects.
> (quoted from Thure von Uexküll 1982, 9[281])

This naturalization of Kant implies two worthwile endeavors as quoted: to extrapolate the phenomenal world's constitution from a naked transcendental subject and to the body in a Merleau-Ponty-like embodiment intution, and to extrapolate its range from our species only and to other species in an ethological intuition. But at the same time, von Uexküll – as is evident from the opening statement in the quote – falls prey to a widespread German temptation at naturalizing the constitutive subjectivism as well, with a result not entirely different from Nietzsche's perspectivism. The semiotic theory just outlined is seen as the constitutive theory of all other possible thought; physical laws of nature, for instance, become mere extrapolations and abstractions in the specific human *Umwelt* – and the *Umwelt* in general being something that arises by a 'projection' on part of the organism. Here lies one basic flaw in von Uexküll's reasoning: it is not possible to take one feature of our *Umwelt* – the organism in question – and hold it outside the constitution, only to let it form the basis for the consecutive construction of its *Umwelt.* The organism is not, of course, able to constitute itself, so two possibilities remain – either the naturalized subject as organism is kept outside constitution (but then it is no longer constitution in a transcendental use of the word, in so far as empirical phenomena (the organism itself as phenomenon) remain unexplained by the constitution), or else the organism was already in an *Umwelt*, namely the scientist's – pushing the constitution one step further back and losing what was gained in naturalization. Here lies a dilemma: either the naturalization of the transcendental subject to organisms is pulled through – or else the constitutive function of the subject is maintained. You cannot have both. A parallel flaw is seen when the Kantian forms of intution space and time are sought naturalized (otherwise with very interesting consequences, see below): the specific signs of orientation in the *Umwelt* (which in themselves form the elements of a sound theory of biological environment mapping) are taken to constitute time and space *as such*. This analysis is undertaken in the following steps. Organizing signs are distinguished from content signs, so that the former set of signs constitute time and space from more basic sign types: local signs, directional signs, and impulse-to-operation signs (*Wirkzeichen*) respectively. The latter, what we would nowaday probably class under the heading of proprioception and kinaesthetics, even permit us to construct the category of space without any external experience, from the perception of our own body and its movements only.[282] But at the same time as all this, we are told that the *Umwelt* is constituted, in turn, by the very *projection* of these signs of time and space onto the organism's outside. But the flaw in von Uexküll's argument is evident here from the fact that 'projection' and 'outside' are *already* spatio-temporal concepts which may not, then, be tacitly presupposed in a time-and-space constitution theory. So before the organism is supposed to construct space from various *Ordnungszeichen*, it is already itself localized in a surrounding space which it consequently invades and imposes order upon by these signs – hence space is presupposed before this organizing which henceforth may not be said to constitute space as such in any foundational sense of the word, merely in an epistemological sense, relevant only for the specific Umwelt in question.[283] Which status von Uexküll ascribes to this common field in which the various *Umwelten* unfold, varies in his work,[284] but he in any case maintains the inaccessibility of that *Ding-an-sich*-like domain. Yet, in *Theoretische Biologie* we read that

> The material from which a foreign Umwelt is constructed consists in all cases of our objectified quality, because no other qualities at all are accessible to us. The only difference to our Umwelt consists in the fact that they have a lower number. [...] A location which lies more to the left to us than another, lies also in the foreign Umwelt, when both locations are present as perception-signs in them, farther to the left, also when the number of location-perception-signs distinct from one another is less than in our world.[285]

Even if perceptual granularity (see below) may vary across *Umwelten*, directions are thus preserved in mapping between them. But this invariance constitutes objective space. von Uexküll evidently agrees: to him, objectivity is correlative to spatial determination, but still this determination is strangely seen as subordinated to the overall subjectivism. But if space is not only objective as understood in relation to a specific *Umwelt* but also across *Umwelten*, it ceases to be completely relative to the subject; only the granularity of its construction remains in some sense subjective. This is also evident in von Uexküll's very methodology: we are not supposed to discover other species' *Umwelten* by psychological *Einfühlung*, but precisely by empirical observation of which impulses may be perceived and of which actions they release – comparing these observations with our own *Umwelt*. von Uexküll's contention that the animal *Umwelt* in question must always contain fewer locations than ours is an evident fallacy that may have two reasons. One is an anthropocentric belief in the extreme perfection of the human *Umwelt* as compared to animal *Umwelten* (hardly probable, given von Uexküll's tendencies towards relativism), the other is a methodological idea that we may only reconstruct in other *Umwelten* what is already in our own. But we may easily observe supersonic hearing or ultraviolet vision in other species, just like we may observe far finer granulations of perceptual space than our own (the sense of smell in many carnivors, e.g.), given the relevant observation and measurement techniques. So the granularity or range of a construed animal *Umwelt* needs not, as claimed, be more gross nor narrow than ours.

The very objectivity in our construction of other species' *Umwelt* presupposes the possibility of cross-*Umwelt* mappings of at least two kinds. One is identity mapping: the mammal that the tick smells via a molecule of butylic acid is the very same creature as we ascribe completely other perceptual properties (if not, we would not be able to construe the mammal inventory of the tick *Umwelt*). Another is task similarity: being able to specify the functional circles of another species requires that we classify the circle in question in analogy with our own functions (cf. von Uexküll's circle types of food, flight, sex, etc.). The first of these mappings requires spatial, indexical invariance (the two *Umwelten* in question share the same underlying space, even if describing it differently topologically, geometrically, metrically, etc.), the second requires semantic, iconic invariance (the two *Umwelten* in question share functional circles with more or less analogous telos and specific actions).[286]

To sum up, there is a constant tension in von Uexküll's naturalized Kantianism, evident in quotes such as

> Every action, therefore, that consists of perception and operation imprints its meaning on the meaningless object and thereby makes it into a subject-related meaning-carrier in the respective Umwelt (subjective universe). (von Uexküll 1982, 31)

At one at the same time, the organism is seen as constitutive of reality as such, on the other, a 'meaningless object' – heir to Kant's *an sich* – is supposed to exist out there ('on') *before* these constitutions. But either constitutive subjectivism or naturalized subjectivism without constitutional abilities – you cannot have both. You cannot keep your subject and eat it too. Flaws of this kind are inherent in any attempt at naturalizing Kantian subjectivisms (the arch-example of course being Nietzsche): in Kant, the subject remains transcendental (which raises lots of other problems, to be sure) and this trap is avoided. To us, the choice is easy: naturalized subjectivism must be chosen at the price of constitution – this permits us, on the other hand, unproblematically to accept indexical as well as iconical cross-*Umwelt* mappings.

WHAT MAY CLAIM TO BE A 'SUBJECT'?

Another peculiarity deserves to be mentioned: In von Uexküll's presentation, the predicate 'subjective' is often connected to single concrete organisms, so that there is a tendency in the concept of *Umwelt* to be understood as something taking place within the (proto-)mind of the single animal. This tendency might be more dangerous in certain von Uexküll interpreters celebrating the dissolution of the universe in a 'pluriverse' of subjective experiences. But von Uexküll's *Umwelt* construction method takes its departure in objective, spatial qualities characterizing behavior and consequently does not imply anything about the quality of subjective what-is-it-like-to-be-a-bat experiences. Thus the word 'subjective' in von Uexküll's Kantian declaration of faith above should not be conceived in any awareness-consciousness-like use of the word. von Uexküll does not exclude the possibility of a research into this domain – presumably by animal psychologists – but he maintains an objectivist stance as to the biological *Umwelt* reconstruction, even if he now and then comes close to using the predicate 'subjective' in a psychological way. Let us here stick to (methodological) anti-psychologism: the decisive feature is the existence of functional circles, just like there is in semiotics, in general, no reason to restrict signs to conscious intentions.[287] The methodological principle is stated in *Theoretische Biologie* but is also present in the *Bedeutungslehre,* for instance in the idea that spontaneous reflexes should be included in the semiotic interpretation of the functional circle:

Even the simple blink-reflex, caused by the eye being approached by a foreign body, does not consist of a mere sequence of physical causes and effects, but of a simplified functional circle, beginning with percepiton and ending with effect.That the functional circle, in this case, does not pass through the cerebral cortex, but makes its way through lower centers, changes nothing. (34)

It thus seems perfectly reasonable to speak of the *Umwelt* as the whole set of types of perceptions-reactions – not necessarily conscious – of the animal. At the same time, it also seems – as von Uexküll also in some cases does – reasonable to speak of the *Umwelt* of a whole species ('the tick'). But, having gone so far, is there any reason not, even more generally, to admit the *Umwelt* of more complicated systems like ecologies, corporations, institutions, economies, or the like? If that be the case, then the 'meaning' of his *Bedeutungslehre* would cease to be something taking place in a proto-consciousness, but rather be an objective structure defined in relation to sufficiently complex systems (probably what the general theoretical biology of the current Santa Fe school nicknames CASes – complex adaptive systems). In the same movement, von Uexküll's theory will be be cleansed for its tendency towards scepticism, as the *Umwelt* ceases to be a projection due to a subject and becomes instead a self-organizing entity involving an organism and its environment.[288]

MEANING AS MUSIC

Another way of addressing the meaning question in the *Bedeutungslehre* is the idea of the apparent harmony in the incredible fine-tuning between different species' *Umwelten* in ecology. The von Baer musical metaphor for the organism and for biological correspondences in general pervades the whole of the *Bedeutungslehre* and is already present in the *Theoretische Biologie*. There are at least two issues at stake here: the idea of biological perfection and the (more implicit, to be sure) idea of categorical perception. In the *Theoretische Biologie*, a strong emphasis is placed on the perfection of every single species. Darwinism is counterargued by the idea that adaptation implies imperfection being pervasive in nature, and evolution is rejected (or its role, at least, minimized very much) by the idea of the perfect fit between an animal and its environment. In this static idea, a series of problems in von Uexküll's position becomes clear. The perfect fit between animal and *Umwelt* of course requires the static nature of the latter. In a changing world, no perfect fit may a priori be assumed, but rather a continuum of differing degrees of fitting, because environmental change naturally will disturb established animal-*Umwelt* fits. The idea of a perfect fit has its correlate internally in the functional circle, in so far as it is assumed that, in simple *Umwelten*, to every *Merkmal* corresponds a *Wirkmal* and vice versa. This evidently goes for the tick and species of similarly low functional complexity, but it seems that as soon as we reach higher animals, a considerable openness of the *Umwelt* softens up the one-to-one mapping between *Wirkmale* and *Merkmale*:

> As soon as body profiles appear as perception-signs, the picture of the perception-world changes thoroughly, for now relative positions in space begin to play a decisive role. [...] Only when spatial differences appear within the perception-sign itself you may talk about a higher stage of perception-world. The presupposition for the existence of objects in the perception-world of an animal is provided by the ability of the animal to construct its own functional rules for action.[289]

The first quote assumes (tacitly, that is) that higher perception worlds become more objective by the introduction of spatial environment mapping; the second connects this higher perception to the possibility of the making up of new functional rules on an ontogenetic time scale, that is, learning. This latter idea in fact connects the objectivity of spatial mapping with the emergence of pure objects (apart from their role in functional circles) and a freedom in action as a result of the tendential dissolution of the very fixed functional circles in lower *Merkwelten*. This remarkable differentiation of *Merkwelt* types reaches its peak in the idea of a specific openness in the human *Umwelt* connected to the appearance of neutral objects. But at the same time, these almost enlightenment ideas run contrary to the dogmatist perfection principle. For it seems difficult to conceive of the invention of new functional circles without assuming some kind of trial-and-error test of them. And the very 'awareness' of neutral objects in an *Umwelt* corresponds to a heightened sensitivity for unknown events and dangers – that is, a breaking up of the perfect fitting-in, as well of *Merk-* and *Wirkwelt* respectively, as of functional circle and *Umgebung*, respectively.

So it seems that in von Uexküll's musical metaphor, perfection should rather be left out (as a necessity, not, of course, as a relative possibility). And with it, the Schelling-like idea of the whole of the universe as one harmonious symphony. What is gained, on the other hand, is the ability to act in a changing environment due to general-purpose perception and actions, not tied to specific *Umwelt* functions – improvisation, to include it in the musical metaphor.

But other interesting sides of the musical metaphor remain. In a central example, von Uexküll considers the pea beetle larvae, being born from an egg in a pea, which makes its way to the outside of the pea, only to drill this tunnel and bite a door in the pea skin while the pea is still soft, so that it later, as a grown-up beetle after metamorphosis, can take advantage of this passage to get out even if the pea is now irreversibly hardened. Furthermore, a certain wasp may take advantage of the existence of this tunnel and lay its egg in the larvae after this tunneling operation, so as to use the larger larvae as food supply for its own larvae, sure to find the same tunnel ready for its exit after having digested its constructor and become a grown-up wasp. Examples like these are impossible by trial-and-error, argues von Uexküll, and he takes recourse to the musical metaphor in order to understand such subtle intercouplings between *Umwelten*. Fine-tunings like this example are examples of a harmony, he rightly claims, and he quotes the famous Goethe stanzas about the pre-established harmony between eye and sun ('War nicht die Auge sonnenhaft...') in order to depict this harmony.

Thus, the beetle is pea-like and the pea is beetle-like in this Goethean reciprocity (and both presumably wasp-like etc.); the whole of nature is fused together with meanings of this kind where *Umwelten* marvellously fit each other locally. While the naturalized subjectivism tends to make it impossible to see beyond the horizon of the *Umwelt*, the musical metaphor makes possible an inference to mend this problem: the reciprocity of *Umwelt* meanings is made possible by the composer of nature's symphony, presumably a Mozartian god whose hands interchangeably play over the claviature of nature's possibilities during the course of earthly time. Even if any evolutionist must immediately reject this far-stretched creationism, there remains something of the Goethean harmony picture even if we secularize von Uexküll's vision and grant evolution its rightful role as the 'creator' of these harmonies. There remains, namely, a gestaltist and hence non-irrational account of the organization of the life of an organism. In a through-and-through vitalist as Hans Driesch, we find the interesting idea that the entelechy, the organizing principle, emerges out of an 'ultra-dimension' in addition to the physical constraints in space and time. But this dimension needs not be the refuge of unknown forces. The melody – arch-example for the Gestalt Theorists from von Ehrenfels, Stumpf and the early Husserl onwards to the Berlin and Graz schools – articulates an organized structure disconnected from the here-and-now of physics and implying a teleological circle foreseeing the last note already by the intonation of the first. Thus – as Merleau-Ponty remarks[290] – this metaphor makes it possible to see the life of the individual organism as a realisation, a variation of the theme, requiring no outside vitalist goal – a variation, we may add, that forms the condition of possibility of the modification of the animal's system of functional circles and hence the acquiring of new habits, possibly to govern evolutionary selection in Baldwinian evolution. Here, the lucky possibilities of Uexküll's melody metaphor go beyond his own static perfections intentions: music may be perfect, but it is far from always the case. Sometimes only dissonant tuning of the instruments may be heard, in other cases false breaks during a melody or more or less successful new variations on a theme; in any case, evolution as a whole might be a symphony, but only if it is one of continuous improvisation.

But there is even more to be gained from that metaphor. We may ask: what are the means used by nature in harmony – of course reconceived in an evolutionary perspective rather than in von Uexküll's religious vitalism?

CATEGORICAL PERCEPTION

Let us take a closer look at some of the implications in von Uexküll's musical metaphor:

> We know from Sarris's experiments that a dog trained to the command 'chair' learns to sit on a chair, and will be on the look-out for other seating-accommodations if the chair is removed; indeed, he searches for canine sitting-accommodations, which need in no way be suitable for human use.

> The various sitting-accommodations all have the same 'sitting-quality' (*Sitz-Ton*); they are meaning-carriers for sitting because they can be exchanged with each other at will, and the dog will make use of them indiscriminately upon hearing the command 'chair'.
> Therefore, if we make the dog a house-occupant, we will be able to establish that many things will have a 'sitting-quality' for the dog. A great number of things will also exist that will have an 'eating-quality' (*Fress-Ton*), or a 'drinking-quality' (*Trink-Ton*) for the dog. The staircase certainly has a 'climbing-quality' (*Kletter-Ton*). The majority of the furniture, however, only has an 'obstacle-quality' (*Hindernis-Ton*) for the dog – especially the doors and cupboards, which may contain books or washing. All of the small household effects, such as spoons, forks, matches, etc. do not exist for the dog because they are not meaning-carriers. (von Uexküll 1982, 29)

There are several interesting lessons to be learnt from this quote. One, central, is that the musical metaphor pervasive in von Uexküll's *Bedeutungslehre* has the further virtue of permitting him to talk about categorization in meaning without explicitly mentioning it, taking us to the issue of categorical perception already touched upon in Chap. 9. It is by now well known that music listening involves categorical perception so that a small frequency environment around a pure tone is heard as that tone, while the frequencies beyond a certain limit is heard as false tones, until the next frequency window around a pure tone (depending on the scale chosen, to be sure) appears.[291] Thus, the tonal system constitutes a discontinuous categorization of the continuum of frequencies. This fact now permits von Uexküll to use music as a metaphor for biological meaning. All the examples in the quote discuss how the animal ascribes a certain tone to a range of different but comparable phenomena, that is, categorizes them. Edible stuff is categorized, just like drinkable, climbable, unsurpassable, etc. stuff. Thus, it seems a crucial but not explicitly admitted fact in von Uexküll's picture of biological nature as a symphony that this harmony is achieved by means of categorical perception. When the moth hears the bat's supersonic echo-signaling sound (which is the only frequency its ear is able to registrate), then the very anatomy of the ear performs the categorization task just like the case with the tick's extremely restricted and precise sense of smell. The same goes for the pea/pea beetle/wasp example: the pea beetle must be able to categorize pea plants in contrast to all other herbs in order to find a place for its eggs precisely there – and more so with the wasp which has to make a specific subcategory out of those peas having already had a pea beetle visit.[292]

The concept of categorical perception has its origin in psychological linguistics and its research in phonetics: how is it possible that human speakers immediately recognize discrete phonological categories in the phonetic continuum? Only by means of categorical perception, weakening the distinction ability inside a phoneme category and correlatively strengthening the distincion ability across category boundaries.[293] In the same way we must suppose that the dog obeying the 'chair' imperative follows a categorization of 'sittable' objects, and it must be assumed that this category is subject to a version of the same basic weakening and strengthening constraints as in human categorical perception. von Uexküll almost explicitly admits the first of these constraints when he says about the single 'dog chairs' that they '...can be exchanged with each other at will'. Thus I would argue that the whole *Umwelt* biosemiotics should be reconstructed in terms of categorical perception (CP). Now, it is long since admitted that higher animals perform categorical perception[294] but von Uexküll's claim that even single cell animals possess functional circles takes categorical perception back to a much more primitive level of biology. In the previous chapter, we argued that the primitive biological cases of categorical perception point to the fact that the procedure must have stereochemical prerequisites already in biochemistry, and von Uexküll's *Bedeutungslehre* points in the same direction in so far as the meaning circle is assumed to take place in simple reflex processes already and hence must have a basis in relatively simple physiology.

On the other hand, von Uexküll's favorite example leads him to draw too simplistic implications from the seemingly omnipresence of meaning circles (and correspondingly, categorical perception) among animals. His tendency to privilege the stimulus-response-like character of the simpler *Umwelten* which couples one *Merkmal* with one behavior (most examples are from relatively lower animals, and the definition of meaning circle maps *Merk- and Wirkmale* one-to-one...) underestimates the openness he admits in higher animals' *Umwelten*. This as a consequence makes it difficult to tell the single object categorized from the category that subsumes it. This idea corresponds to an early mistake in categorical perception studies, that the weakened distinguishability inside a category should be interpreted as a total lack of inside-category distinctions. Thus, it was supposed that all /b/s are heard as /b/s with no sensibility for dialect, personal or other variations. This idea has long since been given up in categorical perception research, and the same ought to be the case in *Umwelt* biosemiotics. Thus, von Uexküll's late contention that animals can not perceive neutral objects (in categorcial perception terms: that it is unable to distinguish properties in it apart from the properties giving rise to its categorization) is dubious (and runs counter to his more open stance in *Theoretical Biology*, see above):

> The unsatisfactory result of this work [American rats-in-a-labyrinth research], despite the most exact techniques of measurement and their most refined mathematical treatment, could have been predicted, because it was based on the false assumption that an animal can at any time enter into relationship with a neutral object. (27, my brackets)

But it is noteworthy that Uexküll himself, in the dog example quoted above, considers the *Hindernis-Ton* of certain mesoscopic objects in the dog *Umwelt* at the same time as he claims that all objects perceived in an *Umwelt* must form part of specific functional circles. But obstacles are not part of any specific functional circle, be it sex, food, water or other circles, but of a more unspecific and open-ended environment mapping loosened from immediate functional utility. Moreover, the general fact that higher animals may not only be fooled but may themselves discover the fooling points to the fact that they do have the competence for distinguishing different instances of a perceived category, that is, distinguishing objects with respect to properties neutral to a given perceptual category. It is hard to see that this is not exactly what perception of neutral objects is about. There is any reason to believe that higher animals thus master the plural grasping of one and the same object by several simultaneous representations, that is, at least both as types and tokens, and possibly also as symbols for some other type (all three are assumably present in as simple a case as the classic Pavlov conditioning: different bell sounds are categorized as one and the same ringing type, all, in turn, symbolizing general eating). One of the central categorical perception researchers, Stevan Harnad, in his conclusive paper in his state-of-the-art *Categorical Perception* anthology, outlines a three-level categorical perception theory built from exactly these three layers: a continuous, pre-categorization representation; a discontinuous categorization, typification, still on the perceptual level; and finally the possible symbolizing of the perceptual types (for instance as linguistic labeling).[295]

Thus, in higher animals the categorical perception in the meaning circle has considerably vaster complexity and variability than in the moth-bat sound case, and the former case is probably much closer to the extreme variability and extrapolability of the human *Umwelt* than to the latter reflex-like relation. If this is the case, then von Uexküll's sharp distinction between animal *Umwelten* and the human *Umwelt* must also be given up. Of course there are crucial differences between the two but it is not the case as von Uexküll in the *Bedeutungslehre* seems to assume, that all animal *Umwelten* are equally closed in contrast to a specific human openness;[296] if this idea is given up, the higher animals' more sophisticated *Umwelten* may be placed on a scale making the evolution of the extreme plasticity of the human *Umwelt* understandable.[297]

To return to von Uexküll's musical metaphor, there is much in it that can easily be subscribed to: the condition of possibility for nature to link up in these strange 'harmonies' between different species' *Umwelten* depends precisely on categorical perception: the perceptual categories form the tones in the metaphor, and it is their categoricity only that permits them to enter into counterpoint between the single *Umwelten*. The fact that different species may entertain the Goethean likeness-relationship between each other must be due to the reciprocal fitting in of categorical perceptions of their *Umwelten* – only in this can this apparently mysterious likeness consist.

But it is interesting to note that the categorical perception implicit in the *Bedeutungslehre*'s melody metaphor has got some important theoretical underpinnings in the earlier *Theoretical Biology*. That book begins with a presentation of space and time in the naturalized Kantianism we quoted in the beginning. We have already dealt with the problematic assumptions in the foundations of this idea, but if we turn to the formal description of the very mappings of space and time, more interesting observations show up. Both space and time are subject to a crucial granulation in the *Umwelt* construction. *Umwelt* space is constructed – unlike Euclidean geometry – from localities, from *Orte*, just like *Umwelt* time is constituted not from timeless nows, but from short time bits, from *Momente*. The size of these *Umwelt* building blocks is of course relative to the senses and the inner life of the animal in question, respectively. But this implies that the perception and action systems of an *Umwelt* imposes on the surroundings a granularity (the smallest *Ort* in human beings is palpable by the fingertips and visible at close range; the smallest *Moment* the one eighteenth of a second corresponding to the number of pictures per second for movies not to blink). The same is the case for the *Inhaltszeichen* filling shorter and larger chunks of time and space where the ability to distinguish implies a least color (respective tone) unity. These distinction sensivities in space, time and quality, correlative to the perception systems in question, become the condition of possibility for more advanced categorization tasks to be fulfilled. They are performed, now, by *schemata.* In his insistence on this figure in the *Theoretische Biologie*, von Uexküll makes his place clear in the ranks of diagram semioticians. He is especially interested (in contrast to Kant) in empirical schemas, so as for instance object contours making it possible to recognize objects. Such contours are described as 'melodies' comprising many orientations-signs, and

By this process, the melody of the orientation-signs is the gestalting power which, however, only reaches consciousness in the 'gestalt'. The gestalting melody Kant called a 'schema', and the hidden art of gestalting in our mind he called the 'schematism'.[298]

Melodies in this general sense – that is, diagrams – and qualities are what is needed to build the objects of the *Umwelt;* and as the melody metaphor suggests, the schemata may depict spatial (e.g. familiar faces) as well as temporal (e.g. a sequence of movements) regularities. It goes – in von Uexküll – without saying that schemata must be subject to the same granularity as time, space, and qualities (being built from these), and in our Peircean notions we will not hesitate to say that diagram types must be distinguished from diagram tokens and that it is precisely the schematism types that make it possible to use the diagram to categorize a series of individual but related phenomena. von Uexküll places perceptual schemas as crucial to the feed-back control of motor action in higher animals (but does not, strangely, see the action's impulse series as a schema itself), so that sufficiently complicated action must be assumed impossible without implying schemata – organizing general diagrams.

The introduction of schemata fulfills the semiotic prerequisites for the meaning circle: a *Merkzeichen* may refer to a *Merkmal*, but also further on to a *Wirkmal*, a *Wirkzeichen* and a *Wirken*. Thus, the 'meaning' in the meaning circle may be located in many different places, according to which phase of the circuit the attention is directed. The smell of a deer may mean prey to a lion, it may mean the thought of hunt, it may mean bodily preparing for hunt, it may mean the hunt itself. The sight of other lions preparing for hunt may similarly mean deer, prey, hunt – as well as, according to other meaning circles, competition for meat, etc. Any part of the schema governing a functional circle may meaningfully refer to any other part of the same circle: this is where primitive signification lies in the meaning circle and the reason why it gives rise to proto-signs.[299] If we successfully subject sensory material to e.g. our soccer ball schema or 'melody', then this *Merkzeichen* may refer to a ball, to a kick, to the feeling of a kick or to a performance of a kick ... In return, the success of this *Wirken* may now be measured on bodily and mental schemata of a good kick ... As soon as the neutral objects permitted in *Theoretische Biologie* (but not later on in the *Bedeutungslehre)* show up, furthermore, the possibility of applying different schemata on the same sensory material appears, thereby interpreting it merely symbolically, that is experimentally: the ball seen as a globe, as an egg, as a sphere, etc. The semiotic strength of the coupling of category schemata with the functional circle is that the epistemological requirements of the first and the pragmatic requirements of the latter become equally responsible for the categorization inherent in the *Umwelt* construction.

But which kinds of categorical perception are at stake in this concert? A certain amount of confusion has plagued categorical perception research as to the very definition of its categories. Are they defined by the boundaries, distinguishing one category from the other, are they rather defined by the single category's prototype (which may again be defined by different means: typical, ideal, average, etc.), or are they defined with regard to earlier encounters with category exemplars? The idea that one of these definitions should outrule the others seems to be on the way to be given up in favour of a recognition that all of them are mechanisms used to different degrees in various categorization tasks. In any case, these different procedures can be collected under the headline of 'reference points' as a common denominator for boundaries, stereotypes, prototypes, ideals, etc. Similarly, as Medin and Barsalou emphasize (in Harnad 1987), there may be different patterns of sensitivity so that some categorizations or tasks may display heightened sensitivity around prototypes, while others may display lowered sensitivity around them (477). A crucial difference goes between boundary reference and the various prototype references, and one of their suggestions is that

> Perhaps one difference is that reference points are more likely to occur at the boundaries of SP [sensory perception] categories than at the boundaries of GK [generic knowledge] categories. To some extent this may reflect greater innate constraints on SP categories than on GK categories. (475, my brackets)

and that boundary categorization is tendentially more relevant for discrimination tasks (like: is this sound a /p/ or a /b/?) while prototype categorization is more relevant for classification tasks (like: is this animal a predator or not?).[300] If this is correct, we should expect biosemiotic categorical perception to display both, but maybe with a tendency to favor the prototype sort (classification is biologically more basic than discrimination). Uexküll's simplest examples (the tick, the moth) are so simple that they are difficult to determine with respect to the boundary/prototype distinction. Of course the auditory sensitivity of the moth has a lower and a higher boundary, but the animal does not in any way distinguish between what is on either side of it (it simply cannot hear what is beyond it); it would probably be more fair to call it a case of prototypical categorical perception. If the moth's sensibility window is larger than the average bat frequenzy range then we could see it as a simple case of prototypicality. A well-known case of innate ideal reference point is the so-called supranormal releasers which refer to facts like that new-born seagulls accept objects of a certain coloring and form (like a beak) as their mother – but the color-form configuration which has the strongest impact on them is different from any empirical seagull and hence forms an ideal reference point of their 'mother' category. Boundary categorization (which of course may appear together with prototypicality in one and the same category) should be expected where there is a crucial behavioral difference in relation to the two categories on each side of the border, that is, it should be expected in unidimensional gradients where a continuous change leads to ethological difference (behavior difference occuring at a temperature threshold, for instance), or when more complex but very similar categories give rise to different behaviors (similar berries, the one being edible, the other poisonous). Thus, we can imagine perfectly naturally occurring 'pure' cases of boundary CP where no typicality attractor needs be present in the center of the categories, just as we can imagine 'pure' cases of prototype CP where the boundaries are fuzzy or simply do not exist (what are the boundaries of phenomena like 'danger' ...). The more specific musings on these issues must be left to further empirical research, but the point is probably clear on the principal level discussed here: both main types of CP are present already on rather primitive biosemiotic levels and reflect at least a double (icon-token/icon-type) or rather a triple information (icon-token/icon-type/symbol) procedure at stake.

CATEGORICAL PERCEPTION AS A GENERAL SEMIOTIC NOTION

Thus, a closer analysis of a contemporary version of von Uexküll's *Umwelt* forces us to acknowledge categorical perception as a primitive notion in semiotics. But if it be so, then semiotics ought to attempt a reformulation of the originally psychological CP concept in non-psychological terms. The occurrence of categorical perception in both human and animal psychologies seems sufficient evidence for it to be a formal notion not only with psychological relevance. In the previous chapter, we tried to trace the prerequisites of CP in (bio-)chemistry, but one could also point to its possible relevance for other complex organism-like systems with highly selective environment relations (institutions, ecologies, economies, social groups, etc.).[301] A provisional attempt at a formal definition could be: *Categorical Perception is the granularity of Umwelt for a complicated system.* The overall reason for such a granularity (each grain giving rise to a CP) would be economy in environment relations permitting cognition. Or rather, the economy in environment relations *is* cognition, or vice versa. If so, CP is a semiotic piece of formal ontology and should be brought in correspondence with the actual strivings in formal ontology at generalizing von Uexküll's *Umwelt* notion; cf. Barry Smith's idea of the *niche* as a general category in formal ontology discussed earlier. It seems a defining feature of a Complex Adaptive System that it must entertain Categorical Perception in relation to its niche so that certain types of niche stimuli evokes highly specific regulatory actions in the Complex Adaptive System, and that certain stimuli are even able to change the CAS's niche relations in general (this would be the formal notion of adaptation).

If this be the case, then this formal conclusion will allow us to draw one further implication for biosemiotics. We have already ridiculed von Uexküll's ardent anti-evolutionism and his support to an irrationalist vitalism or creationism, and an updated version of the *Umwelt* must of course embed it in an overall evolutionary framework. Still, in the Categorical Perception reading of the *Umwelt*, it contains a more delimited and more precise arguments against certain assumptions in the Neo-Darwinist doxa, namely the idea that all order in nature is the result of natural selection. As a primitive order notion, CP in its formal version seems to be a necessary *prerequisite* for life as such, rather than being a contingent result of selection.[302] Thus, it constitutes a germ of order which may not be the result solely of natural selection. Rather, natural selection seems to presuppose CP in order to have orderly alternatives from which to select. In so far, the general biosemiotic CP notion supports Stuart Kauffman's claim that biological order is no product of natural selection, rather natural selection determines which specific order among many suits a certain environment best.

Thus, to conclude, our reconstruction of the *Umwelt* concept places it as central to theoretical biology and biosemiotics in particular and to semiotics and formal ontology in general. This only happens, of course, on the basis of a thoroughgoing revision of the concept, depriving it of its subjectivist, skepticist and anti-adaptionist flaws in von Uexküll's version. It must be cleansed for all traces of radical constructivism in favour of a moderate, realist constructivism: each species constructs its own *Umwelt* all right, but it takes place in a common world which is not in itself a construction. In so far, this moderate, realist construction is hardly distinguishable from self-organization. If nature is indeed a symphony, we must not expect of it the perfection read off of the final score. We should rather see it as continuous rehearsals and improvisations without composers nor directors which nonetheless results in stunning performances and the continuous invention of melodies and harmonies as well as scales. The contribution of von Uexküll's *Bedeutungslehre* to our days' semiotics, then, is to provide the outline of a rational analysis of the alamodic notion of 'embodiment' in terms of a schematically supported functional circle, thereby indicating the biological roots of meaning as connected to such schemata, or, in the ever resounding musical metaphor, melodies.

CHAPTER 11

MAN THE ABSTRACT ANIMAL

Diagrams, Abstraction, and the Semiotic Missing Link

...the brutes use signs. But they perhaps rarely think of them as signs.

(Peirce, Consequences of Critical Common-Sensism 1905, CP 5.534)

von Uexküll's simple distinction between closed *Umwelten* in animals versus open *Umwelten* for man obviously does not hold. But if biological evolution is – among other properties – characterized by increasing semiotic sophistication, then it becomes an urgent question to ask which decisive semiotic change took place, if any, with the introduction of language during the Homo Habilis period in human prehistory: what is the semiotic missing link?

In this context, Terrence Deacon's much-discussed and groundbreaking book *The Symbolic Species* attempts a new solution. Based on a threefold background of arguments – philosophical, neurological, and anthropological, – Deacon proposes that the main event in the animal-man transition is the introduction of *symbols*. Now, as is well known, the concept of symbol is probably one of the most ambiguous notions in the history of thought, and Deacon takes great care to make precise the version of it he finds central. He picks the notion of symbol found in Peircean semiotics which stands apart from many other symbol concepts by assuming the symbol as a complex derivative notion related to simpler sign types included in its composition. The symbol, of course, presupposes the existence of the simpler sign types of icon and index, respectively, so that it makes sense to say that genuine symbols presuppose icons and indices. Icons are signs defined by similarity to the object they refer to; indices are signs defined by actual connection to the object, and symbols, finally, as signs referring to their – general – object by means of habit. These three classes are not mutually exclusive partitions of the field of signs; rather, higher sign types presuppose and include more simple types. Thus, icons form the most fundamental sign type (with respect to object reference), and all higher types presuppose icons. Thus, indices are only possible in so far as they possess iconic qualities: the footprint in the sand is a prototypical index, in so far as the sign refers to the object having caused it, but it also possesses iconic qualities, in so far as the footprint's shape to some extent resembles that of the foot.

In order to interpret something as indexical, so Deacon, a higher-order relation must hold between two groups of icons (in the footprint case, there must be a relation between the group of possible footprints similar to the actual print on the one hand and the group of possible foot shapes responsible for it on the other). This corresponds to conditioned response in ethology, Deacon argues. Furthermore, in order to construct a symbol, a whole group of indices are related by means of indexical relations between their tokens.[303] This relation, internal to the symbol, is now strengthened at the expense of the intial index's object reference. Thus, the symbol is, as a tendency, loosened from the closer object contact found in icons and indices. They are bracketed, allowing the symbol to function on its own in representation and reasoning, but in any specific interpretation of a symbol, its iconic and indexical basis must be reinvoked, including the possibility of new icono-indexical specifications of it.

This forms the base of a fertile criticism of rival accounts conceiving of symbols as atomic primitives, so for instance the physical symbol systems hypothesis, making symbols simple physical units corresponding in a rather direct way to other physical units. With reference to a range of neurological brain scanning experiments which we shall not go into here, Deacon argues that philosophical points of view related to simpler symbol conceptions lack empirical support. This criticism is aimed especially at Chomsky's transformation grammar and related positions claiming the existence of an innate grammar module in human beings, a module which allegedly is completely lacking in apes. Deacon's scanning experiments point to the fact that sufficiently complex linguistic tasks inevitably give rise to very widespread brain activity including several separated parts of the cortex in addition to the well-known Broca's and Wernicke's areas involved in the production and reception of speech. This finding forms a strong argument against the language module hypothesis, and to counter that idea Deacon claims a symbol hypothesis. Symbol use is taken as the distinctive advantage of mankind in comparison to other higher animals. Symbol use – as a complex phenomenon – naturally involves the integration of a large amount of highly different, more primitive brain competences. Hence, such a hypothesis makes the semiotic animal-man transition more continuous so that symbol use merely would amount to the integration of a series of competences already to a large extent present in higher animals. The neural equivalent to symbol processing is neither a specific module of the brain nor the simple size difference of the cortex – but the *degree of integration* of the human brain which is even neurologically measurable in terms of much longer growth period and a larger degree of neuron interconnection between spatially distant parts (so as for instance the cortex and the cerebellum, very important for the automatization of phonetic aspects of speech).

Deacon proposes a both a general and a specific scenario for the evolution of speech. The specific one features a huge amount of anthropological hypotheses[304] and may be left aside in this context. The more general scenario revives the notion of 'Baldwinian' evolution (after the American psychologist James Mark Baldwin from around 1900).[305] He proposed the idea that seemingly Lamarckian acquired hereditary properties could be explained within a Darwinian framework with a support hypothesis: that animals able to learn new behaviors may be able to direct evolution as a result of that behavior – because the behavior in question forces fellow species members to assume it or perish. In such cases, a huge selection pressure will favor individuals most able to learn that behavior: '... *those congenital or phylogenetic variations are kept in existence which lend themselves to intelligent, imitative, adaptive, or mechanical modification during the lifetime of the creatures which have them.*' (Baldwin 1902, 95). This Baldwinian evolution argument is evidently stronger, the more intelligent the organism in question is. Thus, Deacon's idea is that the passage to symbol use is intimately connected to the fast evolution of the human brain during the recent millions of years. He imagines a scenario, in which embryonic symbol use in small humanoid groups kickstarts a process selecting for higher brains with symbol processing capabilities within those groups, thus speeding up evolution's pace dramatically, with our sophisticated symbol abilities as a result. The evolution of the human brain thus does not precede language evolution – rather, the two form two aspects of the same evolutionary process, and in some sense the semiotic a priori structure of language, via peer group pressure among humanoids, acts as a further constraint on brain evolution.[306]

My aim here is to try and render the semiotic aspects of Deacon's hypothesis – with which I basically sympathize – somewhat more precise. The main problem in it is that even the Peircean definition of 'symbol' is probably much too primitive – as well as too general – to explain the semiotic aspects of the animal-man transition. In addition, there are some terminological problems to be sorted out; Deacon claims he uses Peirce's terminology, but as a matter of fact, he undertakes his own reconstruction of it, changing the higher sign types – indices and symbols – in two ways. First, they are rendered compositional with respect to lower sign types, maybe for theoretical economy reasons; and, second, they are rendered more complex than is the case in Peirce's account, maybe in order to make them 'fit' the ape/man boundary better. Thus, Deacon's explanation of the index seems to cover cases which in Peirce's account automatically would be counted as symbols. Pure indices – in Peirce only possible as a limit case – will be tied to the actual here and now, while it is the privilege of the symbol to possess a general *esse in futuro* and thus form a habit, regulating future behavior. But when Deacon claims that conditioned behavior is indexical only, it seems he already includes future regulating features in indices (which are nowhere apparent in the prototypical footprint index). The problem is, of course, that if we accept the ordinary Peircean notion of symbol, then symbolic behavior becomes widespread in higher animals,[307] and the notion becomes unfit for the task of distinguishing between animal and human behavior. Thus, a simple Pavlovian conditioning making the ringing of a bell release the excretion of saliva in dogs is a full-fledged symbol in Peirce's terminology: it is a habit, a regulation of future behavior, and it connects a continuum of possible bell sounds with a continuum of possible eating situations, thus being a general sign with a general object – a symbol. Accordingly, Peirce's symbol concept includes a wide range of subtypes of very different complexity degrees, ranging from simple terms over propositions to arguments – each of these, in turn, including a whole fauna of further subtypes. Thus symbol use is neither as simple as Deacon presupposes (with respect to symbol subtypes) nor as complex as he presupposes (with respect to the simpler sign types).[308]

Deacon consequently adds some further requirements to Peirce's symbol concept in order to make it approximately fit the animal-man transition (to be sure, he maintains that a few higher animals, mostly apes, maybe parrots, may learn simple symbol systems[309]). What he adds is a Saussure-like systematicity of requiring the co-presence of several interlinked symbols, both paradigmatically (implying the systematic difference between selected expressions) and syntagmatically (implying the syntactical organization of combined expressions). Only a system of this kind, he argues, permits ape and man to skip icon- and index-consciousness and their tight connection to the actual world and indulge in the semi-autonomous world of symbolicity, giving rise to the possibility of systematical counterfactual imagination. But as to systematicity, higher animals do possess taxonomies, both in perception (prey types) and communication (warning calls for different predators) so it seems systematicity is not the only key to the problem.

Thus, it is correct that Peirce's symbol definition is a necessary, yet not sufficient prerequisite to give rise to the construction of counterfactual, possible worlds. The symbol notion seems too weak to account for the specific advantage of human semiotics over animal semiotics. More specific distinction within the field of symbols must be sought as responsible for this decisive jump. Peirce himself only rarely considers the question of animal semiotic behavior, but he is, at least, quite sure that animals' abilities are far more elaborate than for instance simple conditioning. Take this late deliberation (c. 1911):

> Some seventy years ago, my beloved and accomplished school-ma'am taught me that human kind, being formed in the image of our Maker, were endowed with the power of Reasoning, while 'the animals', lacking that power (which might have made them dissatisfied), received, each kind, certain 'instincts' to do what was generally necessary for their lives. At least, so I understood her. But when I subsequently came to observe the behaviors of several big dogs and little birds and two parrots, I gradually came to think quite otherwise. For, in the first place, I gradually amassed a body of experiences which convinced me that many animals, perhaps all the higher ones, do reason, if by Reasoning is meant any mental operation which from the putting together of two believed facts leads to a Belief different in substance from either of those two. Once, for example, while I was driving (...) along a country road that was very familiar to me, a setter-dog that I had never seen before raced past me at the top of his speed. In an instant a turn of the road hid him from my sight. 'Poor fellow!' I thought, 'he races after his master in fear of losing him forever.' A moment later, reaching the turn myself, I saw the dog again, not far ahead of me, but at a point where the road branched, and now sitting on his haunches. He was not panting nor showing the least sign of fatigue, but evidently puzzled which branch of the road [to take]. After a second or two, he started off at the same tremendous pace as before, on the more travelled of the two roads, though being the older and harder, it was not very obviously the more travelled of the two. These alternations, - a halt between two utmost speedings, with no slightest symptom of fatigue, - seemed to me to show plainly that the dog had stopped to consider which of the two branches of the road his master had probably taken; and his sudden choice of the more travelled showed that he concluded that his master would probably do as most people, which was a kind of argument: technically called a 'probable deduction', - the commonest reasoning of a general in a campaign, when information is lacking, defective, or conflicting. (MS CSP 672,2–5, my square brackets)

Peirce continues with another example of a parrot fooling a dog named Spitz. Every day, the dog's master would come home and call 'Spitz, Spitz, Spitz!' in order to take the dog for a walk. If, by chance, somebody else came to the door, the parrot would repeat exactly the same yell, now provoking the dog to run to the door – only to be laughed at by the presumably practical-joking parrot. It is of course very difficult to ascertain the amount or character of reasoning taking place in animals from behavior observation, but especially the first of Peirce's examples seems illustrative for a deliberate choice based on reasoning seemingly widespread in higher animals. If we take this observation as being correct, this implies that higher animals are not only at the command of symbols in general, but also of those most demanding and complex of symbols called arguments and reasonings, involving diagrams. Even if not making it explicit, the dog's reasoning must implement a Y-shaped diagram in some fashion or other, making it possible for the dog to reason about which branch of the Y to chose. We may note, moreover, that the dog's situation at the fork road Y also, at least in germ-like form, contains the construction of another possible, counterfactual world (like 'what if my master had gone the other way' . . .). If the more complex parrot example is to be accepted at face value, it even contains the deliberate construction of a possible world *for* another animal, implying a theory of other minds – a type of behavior which seems well-documented in apes wanting to fool fellow apes away from food, sexual partners, etc., but let us stick to the more easy-to-interpret first example. Of course, the systematic exploration of worlds of alternative possibilities requires a stable representation system probably in the form of interconnected symbols – but the example here goes at least to show that there is probably no upper bound to the complexity of symbol types which higher animals have access to as single signs. Animals do reason, and one could probably find cases displaying both abductive, deductive, and inductive argument use, to take Peirce's argument typology. A full-blown process of reasoning involves all of the three of them, and if animals reason, they master all of them – even if not necessarily explicitly, of course. They may guess, infer, and generalize from experience, respectively; they are rational because they are forced to – a Popperian argument – just like us. So the problem must lie elsewhere. The problem simply does not seem to lie in the the complexity degree of single symbol types, at least when measured on the term-proposition-argument complexity scale of symbols. Deacon is probably on the right track when he looks for the coming into place of a systematic interconnection of symbols making it possible stably to construct, evolve, and research possible worlds differing from the actually perceived world. But what makes the jump from a sophisticated reasoning ability and to a system of symbols possible? Deacon does not go into this question, but I think the explanation might be found in the ability to make signs explicit and undertake explicitly controlled reasonings with them. To Peirce, explicit control is even paramount to reasoning proper – which is why animal reasoning despite many similarities to human reasoning only rarely may qualify as proper reasoning.[310] The way Deacon describes his redefinition of the symbol concept actually indicates that Deaconian symbols are in fact symbols subjected to such explicit control. And the issue of explicit control points to another part of Peirce's work, namely his *abstraction theories* focussing upon the possibility of making stably explicit the meaning of a term.

PEIRCE'S ABSTRACTION THEORIES

Peirce developed no less than two abstraction theories in so far as he finds the colloquial use of the word 'abstraction' at his time (and in ours probably as well) refers to at least two separate and autonomous issues. Both are relevant here. One is the mind's focussing ability, the other its ability to make issues explicit – referred to as *distinction* and *hypostatic abstraction*, respectively.[311] As often in Peirce, the two notions stem from different traditions in Mediaeval semantics of Scholasticism.

Even if the abstraction problem recurs over and over in Peirce's work, he never consecrates a whole paper to unfold it, so an exposition must be based on a series of small notes spread in his published and unpublished work. The abstraction problem surfaces as early as 1867, but Peirce's interest in it reaches a peak in the fertile years of his mature theory of signs in the first ten years of the twentieth century. The first part of it, however, is stated as early as in 'On a New List of Categories' (1867, EPI, 2; 1.549)[312] where the trichotomy of *dissociation, prescission,* and *discrimination* is terminologically fixed. The idea is that there are three modes of separation which may be undertaken in the analysis of a phenomenon, going from the most coarse, being able to distinguish different qualities, e.g. red from blue (dissociation), over one able to distinguish what may be supposed to exist without the other, e.g. space from color (prescission) to the most subtle being able to distinguish what may only be thought of separately, e.g. color from space (discrimination). This terminology remains constant in Peirce, and in 'Syllabus' (1903), the three modes are directly connected to the definition of his three categories:

> In order to understand logic, it is necessary to get as clear notions as possible of these three categories and to gain the ability to recognize them in the different conceptions with which logic deals. Although all three of them are ubiquitous, yet certain kinds of separations may be effected upon them. They correspond to the three categories. Separation of Firstness, or Primal Separation, called *Dissociation*, consists in imagining one of the two separands without the other. It may be complete or incomplete. Separation of Secondness, or Secundal Separation, called *Prescission*, consists in supposing a state of things in which one element is present without the other, the one being logically possible without the other. Thus, we cannot imagine a sensuous quality without some degree of vividness. But we usually *suppose* that redness, as it is in red things, has no vividness; and it would certainly be impossible to demonstrate that everything red must have a degree of vividness. Separation of Thirdness, or Tertial Separation, called *discrimination*, consists in representing one of the two separands without representing the other. If A can be prescinded from, i.e. supposed without, B, then B can, at least, be discriminated from A. (EP II, 270).

Furthermore, these distinguishing abilities are what make the very separation of Peirce's basic categories possible. None of them may be dissociated, however, but:

It is possible to prescind Firstness from Secondness. We can suppose a being whose whole life consists in one unvarying feeling of redness. But it is impossible to prescind Secondness from Firstness. For to suppose two things is to suppose two units; and however colorless and indefinite an object may be, it is something and therein has Firstness, even if it has nothing recognizable as a quality. Everything must have some non-relative element; and this is its Firstness. So likewise it is possible to prescind Secondness from Thirdness. But Thirdness without Secondness would be absurd. (ibid.)[313]

This implies that the three categories are interrelated as follows (arrows here meaning possibility of distinction; broken arrows impossibility):

1. <--/--> 2.   2. <--/--> 3.

The categories may not be dissociated.

1. <---- 2.   1. --/--> 2.
2. <---- 3.   2. --/--> 3.
1. <---- 3.   1. --/--> 3.

A lower category may be prescinded from a higher, not vice versa.

1. <---- 2.   1. ----> 2.
2. <---- 3.   2. ----> 3.
1. <---- 3.   1. ----> 3.

All categories may be discriminated from the others.

This makes the definition of the categories depend on a calculus very close to the mereology of Husserl's 3rd *Untersuchung*,[314] discussed in Chaps. 7–8. The three separation modes may be rephrased as 1) the distinction between autonomous (genuine) parts, 2) the distinction separating a founding content from a founded content, and 3) the distinction separating any moment (founded content, or *unecther Teil*) from its foundational basis. Thus, like in Husserl, the separation modes are crucial to the explanation of the status of properties (as moments) and their foundation interrelationship (the fact that color properties are founded on spatial properties, not vice versa), and the separation modes can be seen as the devices necessary for isolating general moments in the phenomenon. Prescission and discrimination are the most significant types of distinction because they entail the possibility of isolating predicates by leaving other properties in an object indeterminate[315] (corresponding exactly to Husserl's eidetic variation which inserts algebraic variables in an object for the properties not considered, cf. the Prolegomena in the *Logische Untersuchungen*). With regard to the semiotic man-animal problem, though, it must be admitted that many higher animals can perform corresponding acts, as is evident from the bonobo Kanzi's ability to understand predicate symbols as related in Deacon 1997. Peirce often calls prescission 'prescissive abstraction' to distinguish it from abstraction proper, or as he calls it, 'hypostatic' or 'subjectal' abstraction.

While the separation types makes possible generalization – by the peeling away of still further properties – and thus are tied to the Aristotelian general/specific/particular triad, the other abstraction type is tied to the abstract/concrete dichotomy. While the first one is objective – in so far as it discerns objective aspects of the phenomenon – the other is subjective in so far as it is tied to epistemology and the anatomy of the process of reason (but also has objective implications, see below).

Hypostatic abstraction is linguistically defined as the process of making a noun out of an adjective; logically as making a subject out of a predicate. The distinction between 'hard' and 'hardness' serves as the prototypical example. The idea here is that in order to investigate a predicate – which other predicates it is connected to, which conditions it is subjected to, in short to test its possible consequences using Peirce's famous pragmatic maxim – it is necessary to posit it as a subject for investigation. This is evidently a completely different procedure than the separation types (even if the two very often occur interlinked in the research process) insofar as the output is not more general than the input. It makes an second-order object out of a predicate, an object which may now be taken as the object for a further investigation. It takes a thought as a thing, in short. Consequently, the operation is recursive, and we may produce an unlimited hierarchy of ever more abstract notions.[316]

In the beginning of the century, Peirce over and over again illustrates abstraction with reference to the well-known Molière joke about the 'Virtus Dormitiva', the dormitive powers, of opium.[317] The joke is a parody of sterile abstractions of Scholastic medicine, of course, and in Peirce's positivist time, it apparently functioned as a general warning against abstractions *tout court*. But Peirce turns the table on this interpretation. He admits, of course, that it serves as an extreme example of an idle and useless abstraction, but still there remains, if we put it under a microscope, as he says, an ever so small step forward in the reasoning process, even in this foolish example. By going from the statement that 'opium puts people to sleep' and to the statement that 'opium possesses a virtus dormitiva', a hypostatic abstraction has been performed. *Something* in opium is taken to have this well-known effect. We know nothing more positively about the workings of opium in the brain, but the hypostatic abstraction now permits us to ask further: in what, more precisely, does this virtus dormitiva consist? It might be that opium just put some people to sleep by coincidence, but the hypostatic abstraction – by substantivizing this ability – asks the question of possible further reasons and structures behind this mere fact. Thus, hypostatic abstraction is a crucial motor in the process of research by positing new *somethings*, new *x*'s, as issues to be investigated.[318] It consists in 'asserting that a given sign is applicable instead of merely applying it', as he says already in 1898,[319] it goes from saying that something is red to the fact that redness may be applied to something, and in doing so, it creates an *ens rationis*, a second-intention, whose truth resides in the fact that something holds for other, really existing things: 'For what is an abstraction but an object whose being consists in facts about other things?' ('Logic of History' (1904), NEM IV: 11) The point of the Molière joke, consequently, is not that hypostatic abstraction is futile, but rather that the idea of such an abstraction being *sufficient* as an explanation is foolish.

In a 1905 manuscript 'Basis of Pragmatism' (Robin nr. 284), Peirce attempts to give hypostatic abstractions a systematic place in his semiotic architecture. In a chapter on the 'Division of Signs', he gives a new trichotomy pertaining to the sign's relation to its immediate object (distinct from the well-known icon-index-symbol pertaining to its relation to its dynamic object). Here, he distinguishes *vague, or indefinite* signs, *singula*r signs, and *general* signs, respectively. The singular sign refers to one particular object, while the vague signs refers to objects which are in need of more precise description in order for their object to be determined, and the general signs refer to a possible continuity of objects, among which the interpreter is free to chose any he likes. Among the singular signs, now, a further subtrichotomy is posited as follows: *hypostatically abstract* signs, *concrete* signs, and *collective* signs, respectively. The latter two refer to singular existing entities and entities built from parts or elements,[320] respectively, while the former are characterized thus: 'The Immediate Object, though Singular in form, is represented as having the logically material character of the Priman, which is the absence of the matter of existence.' (67) Existence is abstracted away to highlight the 'Priman', that is the Firstness, the quality which is then represented as an object. As soon as an abstraction is performed linguistically, it becomes a symbol, so we should place hypostatic abstractions as a specific subtype of symbols. Of course it is possible to refer to abstractions by other means than language – diagrams will be a typical way of referring to them, cf. below – but in these cases a symbolic, general indication of their object will be a part of the sign. Collections, the third subtype, are of course already itself abstractions, and an interesting fact is that also the single existing object for Peirce is an abstraction[321] – it is only possible as a limit case for investigation; to that extent, it is no wonder that hypostatic abstractions are seen as the most simple singular signs.

Hypostatic abstraction is supposed to play a crucial role in the reasoning process for several reasons. The first is that by making a thing out of a thought, it facilitates the possibility for thought to reflect critically upon the distinctions with which it operates, to control them, reshape them, combine them.[322] Thought becomes emancipated from the prison of the given, in which abstract properties exist only as Husserlian moments, and even if prescission may isolate those moments and induction may propose regularities between them (and we have any reason to believe higher animals may perform these two logical operations), the road for thought to the possible establishment of abstract objects and the relations between them seems barred. The object created by a hypostatic abstraction is a thing, but it is of course no actually existing thing, rather it is a scholastic *ens rationis*, it is a figment of thought. It is a *second intention* thought about a thought – but this does not, in Peirce's realism, imply that it is necessarily fictitious. In many cases it may indeed be – as when we make the abstraction of unicornicity – but in other cases we may hit upon an abstraction having real existence:

Putting aside precisive abstraction altogether, it is necessary to consider a little what is meant by saying that the product of subjectal abstraction is a creation of thought. (...) That the abstract subject is an *ens rationis*, or creation of thought does not mean that it is a fiction. The popular ridicule of it is one of the manifestations of that stoical (and Epicurean, but more marked in stoicism) doctrine that existence is the only mode of being which came in shortly before Descartes, in concsequence of the disgust and resentment which progressive minds felt for the Dunces, or Scotists. If one thinks of it, a *possibility* is a far more important fact than any *actuality* can be. (...) An abstraction is a creation of thought; but the real fact which is important in this connection is not that actual thinking has caused the predicate to be converted into a subject, but that this is *possible*. The abstraction, in any important sense, is not an actual thought but a general type to which thought may conform. (Letter to E.H. Moore, Jan. 2. 1904; NEM III/2,918).

The seemingly scepticist pragmatic maxim never ceases to surprise: if we take all possible effects we can conceive an object to have, then our conception of those effects is identical with our conception of that object, the maxim claims – but if we can conceive of abstract properties of the objects to have effects, then they are part of our conception of it, and hence they must possess reality as well (cf. the 1903 *Lectures on Pragmatism*, Peirce 1997, 134). An abstraction is a possible way for an object to behave – and if certain objects do in fact conform to this behavior, then that abstraction is real; it is a 'real possibility' or a general object. If not, it may still retain its character of possibility (just like Husserl in the Prolegomena to the LU states that the law of gravity would not cease to hold even if the last heavy object in the universe vanished). Peirce's definitions of hypostatic abstractions now and then confuse this point. When he claims that 'An abstraction is a substance whose being consists in the truth of some proposition concerning a more primary substance.' (Peirce 1997, 135), then the abstraction's existence depends on the truth of some claim concerning a less abstract substance. But if the less abstract substance in question does not exist, and the claim in question consequently will be meaningless or false, then the abstraction will – following that definition – cease to exist. But 'unicornicity' does not cease being an abstraction just because no unicorns exist? The problem is only that Peirce does not sufficiently clearly distinguish between the really existing substances which abstractive expressions may refer to, on the one hand, and those expressions themselves, on the other. It is the same confusion which may make one able Peirce scholar claim that hypostatic abstraction is a deduction and another – no less able – claim it is an abduction.[323] The first case corresponds to there actually existing a thing with the quality abstracted, and where we consequently may expect the existence of a rational explanation for the quality, and, correlatively, the existence of an abstract substance corresponding to the supposed ens rationis – the second case corresponds to the case – or the phase – where no such rational explanation and corresponding abstract substance has yet been verified. It is of course always possible to make an abstraction symbol, given any predicate – whether that abstraction corresponds to any real possibility is an issue for further investigation to estimate. And Peirce's scientific realism makes him demand that the connections to actual reality of any abstraction should always be estimated: '... every kind of proposition is either meaningless or has a Real Secondness as its object. This is a fact that every reader of philosophy should carefully bear in mind, translating every abstractly expressed proposition into its precise meaning in reference to an individual experience.' ('Syllabus', 1903, EP II, 279; 2.315). This warning is directed, of course, towards empirical abstractions which require the support of particular instances to be pragmatically relevant but could hardly hold for mathematical abstraction. But in any case the step of hypostatic abstraction is necessary for the ongoing investigation, be it in pure or empirical cases.

The pure case corresponds to the second reason for abstraction's central role in reasoning: the role of hypostatic abstraction as the most central operation in mathematics. It is the possibility of making an operation into the object of a new operation, so as to investigate the rules holding for the first operation (e.g. its transitivity, symmetry, etc.). As everywhere, this abstraction procedure is recursive and may form a hierarchy of concepts. Generalization undertaken by prescission is, of course, equally important in mathematics. Mathematics is linked to hypothetical deduction and diagrams in a very tight fashion in Peirce: mathematics is the science that draws necessary conclusions, and diagrams are the vehicles for all deductive reasoning. Deductive reasoning featuring empirical matter must thus imply a diagrammatic, mathematical structure, and diagrammatic reasoning forms the center of Peirce's epistemology: the iconicity of the diagram ensures its structural similarity with its object, the symbol governing it determines the possibility of manipulating it with regard to gaining new information, as discussed in Chap. 4. Both abstraction types play a crucial role in diagrammatical reasoning:

> All necessary reasoning without exception is diagrammatic. That is, we construct an icon of our hypothetical state of things and proceed to observe it. This observation leads us to suspect that something is true, which we may or may not be able to formulate with precision, and we proceed to inquire whether it is true or not. For this purpose it is necessary to form a plan of investigation and this is the most difficult part of the whole operation. We not only have to select the features of the diagram which it will be pertinent to pay attention to, but it is also of great importance to return again and again to certain features. Otherwise, although our conclusions may be correct, they will not be the particular conclusions at which we are aiming. But the greatest point of art consists in the introduction of suitable *abstractions*. By this I mean such a transformation of our diagrams that characters of one diagram may appear in another as things. A familiar example is where in analysis we treat operations as themselves the subject of operations. (Peirce 1997,226)

Thus, the two abstraction types are seminal for diagram formation. Prescission permits to construct a general diagram, bracketing all contingent features of the particular diagram drawing in favor of the features of it to be read as referring to a general property. Only the required predicates is preserved by this prescission procedure. Abstraction allows diagrams to be recursive and to investigate the properties of other diagrams, taken as objects. By these two operations, diagrammatic reasoning performed by rule-bound experimentation on the diagrams is made possible. But it is important to notice in our context that we have no reason to suppose that animals may not make simple diagrams (the road fork's Y), nor experiment upon them (the dog's probable inference taking the more travelled of the roads is such an experiment). But we have no reason to believe that the abstractive making a diagram explicit is a part of higher animals' reasoning abilities.

Seen from a logic point of view, abstraction's character of representing predicate classes makes it pertain to our days' concept of second order predicate logic (which is, of course, unlimited in contradistinction to first order predicate logic, not allowing quantification over predicates, quantifiers, etc. but only over individual variables). T.L. Short remarks (Short 1997, 295) that hypostatic abstraction is identical with 'the transition from first- to second order predicate logic', and he adds: 'It does not follow that every fact about an *ens rationis* is inferable from facts about other things. Second-order predicate logic is not reducible to first-order predicate logic; mathematics could not be done without referring to classes or to other abstract entities.' (296) This makes explicit the purpose of abstractions: they are not only shorthands for information already available at the concrete levels. They may add genuinely new information – corresponding to Peirce's idea that by theorematic reasoning with diagrams (as opposed to merely corollarial reasoning), new information may appear that was not explicit in the construction of the diagrams in question.

A more detailed investigation of hypostatic abstraction must try to analyse its basic subtypes. Obviously, there are a lot of different dimensions along which it is possible to perform hypostatic abstractions. In language, they may give rise to a linguistic variety of semantically different abstract noun types and, more broadly, nominal constructions. I know of no Peirce scholar trying to go this way, but in the Husserlian tradition of pure a priori grammar from the 4th LU, a scholar like Jean-Louis Gardiès presents some ideas on linguistic hypostatic abstraction types. There are, for instance, at least three types of possible quotation-marks (in a wide acceptance of the term), each of them nominalizing the expression in question: 1) The operator 'the fact that...' which forms the name of a state of affairs; 2) the nominalization of a predicate ('redness', 'humanity'); 3) ordinary quotation marks referring to the name of a proposition (or any other element of discourse expression ('or' is pronounced parallel to 'door'), structure ('or' is a conjunction), or content ('or' may mean XOR or it may mean V)). We may add – following Peirce's ideas above – 4) the collection operator forming a set of objects ('my books', 'mankind'), 5) the individual object operator: 'that object as it exists fully determined now and here' (or with any other spatiotemporal or other specification), cf. Peirce's contention that the unique object with all properties completely determined is also an abstract idea. From linguistic tradition, a whole series of 6) verbal substantives permits to abstract a verbal predicate in other ways than ordinary nominalization: a) present perfect ('operating') forming the abstract idea of an ongoing process; b) past perfect ('operated') forming the abstract idea of a process having taken place; c) infinitive ('operate') forming the abstract idea of the process content apart from realization; d) nominalization ('operation') forming the abstraction of the process as a whole; e) nominalization of the agent ('operator') forming the abstract idea of a specific ergative subject for a process; f) adjectivisation ('operational') forming the idea of some other x having to do with the process. From predicate relations with more than one relative, several different roles may be abstracted (from 'give': 'the giver', 'the gift', 'the gifted', 'the given'). Other languages may add still further types (gerundive 'the one that ought to be given something' etc.). But hypostatic abstraction needs not be expressed in nor refer to linguistic entities (even if they support it by abstraction suffixes like '-ness', '-ity', '-ation' and many more, and highly enhance the possibilities for using it). A recurring example in Peirce is the idea of seeing the geometrical line as an abstraction from the trajectory of a particle. Thus, the nominalization act of hypostatic abstraction may also include the spatial 'stiffening' of temporal processes or aspects thereof into objects of an abstract space. All abstraction types probably refer – explicitly or implicitly – to such spaces in which diagrams may take other diagrams as their objects. The description of hypostatic abstraction in terms of linguistic or logical vocabulary should not keep us from seeking the phenomenological basis for it, and the possibility for diagrams of taking other diagrams as their objects (thoughts taken as things) precisely presupposes abstract spaces embedded in other abstract spaces. The list of how this may be achieved and represented is possibly open-ended, given the fact that an abstraction of a given predicate may be attempted with reference to many other already constructed abstract ideas; this open-endedness corresponds to abstraction's homology with second-order logic.

In spite of the fragmentary treatment of these two abstraction types, they play, as is evident, a central role in Peirce's architectonic. In his Carnegie application from 1902, e.g., one of the few occasions when he proposes a systematic exposition of his mature thought, hypostatic abstraction appears already in lecture 4 (out of 36, and long before the introduction of categories, signs, etc.). This is of course because of hypostatic abstraction's central role in mathematics – the possibility of an operation to be taken as an object for another operation, investigating the first operation's properties. And prescissive abstraction is logically prerequisite to hypostatic abstraction: before hypostatic abstraction of a predicate to a subject, a predicate must already be prescinded. This interplay between the abstraction types are rarely treated explicitly in Peirce, but on one significant occasion, he links the two with the animal-man transition problem, worth quoting at length (in the letter to E.H. Moore, Jan. 2. 1904):

> There are two entirely different things that are often confused from no cause that I can see except that the words *abstract* and *abstraction* are applied to both. One is αφαιρεσισ, leaving something out of account in order to attend to something else. That is *precisive abstraction*. The other consists in making a subject out of a predicate. Instead of saying, Opium puts people to sleep, you say it has a dormitive virtue. This is an all important proceeding in mathematics. For example take all 'symbolic' methods, in which operations are operated upon. That may be called *subjectal abstraction*. This use of the word abstract goes back to the beginning of the XIIIth Century while the other use is earlier still. So both are of unquestionable respectability. But they have nothing in common. What I say in treating such subjects I am apt to mean. They have nothing in common. No doubt subjectal abstraction presupposes a certain considerable precisive abstraction in each case; but that was not introduced in making the subjectal abstraction, it was there before. Experience is first forced upon us in the form of a flow of images. Thereupon thought makes certain assertions. It professes to pick the image into pieces and to detect in it certain characters. This is not literally true. The image has no parts, least of all predicates. Thus predication involves precisive abstraction. Precisive abstraction creates predicates. Subjectal abstraction creates subjects. Both predicates and subjects are creations of thought. But this is hardly more than a phrase; for *creation* and *thought* have different meanings as applied to the two. Without precisive abstraction man would not be man; but I can well believe, - indeed, I do think it probable, - that a large fraction of the races of mankind, by no means necessarily very low in the arts, are entirely devoid of the power of subjectal abstraction. (...)(NEM III/2, 917–18).

Lots of interesting ideas are implied in this. Here we find the idea that prescissive abstraction precedes hypostatic abstraction, that the former creates predicates and the latter, in turn, creates subjects. In so far as even simple collections are abstract entities, it follows that this creation process goes on in human thought all of the time and not only in its purified form in the sciences. Everyday reflection is impossible without it; T.L. Short even argues that the self – and correlatively self-consciousness – is an entity inferred by means of hypostatic abstraction from faults in single actions (as the source of those errors).[324] In relation to the semiotic animal-man transition question we find in the passing a reflection upon abstractions' relation to biology: the idea that without prescissive abstraction man would not be man, while many human beings, maybe even cultures,[325] may be in lack of sufficient ability to perform hypostatic abstractions. This is not, it must be admitted, very precise, and we have already assumed prescission to be widespread in higher animals, evident in their ability to associate via qualities (Peirce: 'The most ordinary fact of perception, such as 'it is light', involves precisive abstraction, or prescission' ('Minute Logic', 1902, 4.235)). In any case, if many higher animals may prescind and man not be man without it, hypostatic abstraction seems restricted to mankind, even if maybe unevenly distributed among us (which might in fact be an indication that selection pressure for it is still at work, or has been until recently).

As is evident from the above, this conforms with our general idea: it is the ability to form not symbols in general, but the special symbol type called hypostatic abstractions, that distinguishes (most) men from (most) animals. It is, of course, a very difficult problem to ascertain which mental procedures higher animals are capable of. But it seems reasonable to assume that they master symbols, including arguments, action according to diagrams, and even symbol systems in some rudimentary form, involving huge amounts of generality made possible by prescission – but with no means to extract that generality from sensory experience and to isolate it, control it or experiment upon it. Here, Peirce's 1905 reflection briefly quoted at the beginning of this paper adds the explicit control dimension as an important role for the abstractions to play:

> Pragmaticist. To my thinking that faculty [of language] is itself a phenomenon of self-control. For thinking is a kind of conduct, and is itself controllable, as everybody knows. Now the intellectual control of thinking takes place by thinking about thoughts [cf. the description of hypostatic abstraction in such second intention terms]. All thinking is by signs; and the brutes use signs. But they perhaps rarely think of them as signs. To do so is manifestly a second step in the use of language. Brutes use language, and seem to exercise some little control over it. But they certainly do not carry this control to anything like the same grade that we do. They do not criticize their thought logically.
> ('Pragmaticism, Prag. [4]' c. 1905, 5.534, my brackets)

Man as well as animals are consequently rational beings, probably even necessarily so. Both are involved in a constant series of arguments in a reasoning process involving a whole range of simpler sign types. But what enables man to build up his symbol systems and its resulting more acute and accelerated rationality is prescission and abstraction working together, making it possible to isolate and to make explicit single phases in the ongoing chain of arguments in order to control them, scrutinize them, experiment upon them, combine them, recombine them, and improve them. Animals may possess the same abilities in germ, prescission probably especially so, but the continuum going from animal to man is to be grasped in terms of gradually higher mastering of abstraction. It must also be admitted, though, that this ability greatly enhances man's ability to commit errors, to be fooled, to lie. Of course, higher animals possess all these abilities, but abstraction adds the possibility for the construction of the enormous subdomains of discourse involving counterfactual universes: myth, religion, literature, science whose vast capacity for general truths mirrors an equally large capacity for general fallacies.

This would also conform well to Deacon's Baldwinian assumption: that the behavior-selection feedback in symbol using *Homo habilis* communities acquired an extreme pace measured against evolution's normal velocity. For the active controlling and explicit experimenting on signs makes it possible to develop them significantly within one single biological generation, while the spontaneous historical aspects of language evolution (change in phonetic patterns, etymology, etc.) is a much slower phenomenon, even if still quick as compared to biological evolution. Given this scenario, it seems reasonable to assume that a very strong selection pressure has prevailed against the increased possibility of fallacies, especially against the formal logical fallacies without any empirical content, but also against violations of basic linguistic constants like the subject-predicate structure.[326] All in all, these abstraction operations permit us to construct an indefinite amount of abstract objects, more or less apart from the actually surrounding world of here and now – and it permits us, by the same token, to construct explicit diagrams to bring these abstracta to the test, yielding the amazing increasing insight in empirical regularities as well as in formal and synthetic a priori laws of fallibilistic apriorism.[327]

If this idea is correct, human beings are indeed a symbolic species, even if not the only one. Man is rather the abstract animal.

CHAPTER 12

THE SIGNIFYING BODY

A Semiotic Concept of Embodiment

If biosemiotics is right in claiming the fundamentally semiotic character of biological processes, then this should throw a new light upon the concept of 'embodiment' so fashionable in recent linguistics and philosophy. The semiotic 'missing link' issue of the previous chapter must thus be a special case only of the more general issue of the emergence of semiotically competent body types during evolution: a natural history of the signifying body.

In traditional semiotic thought, however, the body has been almost ignored. If we take structuralist semiotics, we should expect a treatment of the body as of any other concept in language, persuading us that the body differs in different languages, cultures, in short in different semiotic systems and that a study of such different systems will show us as many different cultural representations of the body. No extra-structural constraints are supposed to determine the spectrum of possibilities of body representation.

Thus, the body would be conceived of as a concept subjected to the free arbitrarity of semiotic systems – and no special attention would be paid to the body as a crucial prerequisite to semiotic articulations. This ignorance of the body is about to undergo a complete change – epitomized in the widespread popularity of different versions of the concept of *embodiment*. What is called for is a concept of the body which, in itself, makes evident the basic semiotic competences of an organism – thus, a body concept which entails semiotics. It is to be expected, naturally, that this will give rise to a more fine-grained typology of bodies depending on how complicated and sophisticated semiotic behavior the organism in question is able to indulge in.

This 'embodiment' turn of semiotics which has been underway during the recent decades, has called for new interest in old positions all the way back to Aristotle, as well as a reorientation in actual tendencies of biology and philosophy. Among the former count reappropriations and reinterpretations of the foundations of ethology and theoretical biology, examplified here in Jakob von Uexküll, as well as a renewed interest in phenomenology, examplified here in Maurice Merleau-Ponty's work with its focus upon the body and the 'flesh' as the necessary concept to avoid sterile mind-matter dualisms. Among the latter count the cognitive semantics movement (Eleanor Rosch, George Lakoff, Mark Johnson, Leonard Talmy, Mark Turner, Gilles Fauconnier, etc.) and its emphasis on the 'embodiment' of cognitive structures; the 'complexity theory' around the Santa Fe school (Murray Gell-Mann, Christopher Langton, Brian Goodwin, Stuart Kauffman) and its reinterpretation of classical issues of theoretical biology within the framework of a general theory of complexity; and finally, within semiotics itself, the emergence of 'biosemiotics' (Thomas Sebeok, Terrence Deacon, and the Copenhagen school (Jesper Hoffmeyer, Claus Emmeche) to which I myself to some extent belong as a fellow-traveller).

This paper will briefly present and discuss the body concepts involved in these developments and their semiotic possibilities.

COGNITIVE SEMANTICS – THE BODY IN THE MIND

The well-known American tradition in linguistics and philosophy known as 'Cognitive Semantics' or 'Cognitive linguistics' (Lakoff, Johnson, Turner, Fauconnier, Sweetser, Talmy, Langacker, etc.) has, during the recent 25 years, provided a new view of language, loosening the autonomy of linguistics in order to connect it to developments in cognitive science – using insights from psychology, philosophy, comparative literature, anthropology, neurology, etc. 'The linguistic turn' is rolled back – language is seen as a specific combination of a series of different, cognitive, pre-linguistic competences. In doing so, Cognitive Semantics covers a large field of semiotic issues; thus it constitutes one of the main developments of semiotics from 1980 onwards – even if it most often does not explicitly use the term 'semiotics'.

A very basic tenet in this tradition has been its insistence on the bodily motivation of cognitive, semantic, and linguistic structures. Such structures are claimed to be 'embodied'. This claim is aimed against the formalist and logicist trend in American analytical philosophy and linguistics (especially against the Chomskyan tradition) – that is, against the tacit or outspoken assumption that the bodily basis of thinking beings is irrelevant for the study of thought and language (which may then be studied by purely logical, formal means) – a famous version of this claim is Putnam's old 'functionalist hypothesis' equating mind and brain with computer software and hardware, respectively. Against the implications of this analogy (the brain as a Turing machine able to 'run' any mental program ...), the idea is that the specific architecture of body and brain is fundamental to thought and language. This part of the program has been made explicit especially in Lakoff and Johnson's large tractatus *Philosophy in the Flesh* (1999). A closer look reveals, though, that is it not always completely clear what is intended by the embodiment claim. The following subclaims constitute the details of the embodiment hypothesis.

a) A philosophy must be true to bodily experiences – instead of indulging in A Priori philosophizing. This implies, in turn, two things. It must be open to the results of the various sciences pertaining to bodily experience – psychology, anthropology, linguistics, neurobiology, etc. Furthermore, it must refrain from universal claims, since all facts about the mind depend on the empirical variation studied by these special sciences, cf. the so-called 'experientalism' of Lakoff and Johnson. It is important to note, however, that these two claims are not necessarily connected. The interdisciplinary view of cognitive processes implied by the former does not entail the anti-apriorism of the latter. Rather, a Husserlian idea of a priori structures with necessary conceptual networks underlying the special sciences would easily fit with interdisciplinarity, and it may be argued that Cognitive Semantics often, in fact, involves a priori arguments without admitting it (the embodiment hypothesis could, inter alia, be seen as exactly an a priori hypothesis).

b) Reason and thought use, to a large extent, competences involving the sensori-motor system – in conceptualizing phenomena in terms of spatial and motor relations borrowed from or directly relying upon these systems. This implies the possibility of sensori-motor inferences: structures from these parts of the brain facilitate reasoning determined by spatial and motor properties of the phenomena intended.

This idea is examplified in two core hypotheses of Cognitive semantics: the dependency of language and thought upon two fundamental sets of semantic primitives, 'basic-level concepts' and 'kinaesthetic image schemata', respectively. Both are cognitive entitites which are inherently meaningful due to the fact that we know them from our bodily interaction with the environment (hence, they are not vulnerable to the 'symbol grounding problem' asking about the foundation of the meaning and reference of symbols – they are always-already meaningful). The first idea stems from psychological investigations by Eleanor Rosch and claims that the fundamental concepts in the human mind refer to types of things or actions with which we have a basic, typical motor experience – and of which we can hence form simple schematic image representations. Chairs, tables, cars, houses; walking, talking, sleeping, etc. The basic idea is thus that our sensori-motor acquaintance with the world determines our fundamental concepts – an idea not foreign to an Uexküllian point of view. More abstract concepts ('furniture', 'vehicles', 'movement', 'action', etc.) lack associated specific motor programs as well as a clear schematic images in termes of the specific whole-and-part structures of the phenomenon in question. More specific concepts, on the other hand ('kitchen table', 'Louis XVI chair', 'dozing', 'marching', etc.) are formed as subclasses of the basic level concepts by further specifying the actions and images involved on the basic level. The specific lexicon of basic level concepts is, of course, culture dependent. It has a core, motivated by basic bodily functions, but large parts of it are relative to the perception and action practices developed on top of them in a given culture.

Image-schemata are embodied Gestalts (be they explicit or not) used in perception and reasoning: part-whole, center-periphery, link, source-path-goal, cycle, iteration, contact, adjacency, forced motion, support, balance, straight-curved, and near-far, and a few more, but probably not many more. The idea is that we tacitly know these Gestalts, their structure, and the inferences which they support, from our ordinary bodily activity; they are crucial aspects of what it is to be a body. Their description thus adds to the implicit body theory of cognitive semantics: a body is hence characterized by goal-oriented behavior (source-path-goal); connectedness (container); stable mereological-anatomical structure (part-whole); hierarchical structure – head/body vs. limbs (center-periphery), orientation in gravity fields (up-down) etc.

A crucial implication of these hypotheses is a dissolution or, at least, a relativization of the perception-conception boundary: concepts are motivated (but not exhaustingly determined) by structures in perception and action. A further crucial implication is that these basic bodily experiences provide the starting point for more sophisticated mental activity; thus abstract thought is taken to arise from metaphorical projections of these structures from the basic bodily field and onto other domains more remote form sensori-motor activity. The extensive metaphor theory of this tradition is constructed on the basis of this hypothesis, rendering metaphor an important cognitive tool, giving rise to structural metaphors, each of them underlying many linguistic metaphorical expressions. The structural, conceptual metaphor 'Knowing is seeing' known in many languages thus gives rise to a long series of different expressions like 'enlightenment', 'Can't you see what I explain?', 'Take a closer look at this problem', etc. This gives the implication, in turn, that imagination becomes an important cognitive tool, not only in these conceptual metaphorical projections, but also in the trial-and-error construction of more elaborated conceptual models in thought experiments, so-called 'idealized cognitive models', built from basic concepts, image schemata and layers of mappings between them.

This gives a rough picture of the body concept at stake in cognitive semantics. There are, however, certain problems in that concept. It is not at all clear what the exact extension of these claims is. The immediate – prototypical – body referred to here is, of course, the human body. The general reference is to 'our body and brain', this 'our' supposedly referring to the body and brain of the human species, but the extension of this expression is unclear in at least two dimensions. One is 'downwards' in the animal kingdom: how many of these competences used in the description of this body concept hold for higher animals? We should probably expect higher animals to make use of basic level concepts and kinaesthetic image schema as well – but maybe to a lesser extent the metaphorical extensions of these basic tools.[328] Another is 'sidewards' in a Kantian manner, so to speak: are these claims valid not only for the contingent, empirical human race but for any possible reasoning subject as such? Cognitive Semantics itself gives an explicit answer to this question: no. They claim their theory is empirically built by the investigation of human languages referring to the human body and brain and thus reject all a priori reasoning. Still, a problem remains here. Even if it is easy to agree with Cognitive Semantics that theories of meaning which disregard the body must themselves be disregarded, it is less easy to agree that their theory itself is, in fact, purely 'experiential' and without any a priori assumptions. Some of the assumptions used are taken from various special sciences, true, but this does not entail they have no a priori validity.[329] So the question remains whether it is, in fact, the case that any possible real intelligence must possess some sort of a body, conceptualize in some set of basic level concepts and use some repertoire of kinaesthetic image schemata, etc.? This a priori question is not answered within Cogntive Semantics, but as far as I can see, it becomes a crucial question for generalized semiotics on the one hand and theoretical biology on the other.

Another problem in the theory is that the relation between embodiment and general semiotic competences is fairly underdeveloped. If a body is characterized by instantiating (some of the) Gestalt schemas mentioned, this does not in any way imply that that body is necessarily able to *understand*, still less *represent* such schemata nor build language from them. A bacterium behaves according to the source-path-goal schema (when swimming upstream in the sugar gradient thanks to its 'biased random walk'), it instantiates the container and part-whole-schema by its closed cell membrane, and so on, but this does not in any way imply that it has any functional, not to talk about mental representation of those schemata. Ability to reason with such schemata requires the tacit or explicit mental *representation* of them, separated from actual sensori-motor behavior, due to the kinaesthetic sense and to the existence of sensori-motor integrative body images in the nervous system. Here, Cognitive Semantics is in need of further underpinning of neurological as well as phenomenological work on these issues.

VON UEXKÜLL REVISITED – THE BODY AS FUNCTIONAL CIRCLE

We have already discussed a theory at the root of ethology which has reflected some of these problems in a proto-semiotic manner and which has gained wide renewed interest in semiotic circles during the last decades: von Uexküll's 'Umweltlehre'. The functional circle in von Uexküll provides, in fact, a basic sensori-motor body conception, described in semiotic terms: perceptions and actions are classed as sign types: *Merkzeichen* and *Wirkzeichen*, respectively, all in all making up the 'Umwelt' of the species in question. The body and its surroundings are thus conceived of as correlatively defined entities: a body is a body only with respect to its niche in a specific *Umwelt*, and vice versa. This definition of the body is thus correlated with that of the environment it constitutes. In this conception of the organism, the body per se is conceived as a semiotic device: it is an intrinsic property of a body that it is able to perceive the surroundings through signs and act correlatively through signs. The extension of this definition of the body to what is later called 'endosemiosis' is natural: when one part (the external parts) of the functional circle requires sign use, why not other parts (the internal parts) of it?

We saw that von Uexküll vacillated with respect to admitting the existence of *neutral objects* (not defined by correlated actions) in higher animal *Umwelten*. But the choice for us is easy: neutral objects are necessary, *precisely* because they disrupt the virtually perfect fit between organism and *Umwelt*. The existence of general – underspecified – perception and action diagrams are hesitantly admitted in von Uexküll. And this generality in *Umwelt* relation of any body increases with complexity of the organism and its related perception and action patterns. The perception of neutral objects is, importantly, a prerequisite to learning because learning in some sense consists of nothing but the de-neutralization of neutral objects, drawing them into more complicated segments of functional circles. But in order to be thus invested with meaning, neutral objects, of course, must be phenomenologically present *before* their use in learning. The perfect fit between organism and environment must thus be relativized: life time adaptability presupposes to some degree the perception of neutral objects, not immediately functionally relevant.

As we saw, the musical metaphor in von Uexküll has, on the other hand, a crucial implication which is never made explicit in the *Umweltlehre*. In the same manner as tonally discrete music, the body-surrounding fit is possible only through *discretization* of the continuum of possibilities, both in the perception and the action relation. Perception possesses a highly constrained selection of possible environment stimuli – ranging from simple cases like the possibility of sensing only groups of specific chemicals and to more complicated cases like the necessary limit of discrimination ability in any continuous perception spectrum (visual, auditive, tactile, etc.). In short, perception and action both possess a certain *granularity* which allows them to be pragmatically efficient at the prize of a certain imprecision. This imprecision, it is evident, implies certain limitations – larger or lesser – on the perfection of the organism-environment fit. Both more perceptual precision (which is also energetically more expensive), on the one hand, and more perceptual economy (which is also less precise), on the other, may be favored by selection, according to the specific conditions in the single case. In semiotic terms, this implies that in the functional circle, a tension is at stake in embodied semiosis between semiotic complexity on the one hand and semiotic economy on the other. The 'perfection' discussed by von Uexküll can be nothing but a local optimum (always potentially subject to change due to environmental pressure and change) in the tension between these two possibilities in all functional-circle defined bodies. Thus, the Umwelt concept furnishes semiotics with a basic idea of a body as semiotically defined by the set of its perception and action sign possibilities[330] – which in the right interpretation may potentially serve adequately as a biological underpinning to the often vague embodiment talk in Cognitive Semantics.

MERLEAU-PONTY – THE BODY'S FUNCTIONAL CIRCLE OPENING UP

It is well-known how Merleau-Ponty's special brand of phenomenology took its point of departure in Husserl, enriched by ethology and psychology of perception. This leads Merleau-Ponty to a characteristic third position in relation to mind-matter problems where this well-known dualism and various, more or less one-sided, attempts at resolving it are rejected with reference to the primary position of bodily being and perception. Thus, both rationalism and idealism are rejected as hypostatizations of pure mind, while, correlatively, materialism and traditional naturalism are rejected as hypostatizations of pure matter. The very condition of possibility of mental or idealized structures on the one hand and pure, material objects on the other, is bodily being whose behavior and perception form a prerequisite basis which cannot be further dissolved – cf. Merleau-Ponty's concept of 'flesh' antedating both mind and matter. Unlike many other phenomenologists of Heideggerian influence, Merleau-Ponty never, however, saw such a philosophical stance as alien to science, and it is well-known how he supported this hypothesis by references to biology and psychology of perception. It is probably less well-known how he – to the end of his short life – undertook a major work to support this position by reference to central figures in different branches of biology. These reflections appeared in university courses in the latter half of the fifties and Merleau-Ponty's lecture notes from those courses have only recently (1995) been published – and they add further details precisely to his conception of embodiment. Here, Merleau-Ponty confronts a long series of biological thought ranging from German idealism (Kant, the romantics, Husserl), vitalism (Driesch), behaviorism (Watson), Darwinism, ethology (Uexküll, Tinbergen, Lorenz), and many others – reading these currents in order to interpret their results philosophically and extract the points relevant to Merleau-Ponty's own embodiment phenomenology.[331]

Here, I can run through only some of the main points he addresses. Referring to von Uexküll in particular, Merleau-Ponty takes care to note that the notions of an organism equipped with an *Umwelt* in which to unfold its behavior – 'comportement' – is more basic than consciousness; quite on the contrary, consciousness is but one of the special forms this behavior takes, not unlike the claim of Peircean semiotics. The functional interrelation between organism and surroundings is prerequisite to the emergence of consciousness, not the other way around – an idea in which von Uexküll would probably agree.[332] Correspondingly, with its functional definition, the *Umwelt* is not only to be found at the level of the whole organism; even a single organ could be said to have its own Umwelt (220) – an idea which, of course, opens the 'endosemiotic' issue. The distinction is drawn between lower animals – mere functional bundles – where the Umwelt forms a closed set of reaction types, and higher animals who possess, thanks to the central nervous system, an internal representational mapping of the surrounding world facilitating an open *Umwelt* (225), not unlike our argument in Chap. 10. Thus, higher animals have sensory organs *able to move* (cf. movable snout, eyes and ears, intertwining sensation and motion in quick perception-action searching sequences) which enable them to actively explore and inform the world (225) – this moveability of the sensory organs so to speak short-circuits the functional circle and enables the organism to increase precision and decrease the time and space granularity of the *Umwelt* relation considerably. This implies, correlatively, the 'possibilities of objects';[333] the animal may distinguish its own spatial position, e.g. against gravity; this ability becomes complete only with a proper neural system of proprioception (226) facilitating feed back control of behavior as measured against the interiorized *Umwelt* and its *Merk*- and *Wirk*-components – a proper perceptual world and a behavior world. Only now when the body is, in itself, perceived, the perceptual world becomes possible as an oriented, represented mental map world inhabited not only by stimuli but by independent, body-like objects.[334]

Merleau-Ponty's philosophical interpretation of Uexküll's *Umwelt* concept takes its departure from his basic melody metaphor: the *Umwelt* as a melody singing itself. This image loosens the *Umwelt* from pure, actual, physical time, just like the melody it is aimed towards the future, as it cancels the priority of effect over cause, ends over means, essence over existence. The *Umwelt* thus, in Merleau-Ponty's interpretation, acquires an *ideal* character, not in any subjectivist sense, but as general, opposed to the actuality of the present existence – the *Umwelt* is an ideal structure which, like the melody, persists over and above the vissicitudes of the single moment. Behavior in such an *Umwelt* thus cannot be understood taken moment for moment, but only as a meaningful whole extended in time. Without being defined in actual time and space, it is thus trans-temporal, trans-spatial – almost a Platonic idea, in some sense, but at the same time the result of a process of self-organizing schematization involving organism and environment. In higher animals, the functional cycle of the *Umwelt* loosens ever more from its immediate pursuit of teleological ends and indulges in interpretations of symbols. Merleau-Ponty thus takes sides in Uexküll's fight against himself as to the possibility of the existence of neutral objects and the corresponding freedom in the *Umwelt*. To Merleau-Ponty, this possibility is crucial and points to the possibility of the existence of 'pre-culture' in higher animals (231). Symbols point out of the immediate present to future perception, event, and action possibilities, and the action through symbols permits organisms to perform not-innate complicated action wholes.[335] Merleau-Ponty's fertile reinterpretation of von Uexküll is enriched by the reflective introduction of a long series of other biological thinkers. I shall here restrict myself to those who add to the body concept here outlined. E.S. Russell's idea of 'directiveness' of organic activities is referred for its generalization of the notion of behavior to encompass not only the 'external circuit' of a body's relation with its *Umwelt* – but also the internal regulation of an organism. There is no definite limit between these two (235), and hence it makes sense to talk about bodily behavior already during morphogenesis. This idea thus, moreover, dissolves the strict borderline between the organism and its behavior which should be seen, rather, as dual concepts: 'The body belongs to a behavioral dynamics. Behavior is inscribed in embodiment.'[336] Thus, the relative plasticity of behaviors is seen as an integrated property in the being of a bodily organism as such, both at the external and internal levels – pointing again towards the semiotic notion of 'endosemiosis' for intra-organism semiotic processes.

Adolf Portmann's study of the outer appearance of animals becomes an important source to a beloved theme of the elder Merleau-Ponty, namely the duplicity of perceiving and being perceived – of visibility and invisibility. Parts of an animal's outward appearance may be the chance result of a local process (the shell of a snail) and is thus without interest in this respect, but other aspects of appearance (the skin of the zebra) is the result of non-local processes characterizing the whole animal and its *Umwelt* relation. Thus, these aspects have two characteristica: one, they satisfy a mimetic teleology as if there were a perceptive relation between the two – animal morphology and environment (246). But this apparent teleology, to Merleau-Ponty, cannot be primary but requires, in the first place, a certain freedom of expression which may, in turn, direct the appearance towards utility in some cases, in others not so. Thus, animal appearance has an aspect of 'presentation', of existential manifestation by which the animal makes itself visible to fellow creatures (and invisible to predators, it should be added . . .). In a parallel to the older

Husserl's notion of 'intersubjectivity', Merleau-Ponty thus introduces the notion of 'inter-animalité' as granting an ontological status to the notion of species (this idea probably may freely be extrapolated to the notion of ecosystem): an animal looks in a way which it is itself able to see (in order to be able to be recognized by specimens of its own species): 'l'animal voit selon qu'il est visible' (247).[337] Inter-animality thus requires the constitution and recognition of other creatures from profiles, it requires a degree of understanding of their intention types – this covering species fellows as well as typical prey and/or predators in the ecosystem of the species in question. This inter-animality must further include common aims as well as internal battles in herd-animals – so as to form the phenomenological basis of the well-known sign and communication systems often developed in such species.

The notorious pupil of von Uexküll, Konrad Lorenz, is also made the object of an interesting reinterpretation, notably concerning his controversial notion of 'instinct'. According to Lorenz, an instinct is an innate action series which requires a certain environmental releaser in order to be actualized. This idea has often been interpreted and dismissed as a purely mechanistic idea, but Merleau-Ponty rejects this reading and points to the fact that Lorenz's notions of objectlessness and *Prägung,* imprinting, entails quite the opposite. The instinct in Lorenz is objectless and it thus possesses a ceremonial, ritual excess on top of its possible function. Moreover, the phenomenon of *Prägung* shows that instinct is in many cases incomplete and requires a fill-in from the environment which implies that it entails an open orientation towards the surroundings, especially in complex cases where the actualization of an instinct is only possible with an adjoining *Umwelt* construction, with systematic world elaboration (255). This relative emptiness and openness of instincts is what enables it to become, in turn, a kernel in symbol construction: instinctive action series may be cut off from their – possible – telos and be taken to symbolize quite different phenomena in animal communication (254). Exactly the objectlessness of the instinct makes possible its imaginative reinterpretation as a basis for symbol formation.

Merleau-Ponty's interpretation of his great master Husserl is interesting in this context. Husserl's problem is, Merleau-Ponty argues, analogous to that of Schelling: to find a place for nature in a philosophy of reflection. Hence, two competing tendencies are to be found in Husserl. One, connected to his transcendental philosophy, sees natural objects as connected to the 'natural attitude', the attitude of innocence, characteristic of the unphilosophical observer who must unlearn this naive way of seeing in order to grasp the constitutive, phenomenological stance. The other tendency attempts to understand natural objects as pertaining to different regional ontologies, and the break with the natural attitude is an attempt also to clarify this pre-reflexive stance, in which the natural world is given to us in a passive synthesis: 'Phenomenology rejects the natural attitude and, at the same time, does more than any other philosophy to rehabilitate it.'[338] Nature, interpreted as consisting of pure things, is the correlate of pure consciousness, but antedating this, there is the more original, perceived and lived world, the *Lebenswelt.* This world is the world of the body – the body as the organ of the Husserlian 'I can', so to speak Husserl's version of the sensori-motor *Umwelt* alien to the merely perceptual, action-free world supposed by large parts of the philosophical tradition since Kant. The body perceives the objects not in a detached way, but by considering the motor possibilities implied: 'The object appears to me as a function of the movements of my body.'[339] The body is the privileged place both for my inhabiting the world of things, and, at the very same time, for my perceptions. (107). The body is, as always in Merleau-Ponty, subject and object at one and the same time: the world of things is part of my body. (108) The body is the zero-point of orientation, not only in space-time, but in all normative scales: it is thus the body which founds the very idea of normativity.

As to the existence of other bodies, I grasp them by an *Einfühlung* which is basically corporeal (109): I perceive them as perceiving bodies before I perceive them as thinking, and the latter idea presupposes the former. This bodily relation with other bodies is indispensable for the possibility of the thought of pure things: it is only now that the pure object can be defined as a thing to which other bodies have virtual access: this 'intercorporéité' (109) defines the very access to pure objects. (cf. Dan Zahavi's discussion of intersubjectivity as object constitutive). It is in this relation that I pass from prehuman to human, Merleau-Ponty claims, anticipating Tomasello's 'joint attention'-hypothesis (even if it remains unclear exactly how this 'intercorporéité' differs from the 'inter-animalité' of animals which also possess the ability for empathy).[340]

But this reference of the universe to the body and to humanity, does that not overlook that life might disappear? and what would, in that case, be left? (111). This possible disappearance, Husserl claims, would not entail that the evidence of references would disappear, so we must assume a mute world would remain even in that case. Thus, in *Ideen II*, Husserl may define nature as that to which I have an original and primordial relation '... le seul unique monde pour tout le monde' (112) – the only world for everybody in the world – as Merleau-Ponty jestingly puts it.

Merleau-Ponty remarks the constant tension between this idea and the idea of the break with the natural attitude which it presupposes, and he concludes by saying that Husserl never really resolved this tension which he implicitly admitted, e.g. in his double concept of constitution (constitution 'par actes', through conscious acts, on the one, anti-naturalistic side, and 'latente', on the other, naturalistic side, 112). Obviously, Merleau-Ponty ontologically prefers to underline the latter, pertaining to the 'passive synthesis' while, at the same time, maintaining the methodological necessity of the transcendental 'epokhe' of the former.

All in all, the body concept in Merleau-Ponty's reflections on nature has the following characteristics:

- It refers to future possible states and thus transgresses the pure actuality of physics – and at the same time points to 'real possibilities' as having ontological existence (cf. Peirce's insistence on exactly that notion)
- It is prerequisite to both the subject and object category.
- It has a primary sensori-motor relation to the world, forming an integrated complex of *Umwelt* and behavior.

- It has the possibility of transgressing its own finality in behavior with an 'open' side facilitating mere expression and symbol use – already in the case of higher animals which thus possess the germs of culture.
- Merleau-Ponty, furthermore, envisages the need for the transgression of the body concept in both up- and downwards directions, so to speak: he admits the possible *Umwelts* even of single organs (and cells?) on the one hand, at the same time as he locates the animal in an 'inter-animality' anticipating intersubjectivity.

In comparison to the body concept of Cognitive Semantics, Merleau-Ponty lays the same stress on the primacy of the sensori-motor entanglement of the body, on its gestalt competences, and its imaginative abilities – and in rare moments, the cognitive semanticists also count Merleau-Ponty as a forerunner, if not inspirator. In addition to Cognitive Semantics, Merleau-Ponty, with his Uexküll and related references, extends his body concept to cover higher animals as well, while the specificity of the human body comes into play along with language, intersubjectivity and the appearance of pure objects (even if his notes are not unambiguous on this point – the access of higher animals to neutral objects must provide an important step towards pure objectivity, just like the notion of 'inter-animalité').

COMPLEXITY THEORY – 'AUTONOMOUS AGENT' AS A FORMAL BODY DEFINITION

A recent tradition of thought with huge implications for theoretical biology is the so-called 'complexity theory' of the 'Santa Fe school' (Murray Gell-Mann, Brian Arthur, David Pines, et al.) finding formal regularities of complexity in all domains where a manifold of interacting entities are at play: from solid state physics over biology to economics and sociology. Especially the work of Stuart Kauffman is of interest in our context; even if his work does not contain explicit reflections on the body concept, such reflections can be easily inferred from his general discussion.

He takes his beginning point in the Darwinian idea of the origin of life in a biological 'primordial soup' of organic chemicals and investigates the possibility of one chemical substance to catalyze the reaction of two others, forming new reagents in the soup. Such catalyses may, of course, form chains, so that one reagent catalyzes the formation of another catalyzing another, etc., and self-sustaining 'loops' of reaction chains is an evident possibility in the appropriate chemical environment. A statistical analysis shows that such catalytic reactions may form interdependent networks when the rate of catalyzed reactions per molecule approaches one, creating a self-organizing chemical cycle which he calls an 'autocatalytic set'. When the rate of catalyses per reagent is low, only small local reaction chains form, but as the rate approaches one, the reaction chains in the soup suddenly 'freeze' so that what was a group of chains or islands in the soup now connects into one large interdependent network, constituting an 'autocatalytic set'. Such an interdependent reaction network constitutes the core of the body definition unfolding in Kauffman, and here, as in Uexküll, its cyclic character as the basic precondition for self-sustainment must be noted. He now defines an 'autonomous agent' – a sort of general, formal organism concept – as follows. An autonomous agent is an

> Autocatalytic set able to reproduce and to undertake at least one thermodynamic work cycle. (49)
> (- defined, in turn, as the recurrent release of thermal energy for performing mechanical work)

This definition implies two things: 1) reproduction possibility, and 2) the appearance of completely new, interdependent goals in work cycles. The latter idea requires the ability of the autocatalytic set to save energy in order to spend it in its own self-organization, in its search for reagents necessary to uphold the network. These goals evidently introduce a – restricted, to be sure – teleology defined simply by the survival of the autocatalytic set itself: actions supporting this have a local teleological character. Thus, the autocatalytic set may, as it evolves, enlarge its cyclic network by recruiting new subcycles supporting and enhancing it in a developing structure of subcycles and sub-sub-cycles. This has its exact analogy in economy, Kauffman argues: in eighteenth century England, coal-fired pumps were employed to empty the coalmines for water – and thus made possible more efficient coalmining and industrialization as a whole. Again, we recognize the Kantian description of teleology by locally self-sustaining cycles.

Now, Kauffman proposes that the concept of 'autonomous agent' implies a whole new cluster of interdependent concepts (forming so to speak, in itself, an autonomous agent network on the conceptual level...). Thus, the autonomy of the agent is defined by 'catalytic closure' (any reaction in the network demanding catalysis will get it) which is a genuine Gestalt property in the molecular system as a whole – and thus not in any way derivable from the chemistry of single chemical reactions alone. Tasks and molecules constitute, on this level, a duality:

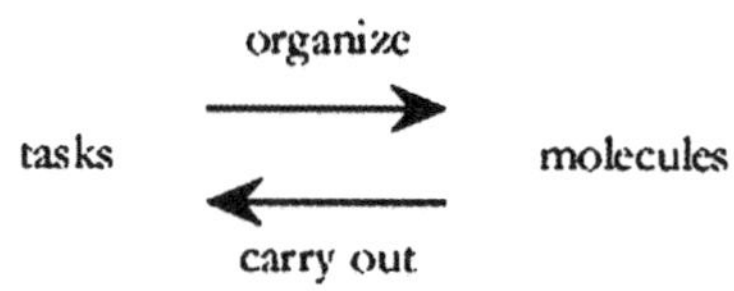

Figure 31.

The work definition here concerns the *governed* release of energy with respect to a task. Thus 'organization' becomes the set of constraints pertaining to an energy release process in the system. To the conceptual cluster around 'autonomous agent' thus also belongs: the emergence of organization consisting of entities measuring relevant properties of nonequilibrium systems in the environment, identifying sources of energy that can perform work. This requirement is, as is evident, co-extensive to what we normally call perception and action – the search for, identification of, digestion of, and putting to use energy sources in the environment (83). Thus, also all of the following concepts pertain to one and the same conceptual network defining an 'autonomous agent': Work, constraints, construction, measuring, energy, information, event, organization (as *closure* of the set of possible molecules, of catalytic tasks). Thus, semantics is first of all defined by teleology – in an autonomous agent, chemical agents (or to be more precise, specific molecular configurations on the surface of macromolecules) can become signs.[341]

It is interesting to note that Kauffman's definitions on the basis of speculative chemistry thus entail not only the Kantian cyclic structure, but also the primitive perception and action phases of Uexküll's functional circle. Thus, Kauffman's definition of the organism in terms of an 'autonomous agent' basically builds on an Uexküllian intuition (even if there is no reference to von Uexküll), namely the idea that the most basic property in a body is metabolism: the constrained, organizing processing of high-energy chemical material and the correlated perception and action performed to localize and utilize it – all of this constituting a metabolic cycle coordinating the organism's in- and outside, defining teleological action. Perception and action phases are so to speak the extension of the cyclical structure of the closed catalytical set to encompass parts of its surroundings, so that the circle of metabolism may only be completed by means of successful perception and action parts.[342]

Of special semiotic interest are certain corollaries to the concept of 'autonomous agent'. Environment perception must, for economical reasons, be semiotic. One aspect of this is the fact that perception must coarse grain the environment and search information on a certain level of stimulus dissolution. Biological evolution makes perception tend towards an optimal coarse graining of environmental information. Another aspect is that 'usefulness' of an object in relation to a given agent only pertains to a few properties – just like a few properties then may play the role of signs for their useful objects. The useful properties and the sign properties may, in many cases, coincide (giving a high degree of interpretational security), but in other cases, the sign properties may merely overlap or even differ highly from the object properties which are sought after, introducing an arbitrary aspect of such signs. A third semiotic constraint in the concept cluster surrounding the autonomous agent concept is the fact that we – as any other autonomous agent – are unable to overview the configuration space of the biosphere and search for regularities therein. The space of all possible biological species is so enormous that it is impossible to access directly in any rational way (and much worse, then, the space of all possible interspecies relations which is larger than $2^{\text{number of species}}$). This necessitates *storytelling* as the only way to access general knowledge about typical event series – storytelling here ranging from innate action sequences which have proved useful for survival, over acquired knowledge about typical event sequence structures in environment interaction, and to human explicit narrating. Narratology thus, analogous to our argument in Chap. 9, becomes an a priori consequence of the vastness of biological configuration space and the a priori impossibility of access to that space. The autonomous agent furthermore introduces the distinction ought/is, correlative to the distinction between task functions and effective causes in an organism, while on the other hand the physical distinction between laws and intitial/boundary conditions is relativized by the circularity and growth potential of the agent. Stephen Jay Gould's well-known notion of exaptation[343] also stems from the inability to access biological configuration spaces: no finite list over biological functions can be predetermined. Thus, invention refers to new combinations of select properties among the indefinite number of properties of a system.

The evolution of autonomous agents is taken as the empirical basis for the hypothesis of a general thermodynamic regularity based on non-ergodicity: the Big Bang universe (and, consequently, the biosphere) is not at equilibrium and will not reach equilibrium during the life-time of the universe. This gives rise to Kauffman's idea of the 'adjacent possible'. At a given point in evolution, one can define the set of chemical substances which do not exist in the universe – but which is at a distance of one chemical reaction only from a substance already existing in the universe. Biological evolution has, evidently, led to an enormous growth of types of organic macromolecules, and new such substances come into being every day. Maybe there is a sort of chemical potential leading from the actually realized substances and into the adjacent possible which is in some sense driving the evolution?[344] In any case, Kauffman claims the hypothesis that the biosphere as such is supercritical in the sense that there is, in general, more than one action catalyzed by each reagent. Cells, in order not to be destroyed by this chemical storm, must be internally subcritical (even if close to the critical boundary). But if the biosphere as such is, in fact, supercritical, then this distinction seemingly a priori necessitates the existence of a boundary of the agent, protecting it against the environment. Kauffman does not go deeply into this, but it is, as a matter of fact, an a priori argument for the existence of cell membranes, skin, fur, plate, etc.

To sum up, the body understood as an 'autonomous agent' has the following implications:

- It defines a conceptual cluster involving reaction chains forming a loop in autocatalytical closure, emergence of organization, perception, action, work, constraints, construction, measuring, energy, information, event, organization, semantics, teleology.[345]
- Thus, the birth of meaning, signs, and intention supposedly take place alongside the self-organization of autonomous agents.
- Narration and story-telling are necessary in the absence of access to the configuration space of autonomous agents and their interrelations.
- The body boundary is implied by the necessity of the sub/supercritical distinction.

In relation to biosemiotics, it is crucial to note that the definition of autonomous agents takes the body concept all the way down to cell level (and maybe below as a limit case, cf. 'autocatalytic closure')

BIOSEMIOTICS – A NATURAL HISTORY OF EMBODIMENT

Biosemiotics, as discussed in the previous chapters, has emerged as a semiotic project during the last decades, involving the initiative of the late Tom Sebeok, and, among others, the Copenhagen school (Jesper Hoffmeyer, Claus Emmeche).

I can not run through the whole set of ideas of Hoffmeyerian Copenhagen biosemiotics here, and many of them have been repeated through the previous chapters, but let me sum up some of its basic assumptions. Independently of Kauffman, biosemiotics sees the cell as equipped with a point-of-view – defined as a 'stable integration of self-reference and other-reference' (Hoffmeyer) – and like in Merleau-Ponty or Kauffman, this does not necessarily imply the presence of consciousness. This definition refers to the self-description by genetic means on the one hand, necessary to pass on the information about the stable structure of the body to the next generation, and, on the other, to the stable representation of the outside within the cell, due to its system of 'perceptions' facilitated by specific chemical receptors. The self-description, in turn, is double, due to life's 'code duality': the digital DNA representation is not the only information inherited, and it is certainly not, as it is often maintained, the only 'cause' of the phenotypical organism. The whole cell structure with different organels constitutes an additional 'analogue code' which is inherited directly (some of the organels may even have their own DNA) in the egg cell.

Membranes are seen as crucial in biosemiotics because defining the inside/outside distinction characterizing all life forms, facilitating the highly constrained traffic across this boundary in the form of signs (perception can, then, be rephrased as 'inner outsides', just as action changing the surroundings into a specific ecological niche can be rendered as 'outer insides'). Moreover, internalized membranes of many different sorts facilitate semiotic processes inside the cell, regulating metabolism. Likewise, in higher animals, internalized skin sensors might have given rise to the part of the important part of the nervous system performing proprioception.

The simplest semiotic process is categorical perception, or environmental granularity (already at biochemistry level) – thus biosemiotics finds the semiotic vocabulary indispensable to describe even biochemical reactions – that is, when they take place in the context of the cell's metabolism. 'Scaffolding' is a concept invented by Hoffmeyer for the general process of stabilizing, channeling, automatizing and sophisticating (by adding new stable possibilities) a segment of metabolism: ranging all the way from cell architecture, organ structure, and to nest building, herd behavior, language, writing. In Hoffmeyer, increasing 'scaffolding' implies the converse notion of 'increasing semiotic freedom' – higher animals may not only be able to recognize tokens as instantiations of types, but also use these, apart from their possible reference to the immediate surroundings, to symbolize, to play, to reason, to argue, to use diagrams. This bodily basis even permits sufficiently complex organisms to make abstractions by treating relations as a thing (metaphorically) – probably only humans are able to do this explicitly and at will.

Biosemiotics in general thus points to the necessity of defining the body concept in constant interaction between biology and semiotics: the body as the minimal biological entity simply *is* a sign processing device. The evolutionary stance of biosemiotics, moreover, urges it to try to define a series of differently abled bodies so that the ability to specific sign use is correlated to a type of body[346] – attempting to charting a 'natural history of the sign' (Hoffmeyer).

Umberto Eco originally proposed the idea of a lower threshold of semiotics, presumably distinguishing human language and sign use from simpler signal systems in biology. This idea has, of course, been challenged by biosemiotics finding instead a crucial threshold between inorganic and organic nature – cf. Sebeok's idea

A biosemiotic Scala Naturae

Searle threshold – conscious, deliberate, intentional communication actions – linguistics
Eco threshold – linguistics and other human sign systems – humanities
the 'biosemiotic missing link' – probably consisting of several sub-thresholds? *Deacon threshold* – symbols? *Lakoff threshold* – metaphors? (- but is there any clear limit between metaphor and concept extension?) *Husserl threshold* – intersubjectivity? *Tomasello threshold* – joint attention? hypostatic abstraction?
Merleau-Ponty threshold – multicellular bodies with central nervous system (and probably consciousness), symbol processing, 'interanimalité' and environment mapping – higher zoology
Uexküll threshold – active information gathering, functional circle – zoology
Lynn Margulis threshold between unicellular and multicellular organisms (plants, fungi, animals) – pertaining to the introduction of stable intercellular semiosis and cell differentiation
threshold between prokaryotes and eukaryotes where the DNA in the former is not yet separated from the protoplasm so that it may much more freely be communicated to other prokaryotes
Sebeok threshold – semiotic processes proper – biology
Peirce 'threshold' – protosemiotic processes – all of the universe

of biology and semiotics as co-extensive. Thus the idea emerges that the crucial task is not to find one lower threshold of semiotics but rather to establish a whole ladder of thresholds of increasing biosemiotic complexity. In addition to thresholds separating the inorganic from the organic world and animals from human beings, we may expect semiotically defined thresholds separating unicellular from multicellular organisms, plants and fungi from animals, animals with/without central nervous systems and correlated environment representations – and probably many more. As we saw in Chap. 11, Terrence Deacon argued (1997) that the semiotic 'missing link' is the transition from iconic and indexical to symbolical signs; a problem here is that if we use Peirce's symbol definition, many higher animals use symbols.

Maybe the semiotic missing link is constituted by several subthresholds lying close to each other which yet await untangling? – the ability of diagrammatical reasoning is probably also shared by some higher animals, the use of metaphor likewise, the existence of intersubjectivity and, correlated, objectivity arguably has its primitive form in Merleau-Pontyan 'inter-animalité'. The use of the special symbol type of argument can also, presumably, be found in many higher animals. As we saw, the best candidate for the semiotical missing link might be a special subtype of the Peircean symbol: the so-called 'hypostatic abstraction', making of some aspect or relation of a phenomenon a new, explicit object. This would be the ability to make *explicit* signs, arguments, symbols and thus subject them to deliberate control and change – and hence facilitating their fast development in culture.

In any case, one of the tasks of biosemiotics must be the construction of a semiotical ladder of evolution, combining basic body types with semiotic ability. On the opposite page is indicated, as a conclusion to the last four chapters by way of proposal,[347] a first outline of such a scale with some of the single steps nicknamed after central authors.

CHAPTER 13

CHRIST LEVITATING AND THE VANISHING SQUARE

Diagrams in Picture Analysis

Most studies of pictures and pictural phenomena take place without any further determination of the concept of picture: the everyday meaning of the word are without further notice taken as sufficient. In many cases this causes no problems, to the extent that the study only involves the description and interpretation of selected properties in the picture in question. For a more systematic gaze, though, it is hardly satisfactory that the very category of picture remains vague. Very often, it is spontaneously assumed that a picture is an object somehow picturing another object. For specialists with some measure of knowledge of art history, such an idea must immediately cause problems: most of the abstract or non-figurative tradition of twentieth century art depicts – for a first glance – no other object. Other spontaneous ideas hold that the picture is a visual entity which some person has produced with the intention of communicating some content – maybe even an especially aesthetic content – or, maybe, a content about which something is claimed, so that the picture has the character of a proposition. The ideas of what is a picture thus flutter around concepts like visibility, intention, communication, similarity, abstraction, reference, proposition. There is, however, some of these concepts which for a closer gaze proves irrelevant for the basic determination of the picture concept. The idea that *visual* pictures should exhaust the picture category is already undermined by the fact that we talk, without further notice, about pictural meanings, iconic language and so forth (without any idea that these notions should in themselves be pictural!), just like synaesthetic phenomena of all kinds point to the fact that the category of pictures does not belong to vision alone. Nothing prevents a sound from picturing another, and even if visual pictures constitute a prominent and prototypical class of pictures, nothing indicates that they exhaust the category. The idea, on the other hand, that pictures should be the expression of the *communicative intention* of somebody, does not seem plausible either on closer scrutiny. Naturally occurring images, mirrorings, echoes, mimicry phenomena and so on point to a broader concept of picture, just like the developments away from purely intentional meaning concepts in the aesthetic disciplines in general. Just like in the case of the work of art, it is not exhaustive or even decisive for our experience of an image that it corresponds to what the sender might intend it to mean. The possibility of discovering important properties in an image which the artist did not intend when he painted it, lies open to us; if this was not the case, the artist might just as well serve as an index of solutions for the aesthetic analysis, and he might ease his own work as well as that of the critic by simply publish a written account of his intention instead of painting. *Similarity*, then, seems to be definitely dealt with already in the art history of the twentieth century – for is it not the case that similarity in pictures is but the result of a restricted, mimetic, and outlived poetics which has long since been obliterated by the avant garde? Has it not long since – supported by deconstructivist and negativist theories – made it clear that behind every tempting similarity hides an abyss of differences? My hypothesis is, however, that it is, despite of all this, in similarity we shall seek the very definition of the concepts of image and picture as such.[348] If we take a series of the archangels of the so-called nonfigurative tradition, e.g. Malevich, Kandinsky, Klee, Mondrian, Newman, Rothko, Reinhardt – it is striking (even if their claims will, of course, not suffice to settle the case definitively) that they themselves by no means suggest that their pictures do not represent anything, quite on the contrary. It goes without saying that they do not represent interiors nor exteriors from the natural or cultural world, easy to recognize for an untrained gaze – but they are taken to refer to more subtle matters, for many of the painters mentioned even to mystical, religious spheres. Even an observer like myself – who has a hard time letting Rudolf Steiner figure as the key to Kandinsky – has the immediate and spontaneous experience that these pictures represent *something*, be it personal mental ideas or emotions, be it realms of ideal geometry, be it that-which-cannot-be-depicted... it does not matter in this context. The decisive thing is that similarity cannot be given up as that which defines a picture in the first place. The iconoclastic tradition discussed in Chap. 3, supposed that similarity was (almost) identity, that it was a dull copy of something given, that it maybe only mimed natural beauty and thus constituted a primitive set of tablets which must be crushed in order to push forward to a radically new and sublime language of art. I will suggest that we – from the vantage point of our time – can see this idea as a special sort of avant garde kitsch. It is the avant garde version of the deer by the forest lake over the couch because it just like the deer – only with reversed value ascription – assumes similarity to be an easy and unproblematic issue.[349] The history of the avant garde demonstrates – quite on the contrary – that the limits to the concept of similarity are very wide indeed. Just like the concept of *œuvre*, of artwork, has only been extended and become still more multifaceted and enriched by the many attacks on it by the avant garde, the concept of similarity has been diversified and extrapolated to realms where we did not recognize it functioning earlier. Especially the concept of the artwork which – however this may take place – gives the single piece of art its definitivity, its closedness, a set of borders, results in the fact that what appears within these borders assumes the character of a sight seen within a frame, a margin, or other boundaries in space and time, that what is seen within these limits appear as representing *something*. The fact that it may very often be quite difficult to determine *what* this something is, does not change the fact that as soon as we are within the realm of the artwork, we are perfectly aware that the work has, in some way, to be decoded. But as the artwork cannot – if it aims at being a good work, that is – rely on conventions alone (this would conflict with the second and in other respects problematic demand of the avant garde: originality[350]), no other resource is left but similarity. The history of art of the twentieth century with its triumph of experiment has thus – quite to the contrary to what is often assumed – been a triumph of similarity.

What does this similarity entail for the definition of the picture? Many artworks which we spontaneously conceive of as pictures (and thus primarily iconical signs), however, also possess indexical aspects (maybe due to their title which may refer to existing phenomena – the portrait painting's title referring to the person portrayed; the landscape paintings to the locality; the indexical reference of the signature to the artist, etc.), as well as symbolical aspects (e.g. by virtue of their use of different conventions for indirect reference to phenomena – the iconographical convention like the lion referring to the apostle Marcus). We may concentrate upon, however, the primarily iconical signs – what Peirce terms hypoicons. They are the class of signs, then, which primarily functions iconically. Here, Peirce's non-circular description of what 'iconical' means remains decisive – his requirement that icons are signs, *by the contemplation of which it is possible to learn more about their object that what lay in the mere construction recipe for the sign.* When possessing the key to an icon, it is possible to take out of it more information about its object than what is explicitly stated in the sign itself. It is for this reason that similarity-based signs seem 'deep', and the definition by 'more information' is, contrary to the intuitive similarity definition, operational to the extent that it may be used as a criterion to determine whether a sign is in fact an icon. The fact that many easily decodable icons exist might give rise to the idea that it should always be very easy to determine the object of an icon; this is, however, a very dangerous fallacy. Icons may, in some cases, be very difficult to decode. Perhaps the object may be determined vaguely or partially only; the decisive thing remains, according to the operational definition, whether more information may be retrieved about the object than what explicitly appears. To grasp more exactly what lies in this operational definition, we must again consider Peirce's triadic subtypology of the category of hypoicons. They fall in images, diagrams, and metaphors, respectively. Images in this restricted, technical meaning of the word are similar to their object due to some simple quality (color, shape, tonality, size...); diagrams are all similarity-based signs which refer to their object by means of some skeletal analysis of the object into mutually connected parts. The diagram consists of a sketch-like anatomy of its object – as the most ordinary examples one may point to function diagrams, cake diagrams, column diagrams, matrices – but also, cf. below, a much larger set of icon types. Metaphor, finally, is the picture type which refers to its object via the intermediary of a third object. When a metaphor does not in an evident way display which diagram it makes use of, it becomes ambiguous and opens up interesting possibilities for poetical use, but this will, from the Peircean point of view, be a marginal, atypical use, even if strongly interesting and fertile.[351] In art, however, this use becomes central. The naked, easy-to-grasp diagram rarely occurs in art – and even in prototypical diagrams, surplus image similarities (in Peirce's special use of the word 'image') often occur – as when a column diagram over the average income in different countries are tinted with the colors of the flags of those countries.

The decisive implication for the status of the picture category in these distinctions lies in the extension of the diagram category, cf. Chap. 4. This category, at a first glance so easy to determine and seemingly bound to lines and arrows on paper, proves much more comprehensive, given the operational icon definition. Peirce gradually realizes this, as when he sees that e.g. algebraic formalisms also have diagrammatical character – to the extent that they may be manipulated in order to display new information not explicitly present. Here, Peirce's non-circular definition of similarity demonstrates its full force: icons are all those signs by means of which more knowledge about their object may be gained. Diagrams, consequently, are that subset thereof which functions by analyzing their object in parts, interconnected by means of rule-bound relations. This entails a series of further determinations of the diagram: it includes all signs by means of which something can be inferred with necessity (which then, in turn, holds for their object) – that is, diagrams include all signs allowing for deduction. As discussed in Chap. 4, this implies that diagrams as a decisive feature possess the possibility of being *manipulated* with the aim of reaching deductive consequences. This manipulability and, correlatively, deductibility, is what makes diagrams icons with the special feature that they may be used to think with. If we consider the less abstract side of diagrams, we see that this holds also for very 'mimetic' diagrams. As mentioned, Peirce never undertook any further partition of the diagram category, and the establishment of a rational typology of diagrams remains one of the most important tasks for actual semiotics to pursue,[352] but it seems obvious that *maps* are involved as a crucial type of diagrams. Maps come, even if we restrict ourselves to geographical maps, in a continuum of widely varying degrees of abstraction, from aerial photos, preserving distances, angles (provided the area depicted is sufficiently small), color and much more . . . and to e.g. subway maps which are neither color-, distance-, nor angle-preserving but only preserves the mutual topological relations between subway stations and lines as points on a network of topological connections. Common to all maps, however, seems to be the idea that a route in the diagram corresponds to a route in the object which facilitates the possibility of certain diagram experimentation on maps (what is, e.g. the highway distance between Paris and Berlin? – this is not stated explicitly in any way on your average map of Europe, neither in the printing process nor in the construction of the map based on triangulations in the landscape, but the information may be easily retrieved by a small diagram experiment using a ruler and the scale of the map).

Maybe my conclusion is evident: all pictures, also in the ordinary art-history-meaning of the word, are also diagrams – primarily *maps* in the general meaning of the term suggested.[353] As I hinted to above, this idea must be seen in relation to Peirce's idea that signs are only signs *in actu*, that is, to the extent that they are used as signs. This implies that if I see a picture hanging remotely on the wall, peripherically in my field of vision where it appears almost as a vague impression, then it is only, of course, an image in Peirce's simple sense. But as soon as I – be it as a spontaneous viewer of the picture or as professional analyst – go closer into the picture, the use of it as a diagram automatically becomes relevant.

I measure distances on the picture plane with the eyes and spontaneously infer information about distances between foreground figures, medium ground appearances and background features in the space depicted; I observe the striking effects of contrast between certain colors and may oppose them to other, non-contrasting colors. I may construct a space,[354] in which I can imagine my body moving around; this very wandering route inside the landscape has the characteristics of a diagram manipulation. I may question, for instance, the artificial, the striking, the aesthetic, the failed or other qualities in the distribution of picture objects on the plane – doing so, I make a manipulation where I imaginatively move around objects on the plane and observe the changes in their mutual relations. It is, in short, impossible to reflect or speculate upon a picture – in spontaneous perception or with the distance of the analyst – without conceiving of the picture as a diagram, manipulating with its parts according to different rules and, so doing, retrieve new knowledge of the objects depicted, be they concrete or abstract.

Now pictures – in the art history use of the word – constitute a special subset of the diagram category. They are visual, they are delimited, they are two-dimensional (2 1/2-dimensional, if we include texture, bas-relief, frame, etc.). They are, moreover, characterized by the fact that they are only rarely accompanied by any explicit symbolic set of rules or conventions governing the possibility for their diagrammatical manipulation (this in opposition to mathematical diagrams). But this does not mean they do not function as diagrams; it rather means that we, by contemplating them, make use of two things: 1) the spontaneous diagrammatical abilities characterizing natural perception, constituting an autonomous intelligence inherent in the visual system (giving rise to facts like that it is almost impossible not to see depth in a perspectively construed picture), 2) the experimental use of diagrammatical abilities taken from other fields of experience. Art is thus diagrammatically underdetermined in a way analogous to the metaphorical underdeterminacy of poetry – which is why it calls for interpretation: many different diagrams and metaphors may call for application in experimenting analytical picture observation. Thus analysis of art (but this goes for any analytical process) involves a crucial abductive component – abduction being the Peircean prerequisite for any gaining of new knowledge. Abduction makes a guess at an unexplained phenomenon by suggesting a general law or state of affairs which would have the phenomenon in question as necessary consequence (other possible laws, regularities, or prerequisites might have the same effect which is why abduction is not necessary but remains a qualified guess). But once an abduction is proposed, then it may be tested by the manipulation of the diagram implied in the hypothesis proposed (we return to the role of ab- and deduction in interpretation in Chap. 16). The outcome of the diagrammatic deduction may then, in turn, be inductively compared with other knowledge about the phenomenon in question, giving Peirce's three-beat motor of reasoning as a result: abdution-deduction-induction.[355] It may thus be discovered de- and inductively whether the abduction in question is fertile; in the opposite case another abduction must be suggested.

CHRIST LEVITATING ...

Let me illustrate this idea of the analysis of pictures by a couple of examples. Some years ago, the Danish art historian Erik Fischer published an exquisite little book on the Danish romantic painter C.W. Eckersberg[356] – including a brilliant analysis of Eckersberg's famous altarpiece from Frederiksberg Church in Copenhagen. Fischer recapitulates the iconographic background of the picture. Eckersberg was asked by the rector of the church to base his picture on John 17.6–19 which is the so-called 'high priest's prayer' in the latter part of the last supper – a topos with, it must be admitted, a rather meager pictural potential.[357]

Fischer now wonders what Eckersberg's seemingly uncomplicated last supper rendering may yield as an illustration of this ecstatic speech of farewell, in which Christ claims not to be of this world. In his search for further content in the picture, Fischer stumbles over the empty seat of Judas in the right foreground zone of the picture and sharply remarks: is it not the case that this stool is positioned in a strangely oblique way in the room? Fischer makes this observation and now adds the abduction that the stool is not only oblique but that this serves a deliberate artistic purpose which is why he now indulges in an experimentative manipulation of central relations in the picture, that is, treats it as a diagram (69–70) 'His empty

Figure 32. Eckersberg's Altar Piece for Frederiksberg Church, ordered 1839

stool is highlighted in the foreground. Not only is it highlighted, it also seems to play a compositional role of its own: its vanishing point (FpII), at the extreme left of the painting, demonstrates that Eckersberg provided his composition with a second horizon, C-D, a unique phenomenon in the history of Post-High Renaissance art. The C-D horizon forms a Golden Section in the part of the composition that is below the J-K line.' First, he constructs the perspectival horizon line as implicated by the lines of the ceiling girders. This diagrammatical deduction rests, of course, on the premiss that the picture depicts a room which has as will normally be the case, a horizontal ceiling with parallel girders. The result is a horizon with the room's vanishing point (FpI) at Christ's right shoulder. He now repeats the same procedure with the seat of the stool and finds that Judas' seat has a completely different, much lower horizon from that of the dining room, indicated by the vanishing point (FpII). This observation is now corroborated by an earlier sketch of the picture, in which Eckersberg again has made this construction of the horizon corresponding to the seat, and Fischer consequently finds that the seat really, measured against the horizon line implied by the ceiling, is skew. This diagrammatical finding supports his abductive hypothesis and now allows Fischer – using another unspoken assumption, namely that the disciples did not know of ergonomically correct seats but used stools with approximately horizontal seats (this unspoken premiss is, of course, supported by the general symmetry of the stool's design) – to conclude that the picture is a case

Figure 33. The painting with Fischer's auxiliary lines showing the double horizon of the ceiling and the stool, respectively

Figure 34. Eckersberg's sketch with Fischer's auxiliary lines

of deliberate use of double perspective with the rhetorical effect of a structural derogation of Judas. His horizon is, literally, far below that of our Saviour: 'The second horizon bears a subtle iconographic message. Judas is *the son of perdition*, so 'his' horizon is placed low in the composition cosmology; his 'vanishing point', at the extreme left, must be seen as an unmistakable indication that he has moved away from harmony toward betrayal.

The line L-M further stresses the expulsion of Judas: it touches the far corner of his empty stool, which is thus once again symbolically relegated to its own sphere at the very bottom of the composition.'

If we now take further Fischer's idea and repeat the same operation, now with respect to the *sides* of the stool resting on the floor (again based on the helping hypothesis that they are parallel), then we obtain yet another horizon line, the X-Y line, this time for the floor of the room. This line, surprisingly, lies even lower than the one construed from the seat. This implies that the stool is almost impossible as a geometrical object – if measured on our ordinary ideas of how stools normally have parallel sides and orthogonal vertices. This implies that the stool stands out as a disquieting object which due to its double perspective and its oblique form irritates our gaze and indirectly insists on Judas's betrayal and disappearance, and in so far this observation yields further support to Fischer's reading. But what if we suppose, now, that the stool is not skew in relation to the perspective of the floor but that it stands, in fact, in a normal way on the floor? If this is the case,

Figure 35. The painting with my auxiliary lines added to Fischer's, indicating the third horizon pertaining to the floor

then the very perspective of the floor deviates strongly from that of the ceiling, and then further perspective anomalies follow. Then the whole geometry of the room of the supper explodes; ceiling and floor yawn and moves away from each other in the direction into the picture towards the horizons. Whether it is the floor that goes downward or the ceiling that goes upward (or both) is of course impossible to decide within the picture's own system of reference, because it gives no hint which may decide which one of the now no less than three possible horizon lines should be privileged as being horizontal. But no matter this undecidability, a remarkable fact remains. Namely the fact that given the tilting floor, Christ is now simply unable to reach the floor behind the table, if we assume that he was created like a normal man with approximately the same distance from top to groin as from groin to feet. This diagram manipulation thus urges the question: What could be the possible reason for Christ to stand on an invisible platform behind the table or to wear extreme Sylvester Stallone boots to appear taller than he actually is? Or what could be the reason for a radically long-legged Christ which I believe we have no further biblical evidence to support? – or, the other way around, a Christ secretly levitating over the floor behind the tablecloth? The left side of the table is mercifully hidden by the sitting disciples and thus does not allow for an additional, corrective horizon construal based on the (allegedly) rectangular, horizontal shape of the table top. Our experimental deduction on the picture read as a diagram thus leads us to infer these facts. Do they corroborate our abduction – that there might lie further information hidden in the multi-perspective structure of the picture? This is the case, indeed, because the scene of the last supper leads directly on to resurrection and ascension – decisive upward movements which are then prefigured in this picture's intricate application of a triple perspective. Christ is here already, during the last supper, characterized by an upward movement which very well points to the claim in the evangelical text base of the picture: He is not of this world. These two diagram deductions – Fischer's and mine – possess an objective validity which does not depend on what Eckersberg's intention (or the picture observer's, for that matter) might have been. On the contrary, our judgment of these diagram experiments depends on their fertility in relation to the given picture and its contextual background. Thus, it is at this point we cease to read the picture as a pure diagram and include the symbolical reference it carries (due to its conventional use as an altar piece and the correlative reference to a specific phase of the passion of Christ). And in both cases, Fischer's and mine, we find in this symbolical framing of the picture as an applied diagram, so to speak, facts corroborating our manipulating deductions. It is not only Eckersberg who failed on a bad day and crafted a malfunctioning diagram (as would be the case if the picture was to be read naturalistically only, and the different additional hypotheses as to horizontal ceilings, floors and seats were correct) which makes an ordinary room twist in surreal lines. Quite on the contrary he has made an ambiguous diagram, open for a triple reading, whose tension between the perspectives of ceiling, floor and seat enables the observer to judge Judas' low status and skew nature, as well as the beginning levitation of Christ pointing forward to resurrection and ascension. Fischer's analysis has, of course, a further support due to Eckersberg's preliminary sketch which is lacking in the case of my additional argument (the floor horizon in the sketch coincides with that of the stool top). But the sketch has only the role of inductively strengthening Fischer's argument; it proves deductively nothing (and presupposes an intentionalistic reading of the picture) but it merely serves the purpose of further support for the probability of his claim. The decisive test of both analyses lies in their inductive continuity with the symbolic dimension of the picture – corresponding to Peirce's notion of the symbolical government of the pure diagram, deciding which conceptual reality it depicts in concrete usage. Here, the category of the artwork plays an important role as that which allows us to detect these hidden similarities in the picture: if the picture was not a work of art with the special definitivity this implies, then we would rather be tempted to dismiss it as a bad piece of student's work bearing witness to a vacillating understanding of the laws of perspective with the result that all parallel lines do not meet in the same horizontal point. Herein, the analysis of an artwork is different from a pictural analysis as we would make if this picture was an ordinary photograph: then we would be forced to other helping hypotheses, so as for instance that the stool of Judas is in fact deform, that Christ stands on a small, hidden platform on a tilting floor, that Christ has been ingested growth hormone by John the Baptist during adolescence, etc. The category of the artwork obliges us with its definitivity to search for the coherent reading rather than the probable reading, in the hope that it will, in the end, convince with its fertility in the final understanding of the work. But pictural phenomena which are not artworks and do not share their strongly teleological composition, do not demand such an integrated understanding; they may as phenomena require a decomposition in parts calling for each their diagram. They are none the less worthy phenomena for that, of course, whereas the work of art will suffer from such inconsistencies if not they may be deemed motivated at some level of analysis. Herein, the object of artistic analysis differs from other objects.[358]

This presentation demonstrates the characteristic shifting between abduction understood as qualified guesses faced with strange observations, deduction understood as diagram experimentation on the picture, abductive helping hypotheses, inductive probability arguments and a final conclusion measured on the symbolical governing of the diagram. This procedure is probably not unique for the analysis of pictures, not even for aesthetic analysis as such. Rather, it is a general heuristic for scientific analysis of object properties not yet fully determined (in that case its behavior might be deducted directly from theory). It is the road of the power of judgment to abductively guess and afterwards to corroborate this guess by deductive experiment and inductive probability support. Still, this leaves the analysis of pictures with a special status, because unlike the case in many other disciplines its very object contains a direct diagrammatic aspect. In most other disciplines it is a task for the scientist to find, formalize and corroborate which diagram may adequately map a given phenomenon. A (sufficiently complicated) picture, however, only *exists* to the extent it is already a diagram.

This implies that the analysis of pictures ought to be interested not only in its connection to visual objects in general or to aesthetic objects in general – but to diagrams in logic, mathematics, maps, algebra, and graphs. Diagrams comprise all things by means of which necessary conclusions may be drawn. Thus, the diagram category implies that all pictures with diagram qualities (which is: most pictures) involve the possibility for logical inferences regarding certain implications for their object. A traditional logical understanding of this fact would probably be that what is pictural about the picture should be seen as a mere heuristic surface over a depth of logical propositions whose content might be more adequately expressed in purely formal calculi. Such an idea would, with a certain displacement, repeat the linguistic imperialism in early picture semiotics where pictures' character of being signs with content was seen as so central that the very visual qualities of the picture tended to be reduced to a peripheral epiphenomenon.[359] To avoid merely substituting a logical imperialism for linguistic imperialism, it is important that this logical aspect of pictures be correctly understood. The logical aspect of the picture is inherent in its very iconic construction, in its very diagrammatical structure, and contrary to the possibility of a logical imperialism of the sort mentioned, a Peircean view would see symbolical calculi as a mere subset of the general iconic realm of logic.

... THE VANISHING SQUARE

This diagrammatic rendering of the analysis of pictures might give difficulties, however, when we face pictures where no evident diagrammatic systems are at stake (as in the Eckersberg case perspective and its analysis of perceived space into mutually connected positions), nor any symbolical content which may govern the diagrammatical experiments – phenomena which were abundantly present in Eckersberg's painting. Let us now, to support for our hypothesis, turn to a diagrammatic case study of a picture without any explicit diagrammatic structure and symbolical content. Malevich's *Suprematist Composition: White Square on White* from 1918 may of course not, with its low-profile articulations, hide a complexity of the same kind as Eckersberg's altar piece; but still it would be a mistake to claim that we view even a picture of this simplicity as a mere image in Peirce's sense of the word.

The meditative calm which may inhabit the viewer by the contemplation of such a work of art was understood by Malevich himself as a direct access to being itself with the acquittal of ordinary reason connected to everyday perception,[360] but even a quasi-mystical experience of this kind does not take place without diagrammatical means. Even the symbolical governing inherent in the painting title is here extremely vague and is delimited to a description of the picture surface in very general terms. Already in this fact, though, we find a hint: there is a delicate tension between the generic 'white square on white' and the specific appearance of the canvas surface. A 'white square on white' might with the same means be presented in infinitely many other ways: the small square might be differently located on the canvas and have any other size as long as it remained smaller than the picture. This immediately opens the picture for diagrammatical manipulation: we may imaginatively let the inner square reduce and grow, move around on the surface. The fact that this diagrammatic exercise would not change anything in relation to the generic title is a point which makes it possible to see the specific picture *itself* as generic, so that the painting in a certain sense represents an infinite set of other pictures of white square on white – an effect which Malevich has no doubt sought in his idea that his painting should display nothing less than the absolute itself. But this does not exhaust the diagrammatical possibilities of experiment on this surface. Ground and figure are only distinguished by a difference in shades of white on the lower limit of being perceptible, making the small square an inkling bluer than the background, against which it luckily, gestaltistically stands forward (which is also referred to in the title's 'white square *on* white') by means of a discrete comic strip contour. This makes possible a variation well-known from Gestalt theory: the *Kippfigur* shifting between object and background, so that the small square changes from being an object in the foreground and to being a hole in the surrounding foreground – a change facilitated by the more blueish tone's more remote character in atmospheric air. This diagrammatical experiment is even basically made possible by the visual system and is not necessarily in need of any conscious effort. A further fact is that the title's repetition of the concept 'white' is contradicted by the two main nuances on the surface, to the extent that

Figure 36. Suprematist Composition: White Square on White
(Kasimir Malevich 1918)

we should understand exactly the same color by the word 'white'. Neither fore- nor background are, in fact, perfectly white, and the 'white' of the title must thus also be generic and refers to white as a class of different whitish shadings.[361] The diagrammatical exercise called for by this fact is naturally to imagine the difference between the two whites disappear and the contour to vanish completely so that the small square simply implodes in the background – which would be the case if both whites were in fact identical. This procedure introduces a subtle play, for the very contour between the two parts of the surface is what introduces a minimum of depth in it – whether it turns one way or the other. If the square in the foreground ceases to exist, the depth of the background remains as a vacuum into which the viewer is sucked. Even if the two surfaces are only minimally distinct, the picture surface thus holds a possibility for a Rothko-like hovering into the picture which lets the – diagrammatically imagined – completely emptied surface have a depth, a void, a nothing with spatial qualities; probably for Malevich an adequate representation of his Absolute. A further diagrammatical experiment is possible with reference to a diagram already present in the visual system before any ascription of content: If the vertically suspended picture is supposed to refer to a zone equipped with gravitation, the inner square immediately assumes weight, and its oblique position related to the frame indicates an instability which at any time may be released, letting the square fall away with the pure, unspotted depth remaining.

Even if the title of the picture also in this case allowed for the construction of a symbolical frame which might, to some extent, orient the picture's interpretation as a diagram, then the overall result is here – as in large parts of twentieth century art – a radical underdetermination compared to the Eckersberg case. This implies that, in viewing such a picture, you are forced, as an observer, to an abductive trial-and-error process, attempting the use of diagrams and consecutive manipulations with no other guaranty than that of perception for one diagram being better than another. An analogy is, again, the use of sophisticated metaphor in much modern poetry which does not allow one unambiguous diagrammatical interpretation: such underdetermination only *augments* the possibilities for diagrammatical manipulation. This diagrammatical diffusity does not, however, contradict the fundamental similarity character of the picture. Even if the surface is only made to refer to a universe of pure geometry[362] – cf. the title – then it *is* an undoubtable relation of similarity, just like the secondary relation which, in turn, with a Platonic gesture lets this idealized universe of geometry refer to the Absolute as such (if we agree to ascribe the Absolute traditional properties like ideality, purity, simplicity, etc.).

What is gained by realizing the diagrammatical character of picture viewing is not least the close relation between picture and thought. It is a corollary to Peirce's generalized conception of logic that thought, even if general, can never leave an intuitive, iconic basis, and the diagram as a category is, as mentioned, Peirce's heir to Kant's famous *schemata* as a meeting place between intuition and thought. This implies that we automatically, spontaneously see pictures as bundles of diagrammatical variation possibilities (some of which are already present in the hardware of our visual system), and we think – when we do so, that is – in and with them by attempting to make explicit these diagrams, to add other and less spontaneously appealing diagrams – and to examine some or all of their manipulative possibilities. This gives the strong insight – for the analysis of pictures – that the contemplation of pictures is ripe with connections to general epistemology.[363] This implies the necessity for the picture analyst to go deeper into the field between arts and diagrams (in a more narrow use of the word), and it further implies the maybe more difficult requirement that he must give up his spontaneous idea that the picture displays what it displays and nothing more. Frank Stella's famous 'What you see is what you see' only remains correct in a diagrammatical light, if 'see' is taken in a broad sense including more than vision proper – namely logical and conceptual insights. Seen as a diagram, the picture is a *machine à penser*, allowing for a spectrum of different manipulations. It is these manipulations the analysis of pictures must reveal, and they are not immediately visible in one glance on the canvas.[364]

CHAPTER 14

INTO THE PICTURE

Husserl's Picture Theories – and Two Types of Pictures

'Things, however, are difficult, and it is easy to run wild . . .'

Husserl (Hua XXIII, 342)

We have investigated the possibilities of diagram manipulation in picture viewing and analysis. A further attempt at surpassing naïve picture understanding is provided by the sections of Husserl's phenomenology in which he outlines a taxonomy of acts and thus describes the picture as an act type and spends extensive discussions on placing it in relation to other act types.

After having put forward the idea of different types of intentional acts in *Logische Untersuchungen* which we touched upon in Chap. 6, Husserl often in the following years returned to the attempt of devising a more fine-grained typology than the one outlined there. The distinction between perception, imagination, and signitive acts of various types had to be refined, especially as regards the imaginative types of acts. In the years from LU and to 1912, but even well into the 20s, Husserl struggled with the elaboration of this branch of phenomenology, and the papers containing these strivings are collected in *Husserliana* (*Hua*) Vol. XXIII with the title *Phantasie, Bildbewusstsein, Erinnerung*, a title suitable for the project because Husserl's investigation concentrates upon establishing a series of subspecies to the imaginative acts of LU where the three types mentioned hold prominent positions.

Most of the papers have a preliminary, discussing character, and even the more thoroughly argued texts often have the character of the investigation of an issue where the conclusion still remains to be made – but in the long run of the repeated investigations, gradually the central problems in the descriptions become clear. A large work dating from the 1904–05 lectures about phenomenology and epistemology deals with the distinction between fantasy and pictorial consciousness in an explicitly experimental way – Husserl investigates a claim he earlier shared but which is now abandoned during the text: the idea that fantasy is a sort of pictorial consciousness (as in Beilage 1 from 1898). A long series of smaller papers, notes and addenda from these years and thereafter are concerned with fantasy versus picture, fantasy vs. memory, contradiction ('Widerstreit') as the basis of pictorial consciousness, empty presentations, etc., and various taxonomies of act types are proposed. The next larger work to follow is 'Modi der Reproduktion und Phantasie. Bildbewusstsein' from 1912 where the discussion about the status of fantasy is attempted solved – now on the base of emerging transcendental phenomenology – by the determination of fantasy as a reproductive modification, based on a distinction between impression and reproduction. This text is also accompanied by a series of *Beilagen*, inter alia about fictional consciousness, 'iconic phantasies', the relation between fantasy and actuality, the different aspects of pictorial consciousness, aesthetic consciousness, etc. Finally, a last group of smaller texts from the years 1918–24 are about the modi of intuition, the relation of fantasy to pure possibilities and to neutralization (taken as the suspension of the act's character of existence claim).

The notion of picture belongs to the most stable elements in this flux of new ideas and typologies making the many texts not only variants but outlines of original and untested ideas. Here we shall extract an outline of the resulting concept of picture – followed by a concrete proposal for a new phenomenological distinction on that basis.

The distinction in the LU between perception and the bundle of less direct types of access to the object, the imaginative respective signitive acts rests on the idea that an indirect approach to the object may take place with or without accompanying pictures. Here, it is taken for granted that pictures, fantasy, dream, memory, expectation, etc. all share the character of being imaginative – and further investigation naturally must attempt a description of the distinctions between these imagination subspecies. When reading Husserl's vast works on these problems up to 1912, it is striking to see that the notion of picture holds, as mentioned, a rather stable definition throughout the period. The determination of the concept of picture by the triad of *picture, pictural object* and *sujet* is constant, just like the relation of similarity, tying them together, and the relation of contradiction which complementarily prevents them from being identical. This double tension of similarity and contradiction provides a constant description of the picture and the type of consciousness assumed to correspond to it. The decisive problem for Husserl during that period is rather the description of *fantasy* which seems to be a sort of in-between, difficult to determine, between the explicit relation of similarity between well-defined, separate objects in the picture on the one hand and the immediate and direct access to the object offered by perception on the other.

Fantasy seems to share the picture's relation to its object, defined by similarly, on the one hand; on the other, fantasy seems, just like perception, to take place directly (even if not referring to any object present) and without any intermediary, thus there is not in fantasy any contradiction or tension between several ways of accessing the same perceptual content as is the case in pictures. Husserl's never-ending reflection in that period has, for that reason, fantasy as its primary goal rather than pictures proper (fantasy along with memory, dream, expectation, etc. which share fantasy's picture-like qualities without any explicit similarity/contradiction relation like in pictures proper). Correspondingly, Husserl's picture concept is narrow, being tied to externally, physically existing picture objects endowed with certain visual or tactical moments. In the perspective of Peirce's broad concept of icons, iconicity naturally involves fantasy which in a Peircean view must be classified as being a type of hypoicon. A further reason for including both fantasy and pictures under the icon concept is the role of fantasy in the grasping of sophisticated types of icons – more about this later.

Let us present the picture theory of 'Phantasie und Bildbewusstsein' where the triad mentioned is outlined:

When we here distinguish between object and picture, it soon becomes apparent that the concept of picture is here double. The object depicted is opposed to a double entity: 1) the picture as a physical thing, as this painted and framed piece of canvas, as this printed piece of paper, etc.
2) The picture as the pictural object which appears through this particular bundle of colors and forms. This is not the object depicted, the pictural sujet, but the concise analogon of the fantasy picture, namely the appearing object which is the representative of the pictural *sujet*.[365](Hua XXIII, 19)

In contrast to many naïve definition of pictures by means of similiarity, this definition – just like Peirce's – avoids the identification of meaning and reference. Against the picture taken as a merely physical object stand not one diffuse meaning, but two distinct phenomena: the object, be it real or not, to which the picture refers, and the pictorality, the 'pictorial object' (Bildobjekt) in Husserl's terminology, which is what makes it possible for the picture to refer to its sujet, to its object of reference. As Husserl sums up, (ibid.):

Three objects are what we have. 1) The physical picture, the thing made out of canvas, of marble, etc. 2) The representing or picturing object, and 3) the represented or depicted object.[366]

This distinction involves Husserl's decisive criterion for pictorality, namely whether a 'Bewusstsein von Differenz', a difference-conscoiusness distinguishing between 2) and 3) is present (20). If 2) did not differ from a perception of the object itself, there would be no picture. If we compare this to Peirce's concept of the icon we find a related idea: a transformation of invariant structure must hold between the two sides of the icon (corresponding here to 2) and 3), respectively), and the case of identity will correspond to a null-transformation where we would not talk about iconicity. A decisive difference between the two, on the other hand, lies in the question of the picture's dependency upon consciousness (maintained by Husserl, not so by Peirce). This implies, of course, that the difference between object and meaning – Husserl's 'Widerstreit' – to him is not only a structural fact in the picture sign as such, but requires a more or less explicitly articulated consciousness of contradiction.

This pictorial consciousness consists in the fact that the picture is constituted by the contemporaneous presence of two different presentations of the object: first, the presentation of the picture as a physical object appearing – second, the derivative act, founded on the former, which takes (aspects of) this presentation as picture of another object which is not, in itself, intuitively given, and which appears only through this presentation. The relation between pictorial object and reference object (what Husserl calls sujet) repeats, on the one hand, the tension inherent in perception between intuitively given profiles and their synthesis into an object, on the other hand a pictorial consciousness must precede it and select certain moments in the physical picture, making them interpretable *as* such profiles. These moments in the appearance of the physical object thus provide the basis of two contrasting acts, one direct and one pictorial. Pictorial consciousness thus involves two sides: the conscious directedness towards an object (31) through a picture – and the consciousness of contradiction which distinguishes between the two contrasting interpretations.

This analysis of pictorial consciousness has a series of consequences:

- It becomes possible to distinguish between those moments in the pictorial object which are similar to the object and allows it to appear, on the one hand, and those moments which do not resemble it or are decidedly in contradiction to it, on the other: 'What functions representationally in the content of the pictorial object is indicated in a special manner: *It produces, it represents,* depicts, *makes intuitive.* It is as if the sujet gazes towards us through *these* features. These features appear first in simple perception and only stand out in it against the other features of the pictorial object, against moments, parts, properties which either decidedly possess the opposite character, that of contradiction to properties in the sujet referred to, or which do not have neither one nor the other character. Such characterless features depict nothing, but it remains undecided how the real object is represented in them.'[367]
- On the basis of this idea, a further distinction between different pictures of the same object can be made – and, more generally, a distinction between different types of pictures using different classes of moments in this similarity relation.
- It becomes possible to distinguish between two different types of use of the picture: one which uses the picture mainly as sort of indexical 'trigger' to refer to the object (what Husserl calls a 'symbolical' use of the picture, cf. his idea of the trigger role of illustrations in geometry) versus a more proper use of the picture which momentarily lets pictorial object and reference object fuse together so that we directly grasp the latter through the former (analogous, in fact, to Peirce's idea of an 'imaginary moment' in icon processing where it is as if the object is perceived directly).
- Aesthetic contemplation of pictures may be distinguished as a special focussing on the pictorial object *as appearance* (where both the picture as physical support as well as object reference are bracketed). In the text in question, Husserl draws the consequence that the aesthetic picture must stand in a clear distance to its object (41), so that the differential character of pictorial consciousness is at its most clear and the bracketing of the object depicted is easiest. It is interesting to note that Husserl's deliberations here touches upon his recurrent wax doll example (where we erroneously take a life-size wax doll to be a human being) where the spectator is spontaneously fooled by the high degree of perceptual resemblance *despite* our clear, conceptual insight that it is not a living human being. The purely conceptual knowledge about the *Widerstreit* is thus not sufficient for the aesthetic feeling, about which we know that '... the picture must distinguish itself *clearly* from reality, that is, purely intuitively, without any help from indirect thoughts.'[368] Pure consciousness of *Widerstreit* must be spontaneous and intuitive, not a secondary product of reflection. Aesthetic contemplation of pictures thus appears as a threshold phenomenon in a continuum: on the one hand it must be distinguished from pictures with unclear, vague, or diminished consciousness of *Widerstreit* where the interest in the object has the upper hand and the attempt

to grasp the object directly is important (cf. 'the imaginary moment'); on the other hand it must be distinguished from merely symbolical uses of the picture where it it is *also* taken primarily as a type of access to the object, but this time with extra sharpened contrast consciousness and without any direct imagination of object qualities.

- Finally, the analysis of pictorial consciousness serves in a recurrent fashion as Husserl's background for his (vain?) attempts at a consistent description of fantasy as act type. The large text which we have here discussed, in this respect plays the role of an intermediary text where Husserl's earlier idea about the pictorial quality of fantasy (maintained in Beilage 1 (108) from 1898 and, even if less explicitly, in the *Logische Untersuchungen*) is finally given up.

The fact that the constant tension between similarity and contrast in the definition of pictures must, in fact, be at stake in two separate relations in the picture as a whole, is finally made explicit in a contemporary note, Beilage III (dated 'wohl um 1904/05') which, given the triadic definition of the picture, clearly poses two separate questions: 'Welches Verhältnis besteht zwischen Bild und Sache? Welches Verhältnis besteht zwischen Bildobject und Sache?' (138).[369] Here, the basic contrast between the picture and the picture object which the sensory appearance provides for intuition is outlined in addition to the already mentioned contrast (similarity, but no total identity) which must lie between the pictorial object and the sujet, the referential object.

The decisive relation of similarity, defining pictoriality,[370] is here analyzed in two steps. First, a relation between the picture taken as a physical object and the pictorial object it makes the observer see. Here, of course, a similarity must be at stake, while the latter is identical with a subset of the moments of the former – subjected, of course, to a different interpretation. At the same time a difference must be at stake because it is a subset only of the aspects of the physical object which participates in the pictorial object. We can add that it is possible to imagine extreme cases like pictures giving rise to entirely different pictorial objects, depending upon which moments are taken to give rise to a pictorial object and in which way (Wittgensteinian duck/rabbits, photographic double exposures, puzzle pictures, partial reflections in glass, etc.). Second, there is a relation between pictorial object and reference object, sujet, state-of-affairs or whatever we call the object referred to. Here holds a to some extent analogous distinction between which moments of the pictorial object are taken to refer to the object by means of similarity, and which not. Pictorial consciousness, to conclude, must consist of two sets of solidary relations of contrast and similarity.

Is it a given thing, now, that the very same moments in the picture's appearance which constitute the pictorial object are identical with those which, in turn, refer to the object? Husserl does not pose this question, maybe because of the fact that it seems immediately obvious that the pictorial relation prototypically involves one and the same subset of moments of the picture (even if, of course, taken in three different ways). Yet it must be emphasized that a vast set of modifications of this basic relation seems possible: different moments of the picture may contradict one another with a lot of different rhetorical effects as the result. The fact that one picture depicts a face which is considerably more oblong than any existing face known must, immediately, be taken as the reference to such a deform or mutated bodily part. If, however, the object reference in some way (maybe by means of a symbolical indication determining the object referred to as a really existing person, maybe by means of pictorial aspects of the very pictorial presentation (the shape or context of the face in the picture makes it clear the figure in question is a certain American president), etc.) is indicated as being a real person, then the oblong character of the face changes from being a real property of the object depicted to being a rhetorical effect in the description of that object, and the picture changes status from depicting an abnormality to being a caricature. The two sets of similarities thus provide the theoretical basis for a vast range of different pictorial rhetorics, depending upon which elements and aspects in the double transformation relation between physical picture and object possess character of reference and which possess character of meaning, to put it crudely. It is on the basis of an idea like this that the Belgian semiotician gang Groupe μ has devised their investigation of different types of transformation between pictorial object and reference object.[371]

The notion of picture is grosso modo kept unchanged in the group of 1912 texts, e.g. in 'Zur Lehre vom Bildbewusstsein und Fiktumbewusstsein' where the same triad of Bildding, Bildobjekt, Bildsujet is presented (489). Yet, an important change has taken place: where pictures were earlier presented as a derivative structure founded on perception, the contradiction inherent in pictorial consciousness now has the character of 'Hier streiten zwei *Wahrnehmungen*, bzw. zwei Wahrnehmungsobjekte' (487). This is *not* the case in *all* pictorial consciousness, now, only when pictorial consciousness is founded on a perception. In that case, the fantasy picture may be 'saved' as a special case of picture with no direct perceptual basis – but then in favor of giving up of the definition of pictorial consciousness by the duplicity of similarity and contrast. This change is deeply connected to the whole transcendental *Kehre* in Husserl's thought which is evident from 'Phantasie als Durch und durch Modifikation. Zur Revision des Inhalts-Auffassungs-Schemas' from 1909. Here, Husserl begins by asking to the source of the ever failed attempts at understanding the relation between perception and fantasy. The fallacy is localized in the distinction between 'Auffassung' and 'Auffassungseinhalt' which makes seem that one and the same content should be given for two different apprehensions in the two act types. This is wrong, instead '*Rather: "Consciousness" consists all the way through of consciousness, and already sensation as well as fantasy is "consciousness"* '[372] Where the content was earlier seen as an entity independent of consciousness which was subjected to different act types, now consciousness is closed around itself as a consequence of the transcendental reduction (or as its prerequisite?). This in contrast to the earlier doctrine where sensation and sensorial content was one, but on the other hand distinct from perception as conscious act. Now, sensation and perception are placed on one and the same, conscious level.

The earlier doctrine is, in this context, represented by Beilage XII ('Empfindung – Phantasma und die ihnen wesentlichen "Auffassungen"', from 1904/05) where the idea, quite on the contrary, is that sensation and fantasy as contents are not, in themselves, conscious, but only becomes so in perception and fantasy as acts which take the sensation and the fantasy, respectively, as objects. The difference between these two texts thus display very well the theoretical development from object to noema in the field of imagination types especially.[373] The reason why conscious acts here must direct themselves toward non-conscious contents is that the opposite is taken to lead into regression: the fantasy imagination is not itself a fantasy, and if it were, then it would require yet another imaginative act in order to grasp that, another, in turn, to grasp that, etc. But then, on the other hand, so goes the criticism of this earlier doctrine, the content must in itself be marked by the difference between sensation and fantasy, for we can not perceive a fantasy content nor vice versa: phantasize about a perceptual content. We may only make founded second-order acts like the perception of a fantasy imagination or the phantasizing about a perception.

Accordingly, we may quote two of the more elaborated taxonomies in order to illustrate the development in Husserl's thought. In Beilage IV (from 1904/05, 39), the typology is rendered in the following Porphyrian tree:

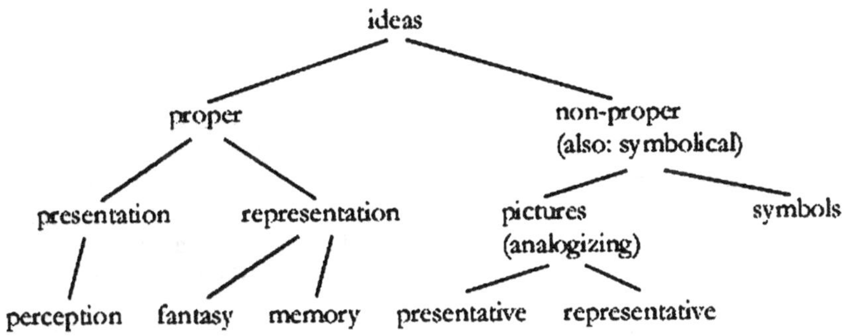

Figure 37.

The two subcategories under the picture category refers to pictures founded in perceptions and pictures founded in phantasies (picturing something not-remembered in fantasy), respectively. As is evident, fantasy is here taken to be representative (because it represents something not present), but not analogizing like the picture. It is clear that the type of representation here referred to (comprising fantasy and memory, differing from picture as well as from symbol) has a special character. It is construed as a derivative and secondary access to the object as compared to perception, but is opposed, on the other hand, to improper (uneigentliche) *Vorstellungen*, which require outer support in the form of pictures or symbols.

If we jump five years to 1910, we find in text 12 ' "Empfindung", Erinnerung, Erwartung und Phantasie als Modi des Zeitbewusstseins. Bewusstsein als Zusammenhang' a completely different taxonomy, motivated by the transcendental *Kehre*. The point of departure is now the molecule of time consciousness involving retention-presentation-protention. While presentation is characterized by

Gegenwart, presence, all three aspects of time consciousness may be subject to an assertive[374] repetitive consciousness ('Wiederbewusstsein') giving rise to three new categories: memory, perception and anticipation (Erinnerung, Empfindung, Erwartung), called forth by re-presentation, co-presentation, and for-presentation, respectively (Wiedervergegenwärtigung, Mitgegenwärtigung,Vergegenwärtigung). Finally, these three categories may be modified, in turn, by a non-assertive act giving rise to fantasy (which must fall in the same three temporal classes). This gives us something like the schema below (referring only to the left part of the former tree dealing with *Vorstellungen* proper):

Non-assertive Representation	F A N T A S Y		
Assertive Repetitive Consciousness	Memory	Perception	Anticipation
Original Consciousness	Rentention	Presentation	Protention

This paper virtually explodes (quite explicitly) of unsolved questions: what then with picture and symbol consciousness? They must now be the result of still further modifications – symbol consciousness as the mere associative linking of two conscious states, and picture consciousness (presumably) as the analogizing such linking. In this context, the overall principle now is the associationist credo '*Bewusstsein ist immer Zusammenhang und notwendig Zusammenhang*' (291) – consciousness is always connectedness and necessarily connectedness.

If we go on, however, to 1912 (in the large 'Modi der Reproduktion' and contemporaneous texts and fragments), fantasy again approaches pictorial consciousness, now in a reverse movement, so that the fantasy concept tends to embrace the picture concept rather than the opposite. The notion of fantasy must be generalized ('Reproduktion und Bildbewusstsein', c. spring 1912, 475), and two types of modification by *Vergegenwärtigung* – Husserl's word for representation – are presented: perceptive and reproductive, respectively (corresponding to fantasy proper and pictures). This happens based on the argumentation that the distinction between assertive and non-assertive is independent of the distinction between picture and fantasy and, more generally, independent of the distinction between impression and reproduction. These distinctions may be freely combined. This idea corresponds to that maintained in *Ideen* I, §§109–15. Thus both assertive and non-assertive pictures and phantasies exist, and the distinction from 1910 where phantasies were seen as non-assertive is given up.[375] On the other hand, pictures and phantasies now fall in the same category, both being reproductive acts in contrast to the impression of perception. The picture (perceptive or 'eikonische Bildlichkeit') now has the same structure as reproductive fantasy (as against perception and simple fantasy), both may be (non-)assertive (384), both are pure *Darstellung* and may, in turn, give rise to real propositions as well as fantasy propositions: 'I follow a play in the theatre, or I contemplate a painting. Here differs, just like in reproductive fantasy, the ideas, perceptions, judgments, emotions etc. which are produced in the picture (pictorially produced) from the actual ones called forth in me, the observer. Actual is the very production itself.'[376] – depending upon whether the acts in question are taking place *in* fantasy (picture) or has fantasy (or picture) as their content.

The reintegration of the two may also get an expression as the following: if the sujet is lacking in the fundamental triad of the picture (picture – picture object – object), the result will be fantasy (in a narrow sense, we must assume) or memory ('Zur Lehre vom Bildbewusstsein und Fiktumbewusstsein', c. 1912, 489).

The result of this re-approachment of picture and fantasy – just like in the phase prior to 1905 – is, consequently, that the *Widerstreit* is dampened to be a mere possibility in pictures: 'But only where a pictorial consciousness is based in a perception situation (where a perception does in fact form part of a perceptual context but is not embedded in it unanimously) or where a pictorial consciousness in the same manner is embedded in a memory context do we have pictorial consciousness connected to contradiction; all this must be grasped in a more precise manner. I must protect myself against unsuitable generalization: as if picture and contradiction necessarily belonged together.'[377] Contradiction now is found in the special case of perceptual pictures and memories only, while it is not present in pictorial consciousness (comprising fantasy as well) in general.[378] Six years later, this continuous vacillation in the relation between picture and *Widerstreit* is sought solved by a lucky idea: a distinction between real and merely possible *Widerstreit*: 'One must distinguish between the *cases of real contradiction* as in the case of actorial presentation, and the cases of *potential contradiction* which depend on environmental intentions which hang on to the object intuited but which must first be 'developed' in order to lead to real contradiction.'[379] Contrast needs not be explicit, it may lie as a possibility in the relation of the picture to its context (that is, the Widerstreit of the picture, fantasy, memory against the actual world of perception). This was anticipated already in the distinction between actual and potential assertion in the *Ideen*. We shall return to possible consequences.

The reason for the new subsumption of pictures under fantasy as a special sort of perceptive reproduction is to be found in the insight that both forms may include real moments of perception within them – and thus constitute different combination possibilities between perception and reproduction. The problem is first of all apparent in fantasy: Husserl observes that fantasy very often takes its point of departure in matter from perception which is then recombined, transformed or otherwise changed. Later he goes so far as to say that it is dubious whether pure fantasy without any connection to acts of experience does exist at all (1918, 509).[380] With a recurring example – the reading of fairy-tales – Husserl argues that here factual information is mixed with fictitious phenomena into a fantasy, a quasi-reality (379) with the consequence that the *Widerstreit* is bracketed (382).[381] Generally, 'quasi'-concepts in this period become a central way to present the broad fantasy concept's way of functioning. In fantasy, I am quasi-attentive; what I invest in a fantasy is a quasi-I who quasi-perceives and quasi-judges. The phenomenological basis for Ingarden's theory of literature with its emphasis on quasi-judgment is outlined in these strains of Husserl's work around 1912 (see Chap. 17 – we can add that Ingarden in just this period began as Husserl's student and assistant).[382] The description by means of the 'quasi'-operator may thus without problems be extended to pictures proper which may now be defined as '*quasi* Setzung der perzeptiven Anschauung (Bildanschauung).' (468) of which holds that 'Jedes echtes *Bild*, perzeptives Bild, ist *quasi* Auffassung und nicht etwa Intention, nur herabgedrückt.' (399) – pictures have a quasi assertion of perceptive intuition and are thus only quasi experience.

This large text from 1912 ends with a chapter on terminological clarification whose main distinction goes between passive sensoriality and the whole series of spontaneous activities subsequently built upon it (a constant, dating back to *Logical Investigations*, now in new clothing). This clarification, however, also leads to a new series of questions. The overall problem seems to be that Husserl now has a rather productive combination calculus of many different modifications and intentions – without this combination, on the other hand, being completely free. While the later texts (from 1909 and clearly from 1912 onwards) build upon the transcendental *Kehre*'s closure of consciousness as well as a weakening of the *Widerstreit* as a criterion, they also add lots of interesting new insigths and ideas: the introduction of anticipation in the context of imagination (that only memory should be covered by the picture-fantasy problem seems strange), the close relation to time consciousness, as well as a series of sophistications of the description of the aesthetic *Einstellung*.

We shall here go on to assume the *Widerstreit* as a basal criterion, even if it for us must be freed from its tie to the consciousness concept and be relativized into a merely *possible* consciousness of contrast (just like Husserl himself proposed, tied to the merely possible assertions implied in parts of every picture [383] – see above). The contrast or contradiction between the three poles of a picture is a purely objective relation, and the idea of any explicit consciousness of contradiction runs into problems like that of illusions: perceptions which are fooled by a picture (like a trompe-l'œil when it in fact works). Here, the *Widerstreit* is temporally realized (just like in fantasy), but what did the gaze actually *see* during the time span before the *Widerstreit* was discovered? Then, there was not – according to the picture definition by *Widerstreit* – yet any picture at stake, but it would be ridiculous to claim, on the other hand, that it is a case of ordinary, direct perception, because a non-existing object is what is seen. The formation of a new and strange category to absorp illusions may be avoided by assuming that the picture *did* function as pictorial object – but now without the first, founding sensation became conscious as such. The perception so to speak began with an 'imaginary moment' without contrast consciousness. But the question of whether explicit consciousness is a necessary prerequisite to contrast can be generalized to pictorial consciousness as such. A similar argumentation may be made regarding subliminal perceptions (of which Peirce was one of the first experimental investigators): the fact that it is possible for the mind or brain to process pictures (measured by the fact that they inflict upon behavior) without they reaching consciousness at all. It would be meaningless to say that such pictures are not pictures only because no explicit picture consciousness can be traced. We shall therefore argue for a modified version of Husserl's plastic 1918 suggestion and keep the contrast definition as central – only, it must be generalized to a merely potential contrast.[384]

Let us leave now this long philological discussion with the aim of outlining the most central proposal for the solution of the fantasy/picture problem. We can say that, to Husserl, that discussion never ended. The latest text in *Hua* XXIII (Beilage LXV from the middle 20s) ends, significantly, with the re-emergence of the *Widerstreit*, now accompanied by a question mark: 'Is there not contradiction in all fantasy, also pure fantasy? In perceptions, memories, anticipations. I inhibit all positing of a world. But does not every fantasy possess some place, something, which contradicts it?'[385] Potentially, yes.

Let us now focus upon a series of different interesting detail ideas regarding the picture especially.

PICTURES AND SENSES

An important issue in Husserl's account deals with the basic relation between picture and sensation. The prototypical picture in Husserl is always visual, but Husserl emphasizes in the early period en passant (Beilage III, 1904/05, footnote, 138) that pictures may exist only in the sense modalities of vision and touch – presumably for the reason that any picture must possess a partially contrast-evoking sensorial skin – which these two senses should be the only ones being able to give rise to. Also in this dimension, Husserl's picture category is more narrow than Peirce's (hypo-)icon theory which does not only cover all senses but also abstract, ideal images. But Husserl does not give any closer argumentation for the idea that only these two senses should be picture generating. It seems hard to ignore the picture potential in, e.g. the classical effects of radio theatre: a man in the studio repeatedly bites noisily in an apple, and in the living room of the listener footsteps in snow are heard . . . But you don't need to go that far: if the visual gestalts of the actors on stage pictorially display the figures of the drama, then their voices supposedly fulfill exactly the same task auditively? Husserl seemingly has not thought the issue through which also seems to appear from the fact than in later discussions (cf. below) he, without further notice, may list auditive examples like the relation between different versions of Beethoven sonatas and the intention of the composer. If we expand the question to cover not only pictures as they appear directly for the senses, more problems appear. The sujet which the picture depicts needs not be an individual object,[386] such as might be the spontaneous idea, and the later Husserl is, in fact, ready to admit that the picture as such is an ideal object (537). This opens up the vast issue of the relation between pictorial and fantasy consciousness to the possibilities inherent in categorial intuition and *Wesenserschauung*[387] to grasp ideal objects. In the LU, the distinction between perception and imagination, so important for the intuition of empirical objects, became irrelevant for categorial intuition. Here lies one of the most important lacunae in Husserl's theory – stemming from the spontaneous idea that the objects of pictures and phantasies are necessarily individual. But ideal objects may – as we know from the 3rd investigation – be co-intended as aspects of individual objects, and hence it would not seem strange, rather the opposite, if these aspects might be isolated for access in pictures and phantasies. Rather it could be said that the quasi-character of the picture entails that its object can never be thoroughly determined in all aspects by pictorial means alone (cf. Ingarden's related idea of 'schematized aspects' referring only to certain properties of their object, see Chap. 17), and that the picture for this reason possesses ideal character, just like the later Husserl should maintain. Peirce has, as we have seen, a ingenious solution to this problem in his concept of the diagram.

HUSSERLIAN AESTHETICS

A recurrent feature in Husserl's treatment of the picture/fantasy problem in the whole period is that he tries to develop aesthetic corollaries from it. His basic aesthetic ideas are taken from the definitory triad of the picture and it is contained in the simple idea that the aesthetic attitude highlights the appearance of the picture object, while the sujet, the object of reference, loses interest – cf. Beilage VI (c. 1906) where the interest in the reference object is opposed to the aesthetic interest in the appearance of the picture object. This approaches aesthetic contemplation to fantasy in a narrow sense, to theoretical reflection, as well as to neutralization as the non-assertive presentation of an object or a state-of-affairs, as the later Husserl remarks. In the first round, Husserl seeks to determine aesthetics by the pleasure of appearance contemplation (ibid.), but he constantly returns to the attempt at further understanding of what is special in the aesthetic interest in appearances.

In Beilage IX (c. 1905, 159), the ambiguous quality of aesthetic apperception is highlighted. Husserl comes to this conclusion from the simple nesting of picture consciousnesses within one another. The reproduction of a work of art, e.g., involves two separate pictorialities (the etching as a picture of the original painting vs. the etching as a picture of the object of the original painting – the sonata as Beethoven intended it vs. the sonata as the expression of certain emotions). It is hard to see precisely how the conclusion follows from these premisses, but we can observe that if a picture has, in this way, two nested objects, then it has also two nested appearances. Aesthetic interest may then be directed towards each of them – and, what is more, it is not, in general, predictable which level of appearances aesthetic intention may chose to focus upon (the picture as color, the picture as form, the picture as the result of an expressive act, the picture as the reproduction of objects in a 3-D space, the picture as the reproduction of certain recognizable objects in a 3-D space, etc.).[388] Aesthetic contemplation may not only highlight one or more of these strata of appearance, but also focus upon their possible interaction. In addition to this, a classical demand in the contemplation of art is maintained ('Forderungen, die die Teile des ästhetischen Ganzen gegeneinander üben ...', ibid.) which must also involve these different levels of appearance as aspects of the wholeness of aesthetic experience – an idea of course taken further in Ingarden's idea of a 'polyphony' of the different levels of the work of art.

In addition to this, it must be emphasized that this preliminary formalism with its bracketing of the referred object also must be modified by object properties. In the second picture theoretical assault period ('Modi...', 1915, 390) Husserl claims that '*The content of the object itself is not aesthetically insignificant.* It is not the same whether it is an emperor or not, if it is a significant or an everyday event, etc. Is it then about a co-sounding of effects of emotions (awe, affection)? But also something else: Any objectivity which motivates existential pleasure or phantasizes *quasi*-pleasure. In itself, this pleasure is not aesthetical. But aesthetic pleasure which is dependent on the mode of appearance may connect itself with this pleasure (as an actuality) and the whole has the character of a heightened aesthetical pleasure. Stilleben. The interaction between actual pleasures or *quasi*-pleasures (in nature: pleasure with the fertile fruit trees, fields, etc.) and suffering and other actual positions taken is, however, in itself a main piece of real aesthetical pleasure. So also this belongs to the "mode of appearance".'[389]

Not only the play in between the levels of appearance, but also the interplay between appearance and object – form and matter, if you like – belong to the appearance of the aesthetic, while the aesthetic contemplation at the same time brackets any existence claim as to the matter side, hence the connection to 'quasi'. It is important to maintain that Husserl in these brief determinations thus distinguishes between the aesthetic level (tied to the interplay between forms of appearance and matter, as well as to the emotions involved herein) and the far more comprehensive category of the fictive which is characterized merely by the fact that the distinction between assertive and non-assertive is suspended – cf. the quasi-judgment characteristic of fictional literature in Ingarden's theory of literature. And it is equally important to underline that this aesthetic interplay does not, necessarily, appear explicitly as thematic, conscious content and object of attention (392). But these basic aesthetic determinations hold, without further notice, for picture as well as fantasy; they even hold for aesthetic aspects of the theoretical interest (ibid.): the pleasure of theoretical insight, the beauty of a mathematical proof.[390]

Fictionality, pictoriality in a sufficiently broad sense as to include theoretical knowledge, thus seem to be a prerequisite (but hardly sufficient) to aesthetic contemplation. But a decisive feature in fictivitiy and indeed in representation in general is that it – contrary to empirical reality – is endowed with what Ingarden later named 'Unbestimmtheitsstellen' which in Husserl's brainstorm is merely called *Unbestimmtheit* (1922/23, 561), given by the fact that the pictorial object only represents the object in certain selected aspects and in other respects appears empty: 'Auf die Frage, was der phantasierte Zentaur am Phantasiemorgen essen wird, mit wem sie unterhalten oder kämpfen wird, gibt es keine Antwort.' (1918, 523) – nobody may know what a fantasy creature had for breakfast.[391] The fact that all fantasy and picture activity has this character (not to speak about signitive intentions; nine tenths of Ingarden's literary theory consists of a phenomenological linguistics) is in some sense a simple consequence of the definitions Husserl gave already in the years after the *Logische Untersuchungen*, but these implications are made explicit only more than ten years later. A related fact which Husserl also only thoroughly realizes in this period is the crucial ideality of the picture – exactly because the picture shares the empty slots, it is so to speak endowed with those algebraical variables which already in the Prolegomena to the *Logische Untersuchungen* were characteristic for ideal objects and which in *Erfahrung und Urteil* opens the epistemological possibility of their determination by means of 'eidetic variation'.[392]

Husserl's argument simply concludes that as fantasy does not present reality, it may only present possibilities. And the possible is, as against reality, not individual, it is ideal, on some or other level of generality. About the picture of a mountain, Husserl claims – as against the real mountain, existing in time – that 'Dieses "Bild" ist ein *idealer* Gegenstand (nicht etwa ein Reales in der Zeit Dauerndes), der Berg bietet dieses Bild dauernd, aber es selbst ist kein dauerndes.' (Beilage LVIII, c. 1917, 538). But if the object of a picture is ideal, it can not be real, but merely possible. The ideality of pictures thus refers to pure possibilities, and fantasy in general is unable to imagine individual entities: 'Thus fantasy can not really represent any individual, it gives "something" which is shaped in the form of an individual and which is only made intuitive with its (in relation to individuality) undecided meaning content.'[393]

ICONICITY BETWEEN HUSSERL AND PEIRCE

In the very last text about pictorality from the 20s, Husserl surprisingly approaches many of the central ideas in Peirce's icon concept (the icon's possibility, vagueness, and ideality, icons as the base of communication, their connection to predicates, their lack of conformity to the principle of contradiction, even continuity . . .):

- Vagueness: In 'Reine Möglichkeit und Phantasie' (c.1922/23, **X**), the 'als ob seiend', the 'quasi' of fantasy is described thus: -'Possibilities, individual pure possibilities are flowing.'
- Non-observance of the Principle of Contradiction: imagination gives – 'Infinitely many possible realities, which are mutually inconsistent, etc.' (Bilag LXII, c. 1918, 566) – an infinity of possible realities which are mutually exclusive.
- The predicate character of pictures -'*Predicate* is something identical which may be identical in many, in "manifold", which appears identical in changeable and different possibilities and may determine the concrete.'
- The picture as a prerequisite to any communication of meaning: fantasy as pure possibility is a presupposition for the communication of signification between subjects as intersubjective fantasy, because the subjects involved refer to the same 'intersubjektiv identisches.' (568)
- The predicates of fantasy as something which may be completed by the filling-in of a subject: -'And correspondingly I can think of some subject which supposes this on the basis of its facts.'

– Even an idea analogous to Peirce's metaphysics of the continuum with the loss of individual identity is hinted at, when Husserl says '... when I, in an exemplary manner, run through the extension as pure extension, then it does not occur to me to intend the respective subjective, individualizing differences. If I imagine one single centaur, then I do not mean "it" as imagined exactly by me. Should we say that the singular of a pure extension (of an extension of pure individual possibilities) that the exemplary single is, in itself, already a general, an *identical which, in turn, has a pure extension of possibilities*?' (Our italics). The pure extension of a predicate does not refer to individual objects but to a general object which, in turn, refers to an extension of pure possibilities.[394]

This ideality of fantasy and picture is also developed in relation to aesthetic issues. With respect to decisive references, literature is, e.g. principally undecidable:

In the horizon of undecidability which the given world and time has for everybody, the poet places himself, deciding it, filling it in with Gestalten.

Reader, poet in a world and time. Two extremes:

a) The given world and time may be so fully determined as our surrounding world right now is (not the real world). For instance today's Berlin determined in that way in which it is for us and even for the Berliners themselves. b) Extreme counterexample. Once upon a time in some fabulous country, in another time, in another world with quite different animal beings, even other natural laws, etc. Between these two extremes all art develops.[395]

Correspondingly, art goes forward between depicting something existing (by pictorial consciousness) and producing pure fantasy worlds, between (empirical) realism and (imaginary) idealism. Here, numerous intermediary possibilities may be developed. It can immediately be added that this theory of genre in spe (for all art forms) has its phenomenological basis in the very observation of pictures, namely in its two potential phases, with and without pictorial consciousness, respectively. Also this idea dates back all the way to 1904/05 when Husserl made the distinction between the observation of the picture endowed with *Widerstreit* on the one hand, and the direct grasping of the object through the picture on the other, what was later recognized as a phantasizing relation to it, corresponding to Peirce's 'imaginary moment'. The *Widerstreit* is simply suspended when I focus upon the sujet. (467, 1912).

The possibility grasped may of course be more or less 'pure', depending on whether it involves individual reality (548); this also lay implicitly already in Beilage V where a distinction was made between genuine pictorial consciousness which truthfully sees what is alike in what is alike, and different grades of 'impure' consciousness, all the way down to when the impurity is present to so large an extent that we can no longer grasp the sujet through the pictorial object as if it was in some sense present itself (143).

This continuous variability in the degree of resemblance between pictorial object and sujet may be discussed in the light of Husserl's different concepts – or rather, metaphors – for more or less adequate perception. The recurrent concept of covering ('Deckung') is used already in the *Logische Untersuchungen* as a concept for adequate perception, and in Beilage V, this concept is also used about such 'impure' cases where we only remotely recognize the sujet in the pictorial object – e.g. in a rude children's drawing. We can grasp that it is *meant to* depict a human being, even if we do not *see* the human being in it. 'Or maybe yet a little bit. We have covered the men, the two of them cover each other, but give widely apparent differences. The difference is so crude that we no longer feel disturbed by the consciousness of falsity by our tendency to identify, on the contrary, the tendency to identify is lacking and because of the huge distance we find the representation ridiculous.'[396] It is, however, hardly the only reaction possible; we may also find such a rendering of the object practical, economical, diagrammatical. The decisive thing, however, is the metaphor of 'covering' which allows us to grasp the difference: we sort of place the picture 'on top of' the object, we map the one onto the other and by doing so we grasp the degree of difference between them. The interesting thing here is that this discontinuous idea of covering in the same text intermingle with a continuous idea of 'merging'. The example with the children's drawing is further analyzed as two intentions depicting the same object which are connected and merge into a synthetic unity. This results in a 'Übererinsstemmungsbewusstsein' displaying similarity where 'eins in der andere *verfliesst* und doch wieder als anders empfunden wird'[397] (our italics). But this similarity is triggered by a continuous deformation or variation process rather than a 'covering'. Maybe one could say that the two of them are necessary as simultaneous procedures to keep similarity as well as difference, the flowing together as well as the covering? The continuous variation is tied, of course, to the possibility of unfolding relations lying implicitly in the picture, cf. below.

PICTURES WITHIN ORDINARY PERCEPTION

Let us finally use the results of this discussion of Husserl's concept of pictures to argue for a new phenomenological distinction between two types of pictures, in the first approachment regarding visual pictures of a non-moving kind. Our point of departure will be a couple of Husserl's fragmentary ideas. In Beilage VI Husserl wonders why it is that we have the insight of everyday phenomenology that nature is able to be spontaneously given as a landscape and thus function as a picture. How is that possible?

> Why does nature, a landscape, function as 'p i c t u r e'? A remote village. The houses 'small houses'. These small houses have a) a changed size as against the houses as we normally see them, b) a lessened stereoscopy, changed colours, etc. They are just like pictures conceived as toy houses. The same with human beings: dolls.[398]

The picture thus lies enfolded as a possibility *within ordinary perception already*. When the gaze assumes a certain distance to things, they lose stereoscopic depth as well as size – and, we could add, they also lose immediate corporeal relevance to the perceiving subject. It implies that already the regard at a distance implies a certain neutralization (as well known from everyday idioms like '... seen at a distance ...', '... seen from Sirius ...'), from where the things assume an objectivity which is, at the same time, the basis for fantasy variations (the houses turn into dolls' houses). This is, in fact, a grounding of the possibilities of picture and fantasy *already* in the 'schlichte' perception which is apt to counterargue all claims that the competence of producing and understanding pictures should be something conventional requiring learning, analogous to language acquisition.[399] The remote gaze is, in fact, a double gaze complete with *Widerstreit* and all what follows: we may shift between seeing the remote houses as real houses at a long distance and as dolls' houses very close; we can not see both at once. It is, so to speak the naturalness of puzzle pictures for perception which founds the inherent pictorality of nature.[400] These two ways of seeing are, of course, correlated to to corresponding sets of actions: the neighboring village I can walk to, trade with, exchange women with, wage war against, etc.; the dolls' house village I can play with in an action space of fantasy. This corresponds furthermore, as Husserl will later say, to two different egos involved in the two different attitudes. Thus, there is an *Ichspaltung* already in perception.[401] There is nothing tragic in this splitting of the ego, however, quite on the contrary it corresponds to a panel of possibilities for the subject which may dress up in different fictitious roles, all depending on the character of the picture.

Now, let us take a closer look on the subject's relation to fantasy and picture (nr. 16 (1912), 467): 'I may well also 'phantasize myself into' the picture. This can only mean that I extend pictorial space over me and my surrounding space and include myself – with the exception of the real things which I see – in the picture whereby I bracket by actuality; I myself then become a modified I, without positing. Then my participation is that of a depicted observer (it belongs to the pictural object), not that of a sympathetic observer outside the picture.'[402] This means that, in the phase of the 'imaginary moment' I perform a more or less conscious, fictitious splitting of the ego and appears as a fantasy figure *inside* the picture. When the split in the picture stops, the observing subject is so to say split instead and a phantasized ego enters into the object in the imaginary moment. This ingenious observation gives a new meaning to the traditional embedded observer in romantic painting (Friedrich) which merely makes explicit and thematic a relation which has a broader phenomenological generality: when we observe a landscape picture, we automatically already – even if most often unthematically – find ourselves in the midst of it. When we see the dolls' house village, we already are in it. But given this double character of landscape, a whole bundle of different experimental possibilities open themselves for the fictitious observer who is now placed in its foreground. He may undertake his walk towards the village in the distance. Or he may reach out his arm and begin playing with the dolls' houses in it. The real observer's splitting off of a fiction observer immediately gives rise to the possibility of the splitting off of a second order fiction observer. But we are not threatened by a regress. The doubling will rarely be repeated more than a few times.

But we can speculate – which Husserl does not – what *other* competences may be ascribed to the subject finding itself inside the picture – apart from mere observation. The further evolution of bodily phenomenology since Husserl (from Merleay-Ponty to Rosch, Lakoff, etc.) will highlight, of course, the fictive observer as a body possessing a whole series of characteristic ways of action which may be played through in fantasy action in the picture. As regards the picture formally conceived of as a sign, the Peircean viewpoint will maintain that any use of a picture which rationally distinguishes its parts and their interrelation and indulges in experiments with those interrelations, is diagrammatical. Thus, we must assume that the *observing fiction observer may investigate the picture due to those diagrams he may construct on the basis of bodily action possibilities* – or, in any case, those of them which is actualized by the inherent action possibilities offered by the picture in question. Diagrammatical experimentation with schemata is thus not a privilege for maps and scientific models only, rather, these media are made possible (genetically, of course not a priorically) by the more primitive competence of gaining access to a landscape by means of a body.

Peirce's diagrams, as we know, allow for an experimenting manipulation which corresponds to hypothetical situations allowing that an information which is not explicit in the picture, is extracted and made explicit. Husserl does not develop any corresponding idea, but certain passages in him seems to suggest that he vaguely realized the possibility. Thus, in Beilage XXXIV, he says about the synthesis of ideas that it may possess the character of '... new formation of such syntheses and analyses which separates partial appearances from the unitary appearances with the partial meanings put into synthesis etc. In the explicit, new matter (kernels) appear. Thus *explication of what was concealed in the formerly given appearances.*'[403] New formation of syntheses may produce matter which was hidden in the given appearances – but this is, exactly what characterizes diagrams as synthetical signs. When the fictional observer in the picture makes such new syntheses experimenting with the matter of the picture, he makes diagrammatical experiments with parts of the picture using his spontaneous bodily relation to the pictorial scenario. Thus, this ability in picture observation unites Peirce and Merleau-Ponty.

This idea gives rise to a long range of different possible ways of contemplating pictures, depending on which bodily stance is adopted – a yet unmapped, or at least only vaguely mapped region in traditional theory of pictures. In the next section I shall try to cover one dimension (among many possible) in this new field.

PICTURAL ACCESSIBILITY

It is well known how the opposition between abstaction and figuration which dominated large parts of early modernism in painting long since seems outdated or irrelevant. No partisans of one or the other any longer roam – and if a few should remain, they are unable to put forward their ideas without a moldy taste of retro or of arbitrary or irritating politicizing. This has the simple reason that abstract and figurative is no longer (nor have they ever been) oppositions, but rather extremes on a continuum of possibilities – corresponding, but not identical to Husserl's continuum between the timelessness and the timeboundness of art. An abstract painting is still depicting, more or less, this lies in its basic character of being an idea containing slots of indeterminacy. Minimalism and concretism from the 60s onwards might point towards another tension, which was independent of the depicting qualities of the artwork, but rather involved that function on the one hand and the direct perception of the picture as a physical object on the other: the tension between abstract and concrete. One and the same work may be grasped as abstract, that is depicting something, even if on a general level, and as concrete, that is, depicting nothing but its own matter and form, present for the observer and cut loose from any pictorial signification or reference. During a certain period also this tension gave rise to politicizing partisan activity for one of the other of these possibilities – but this in vain, as the two possibilities in many cases are just as much properties of the observation as properites in the artworks. Of course, it remains to be the case that different criteria must be expected for what counts as a successful artwork, depending on the two attitudes – but it is very difficult to seen which arguments should conclude the strife for one of the attitudes being more aesthetically authoritative than the other – no matter what avant garde partisan myths one or the other may serve. The fact that partisan wars of this kind may be ceasing is one of the good consequences (hopefully, that is) of the victory to death of avantgardism as an ideology in the art institution: art is not politics and to be artist or critic is not to be a partisan.

This leaves, on the other hand, the field open for more sensitive phenomenologial investigations distinguishing different basic possibilities in the picture – without having to choose one or the other as a partisan. A distinction for which I shall here argue has its basis in Husserl's determination of the picture as founded already in ordinary perception (cf. above) – and his idea that picture perception installs a fictitious observer within the very space of the picture. These determinations give rise to the distinction between pictures which immediately opens up a space for the perceiving body to move around in – as against pictures which do not open up such a space. The prototype of the former is of course landscape pictures which the observer digests with the abilities of the wanderer. The picture is mapped by the observer moving along one of the possible routes which the picture displays. On the other hand, lots of pictures do not offer such possibilities. The prototypical picture here is probably the portrait where a face typically provides the main fraction of the picture plane and where the background only rarely challenges the observer to bodily investigation – so much more as the observer is kept still by the presence of the person portrayed, often even by his or hers insisting gaze, and kept at bay by the no-go zone of bodily distance. We speak of prototypes here, of course, the tension between these two types defines here, like in many other cases, a continuum where we may expect a huge zone of intermediaries of many kinds where tempting mountain walks call behind Mona Lisa or where, conversely, all of the central field of the landscape picture is occupied by monolithic tree groups and makes access difficult for the walking or floating body.[404]

It any case it is important to notice that this distinction between accessible and inaccessible pictures is orthogonal to e.g. the continuum between abstract and figurative. A first idea might be to identify the figurative picture with accessibility and the abstract one as inaccessible, but this is not the case. The portrait, as mentioned, but also the Stillleben, the scientific planche or diagram, and the close-up are picture types which do not, or at most marginally, permit the crossing routes of the wandering body. It is not, it is true, precluded for the imaginary body to change form into a fruit beetle and undertake the six-feet walk between the exquisite volumes of the Stillleben and conquer the fruit bowl as a mountain range by the filigran-like diving into the labyrinths of rot of the grapes – but still it remains a marginal and derivative praxis in relation to the immediate observation which rather strikes an evaluative pose on a distance equipped with an imaginary fruit knife. Many abstract pictures have, of course, the same kind of rejection of bodily movements. This holds, of course (but not only) for the strong tendency in abstraction to highlight the surface and the two-dimensionality of the picture, very often coupled with a a concretist emphasis on the material character of the picture plane. It is hardly possible to access the typical Warhol because already the fat layer of silk print gives the picture a fundamental and abstract two-dimensionality, no matter the figurative character of the motif. It is hardly possible to access a Mondrian, even if it ever so much displays the traffic of New York's street grid seen from a skyscraper (the top achievement will be following one of the stylized cars on a restricted, 2-D ride). Or a Barnett Newman where one must push the eye to its limit for seeing the vertical lightning bolt as an opening or the two surrounding planes as scenographic pieces it is possible to traverse. But many different abstract pictures are indeed open to access in this use of the word – not only the later Monet where the spaces depticted has long since freed themselves from any direct reference to the water lilies, but also Kandinsky where the single elements in the picture often float in a diffusely defined, but undoubtable 3-D space. Or Rothko, where the titillating bichromatics often suck like a pipe leading directly to an out-of-the-body or light-after-death experience of occultism. Pollock, where you float around in intergalactic dust, if it does not devaporate into oatmeal. The reason behind these different spatialities is of course the different formal means by which the imagined space in question is constituted – and they do not depend on whether the objectivities depicted are recognizable in any figurative sense of the word.

These means to space construction constitute a long list of artistic devices with each their rules and structures:

- Dedicedly elaborated perspectives with vanishing points on a horizon facilitated by parallel lines in the motif naturally immediately opens the picture plane and erects a space, but far less than that may suffice
- A hint of a horizon line may be sufficient
- A light-dark opposition may construct space, as a tendency letting the lighter parts lie closer to the observer than the darker ones
- A figure-ground distinction which lets the ground draw back and the figure jump forth: this distinction may give rise to whole foreground-middle ground-back ground hierarchies, so that uncut pregnant forms immediately appear as standing before forms whose pregnant contour has been broken 'by' the foregrounded forms

- A differentiation in sharpness drawing the sharply contoured phenomena into the foreground at the expense of more diffuse phenomena relegated to the background
- An articulated grading of the light fall on a surface may let it define its spatial location and, e.g. assume an oblique position from the foreground into the background
- An articulated patterning of increasing tightness in a texture or a pattern where the less tight part turns toward the observer and the more tight parts away from him
- 'Expressive' lumps and bits of color which in addition to their concretion has a tendency to occupy the foreground of the picture, differing from the thinner, more delicate layer of color disappearing into a remote dim
- Bright, violent color valors claim foreground positions while more discrete nuances vanish behind them because of the fact that light intensity falls proportionally with the square of the distance – and atmospheric air adding a further dampening of color intensity
- The related fact that atmospheric air reflects blue light to a larger extent than red light implies that colors from the red end of the color spectrum seem, as a tendency, closer than colors from the blue end
- The upper part of the picture automatically offers itself as celestial and more remote than the lower parts' pieces of middle and foregrounds, due to the fact that we live on the surface of a planet endowed with gravity

Any of these devices may, of course, be counteracted by one of the others, and, in figurative pictures moreover by the mutual relations of recognizable items (the matchbox close, the car remote, if they should not be reinterpreted as high-rise and toy mobile, respectively). But in general, these devices make possible that the body may spontaneously access a large spectrum of abstract paintings which only very remotely or in a very stylized way could be said to possess figurative content, if at all. If some of these effects work against each other, a further effect may result because they do not necessarily simply annull each other – the result may be spaces which remain undeniably spatial but contain zones hard to define, or have a composite character with *kippfigur* – like changing possibilities of spatial interpretation of certain parts or contrasting interpretations of different parts.

Very often, we probably have a spontaneous urge to interpret such spaces as landscapes or exteriors in some generalized use of these concepts, even if they are not inhabited by any figurative landscape effects at all and do not even have to have any horizon. What does the body do when accessing such pictures? The prototypical access is, of course, that of the wanderer, investigating where you can move by foot along walks, roads, stairs, openings, doors, etc. in the picture – and, furthermore, imagining the changing views while being subjected to these transports – most often involving phantasizing due to the slots of indeterminacy in the picture.

But the plasticity of fantasy gives rise to a series of other variations – where the experimenter changes the very rules of the spatial picture as diagram. First, the observer may continuously change his size (and thus crawl upon the pyramid of lemons in the fruit bowl or, conversely, with the steps of a giant gain access to the remote city on the horizon). Second, a series of subcategories must be distinguished, after which modus of movement is allowed for the searching body when accessing pictural space. In certain pictures, you morph into an air spirit and fly over cities and abysses, in short assume properties intermediary between bird and deity, experiencing the whole of 3-D space offered by the picture. Another type of change in movement mode is tempo which (most often) may be speeded up so you hastily can cover different possibilities in the picture, still another will cover other changes in movement type: maybe you not only walk or fly, but sail, swim, drive, etc. – just like an experiment may change the types of hindrances accepted (you walk unhindered directly through a bush, or a house, or a mountain).[405] Thus, we must claim that Husserl is wrong when he (text 12, 1912, 476) tries to understand the character of *Stimmung* in the landscape painting and claims that this has nothing to do with the fictive observer: 'Such represented atmospheres, feelings, etc. do not presuppose a co-representation of the observer, even if he enters action in his own way. More precisely, I – with *this* atmosphere – do not belong to the picture.'[406] Quite on the contrary, our hypothesis would be that the *Stimmung* of a picture is highly dependent upon the bouquet of possible peripathetical action possibilities determining which kind of access to the pictorial space is possible (I do not claim, of course, that the *Stimmung* depends *only* on this). Take as a commutation test a picture of a forest wilderness with no pathways appearing in it. Its *Stimmung* will change abruptly if a narrow, but well-trodden walk is introduced in a small corner of it, leading through it. As an addendum to this chapter, at series of different picture examples are added with differing degrees of accessibility and different positions on the abstract/figurative continuum.[407]

The hypothesis that this wandering access (and its different possibilities and impossibilities) play so huge a role in the understanding of pictures is based on several assumptions. A genetical – and non-phenomenological – assumption is that we, as a mammal species, must possess the basic competence of orienting ourselves in a set of surroundings, find our way, trace prey, avoid predators, deal with species fellows – which implies a high degree of competence presenting qualified guesses of what may hide in the indeterminacy spots of a given set of exteriors. An a priori explanation, a phenomenological essence description (which needs in no way to conflict with the former, as they refer to empircial and ideal regularities, respectively) will point to the idea that understanding is necessarily diagrammatical – it includes maps, diagrams, schemata on a spatial substrate which are dealt with by moving around in an abstract space of possible spatial behaviors and by changing such an abstract mapping. Cognitive semantics has pointed to the fact that such mechanisms are widespread in human behavior and ordinary language. Thus it is no coincidence that it is the case in both concrete, metaphorical, and abstract landscapes that it is possible to run wild. As Husserl – exactly in relation to the question of pictures – metaphorically said:

DIE DINGE SIND ABER DIFFIZIL, UND MAN GEHT LEICHT IN DIE IRRE ...

APPENDIX: PICTURE ACCESSIBILITY EXAMPLES

The following ten pictures demonstrate some of the different combination possibilities between abstract/figurative on the on hand and accessible/inaccessible on the dimensions of the other

Figure 38. Wooded Landscape.
(Gillis van Coninxloo, 1598)

While prototypical landscapes provide a bundle of pathways for the viewer, Coninxloo is one of the first to close these roads, positioning the observer lost in the midst of dark forests. Despite a huge degree of figurativity, we make a considerable way along the axis from accessibility to non-accessibility. We have a deep 3-D gaze into the thicket, but no path is taking us there.

Figure 39. Abstractes Bild.
(Gerhard Richter, 1991) © Gerhard Richter and Marian Goodman Gallery, New York
The later pictures of Gerhard Richter are arch-examples of a virtuoso playing on the axis between accessible and non-accessible. In these almost completely abstract paintings, with color drawn over the surface with wide paint scrapers, a dim haze or thin curtains or spyholes in a surface hint at an underlying space to move into.

Figure 40. Untitled #223.
(Cindy Sherman, 1990) © Cindy Sherman, Untitled #223, 1990, color photo 58 × 42 inches, Courtesy of the Artist and Metro Pictures
The portrait is presumably the prototypical case of a non-accessible genre. The insisting gaze, or, like here, the intimate occupation of the person portrayed maintains a no-go-zone. The acute difference between the foregrounded person and a background losing depth, specificity, and meaning, approaching uniform darkness, prevents the body from entering the picture.

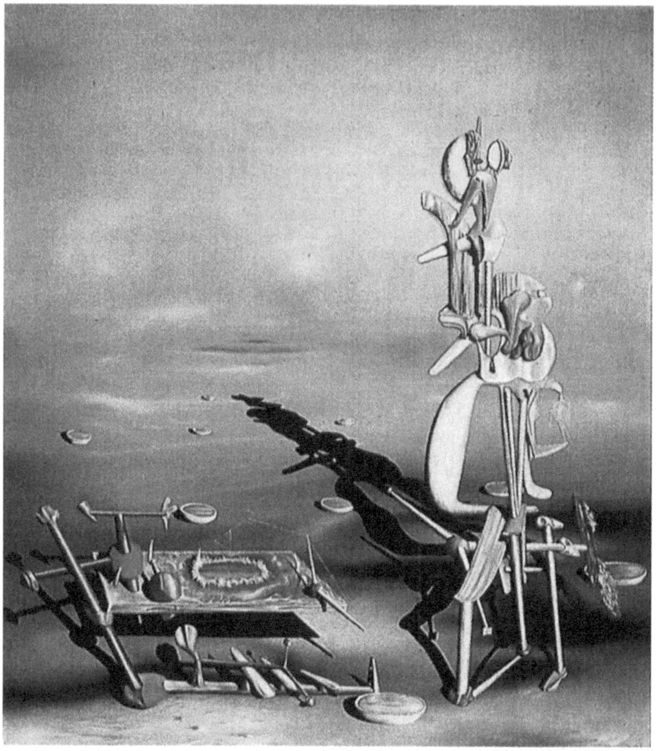

Figure 41. Infinite Divisibility.
(Yves Tanguy, 1942) © Yves Tanguy, Infinite Divisibility, c/o Beeldrecht Amsterdam 2007
An almost too clear-cut illustration of the fact that the axis of accessibility does not coincide with the abstract-figurative axis is provided by the work of Tanguy. Here, we find ourselves in an undoubtable 3-D space, complete with ground, sky and object shadows, opening up for a horizon-seeking wanderer – but where the represented objectivities are abstract.

Figure 42. Press photo from the England-Hungary 3–6 match
(Wembley, London, 1953 [photographer unknown])
The sport photograph illustrates how the ability to read a picture based on the imagined movement of the observer's body may, in many cases, presuppose a massive amount of tacit knowledge on the part of the observer. To understand and appreciate a football picture you must not only know the rules, but also typical playing situations so you can continue the movie the next decisive seconds by the identification of yourself with one or several of the playing bodies, maybe trying different possible developments.
Your stretch your leg a little more than the front player depicted and you reach the ball.

Figure 43. Still-life with Wine Glass and Silver Bowl
(Pieter Claesz, undated)
The still-life genre of course explicitly depicts a stiffened tableau. On the figurative level, it gives the viewer the role of observer of an arrangement organized by others who may have regretted and left. Without further notice, you can sit down by the table and imagine sipping the wine, but there is no reason not to metamorphose into an imaginary fly and take a more ambitious round-trip in the inviting space produced.

Figure 44. The Lu Mountains
(Wang Hui, 1692)

In the Chinese landscape tradition the network of pathways in a 3-D thicket is a frequent topos. In a Husserlian sense, the movement possibilities organized by a picture are most often present in an unthematized way. They are temporal 'back sides' of the frozen moment of the picture. The exception to this rule is roads and pathways – here somebody has walked before you – which explicitly thematize the possibility of movement transgressing the crystallized moment of the picture. The depiction of wanderers may make this possibility actual.

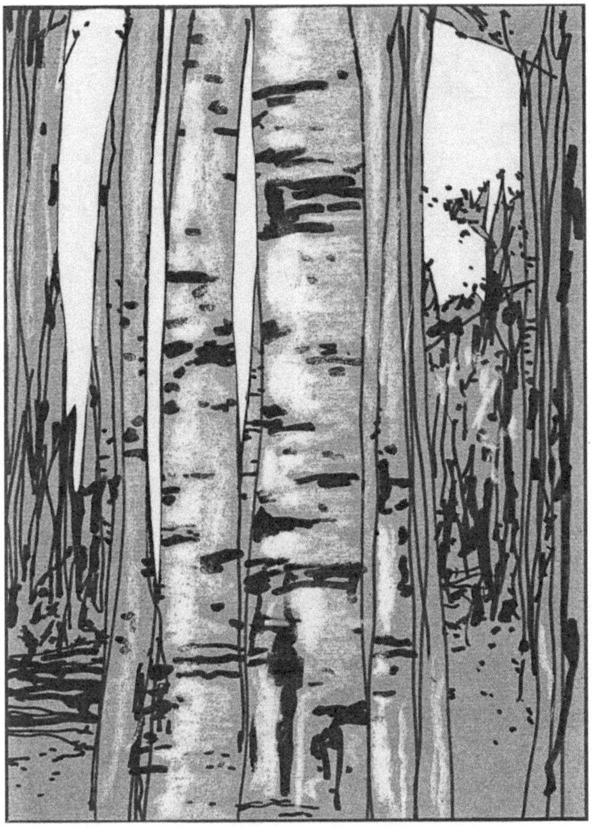

Figure 45. Bare Træer
(Per Marquardt Otzen: Drawing from the anthology *Bare Træer* ['Trees Only'], Copenhagen 1994)

The prototypical path to follow lies in the middle of the field of vision. By blocking this part of the picture, the possibilities of access is preserved because you can enter through one of the side routes. Still, a forceful reduction of movement possibilities is notable, and a picture such as this is a compromise between accessible and unaccessible pictures.

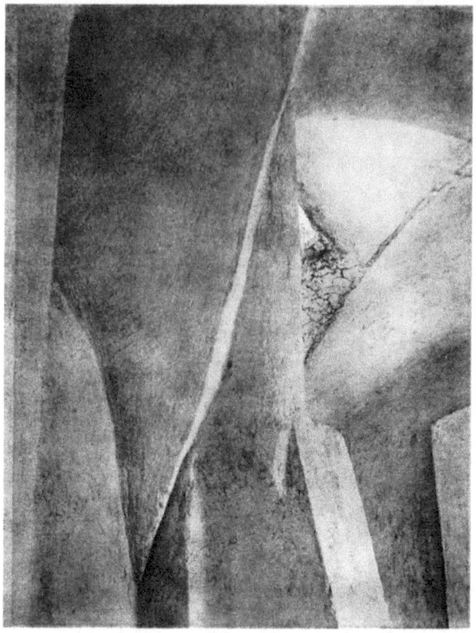

Figure 46. Church Interior. Tepotzotlan
(Tina Modotti, 1924)
A completely figurative picture showing the ceiling in a church loses 3-D interpretability and thus assumes a both abstract and non-accessible character

Figure 47. Park
(Gustav Klimt, before 1910)
Despite the fact that 90 percent of the picture surface seems unaccessible and is vacillating between figurative and abstract and is only pushed into the former by the bottom line of the picture, this painting is eminently accessible. A rich bundle of ways opens into the park landscape at the bottom left

CHAPTER 15

SMALL OUTLINE OF A THEORY OF THE SKETCH

It is striking so little space has been devoted to the *sketch* in pictorial semiotics, measured against its central role in the process of creation in many pictures as well as against its central role in recent art history – not to talk about its role as a general tool of thought and memory.

Here, we shall attempt to draw some preliminary lines to a sketch theory. For a crude art history overview, it seems clear that before modernism, the sketch plays the role mainly as an introductory phase in creation, as a working tool on the road towards the finished work. Presumably, it seems just as clear that one central effect of modernism has been to focus upon various features of the sketch, isolate them, cultivate them and see them as just as essential – or even more so – than the finished work of art. This has lead to the fact that many genres of modernist painting and drawing share qualities with the sketch – as well as to the fact that the sketches inherited from earlier phases of art history have been subject to a revalution so that they in many cases are seen as more interesting than the finished pieces of work which they gave rise to. And it is probably, finally, equally clear that now, when we turn back to form an overview over modernism's different currents, the sketch can not claim neither more nor less prominence than the finished work of art – the polished, overworked piece and the raw, unfinished fragment now appear as parallel possibilities, and there is hardly any point in attacking one of them on behalf of the other – so much more because the artistic reverence for the sketch or the fragment with a strange necessity makes *it* into a work of art.

We shall here conceive of the sketch in the light of the Belgian Groupe μ's pictorial semiotics as it has been presented in the treatise *Traité du signe visuel* (1992). They here present (136) a triangular model for visual signs in general:

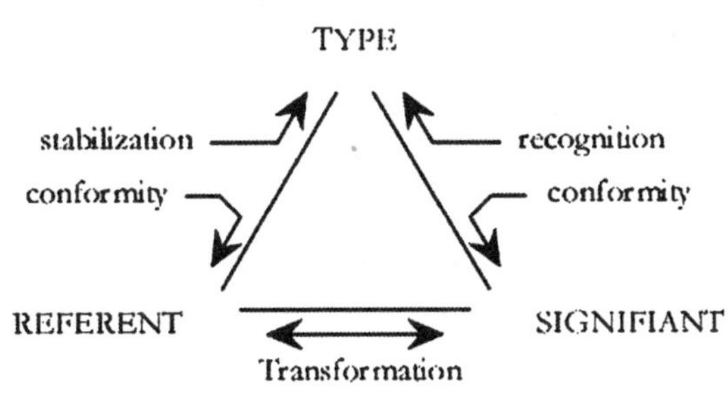

Figure 48.

The overall architecture of the model is hardly shocking for any experienced semiotician; what is at stake is an updated version of that set of distinctions between sign, meaning, and object (here, significant, type, and referent) which in various variants date back to antiquity. Yet, Groupe μ's version of it contains some decisive accents. When the category of meaning is here rendered by the concept of 'type', the group explicitly refers to a *visual* type. This aspect of the Group's visual sign concept is explicitly turned against the linguistico-centrism of much pictorial semiotics since the 60s, partially evident in the import of linguistic terminology in pictorial semiotics, partially, and more dangerously, in the accompanying assumption of the primacy of language so that all visual types presumably can be described exhaustively in ordinary language or in the metalanguage of theory. By emphasizing the visual type, Groupe μ points to the inexhaustibility of the visual sign: it is no superficial vehicle for the communication of a more proper, linguistic or symbolic content; it has in itself, already on the visual level, stability thanks to the typicality of its content. It is, of course, hard to exemplify in text, because ordinary language in many cases has an expression ready to cover a related content, as soon as any visual type for some reason has demonstrated its interest – but a good example might be those spheres with a marked equator that occur in many of Magritte's paintings along with easily recognizable everyday interior. Magritte scholars often refer to them as 'bells', but they cannot be identified with any ordinary utensil and remain enigmatic objects in the context of the space of the picture. This does not, on the other hand, hinder them in begin strikingly easy to recognize from one painting to the next; here we have a purely visual type without any corresponding denomination in ordinary language (and even if the 'bell' of the art historians should become widely known, it would not, of course, exhaust the specific visual typicality which permits them to be recognized – not any old bell will do). Of course, the routine recognition of both concrete and abstract everyday objects is to a very large extent made possible precisely by the existence of visual types: when we see and recognize a bicycle, a chair, a car, etc., we do so because we recognize its visual appearance – only on that basis do we add our linguistically articulated knowledge and identify that object by means of our linguistic denomination.

But this visual type is not determined by the linguistic etiquette, quite on the contrary it functions, in many cases, as a prerequisite to the linguistic denomination.[408] The concept of visual type thus points to the fact that there is a strong pre-linguistic generalizing intelligence in perception able to extract types of (series of) particular visual impressions, and able to keep these types stable, able to vary these types so that different concrete phenomena become visually understandable by being categorized as tokens of that same type, and able to compare and distinguish different visual types.

The other special property in Groupe μ's version of the classical semiotic triangle lies in the appearance of the term 'transformation' at the basis of the triangle. It refers to the fact that visual signs function by means of similarity, that they are Peircean 'hypoicons' – and this implies that some or other property-saving transformation holds between the signifier of the sign and that object it refers to. What is kept invariant in the transformation between the two is exactly the type. A large and useful part of Groupe μ's evolution of its sign concept is the listing of how many different visual transformation we have at our disposal when gestalting visual signs. The fact that visual similarity is no simple property, as too often assumed by many,[409] is demonstrated by the many different geometrical, analytical, optical, kinetic and other transformations which permit one sight to resemble another.

The sketch as a sign is characterized, now, by making use of a special selection of transformations making aspects of the object *discrete*. Groupe μ does not explicitly treat the sketch, but they remark *en passant* (during their development of the rhetorical figures made possible by the so-called analytical transformations): '... your can subdue certain lines so as to leave only those judged important. In that way you get the *sketch* which provides a synthetic vision of the subject and often delivers regulating layouts which may be used for the correct interpretation of the final work.'[410]

This implies that the sketch presupposes an analytical transformation of the object to a structure of lines. The Group here takes 'analytical' to refer to the differential calculus which (among other things) permits to decide where a mathematical function has its maxima and minima (where its first derivative equals zero). But if we envisage the surface of an object as the graph of a function, then it is precisely these points of its surface that interests a draughtsman, because they give rise to lines in the sketch's analysis of them: contour, folds, etc. The drawing thus provides an analytical transformation of the seen object, reduced to a set of lines. In addition to this, a further analytical transformation follows on the level of color – by discretization typical colors of the object may be isolated, adding a new set of lines distinguishing e.g. areas of light and shadow, areas with different color, different texture, etc. Whether these colors and textures are added to the drawing or not, this set of transformations adds another set of possible lines for the draughtsman. The spontaneous extraction of lines from the perceived object is the basis for the construction of the sketch. As emphasizes the Group, the retina is constructed with the aim of extracting lines separating qualitatively different zones in the perceived (typically contours, of course, but also borderlines distinguishing different colors or textures in one object), because the visual cells of the retina are connected in a network so that two connected visual cells getting approximately the same impression tend to annihilate this small difference in a phenomenological *Verschmelzung*, a merging into one continuous colored surface, as the result. If,

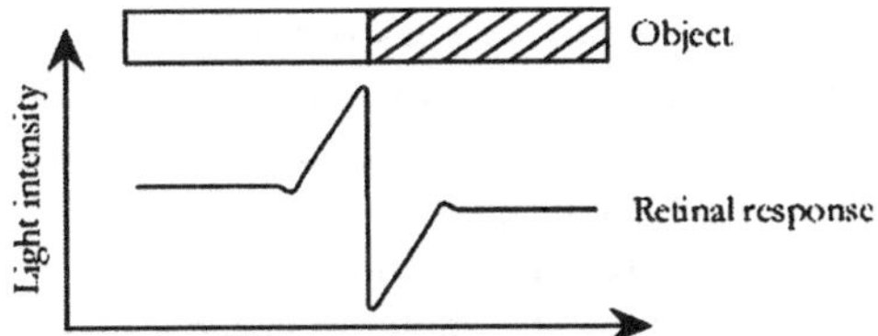

Figure 49.

contrariwise, two close visual cells receive strongly deviating input while each of them at the same time belong to a cell-community with related inputs, the network reacts in a characteristic curve (fig 49; Groupe μ, 66).

The radicality of the edge is 'exaggerated' by the eye which thus in its very construction is aimed at analyzing the visual object into a structure of lines. Given this inventory of lines in the phenomenon, the draughtsman may now make a further selection among them and obtain a sketch. Whether he does this in order to get a first structuring of his work which later will be fully orchestrated with detail depth coloring, etc. – or whether he does it for its own purpose, is not significant for our basic observation. On the other hand, the draughtsman may *add* further lines in the sketch as a trial-and-error attempt at reproducing the line structure offered by the visual system – he may indicate a number of strokes in a zone where the final line would appear only if the drawing is taken further than the sketch phase. Thus, we may distinguish between different sketch techniques:

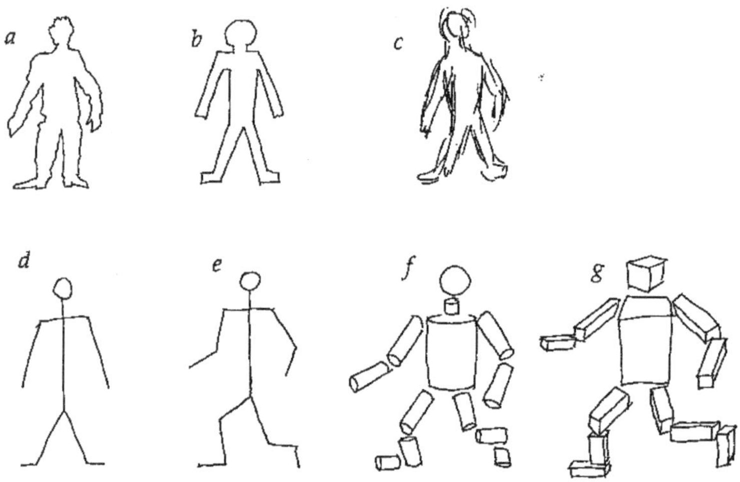

Figure 50. (a), contour (b), stylized contour (c), contour doubling (d), stylization of oblong object into lines (e), d) plus line bendings as symbol of functional bending (f), analysis of the object into cylinders (g), and analysis of the object into prisms

Here is selected seven typical among many different possible sketch techniques. They may be combined, more may be added, but all of them build on the simple set of lines in the visual phenomenon synthesized into an object. We must add that the basic definition of the sketch by the analytical transformation seems too simple. In many cases, the sketch adds what Groupe μ call *stylization* which systematically exaggerates and suppresses different features with some pragmatic aim of making certain aspects of the type more easily read than others. Stylization is strictly speaking another set of devices than sketching proper, but in the ordinary perception of the sketch both procedures occur, because the sketch itself builds on a hasty stylizing of the object: it does not reproduce the exact contour of the object, but a simplified, stylized rendering of it. Among its techniques count the following: the subsumtion of the lines drawn under a certain and limited class of types (most often straight lines and a few curves), the subsumtion of angles under a few angles with selected, discrete values (45 and 90 degrees, e.g.), the 'stretching' of lines to continuity, even rectilinearity, exaggeration of symmetry, etc. (Groupe μ, 365–68). All these transformations may be found in the examples above, with the addition of the subsumtion of volumes under a certain selection of simple types (e.g. cylinders, spheres, cubes, straight lines, etc.). The uniformization of colors and textures to be subsumed under a small selection of types of these is another possibility which we can not here illustrate.

The single sketch technique selects its set of lines with a semiotic purpose, determined by whether the aim is the precise or approximate appearance of the contour, whether (aspects of) the construction of the object, its function, its relation to other objects, etc. is the goal. The sketch thus possesses a rich variation of different transformation possibilities with a corresponding selection of lines, determined by which properties in the object is to be represented. Sketching and stylization may, on the other hand, be driven so far that the effect becomes, in fact, diminished readablity rather than increased.

In addition to this comes the possibility that the scarcity of the sketch can be utilized to point to indeterminacy, both in the object depicted and in the perceiving subject. The visual sign's being a type makes it general and implies that it may subsume several possible objects under it. This effect may, of course, be countered by the appearance of indices in the picture, disambiguitizing who or what the sketch refers to. This typicality is strengthened in the sketch case, the sketch being typically even more general than the more detailed visual sign. With reference to this (deliberate) indeterminacy in the object, the sketch may provide a Peircean diagram of the object. In a sketch like the following:

Figure 51.

this aspect is evident, just like in construction diagrams, function diagrams, etc. which are sketches selecting certain sets of disambiguous information of the object depicted. It is striking, however, that this indeterminacy may, in artistic contexts, be turned 180 degrees against the observer. Maybe he is a cool, technocratic, schematic person who not only regards isolated diagrams with precise purposes this way, but perceives his surroundings in general in this formalized and sterile way? Or maybe he is a confused marginal existence to whom objects only appear as vague clouds? Even more striking is the interpretation of subjective vacillation of the sketch in cases like the following:

Figure 52.

the many competing contours may be given several different interpretations: maybe they are a series of still more precise approximations to the object – or they bear witness to the fact that the object is hard to grasp in a precise way, maybe because of a very faceted morphology (a face), or maybe due to the fact that the object is moving or in a process of change. In the same way, the missing completion of the contour hints into whole, completed objects is ambiguous: maybe a process in development is hinted at here. These interpretations of the sketch is still objective – but in many cases they are spontaneously supplied with subjective interpretations: it may the instability of the perceiving subject which makes it difficult to focus precisely on the object. This instability, in turn, may be ascribed to properties in the perception apparatus (it is difficult to keep eye and head completely still, difficult to synthesize the different profiles of complicated objects from different points-of-view) or to the more emotional or volitional instability in the subject (an angry, upset, ecstatic, traumatized, drunk, confused person has more difficulty in concluding precise perceptions successfully or drawing correct sketches) – so to speak an impressionist versus an expressionist field in subjective imprecision.

We shall assume that the objective use of the sketch and its obvious connections to epistemology is phenomenologically primary (the sketch as a proto-diagram). But the artistic sketch, even if it has its root in this objectivity, should not necessarily be measured on the sketch as a diagram. As always, when art takes over a device, the aesthetic judgment criteria of art itself take over as well, and from this point of view the subjective and objective aspects of the sketch yield equal possibilities for further aesthetic use.

To conclude, we shall propose the distinctions between 1) sketches depending on which lines in the the object or in the understanding of it are highlighted, and with which semiotic aim – and 2) between objective and subjective uses of the indeterminacy assumed by the visual type in the sketch.

CHAPTER 16

WHO IS MICHAEL WO-LING PTAH-HOTEP JEROLOMON?

Literary Interpretation as Thought Experiment

Among the many problems in literary scholarship, an institutional one stands out. During the recent decades, literary studies have been literally swamped with theories imported from a long range of disciplines – to the extent that in an American context, courses are being taught under the naked headline of 'Theory'. This involves all brands of hermeneutics, structuralism, formalism, post-structuralism, critical theory, ecological criticism, gender-class-race, post-colonialism, social constructivism, new historicism, psychological theories and sociological doctrines of all kinds – plus the 'post' or 'neo' versions of all these – and many more. Potentially, this import of inspiration has – in lucky moments – seemed to make literary studies the very focus of the humanities where new syntheses could emerge and new understandings of the humanities at large find their place. But more often than not, the plethora of theories has given rise to bewilderment. And specific strategies have evolved to cope with the confusion. One is the internal splitting-up of literary studies into sub-branches where each scholar picks one theory and spends his time in happy isolation in his office, caring little about the equally narrow-minded research taking place in the neighboring offices. This Balkanization of literary studies, however, is nothing but the recipe for their dissolution as scientific endeavor. Another strategy is a little more promising: the supermarket-like idea that you pick your theory according to the text you tackle. In some cases, formal aspects of the text are intriguing, in other cases the text's embeddedness in social contexts, in still other, the thematic content – and so on. Still, this strategy has its drawbacks: the dissolution is only transported into the individual scholars' head, if he does not undertake any reflection as to the interrelation of this optional menu of theories.

What is lacking is a precise idea of the very role played by theory in the interpretation of literary texts. In particular, any consciousness of the difference between *method* and *theory* is all but absent in the theoretical supermarket. Very often, the two words are used interchangeably, presupposing a very direct relation between theory and text, so that a theory should, in itself, contain a method for its being put to use. But this relation is *not* direct. For between text and theory comes method. Methodology in the scientific treatment of literary texts may encompass highly different procedures: research in readers' response, research into the life of the author, into the period, genre, institution, culture, or any other context of the text, undertaken with interpretative or statistical, a priori or empirical means. But one basic method is common to all the different schools and theories mentioned, a method so simple that it tends to be overlooked: that of *reading* and *interpreting* the literary text itself. It forms the core of all other methods used (and, in turn, in all theory application). This chapter investigates this reading process as a special case of pragmatic epistemology as laid down in Peirce's semiotics. It is important to underline that this investigation is pre-theoretical in a precise sense of the word. It presupposes, that is true, certain theories of the process of gaining knowledge. But it does not presuppose any choice as to a preconceived theory of the nature of the literary text. Quite on the contrary, it tries to describe the place for such theories in the interpretation process.

AB-DE-IN: THE THREE-BEAT MOTOR OF INVESTIGATION

As Peirce analyzes any process of interpretation in three interconnected phases determined by the three argument types abduction-deduction-induction, in that order, it comes as no surprise that literary interpretation may also be so analyzed. A more detailed investigation will show, however, certain particularities as to the structure of literary interpretation.

Peirce's pragmatic semiotics in its mature version involves, it is well known, three levels, with changing titles through the development of his thought. A common way of expressing them is the distinction between Speculative Grammar, Logic, and Methodeutic (cf. Chaps. 2–3). Logic is the study of truth-preserving inferences and arguments and their interrelation – not unlike our days' definition, but still covering argument types nowadays rarely counted as purely logical, so as for instance abduction. Speculative Grammar is sometimes called also semiotics or stecheiotic. It is the study of the constituent signs of those arguments, in short, all instruments which may be used to express thoughts. Methodeutic, on the other side, is the study of how to combine arguments in order for knowledge to progress. In our day we would probably call it epistemology or, more pragmatically, heuristics. The picture is blurred, furthermore, by the fact that Peirce often calls the sum of all three Semiotics or Logic in a broad sense. The decisive point here, however, is that Peirce's logic and semiotics are conceived in close relation to an epistemology focusing upon the growth of knowledge (thus emphasizing the context of discovery rather than, or at least to the same extent as, the context of justification, to put in in Reichenbach's terminology). Van Heijenoort's and Hintikka's classical distinction between two lines in the development of logic is very apt here: on the one side we find logic-as-language (Frege-Russell-Quine), on the other logic-as-calculus (Boole-Peirce-Schröder).[411] Even if many of the core results of the two lines are the same (such as logic of relations, quantification, formal representation all having been invented independently by Frege and Peirce), the overall conception of the task of logic differs. In logic-as-language, logic is conceived as a formal language in which any possible claim can be expressed, and it thus covers the whole universe because it involves the very possibility of all what can be said. In logic-as-calculus, on the other hand, the emphasis is on the role of logic in the ongoing reasoning process – and logic is seen as problem-solving, pointing to a multiplicity of problems in different contexts and thus maintaining the intimate connection between logic and epistemology. Peirce clearly belongs to the latter tradition, and this is evident in his epistemology where the specific abduction-deduction-induction configuration of arguments expresses his pragmatic maxim as his fundamental requirement of sound investigation. Before turning to the details of the Abduction-Deduction-Induction reasoning cycle, let us look more closely on Peirce's original contribution in his doctrine of Abduction.

ABDUCTION: LOGIC AND CREATIVE PERCEPTION

It is well known how Peirce adds to the traditional two forms of inference, Deduction and Induction, a third, Abduction.[412] While deduction is a necessary inference, and induction a probable inference only, abduction is a merely possible inference. At a closer look, it is rather two aspects of what is traditionally seen as Induction which are now separated into Abduction and Induction, namely the proposal of a hypothesis and the empirical testing of it in a number of cases, respectively. The more precise determination of abduction changes over Peirce's lifetime and is, as such, the subject of some controversy. Let us here briefly run through the most crucial among the different aspects of abduction.[413] In ('Deduction, Induction, and Hypothesis', 1878, EPI, 188; 2.623), Peirce gives an early description of the character of the three types of inference, based on different configurations of the rule-case-result triad known from the ordinary syllogism (abduction appearing here under the name of 'hypothesis'):

DEDUCTION.
Rule.–All the beans from this bag are white.
Case.–These beans are from this bag.
∴ Result.–These beans are white.

INDUCTION.
Case.–These beans are from this bag.
Result.–These beans are white.
∴ Rule.–All the beans from this bag are white

HYPOTHESIS.
Rule.–All the beans from this bag are white.
Result.–These beans are white.
∴ Case.–These beans are from this bag.

Here, the conclusion to the abductive argument is a possible statement only: the white beans in the Result may, of course, stem from a lot of other possible sources. As Hoffmann (2002, 254) points out, Peirce's distinction between two premisses in Abduction (and Induction as well) is irrelevant, because they can be simply combined into one, and later on Peirce himself states that the premiss is always only one, however many parts it may be composed of or may be analyzable into. But the conclusion may have highly variable degrees of generality, and in some sense it is the very choice of Rule, more or less general, which forms the interesting part of abduction. Thus, the Case presented here as the conclusion is only reached on the basis of the selection of a Rule – among many possible – which gives the desired

Result. Later Peirce tries to take this into account when he attempts to formalize the logical form of abduction as follows (in his 1903 lectures on Pragmatism):

> Long before I first classed abduction as an inference it was recognized by logicians that the operation of adopting an explanatory hypothesis... which is just what abduction is... was subject to certain conditions. Namely, the hypothesis cannot be admitted, even as a hypothesis, unless it be supposed that it would account for the facts or some of them. The form of inference, therefore, is this:
>
> The surprising fact, C, is observed;
> But if A were true, C would be a matter of course,
> Hence, there is reason to suspect that A is true.
>
> Thus, A cannot be abductively inferred, or if you prefer the expression, cannot be abductively conjectured until its entire content is already present in the premiss, 'If A were true, C would be a matter of course'. ('Harvard Lectures on Pragmatism', 1903, EPII, 231; 5.189)

Here, the Result becomes the surprising fact C to be explained, and a Rule A to explain the fact is what is sought, and the conclusion is neither A nor C as explained by A (as was the case above), but the possibility of A being true (because it accounts for the appearance of C). But, as Peirce himself is the first to admit, the presence of A in both the premiss and the conclusion causes problems. It implies that this logical form does not in any way account for the *process of adoption* of the hypothesis (but only for the *reason to believe* in the hypothesis, namely that it does in fact imply that fact). Hence, the very act of finding hypotheses is a prerequisite to the judging of them in this logical form – as well as the act of comparing the many possible different candidates for A. It is important here to underline that the amount of possible As are in all cases indefinite (in the above example, other possible As could be: 'these beans have been painted white by the experimenter', 'these beans have mutated into a white variant', 'these beans come from a bag of chalk', etc.), so a crucial task in the economy of investigation is to focus upon the candidates for A which are in some sense plausible, given the context.

Here, the role of abduction is to find a hypothesis to explain a surprising case – and the role of induction, conversely, becomes to test a hypothesis against facts in a quantitative investigation.[414] While abduction goes from case to rule, induction runs the opposite way from rule to cases. Between them, deduction intervenes in order to develop testable consequences of the hypothesis reached by abduction.

Abduction takes its point of departure in certain facts and proposes a (more) general, ideal explanation of these facts. Deduction investigates certain ideal consequences of this explanation. Induction, finally, takes us back from the ideal explanation and its consequences to facts, comparing explanation and further facts and judging the former on that basis:

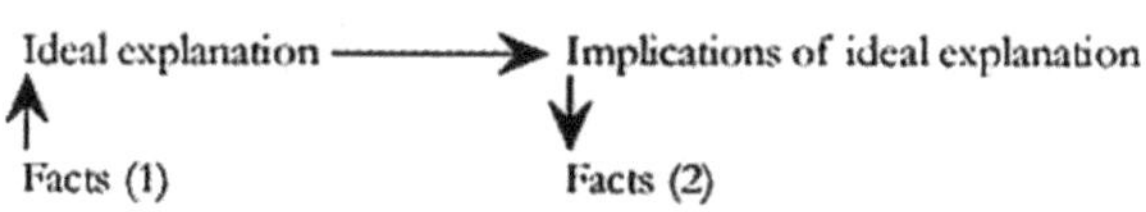

Figure 53.

The problem remains evident in the fact that while Peirce contends that there is substantially no information in the conclusion which was not already (albeit, maybe, implicitly) in the premisses, on the other hand he claims that Abduction is the only source for new information in the reasoning process (deduction and induction providing no new material). But hence, this innovation cannot stem from the logical form of Abduction.

But if the logical form of abduction does not suffice to explain the adoption and selection of hypotheses, then how is that to be explained? Abduction is a 'qualified guess' – but what is it that qualifies the guess and makes it differ from mere fumbling in the dark or mere mechanical testing of an infinity of equally possible hypotheses? In the same manner as induction in which non-logical material is introduced in the empirical testing, abduction includes non-logical material in addition to its logical form. In both cases, the material stem from perception – but in what sense? As Hoffmann argues, this becomes possible to see if we draw a distinction between the logical form of abduction of the 1903 lectures on the one hand and the 'creative' act of perception of the surprising fact on the other, leading to the adoption of a hypothesis.

In the same lecture, Peirce discusses 'perceptual judgments' (given a red chair in the visual field, we may make the proposition 'the chair is red') as an extreme case of abduction. Perceptual judgments differ from prototypical abduction in one crucial aspect: they are not subject to conscious self-control and hence cannot seriously be doubted, even if they are not necessary deductions. As all matter in the conclusion of an argument must be contained in the premisses, perceptual judgments hence must rest upon logical matter being present *implicitly* (or, in the psychological instantiation of the inference, in the mind outside the reach of consciousness) in the premisses:

Ultimately therefore it must come from the uncontrolled part of the mind, because a series of controlled acts must have a first. But as to the logical form, it would be, at any rate, extremely difficult to dispose of it in the same way. An induction, for example, concludes a ratio of frequency; but there is nothing about any such ratio in the single instances on which it is based. Where do the conceptions of deductive necessity, of inductive probability, of abductive expectability come from? Where does the conception of inference itself come from? That is the only difficulty. But self-control is the character which distinguishes reasonings from the processes by which perceptual judgments are formed, and self-control of any kind is purely inhibitory. It originates nothing.

Therefore it cannot be in the act of adoption of an inference, in the pronouncing of it to be reasonable, that the formal conceptions in question can first emerge. It must be in the first perceiving that so one might conceivably reason. And what is the nature of that? I see that I have instinctively described the phenomenon as a 'perceiving.' I do not wish to argue from words; but a word may furnish a valuable suggestion. What can our first acquaintance with an inference, when it is not yet adopted, be but a perception of the world of ideas? (EPII, 233; 5.194)

The creative side of abduction, hence, is a sort of *Wesensschau*, it is, as Hoffmann says, an act of 'creative perception' (264). In an unused section of the lecture, Peirce continues:

A mass of facts is before us. We go through them. We examine them. We find them a confused snarl, an impenetrable jungle. We are unable to hold them in our minds. We endeavor to set them down upon paper; but they seem to be so multiplex intricate that we can neither satisfy ourselves that what we have set down represents the facts, nor can we get any clear idea of what it is that we have set down. But suddenly, while we are probing over our digest of the facts and are endeavoring to set them into order, it occurs to us that if we were to assume something to be true that we do not know to be true, these facts would arrange themselves luminously. That is *abduction*... (EPII, 531)

The creative perception spontaneously organizes the facts in a pattern.[415] It is, in fact, a case of gestalt perception in which a certain interpretation is inherent in the very organization of the perception. Several such organizations may be possible (just like a finite number of measuring data may give rise to an infinity of different mathematical curves accounting for them) which are to the same degree motivated by the data implied. Peirce gives as an example ('Lectures on Pragmatism', 1903; CP. 1.183) a drawing by his father which may interchangeably be perceived as a wall of round stones or as the continuous drawing of one line – in fact, a classic example of the gestalt phenomenon of 'Kippfigur' well-known from Neckar's cube or Rubin's vase.

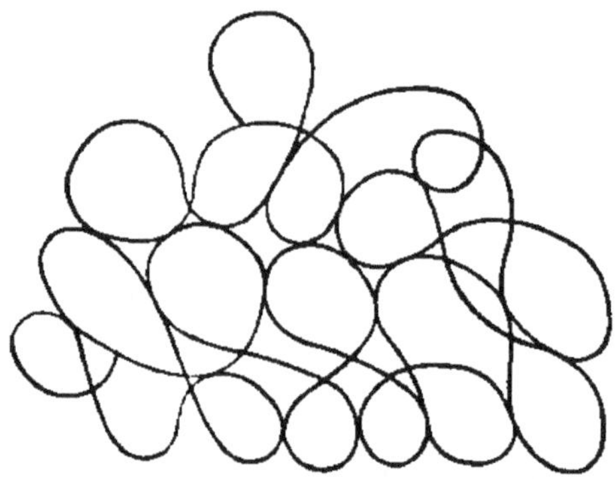

Figure 54. A wall of stones or one continuous line?

Peirce's doctrine on perceptual judgment from around the turn of the century thus unknowingly enters the contemporary debate as to the status of such gestalts. Both the Berliner and the Grazer gestaltists maintained, as against von Helmholtz's reference of such phenomena to *unbewußte Schlüße*, that no act of logical judgment was required to form such complexes. Peirce's doctrine of perceptual judgments might at a first glance seem to side with von Helmholtz. According to him, all thought and sensation are inferences, and it is impossible to reach pure sensory data before logical interpretation of sensation. Any given perception is part of the ongoing perceptive process, resting on former perceptual judgments. Still, the detail of Peirce's doctrine rather seems to give him an intermediate position between von Helmholtz and the gestaltists: the inferences at stake in perceptual judgments are of a peculiar type – they take place without any control and are immediately given without any possibility of doubt; this spontaneous, in-built character takes them closer to the properties of the gestaltist account:

That part of the conclusion which inserts the wholly new element can be separated from the rest with which it has no logical connection nor appearance of logical connection. The first emergence of this new element into consciousness must be regarded as a perceptive judgment. (EPII, 232; 5.192)

Here, it is connected to Peirce's crucial observation that abduction is the only inference that is able to introduce *new* matter in the investigation process – deduction only being able to flesh out what was implicit in some assumption, and induction merely being able to test the implications of a hypothesis against further facts. As this innovative character of abduction can not stem from its trivial logical form, it must stem from this character of 'creative perception'.[416]

Still the problem remains, in the abductive reasoning, of how to select potentially fruitful hypothetical organizations of material without having to represent and go through the whole series of merely possible such hypotheses.[417] Our discussion of literary interpretation may serve as a clue to this question.

ABDUCTION-DEDUCTION-INDUCTION

In the methodology governing the ongoing investigation process, Abduction is connected to Deduction and Induction as phases to be processed in that order. Abduction gives rise to the formation of a hypothesis. Deduction now takes over and seeks out some more or less complicated necessary implications of that hypothesis. Abduction took us from facts to hypothesis, that is, from the empirical world to an ideal world. Deduction works in this ideal world – the only sphere where necessity is possible – and traces certain ideal consequences in the model so proposed. Induction takes these consequences and returns to the world of facts, seeking out further empirical data which may (or may not) corroborate these new consequences of the ideal model. If these data support these consequences, it is to be taken as a sign that the hypothesis proposed has a certain probability of being true (of course, other possible hypotheses might have the same ideal consequences and hence have the same validity in the test). In both cases, the stage is set for a new abduction. If induction confirmed the hypothesis, abduction may now refine or sophisticate the hypotheses and allow for further deductions from the model to be tested. If induction falsified the hypothesis, abduction must start all over again by proposing a reformed or wholly new hypothesis – by the molding of it so as to avoid the pitfalls recorded in induction.

In this way, the Ab-De-In sequence forms a cycle to be run through over and over again – so to speak Peirce's version of Gadamer's well-known 'hermeneutic circle' in text interpretation. As implied by the Gadamerian term of 'Vorverständnis,' it is also the case in Peirce's circle that no pure pre-theoretical beginning is possible. The 'surprising fact' taken as first premiss in Abduction is, of course, only surprising as measured against certain expectations of a general, that is, theoretical kind, be they explicit or implicit. Peirce thus could be said to antedate Gadamer by more than half a century – but his account also adds a host of crucial details to the interpretative circle. Peirce's ab-de-in cycle is nothing special for science or literature, it is more or less subconsciously run through countless times every day in everyday reasoning where its single phases most often do not pose any serious problems which merit explicit reflection.

Having abducted a hypothesis, Deduction takes over. But what is Deduction supposed to do with the hypothesis? Peirce famously distinguishes between two types of Deduction, Corollarial and Theorematical, and it is due to Hintikka that the contemporary relevance of this distinction has been highlighted (cf. Chap. 4). But even before we consider them, there is an often overlooked aspect of Deduction's relation to a Hypothesis: it does not follow by necessity *in which aspect* of the Hypothesis a Deduction should take its point of departure. Beginning with one proposition a whole series of different necessary inferences are possible, the more complicated the Hypothesis is, the more necessary consequences. If the abductive hypothesis claims, e.g., that the facts in question constitute a circle, then we might be interested in finding the necessary tangents or radii of that circle, or we may be interested in finding inscribed or circumscribed polygons of that circle . . . all of which possibilities follow by necessity from the concept circle. Deduction, in short, is no algorithm; rather, it is the condition of possibility for the construction of different algorithms, each of them consisting of different series of necessary steps. Thus, even in the beginning of Deduction, something like an Abduction in the ideal realm must take place: in which direction do we want to test the hypotheses? Which aspects of it would we prefer to take as basis for consequences to be tested?

Deduction deals with ideal phenomena and thus always has a more or less explicit mathematical character – and in short, undertaking a mathematical proof, we most often have an idea of the theorem we seek a proof of, so we select implications of our hypothesis which in some way seem similar to or oriented in the direction of what we intend to prove. This, Peirce claims, may be undertaken in two different manners: Corollarial and Theorematical. The former sets up the whole of the hypothesis, maybe in a drawing or maybe by a synthesizing gaze on all components of the hypothesis – and then the conclusion is directly read off of the hypothesis as a whole. The latter is more complicated. It requires the addition of new material – construction lines of Euclidean geometry is Peirce's prototypical example – and on the basis of this construction some manipulation or experiment with the hypothesis-plus-construction is undertaken. Very often, the intriguing or challenging part of the proof is the abduction allowing to find the right constructive additions to the hypothesis. This idea sets Peirce apart from the algorithmic or formalist ideals in mathematics where intuition should be bracketed and mere symbol manipulation take over – and, on the other hand, approaches him to a Kantian view of mathematics where construction is a necessary part of mathematical proofs. This 'theorematic' part of deduction is what makes it possible for mathematics to be at one and the same time completely deductive and ripe with new discoveries. In deductive inference and diagrammatical reasoning – in Peirce the two are identical – it is possible to observe generality directly, because we objectify our knowledge-gaining apparatus. Diagrams may be represented in material figures (which, it is true, require an idealizing reading) and thus avoid the subjectivist pitfall of the mind as a closed cage of representations. Diagrams are, at one and the same time three things: structures of the ideal hypothetical world discussed, objective material structures open for intersubjective scrutinizing, and subjective representations. None of these aspects can be missed: even if material diagram tokens require an idealizing reading to be understood as types, it is important to see that this reading, conversely, may need the support of the material diagram tokens. The fertility inherent in the multiplicity of possible chains of necessary deductions springing from a hypothesis reaches its peak point in theorematic reasoning. Which construction and which experimental manipulation can be undertaken on the basis of a given hypothesis is an open issue – augmented by the inexhaustibility of mathematics – and of the world of ideal objects in general.[418]

Induction, then, has to test the deductive consequences of the hypothesis proposed. This, not surprisingly, may be done in three different ways.[419] Peirce's shortest presentation of them counts Crude, Quantitative, and Qualitative Induction, respectively (Manuscript 'G', c. 1905, 2.755). Crude induction is the simplest version which he sometimes nicknames pooh-pooh. It is the argument that as x has never been observed, then x is unlikely to occur at all. It is, of course, an extremely weak argument, but still, according to Peirce, extremely important because it is put to use all of the time when, without saying it, it is supposed that the ordinary experience may count as default value. It must, however, give way to the slightest sign of any positive evidence to the contrary. It is the argument that all swans are white, as no black swans have been seen. As it turns out, it plays a crucial if mostly neglected role in literary interpretation. Quantitative induction is the investigation of a hypothesis by controlled sampling in a finite set of investigations, to be treated statistically. This – strongest – inductive argument plays rarely a role, if any, in literary interpretation. Finally, Qualitative induction is applied when the conditions for Quantitative Induction are not present (when there is no finite set from which to select a sample). It is an intermediary between the two others (also in strength):

> It consists of those inductions which are neither founded upon experience in one mass, as Crude Induction is, nor upon a collection of numerable instances of equal evidential values, but upon a stream of experience in which the relative evidential values of different parts of it have to be estimated according to our sense of the impressions they make upon us. (ibid. 2.759)

In normal scientific investigation, Crude Induction merely plays the role of background screening, almost automatically weeding out all sorts of logically possible, but materially fantastic hypotheses. The inferred consequences of the hypothesis chosen are tested by means of Quantitative and Qualitative Induction, respectively. In the experimental sciences, Quantitative Induction takes place in relation to an experiment devised by the hypothesis. The results of the experiment are now compared to the deductively predicted results, and statistic evaluation of a number of results may lead to the conclusion of verifying, falsifying, or, most often, relativizing or refining the hypothesis. Qualitative induction is a weaker sort of test dispensing with quantitative measurability; we shall return to it below.

As is evident, Peirce's Ab-De-In Doctrine forms an early version of the well-known Hypothetico-Deductive method. It adds, however, to this, the deliberations on the abductive choice of hypothesis, the diagrammatical account of deduction, and the admission of non-quantitative inductive testing. Let us try to conceptualize literary analysis in terms of Peirce's apparatus.

LITERARY INTERPRETATION

To a large extent, literary interpretation shares the features of ordinary text interpretation, e.g. of ancient documents, or of more or less complicated pieces of everyday writing. Literary texts, however, add certain dimensions to interpretation as such. Literary texts are prototypically *closed* in the sense that they constitute a work, an œuvre with strictly defined boundaries. These boundaries are not absolute, but they constitute a first barrier for and constraint upon interpretation. Interpretation ought to go as far as possible within textual boundaries and establish the internal systems, meanings, coherences, claims, forms, aesthetics, world view, or whatever interpretation is looking for, before it crosses the boundaries in order to search for the answer to unsolved questions in the text in its context. This is not the case in non-literary texts where we immediately go outside the text to solve puzzles unanswered by the text – if a text lacks information on some point, we do not hesitate to supply it from elsewhere. Of course it is possible to read ordinary texts from a 'literary stance' trying to establish an internal structure before adding outside material, and the conceptual distinction between literary and non-literary texts should not lead us to assume that empirical texts fall in two strictly separate classes; rather, there is a continuum between prototypically literary texts and prototypically 'ordinary' texts. The prototypically literary text's closedness, however, urges us to read it in a *generalized* manner: not only for the particular content it seems to convey, but interpreting it as pointing to some more general claim than is evident on the text's surface. Its closedness and its supposed generality are two sides of the same coin. This is what elementarily characterizes literary interpretation: its generalizing intention aimed at a text seen with a certain closure. Literary interpretation prototypically presupposes that the text carries a general content of which the particular contents of the text is but an instantiation.

No matter what more specific theory the reader may entertain about literature, this interpretation follows a certain method. The beginning of interpretation is nicely caught in Freud's classic description of the analyst's mind during analysis, that of 'gleichschwebende Aufmerksamkeit', floating attention[420] – aimed at nothing specifically but listening in order to discover connections in the patient's speech. The same is the case in literary interpretation. Its base is an ordinary reading of the text, including the passions, emotions, thoughts, observations, pleasures or pains it has given rise to on the part of the reader. It is on the basis of this immediate conception of the text that the 'gleichschwebende Aufmerksamkeit' sets in. It does not look for anything specific, that is, it is equally open to remark phonetic resemblances or oppositions, prosodic repetitions, graphic patterns, semantic overlappings or ambiguities, rhetorical similarities or contrasts, thematic recurrences, narrative developments, metaphorical structures, organization of space-time, different dia-, idio-, sociolects, genre structures – not to mention the possible cross-mappings of patterns between all these and more. In short, all sorts of correlations in the text may be picked up by the first phase of literary interpretation. They constitute the 'surprising facts' of Peirce's epistemology, surprising, that is, in contrast to an 'ordinary' reading of the same text. They call for the selection of a hypothesis – an abduction. This is not a special hurdle to pass, rather abductions spring forth all of the time during this phase of interpretation, most of them so obvious that they need not be explicit. There may be a striking amount of imagery pointing to death in a certain lyrical poem – an abduction now may state the hypothesis that the poem is not only about a pastoral scenery like it immediately seems to be the case – but that it also carries a metaphysical reflection about mortality. This is a striking fact of surprise: a landscape description does not normally entail a death description. Now, the hypothesis erects an ideal model: the poem contains a reflection of mortality.[421] What would we expect from a reflection upon mortality? The theme of mortality has a lot of aspects, and each of them has certain necessary implications (prototypically – all other things equal, of course). Death implies silence and stillness, the ceasing of any projects, a grave implies rotting flesh populated by worms, the passing of a family member implies mourning relatives, etc. A piece of cultural knowledge adds further possibilities: death may be represented by hourglasses, scythes, skulls, unused instruments and books, cut flowers, etc. – the whole imagery of the *memento mori* genres in literature and painting.[422] Are any of these many deductive consequences of the hypothesis chosen corroborated by the text? Here induction enters the scene. Crude induction plays a large role in the interpretation of literary texts, because we can assume that in the absence of any signs pointing to the contrary, everything unmentioned possesses a default value.[423] The role of crude induction is so much larger in literary texts, because there is no other access to the textual universe than text itself – this in contradistinction to ordinary texts about empirical subjects where there always is an infinity of possible information about a given subject, and hence the possibility of counterinformation showing up. The literary text is, even if voluminous, finite, and if nothing ever is told or implied about the main character's legs, we can safely – even more safely than in real life cases – assume that he has two legs. This omnipresence of crude induction in relation to literary text forms a first step in the text's generality, because the corresponding 'Unbestimmtheitsstellen' refer to an infinity of possible extensions (the legs may be long or short, with much or little hair, etc. – they remain general legs until further described). Crude induction does not, of course, tell us anything about more specific hypotheses like the one about death in a lyrical poem. Such hypotheses call for Qualitative Induction. Let us here take a closer look at how Peirce describes it:

> Qualitative Induction consists in the investigator's first deducing from the retroductive hypothesis as great an evidential weight of genuine conditional predictions as he can conveniently undertake to make and to bring to the test, the condition under which he asserts them being that of the retroductive hypothesis having such degree and kind of truth as to assure their truth. In calling them 'predictions,' I do not mean that they need relate to future events but that they must antecede the investigator's knowledge of their truth, or at least that they must virtually antecede it. I will give an illustration of such 'virtual antecedence.' Suppose that to avoid wasting a great deal of time upon a hypothesis which the

first comparisons with the facts may show to be utterly worthless, an investigator of a certain conjecture draws up and resolves to follow a well-considered initial program for work upon the question, and that this consists mainly in working out and testing as many consequences of the hypothesis as he can work out by a certain mathematical method and can ascertain the truth or falsity of at a cost of not more than $100 for each. But suppose that among the half dozen predictions to which that method will carry him, there, quite unexpectedly, turns up one whose truth has long been known to him, though it is a surprise to him to find that it is deducible from the hypothesis under examination. What course does sound logic impose upon him under these circumstances? The answer is that he must reexamine the process of retroduction that suggested the hypothesis; and if the fact that is now repredicted in any degree influenced that hypothesis, it has had its due effect, and must not be used again. But if not, will he then be free to use the prediction if he likes? Not at all: the validity of his Qualitative Induction will be found to depend upon his following a rational and decisive method; he has no more right, but rather less, to favor the inductive rejection of the retroductive suggestion, than to favor its inductive adoption; and he is bound, as a man who means to reason as honestly as the imperfections of his nature and training will permit, to admit the true prediction into his counsels. The predictions must eventually be so varied as to test every feature of the hypothesis; yet the interests of science command constant attention to economy, especially in the earlier inductive stages of research.

Having made his initial predictions the investigator proceeds to ascertain their truth or falsity; and then, having taken account of such subsidiary arguments as there may be, goes on to judge of the combined value of the evidence, and to decide whether the hypothesis should be regarded as proved, or as well on the way toward being proved, or as unworthy of further attention, or whether it ought to receive a definite modification in the light of the new experiments and be inductively reexamined ab ovo, or whether finally, that while not true it probably presents some analogy to the truth, and that the results of the induction may help to suggest a better hypothesis. ('G', 1905, 2.759)

Here, Peirce outlines how Qualitative Induction may fail: if one of the consequences is recognized as true by the interpreter *for other reasons* than the argument in process or facts presented in the induction, then he must reconsider the hypothesis – maybe it was this truth, accidentally occurring among the implications of the hypothesis, which lured the interpreter into accepting the hypothesis in the first place, and not its fertility in the ongoing interpretation. In our example, this would correspond, e.g., to a case where the interpreter has a certain obsession with Freud's concept of a death drive, regards it as one of the necessities of death, and assumes the hypothesis for this reason (rather than because of observations in the text). If such 'false' predictions are avoided, then, the remaining predictions should now be inductively tested. Are any hints of rotting flesh, worms, silence, emptiness, sorrow, etc., or skulls, cut roses, crosses, hourglasses, etc. found in the text? If there are, the hypothesis may be considered strengthened, if not proved, and in any case it may be refined, now maybe concentrating upon a special aspect of or conception of death, which may then be further investigated through new deductions and inductions. A very important implication of this Ab-De-In cycle in interpretation is the fact that only the two of the phases deal directly with the text, namely Abduction and Induction, each in their direction, so to speak:

Abduction seeks a theory. Induction seeks for facts. In abduction the consideration of the facts suggests the hypothesis. In induction the study of the hypothesis suggests the experiments which bring to light the very facts to which the hypothesis had pointed. The mode of suggestion by which, in abduction, the facts suggest the hypothesis is by resemblance, [...] the resemblance of the facts to the consequences of the hypothesis. The mode of suggestion by which in induction the hypothesis suggests the facts is by contiguity, [...] familiar knowledge that the conditions of the hypothesis can be realized in certain experimental ways.

('On The Logic of Drawing History from Ancient Documents', 1901, EPII, 106; 7.218)

Abduction selects a hypothesis able to account for the facts in question – induction seeks confirmation among facts of the deductively 'enlarged' hypothesis. Both share the comparison between facts and hypotheses, and it is easy to see that in the want of the deductive phase, the two of them might easily mirror each other so that any hypothesis might only bring into its horizon suitable facts, so that such 'theory-laden' facts might blind the interpreter from admitting counterhypothetical facts. Here, the intervening deductive phase is extremely important. It turns, momentarily, away from the facts of the text and contemplates only the hypothesis handed over from Abduction. It then considers it as an ideal model and draws a bundle of necessary consequences of it, completely disregarding its background in the text for a moment. This is where theory of any kind may step in. The hypothesis is, in itself, theoretical of nature. In the grasping of its consequences, theorematic reasoning allows for the introduction of new variables, new 'construction lines' in order to facilitate new experiments which may uncover more of the possibly implicit content of the hypothesis. This is the place where theoretical assumptions of all kinds may enter the methodological circuit of interpretation. If I hold certain theories about death – for instance Freud's already mentioned speculative idea of the Thanatos principle – they may be invoked as additional material in order to find what further consequences may be drawn from the idea that the text in question deals with death. Freud's theory, for instance, implies that there is a drive in nature and psychology to reach the lowest level of energy and tension. Consequently, lowering of tension may be ventured as a possible deduction from the enlarged hypothesis, and this implication may now be inductively tested. It is of crucial importance here to distinguish between the two motives for theory import: theory as explicit 'construction lines' imported in the diagrammatical reasoning interregnum, or theory as unrecognized, implicit guide-lines for the selection of the hypotheses in the first phase. In the former case, theory may be so to speak controlled or checked by induction; in the latter case not so, because it silently governs the very selection of hypotheses to be checked. This constitutes, in fact, a difference between interpretation and overinterpretation.[424]

It is, indeed, a subtle difference. For is it not the case that theoretical preconceptions are at stake already in the early abduction phase of 'creative perception'? If the 'creative perception' leading to the death hypothesis in the first place is a perception of the poem's landscape scenery as strangely barren, dark, and desolate, then is it not the case that our theoretical predispositions play a role in our selection of that hypothesis among many other possible hypotheses to be tested? It is, of course, the case. Our arsenal of theoretical devices plays its role twice, both in the pre-explicit choice of hypotheses in the creative perception phase of abduction and in the experimental phase of deduction. What is important, then, is that theory's role in the former case is not that of a presupposed truth, but rather that of an assumption, a way of viewing things experimentally.

The deductive phase on a certain distance to the text may be compared to Ricœur's idea of a so to speak structuralist phase in any ever so hermeneutical text interpretation with its 'Einfühlung' – a phase in which the structures of the text are objectified and, in the passing, seen as beyond any author intention, as an autonomous, ideal system equipped with their own internal regularities. Diagrammatical reasoning makes possible the explicit discovery of truths hitherto being only implicitly present in a diagram. This is why this phase of interpretation is so important: it is here that the not immediately obvious may be revealed.

In our prototypical presentation, we have artificially pretended the interpretation to take place in the course of one Ab-De-In circuit. This is, of course, a simplification for the sake of explanation. In real interpretation, the circuit is traversed any number of times, and each new text observation gives rise to yet another hypothesis-deduction-test series. There is reason to believe that any initially assumed hypothesis will either be refuted or refined during the ongoing interpretation. This does not imply, however, that interpretation following this method, will invariably end in one very subtle and many-faceted hypothesis. A literary text is a huge, complicated phenomenon, and there is no a priori reason to methodologically assume that its crucial features will in all cases sum up in one hypothesis, be it ever so complex. Of course, hypotheses may endlessly be combined by 'and' to yield huge combinatorial hypotheses (yet without any further internal cohesion), but we may easily expect that, in some cases, different text observations give rise to contradictory hypotheses which may both be corroborated by different inductive test observations. In that case, interpretation will have to conclude that an ambiguous text is under investigation.

JEROLOMON REVEALED

Peirce himself did not, of course, test his epistemology in literary interpretation. There exists, however, a very interesting quote where he presents a piece of fictitious interpretation of a fictitious homegrown example:

Suppose, then, that, being seated in a street car, I remark a man opposite to me whose appearance and behavior unite characters which I am surprised to find together in the same person. I ask myself, How can this be? Suppose I find this problematic reply: Perhaps he is an ex-priest. He is the very image of such a person; he presents an icon of an ex-priest. Here is an iconic argument, or abduction of it. Secondly, it now occurs to me that if he is an ex-priest, he should be tonsured; and in order to test this, I say something to him calculated to make him take off his hat. He does so, and I find that he is indeed tonsured. Here at last is an indication that my theory is correct. I can now say that he is presumably an ex-priest, although it would be inaccurate to say that there is any definite probability that he is so, since I do not know how often I might find a man tonsured who was not an ex-priest, though evidently far oftener than he would be one. The supposition is, however, now supported by an inductive induction, a weak form of symptomatic or indexical argument. It stands on a widely different basis from that on which it stood before my little experiment. Before, it rested on the flimsy support of similarity, or agreement in 'flavor'. Now, facts have been constrained to yield confirmation to it by bearing out a prediction based upon it. Belief in the theory rests now on factual reaction to the theory. Thirdly, while the man's hat is off, I read in the crown of it a name that has been pasted into it. I have no doubt whatsoever that it is the man's name. I do not go into the question of how I come to be so confident of that. As long as I have not doubt, it matters not how doubt came to be destroyed. I get out

of the car, and go to call upon the chancellor of the diocese; and that he will tell me the truth I equally believe implicitly. I ask the chancellor, 'Who is Michael Wo-Ling Ptah-Hotep Jerolomon?' (Pardon my nonsense.) He replies, 'he is an ex-priest.' 'Is he the only man of that name?' 'No, there are, or may be, fifteen. Fourteen of them reside in this town and are ex-priests. The fifteenth went, twenty years ago, to High Tibet, and has never been heard of since.' It thus appears that the name read in the hat, thought having no striking 'flavor' of ex-priest about it, nor any such causal connection with the man's being an ex-priest as was the tonsure, yet in consequence of this knowledge becomes a symbol of the man's being an ex-priest; for a symbol is a sign which becomes significant simply by virtue of the fact that it will be so interpreted. So, it might conceivably have been an accident that the man was tonsured, but now that the name Michael Wu-Ling Ptah-Hotep Jerolomon signifies for me a probability on that ground alone is over fourteen to one that he is an ex-priest. There is no escape from that. It is what I consider myself certain of. It is only a probability. Yet now, fourthly, combining the arguments into one mixed argument, and considering, what is logically relevant, that I have no serious stake in the question, I am satisfied to consider the mixed argument as proof, and to dismiss the question until it may acquire more importance. (Although the illustration is silly, it all the better covers the case.)
(MS CSP L 75 E, 163–73)

Abduction: this man is an ex-priest. This idea is based on the surprising fact that he does not look like an ordinary person, but looks like a cleric.
Deduction: an ex-priest probably has a tonsure
Induction: test showing he has, in fact, a tonsure

Now the original hypothesis is strengthened by a further observation: the name Michael (etc.) Jerolomon in his hat (resting on the crude induction that a name in a hat is most often the name of the person wearing it)

Abduction: this ex-priest is Michael Jerolomon
Deduction: he is probably from this town's diocese
Induction: a call to the diocese reveals 14 priests of that name, against whom only one with that name has left the town for Tibet

The inquirer, of course, ought to have compared this finding with the telephone directory of the town (maybe there is a total of 3000 Michael Jerolomons living there) before he concludes to the 14/15 probability of the man being a priest. But suppose he knows that name is not widespread in the town – then the initial hypothesis seems corroborated with a degree of probability bordering to absolute certainty.[425]

INTERPRETATION PERSPECTIVES

Peirce's example is an – albeit fictitious – example of 'ordinary' interpretation. We already remarked the crucial difference between literary and 'ordinary' interpretation: in the former there is a crucial distinction between internal and external text observations. The interpreter, in the first phase, seeks to establish his reading with internal text observations only. This does not, however, imply that he excludes all of his encyclopaedic knowledge about the world while reading. When a text speaks about the Eiffel tower, the reader automatically assumes all his general knowledge of that edifice – that is, until further notice. This is part of the 'crude induction' in literary interpretation. But as soon as any signs in the text hint at the idea that this is not in all respects the 'normal' Eiffel tower referred to, the reader must bracket (parts of) his knowledge of that tower and admit the text's right to alter reality. In so far, the Ab-De-In cycle of interpretation makes use of the reader's encyclopaedic knowledge all of the time, even during the internal 'close reading' phase. Quite another issue is the text conceived no longer as a sign, as a piece of communication, but as a product of its own extratextual context – be it the author, the literary institution, the period, etc. Here, the text is seen rather as a symptom of other phenomena in reality. Here, the introduction of knowledge about the author's biography, the social or political conditions present at the time the text was written, etc. should be introduced with the same care as theoretical assumptions during the Ab-De-In cycle. They may be introduced as further motivations for a hypothesis or for the refinement of it, or they may be introduced as further fact corroborating an inductive testing. But just like the case about the danger of overinterpretation in theory introduction, they must not run counter to textual observations which thus retain a primacy over theory as well as context. Any attempt to directly deduce the text from (parts of) its context will, for this reason, be fallacious. This primacy is not absolute. It may be relativized many times during the ongoing interpretation process when the introduction of external factors is called for: to understand the text as a replique in a conversation, to understand it as aimed at a specific type of reader, to understand it as reflecting certain conditions in the life of the author, to understand it embeddedness in the political culture of the period, etc.This relative primacy of text observations is, so to speak, the rational kernel of the autonomy doctrine of new criticism close reading.

The Ab-De-In cycle of interpretation can be compared with related processes in Peirce's philosophy. One is its very center, the pragmatic maxim from 'How to Make Our Ideas Clear', and, in turn, its basis in laboratory experiment. The pragmatic maxim, as is well known, claims that the content of concepts should be clarified by considering which conceivable effects the object of that concept could have – then this will be the whole of that concept's content. But how does one investigate the conceivable effects of a concept? This should be done, of course, by a process of the same form as the Ab-De-In cycle. The pragmatic maxim is explicitly motivated by the structure of scientific experiment: a hypothesis is thrown forth (abduction), an experiment is constructed in speculation which will give such and such results if the hypothesis is true (deduction), the experiment is performed and supports or falsifies the assumption (induction). On the other hand, as is always the case in Peirce's philosophy, nothing is in logic which was not always already (albeit less controlled) in nature. So the Ab-De-In cycle has its naturalist counterpart in the trial-and-error of Uexküllian functional circles and maybe even – so Peirce – in process of evolution: variation (abduction), species consequences of variation (deduction), and natural selection (induction). Thus, on Peirce's account, scientific experiment, the clarification of concepts, literary interpretation, and biological evolution share the same fundamental trial-and-error structure – involving the crucial diagrammatical experimentation phase introducing ideal objects. The idea that interpretation so to speak has a biological structure is be relevant for the nascent science of biosemiotics (cf. Chaps. 9–12).

All of these processes might be collected under the headline of 'thought experiments' – assuming, of course, that 'thought' is here taken in an objective idealist sense rather than in a subjectivist sense. Thus, literary interpretation as thought experiment is intimately related to semiotic processes in nature and culture, while owing its specificity to the special properties of its object, the literary text.

CHAPTER 17

FIVE TYPES OF SCHEMATIC ICONICITY IN THE LITERARY TEXT

– *An Extension of the Ingardenian Viewpoint*

> When the universe of discourse relates to a common experience, but this experience is of something imaginary, as when we discuss the world of Shakespeare's creation in the play of Hamlet, we find individual distinction existing so far as the work of imagination has carried it, while beyond that point there is vagueness and generality.
>
> Peirce 'Multitude and Number', 1897, 4.172

We have have already encountered Husserl's theory of fictionality (defined by the 'quasi' attitude) as well as his gradual realization that fictions and pictures possess an ideal quality – to the extent that none of them are fully determinate and thus necessarily contain *Unbestimmtheiten.* These observations form the point of departure for his famous pupil Roman Ingarden who is well-known for being the phenomenologist who earliest and most thoroughly worked out the foundations of a phenomenologial aesthetics in general and a phenomenological theory of the literary work in particular.[426] What is less well known is that at the same time as Ingarden developed these possibilities within Husserlian phenomenology, he undertook this enormous work in order to correct what he conceived as a fallacy in the heart of phenomenology, namely Husserl's transcendental turn towards transcendental idealism.[427] Ingarden saw little difference between Husserl's understanding of fictitious objects after the 'quasi'-mode on the one hand, and the reformulation of his theory in *Ideen* where the object pole of phenomenological acts under the concept of *noema* was made a consciousness-immanent entity, on the other. So transcendental idealism, in Ingarden's account, treated the whole world as a fiction. The initial impetus for Ingarden's obsession with phenomenological aesthetics was thus to develop it to an extent where the crucial difference between real and fictive objets became evident. Thus, Ingarden's aesthetics, beginning with his chef-d'œuvre *Das literarische Kunstwerk* (1931) and continuing with *Vom Erkennen des literarischen Kunstwerks* (1938, 1966) and numerous other aesthetic investigations, has a double aim: one, of course, is to found an aesthetics on a phenomenological basis, the other, to provide ontological arguments for philosophical realism – such as it is finally laid out in the unfinished three-volume work *Der Streit um die Existenz der Welt* (1966–74).[428]

The literary work is thus placed as non-real as opposed to ordinary reality – and this crucial idea is argued on the basis of two observations: the *objects* to which it refers are quasi-objects merely, referred to by quasi-judgments. And the literary work *itself* is non-real, albeit in another meaning of the word: it is general, schematic, and (in a certain sense) ideal. While the former of these reasons is what allows Ingarden to counter the basis of Husserl's idealism, the latter is what forms the basis of his own theory of the literary work, a theory which is, in many respects, the direct heir to Husserl's own theories of fictions and pictures (cf. Chap. 14).

Here, we shall discuss Ingarden's detailed theory of the literary work with emphasis on the implications of his ideas about the *schematic* character of the work – or, to put in a Peircean vein, the *diagrammatical* character of it.

Ingarden's account for the structure of the literary work contains, as is well known, four *Schichten*, four strata, levels, or layers: the level of word sounds, the level of meanings, the level of represented objectivities, and the level of schematized aspects. As is evident, the three of these levels roughly correspond to aspects of the sign which have been discussed at least since Aristotle: its expression, its content, and its reference. The fourth stratum, the level of schematized aspects is not part of average sign definitions – but stems from a phenomenologically classical idea, namely Husserl's observation in *Ideen* that perceived objects may never be grasped in their totality but only through one or several out of a huge variety of possible 'profiles' or aspects, dependent on which point-of-views they are perceived from. The basic set of conceptual tools in Ingarden's account is thus surprisingly general and deals with perceptual and linguistic issues not specific to literature. Thus it is no wonder that something like 90 percent of *Das literarische Kunstwerk* is spent on the construction of an ambitious phenomenological linguistics – the main part of whose results covers any language use whatsoever. When Ingarden talks about 'the literary work', it thus also includes e.g. scientific treatises, and only a minor part of *Das literarische Kunstwerk* is spent on outlining the specificity of the literary work *of art* – in *Vom Erkennen* he further tries to single out the precise difference between scientific texts and literary works of art. When it comes down to what distinguishes a literary work of art from any old literary work, the answer is threefold, added on top of the linguistic theory: it is the special use of schematized aspects in order to making intuitive the represented objects; it is the polyphony between the ways the four levels are articulated in the particular work – and it is the evanescent 'metaphysical qualities' which the work – if successful – is able to manifest: the humorous, the tragic, the merry, etc. This is not to say, however, that the pheno-linguistic doctrine he presents as its foundation is trivial, quite on the contrary. Let us run trough the single levels of the construction.

The level of word sounds is probably the least problematic – even if an already classic criticism adds to it the fact that most literary works have their privileged form of existence as printed books, so that the graphic form properly ought to be included in this level. This is, however, possible without major problems. *Das literarische Kunstwerk* is written simultaneously with the first articulations of linguistic structuralism in Prague (which was, through Jakobson, heavily influenced by Husserl's

Logische Untersuchungen just like Ingarden), so it is no wonder that Ingarden's account for the level of word sounds bears resemblance to the structuralist idea of the duality between phonology and phonetics. To Ingarden, word sounds are, just like the other parts of the work, general – they do not consist of the physical acoustic sounds which are uttered in any single, inner or outer, reading of the work (nor of the physical instances of grapheme series constituting the correlating book, for that matter). The level of word sounds consists of *types* – the word sound type is constant from one use to the next, and thus different from the 'phonic material' in which it incarnates in each single case of expression. This 'ideality' (Ingarden would not use that word in this connection) or generality of the word sound of course has functional reasons. The word sound expresses one and the same meaning and hence must be as identical as the meaning itself. (38) On this proto-structuralist basis, a series of further ideas are added: a distinction between 'lifeless' as opposed to 'living', 'vibrant', 'powerful' words, of which scientific terms serve as an example of the former, while the latter have certain onomatopoetic qualities: '... the characters which arise from the purely phonetic 'Gestalt qualities' and move from the meaning onto the word sound reveal themselves more clearly in the word sound and color it, so to speak, with the quality that is characteristic of the designated object...' (44) This poetic function – to use Jakobson's terminology – plays a crucial role in the contribution of the word sound level to the polyphony of the whole work as well as in Ingarden's account for the schematized aspects. Phonetic formations of higher orders include all the well-known effects of rhythm and tempo, melody, rhyme, assonance, tone, etc. and the fact that they are all part of the contribution of the word sound level to the specific polyphony of the work of art is hardly surprising.

The analysis of the meaning level takes up a huge part of Ingarden's doctrine, and it indeed forms an ambitious outline of his phenomenological linguistics with its base in Husserl's 3rd and 4th logical investigations. It forms a first outline of what was to become Ingarden's ontology in *Streit* and provides the clue to language (and hence, literature) as a schematic construct. Like in Husserl's 4th investigation, the two autonomous meaning units are the noun and the sentence, respectively – while the syncategorematica, Ingarden's 'functional words' do not possess an autonomous meaning. And like in Husserl's 5th investigation, a composite act structure is taken as the basis of intentional directedness. Its version regarding the account of the meaning of names differs, however, significantly from Husserl's version: it consists of 1) the intentional directional factor (the reference being singular or multiply oriented (single-rayed or multi-rayed), being actual or potential, being stable or variable, etc.), 2) the material content (the qualitative constitution of the object), 3) the formal content (the ontic status of the object – it being a substance, a thing, a quality, a state etc.), 4) the moment of existential characterization (whether the object is real, ideal, etc.), and in some cases 5) the moment of existential position (whether the object is a fact or a fiction). Verbs, on the other hand, are completion-requiring – not unlike Peircean rhemes.

An important notion here is the distinction between the actual and potential 'stock of meaning' (84). Any specific use of a word actualizes an aspect of its ideal sense – and the other merely potential aspects of meaning may now be indirectly present in intonation in spoken language, and, in written language, by means of reference to the 'system of meanings' to which the word belongs, and by means of the localization of the word in different contexts, each of them selecting among the possible stock of meaning. (88) These indirect ways of indicating possible meanings Ingarden calls 'suggestions' or 'readily potential meanings' as opposed to possible meanings not so indicated – in the course of a work, a word may collect many such suggested possible meanings. This is supposed to cover both material and formal aspects of meaning – and this whole idea becomes important in order to understand Ingarden's difficult notion of 'schematized aspects'.

First, it is this outline of a linguistics which permits Ingarden to ask the questions leading to his basic idea of the ontology of language and of literature. For such meanings of words, sentences, and sentence complexes, are they not ideal objects? In a certain sense, the most decidedly are. They are identical from one use to the next (as a prerequisite to the possibility of understanding), this identity is not numerical (*Faust* remains one work, whether it is printed in a circulation of 2,000 or of 200,000), their kind of existence is not physical (if it was physical, then each single copy of *Faust* would be a different work), it does not exist at a single, specifiable location in time and space. This was what lead Husserl to assume the ideality of meanings (and furthermore to outline the (later abandoned) species theory of meaning in the *Logische Untersuchungen* with the idea that the meaning of an expression is simply the species of the correlated act). Ingarden admits all these arguments against meanings and literary works being real objects. But still they can not be ideal objects in the same sense as logical truths, geometrical objects, or mathematical regularities nor material essences – for they come into being at certain specific points of time in history, they may disappear again, and, unlike these genuinely ideal objects, they invariably depend on subjective intention for their existence: they require the existence of speakers, authors, and readers. If they were really ideal objects, they would not come into being at specifiable points and would, in some sense, have timeless existence, only waiting to be discovered by the author. Hence they form prime examples of Ingarden's third ontological category, in some sense intermediary between ideal and real: *pure intentional objects* (100). They are heteronomous, because they depend on subjective intentions for their existence, they require physical support for their existence, and they may refer, in turn, to any kind of objects, be they real, ideal, or imagined. It is they that generally constitute the basis for the reference of propositions. In ordinary cases, the pure intentional object so to speak remains invisible, because the object referred to is seen through the correlated pure intentional object. Only in cases like faulty reference or fictitious reference does the pure intentional object become visible as such.

Even if pure intentional objects are ultimately dependent on the intentionality of acts of consciousness, the ontic relativity of the objects represented does not refer directly back to such intentionality, but to an intentionality immanent in the units of meaning – constituted by the fact of communication and of its necessity to be able identically to reproduce a meaning from one occasion to another. This shift implies an advantage as well as a loss. The advantage is that meanings, unlike e.g. perceptions, cease to be completely subjective formations and thus become intersubjective – the loss is that the lack of richness and completeness, dependent on single intentions, makes of the meaning a 'skeleton', a 'schema' only. (126–27) Thus the schematic character of the literary work lies already in the basic linguistic schematization of meaning. To say that schematization is a loss, however, might be seen as a strange *façon de parler* – it should rather be seen as a gain, because it is schematization that permits thought economy, plasticity of reference as well as experimentation possibility. These enormous benefits, it is true, are bought at the prize of a loss of richness.

Pure intentional objects imply a special use in Ingarden of the phenomenological notion of *Sachverhalt* – state of affairs. Thus he claims that the pure intentional objects constitute in themselves states-of-affairs – no matter whether any real or ideal correlate exists or not. On the other hand, such correlates are, if they exist, states-of-affairs in themselves.[429] The former use of the concept of states-of-affairs thus gives rise to subtle determinations like the following: '...the states-of-affairs created and developed by the sentence is transcendent with respect to the sentence content yet, according to its essence, belongs to it.' (116) Such 'purely intended states-of-affairs' differ from states-of-affairs pertaining to an 'ontic sphere' (real or ideal) being autonomous in relation to the sentence, and even if their existence is heteronomous in relation to the sentence and finds its ontic basis in sentences, they are, on the other hand, transcendent in relation to them.

This forms the basis of the next level, the level of represented objects or objectivities. They form the only 'thematically apprehended' stratum of the work (217) – that is, the whole edifice apart from the objectivities are only implicitly part of the immediately experienced work. The reader focuses upon the objects dealt with in the work. The object[430] is, of course, not directly grasped from the sentence meanings of the text. These sentence meanings refer to, as we have seen, states-of-affairs, and it is these states-of-affairs which permit, in turn, the access to the objectivities represented. The purely intentional sentence correlate is not – unlike the existing state-of-affairs it may refer to – completely determined (142). As state-of-affairs, it is already schematic – or ideal in the sense of the word that it may be repeated identically. This implies, on the other hand, that it may be ambiguous – it may, for instance, contain mutually contradictory parts which gives it what Ingarden calls 'opalescence' or 'iridiscence', fluctuating indeterminability.[431] The status of the object level depends on the special character of the propositions realized in sentences on the meaning level. They organize a continuum from referring only to purely intentional objects in one end and to referring to autonomically existing objects in the other. The correlative propositions Ingarden names 'pure affirmative' and 'genuine judicative' propositions, respectively – the former roughly corresponding to Meinong's 'assumptions' ('Annahmen') which do not involve any existence claims.[432] On this continuum, the quasi-judgments characteristic of the literary work are located.[433] The 'ontic setting' of a real world of genuine propositions is maintained, but the 'matching intention' (between pure intentional and existing objects) and 'identification' (with a sphere of existence) are left out. The intentionally projected states-of-affairs thus form 'their own world' – which may deviate to larger and lesser degrees from the real world. The purely intentional states-of-affairs already belong to the object stratum (188) – they form so to speak the bridge from meaning stratum to object stratum: 'Since the same object can be revealed in various differently constructed states-of-affairs – since the state-of-affairs are like many windows through which we can look into one and the same house, each time from a different standpoint and from a different side, into another part, or finally, for a second time through the same window – a certain cleavage occurs in the 'stratum of objects' of the literary work. In their representation function the states-of-affairs are that which represents, while the objects constituted within them are the represented. But since the state of affairs is at the same time something which belongs to the proper ontic range of the object (constituted within it), this representation is in the final analysis a *self*-representation of the object in what belongs to it.'

We give this long quote because it pinpoints the difficult relationship between meaning and object levels – and at the same time involves the fourth and last level, the stratum of *schematisierte Ansichten*, of 'schematized aspects'. We so to speak look through the states-of-affairs (cf. the window metaphor) in order to reach the represented objects 'and have them as *given*.' (191), while the medium itself is usually not thematically in the foreground, but remains in the background. Another metaphor for the same is that the different states-of-affairs so to speak capture the objectivities: '... the states-of-affairs, figuratively speaking, merge into a 'net' in which the given object is "ensnared" ' (157), or that it is as if 'a beam of light were illuminating a part of a region, the remainder of which disappears in an indeterminate cloud but is still there in its indeterminacy.' (218)

This implies a delicate relationship between the states-of-affairs and the objectivities. The latter is constituted out of a manifold of connected states-of-affairs. The way they connect these states-of-affairs is seemingly one of continuous merging: the limits between the single states-of-affairs pertaining to the same objectivity are artificial, and they are fused into one complex. This process thus forms a formal equivalent to Husserl's famous analysis of perception and the synthesis of the perceived object from profiles; the object, on its side, being able to give rise to potentially a continuous infinity of profiles. But unlike the case in perception, the basis of the synthesis, in the work, does not form a continuous manifold. Rather, this manifold consists of single, isolated sentences with each their meaning and state of affairs correlate – which implies that the object, in Ingarden's metaphor, is so to speak 'dispersed, like a ray of light in a prism, in a discrete manifold of distinct, though connected, states-of-affairs.' (159) – even if this gives us the possibility of selecting a varied range of different, but still only individual states-of-affairs (198).[434]

Thus, represented objectivities – that is, pure intentional objectivities – differ from real objects in a radical manner. Ingarden examplifies this in his interesting accounts of represented time and space which both differ from real as well as perceived time and space by being schematic because constructed from a finite number of meaning units. While real objects – unlike the case in Peirce – are supposed to be unanimously and universally determined, with all determinations fitting consistently into one whole, and being absolutely individual, so that ideas in them are incarnated or individualized at the lowest possible level of differentiation – pure intentional objects, on the other hand, are schematic and thus not thoroughly determined, they are not necessarily consistent, and ideas in them are not necessarily individualized.[435] This implies, all in all, that purely intentional objects are schematic. This gives rise to the famous *Unbestimmtheitsstellen*, the spots of indeterminacy, or, as Wolfgang Iser would later call them, the *Leerstellen* – of which there is an infinity (or, as Ingarden would later say, a very large number). The supposedly real object in a quasi-judgment hence involves that '... only a formal schema of infinitely many spots of indeterminacy is projected, but almost all remain unfilled.' (250). This forms the basic reason why the literary work is 'in principle incomplete' and thus in need of further supplementation (251) – but it is important here to stress that nothing specifically literary has yet been said by this. It goes for any purely intentional object as such and thus for all objects projected by meaning, hence by all language and all (supposedly) sufficiently complex sign use. The reason why we do not experience these spots while reading is threefold: the represented objects are 'visible' only through the aspects determined by the choice of meaning units; some of the spots are covered by schematized aspects, and, finally, the reader supplies his own selection of concretizations and thus in a sense 'goes beyond' what lies in the text (this giving rise to the important distinction between the literary work and its concretizations which must not, in turn, be confused with their product: the reader's psychological experience). The flip side of this is the possibility for the work of making radically *new* objects: objects of impossible objects or events, transgressing regional ontological boundaries, making inconsistent states-of-affairs – even until 'two different worlds are struggling for supremacy.' It follows from this that Ingarden's final aesthetic criterion of polyphonic organicity is not to be taken for any sort of glib harmony, as it has sometimes been maintained as an argument against him – quite on the contrary, such clashing phenomena on the object level may constitute part of the work's aesthetic quality. The reader's activity in selecting concretizations of such spots forms a major issue in Ingarden's follow-up work *Vom Erkennen des literarischen Werkes*.

That such a concretization is performed in reading the literary work follows from the fact that we most often do not experience its objectivities as indeterminate; we experience them actually as analogues to 'whole', naturally perceived objects. It is as if the 'net' of states-of-affairs catchting the represented objectivity merges into a continuous sheet, meticulously following the shape of the object. How is the finite set of purely intentional states-of-affairs in the literary work to some extent 'restored' to yield something analogous to a perceived object?

The reason for this lies in the fourth level, that of schematized aspects. Shortly described, this stratum is that which makes it possible for us to reach intuitively satisfactory, perception-like syntheses of the objects represented. Still, its basis and ways of functioning belong to the most difficult and least clarified parts of Ingarden's doctrine. An otherwise diehard Ingarden fan like René Wellek simply bluntly dismisses the existence of such a level and subsumes it under the represented objectivities.[436] Already in the account for the synthesis of purely intentional states-of-affairs, two important ideas stand out. Ingarden distinguishes three types of states-of-affairs, pertaining to appearances, essences, and occurrences, respectively (*Soaussehensverhalten, Soseinsverhalten,* and *Geschehensverhalten*). In a few pages (192–96), Ingarden outlines a trans-phrastic theory of the interlinking of such states-of-affairs over the sentence level: while the essence states-of-affairs constitute a kind of kernels, the appearance states-of-affairs allow the connection of them to 'a possible subject', while the occurrence states-of-affairs with their event quality permit the interlinking of essence states. The occurrence states '... have the capacity to exhibit the objects partaking in the respectively developed occurrences to a much greater degree than do pure states of essence.' (195) – an almost pragmatic (in Peirce's sense) idea: effects are what matters in the understanding of an object; states of occurrences 'to a certain extent *exhibit themselves*' (ibid). They thus facilitate the formation of a system of 'realized mutually interconnected states of essence' (ibid.) and contribute to the 'representing, and frequently also to the exhibiting, of entire segments of the world of objects.' (196). Even if the three state types mentioned may thus form an interconnected web of states (governed, of course, by all kinds of narrative, thematic and other gestalt structures on the object side, we may add), they still not suffice to mend the finitude of state-of-affairs nor enter into full intuitive givenness like in sense-perception.[437] This 'intuitivization' (197) in imagination is channeled by the interconnected states-of-affairs – and by the schematized aspects.

'Aspect' of course refers to Husserl's analysis of perception and the objects presenting themselves through a continuous manifold of perceived profiles, 'Ansichten' or aspects – so that there is a '... strict affiliation between every perceptually given property of a thing and the manifold of aspects, strictly ordered, according to rules, in which the given property appears.' (262). This implies that the very idea of a 'property' is, in itself, a certain 'idealization' of the set of profiles, it is in itself a '*skeleton, a schema*, of concrete, flowing, transitory aspects' (no two concrete aspects being identical – while one property may be identical from one object to the other – so already in the concept of invariant property a high degree of generality is present). This property obtained by idealization *is* the schematized aspect – and the single, concrete aspects are now variations within the limits set by the schematic aspect to which they belong. In the work, they have the 'basis of their determination and, in a certain sense, their potential existence in the states-of-affairs projected by the sentences or in the objects represented by means of the states-of-affairs.' (264) – in any case, their immediate base in the work is the states-of-affairs. Such schematized aspects lie predetermined in, e.g. a novel's representation of Paris, and must be filled in – to a larger or lesser degree – by the reader. Just like the case with perceived objects, all schematized aspects in which they may be given belong to the represented objects – but, and this is Ingarden's crucial hypothesis at this point – they are potential only and in order for some of them to be evoked and reach a degree of intuitive givenness, other factors than the represented objects must intervene. Some of the schematized aspects of the represented objects are to some degree 'held in readiness', prepared for actualization. The factors determining this partly lie in the states-of-affairs: the thus-appearance of objects and the exhibiting as against the simple representation of the object. Such aspects thus lie in between potentiality and actuality (obviously, the Peircean notion of 'real possibility' of tendencies and patterns is relevant here). And they partly lie in the well-known figural tools of poetic language: metaphors, images, similes, etc. And finally they lie in Ingarden's favorite (but rarely concretized) example of special word sound patterns, onomatopoietically suggesting certain schematized aspects of an object to be selected, or word sounds referring to specific contexts of linguistic use (that is, referring to different Bakhtinian discourses).

The common ground for these only sketchily described bases of the ready-making of certain schematized aspects is as evident their *indirect* character. Direct reference to schematized aspects, on the contrary, immediately makes of the aspect an *explicit* object (which may then lie open for study in different ways), but this does not result in the (partly) intuitive givenness aimed at, because the object originally referred to will disappear and the aspect taking its place as object instead (this, in fact, will be a case of Peircean 'hypostatic abstraction'). The objects are what is described, but to some extent this description takes place *through* the schematized aspects held in readiness. This forms, as it were, Ingarden's version of the old Hollywood motto: *show, don't tell* – the latter being equivalent to the making explicit the aspects held in readiness. But the aspects themselves are 'jumpy', just like the interconnected web of states-of-affairs (again, as against continuous, perceptual aspects) – and most often only a few selected aspects are held in readiness. Moreover, they are actualized in imagination only. This is why the degree of vividness of actualized intuitivization changes all of the time during reading – sometimes objectivity stands out rich with details, other times it shrinks to a skeleton-like structure, and only the important, implicit 'stabilization' of aspects prevents us from thematically realizing this highly changing appearance of the object, thereby shattering the truthfulness of the objectivity represented. The same goes for the 'murky cloud' from which the actualized aspect emerges (as against the perceived foreground object's embeddedness in its perceived background surroundings). By analogy, this aspect doctrine may be extended to cover our access to mental states in other subjects by means of 'internal aspects' making the 'soul' referred to appear in a certain determinate state.

A certain enigma seems to be at stake here: the finite patchwork of states-of-affairs was in need of continuous supplementation in order to be able to intuitively coalesce in a simulation of real perceptive experience. But how should this be accomplished by means of *another* finite set of patches, not even integrated to the same degree as the states-of-affairs, in the form of schematized aspects? It is as if we expected to get a continuous sheet out of the superposition of two patchwork blankets. Continuing in this metaphor we might say that a superposition of two patchwork quilts would ensure that we encounter a continuity at most points (if the two are suitably arranged), if not in the one patchwork, then in the other. So we would so to speak float along on shifting layers of continuity, and the bumps between the states-of-affairs would be (most often) eased by continuity in a stabilized schematized aspect...

In any case, the contrast picture which Ingarden paints of a text deprived of schematized aspects is frightening: the represented objects would be purely blankly intended in the reading and merely thought in a completely unintuitive manner – they would be empty, purely 'conceptual' schemata, and one would never have the impression that one was dealing with a 'unique, live quasi-reality.' (277) Thus, the stratum of schematized aspects plays a major role in the literary work of *art* – in fact, this forms part of Ingarden's main description of the artistic character of the work, along with quasi-judgments and metaphysical qualities, while all of the other features of the work, including the spots of indeterminacy, remain characteristic of any linguistic or other utterance whatsoever. Ingarden's chapter on the role of schematized aspects in the literary work is short, however, and strangely poor. The basic points remain that the aspects adorn the objects with real-life-like vividness and richness – even to the extent that they may co-constitute them – and that the aspects add to the polyphony of the whole work their own decorative contribution.

This brings us to the final set of ideas in Ingarden's account: the aesthetic qualities of the work as a whole. One side of this is the said polyphony of the strata, merging into one aesthetically pleasing organic whole (including aesthetically pleasing tensions, contrasts, paradoxes, inconsistencies, impossibilities, etc.). The idea is here that the contributions from the single strata supplement each other in some way, and it is probably the case that no further precise determination can be made of what makes such a whole aesthetically successful in general. The other side of it deals with the famous 'metaphysical qualities' which are essences that pertain not to single objects but to whole situations or events – let us give Ingarden's often cited example list: 'the sublime, the dreadful, the shocking, the inexplicable, the demonic, the holy, the sinful, the sorrowful, the inescapable brightness of good fortune, as well as the grotesque, the charming, the light, the peaceful, etc.' As was evident in the simpler case with the aspects, these qualities are not in any way the direct object of the literary work. They are qualities which in some sense pertain to the object level (even if colored by the whole of the work's polyphony), but the whole development and structure of the work in some sense embody these qualities and allow them to appear. They should not, of course, be confused with the psychological reactions they may arouse in the reader. In his description of these qualities, Ingarden's dry, phenomenological tone almost gives way to a vitalist, romantic sweep: the metaphysical qualities have a character of 'revelation' which puts them into contrast to the 'gray, faceless, everyday experiences' (277) – and they do not allow purely rational determination.[438] They may be experienced only in appropriate situations, on which they must be presumed dependent, and then reveal a 'deeper sense' of life (278). The object stratum of the work being heteronomous (dependent upon the meaning level), the qualities associated with this stratum can not be fully realized, but still they may be 'concretized': 'they simulate their own realization' as Ingarden says (294) – which is what permits the reader, on the other hand, to contemplate them relatively calmly, in the well-known aesthetical distance.

THE EXTENSION OF THE SCHEMATIC ASPECT CATEGORY

Now, a whole series of questions remains as to the relation between the synthesized objects caught in the net of states-of-affairs – and the schematized aspects, tightening this net and making it pseudo-continuous. Let us take a look at some of the main problems.

Some of the problems are indicated, more or less directly, in Ingarden himself. His concept of concretization of course refers to the filling-in, by the reader, of spots of indeterminacy along with the actualization of aspects held in readiness. This concretization is what provides the reader's experience with a more rich, detailed, perception-like version of the work than the schematic construction explicitly present at the meaning level, and it requires the active participation of the reader's imagination. Still, we should not think that the concretization has the character of rendering the objectivities in the text fully determined and with full perceptual clarity. The case is rather that a very schematic version is substituted by a less schematic – but still schematic – version. Much of the filling-in must be assumed to take place by Peircean 'pooh-pooh'-induction tacitly inferring the 'standard' or 'normal' version of things or aspects not explicitly mentioned in the text – but this 'normal', default version has, in itself, general or schematic character and, moreover, this filling-in is by its very nature ordinary and un-pregnant. A related issue is the fact that far from all spots of indeterminacy may be filled-in. In some cases, furthermore, the possibility must be regarded that too much filling-in will spoil the artistic effect (Voltaire: the secret of boring is to tell everything), if for instance the filling-in removes an ambiguity or a generality which formed part of the aesthetic polyphony (while in other cases, the removal of ambiguities may be required and, on the contrary, add to the polyphony).

Ingarden (1968) thus envisages two different ways of reading: one – close to the spontaneous reading – concretizing, as opposed to a more clinical, analytical reading which consciously seeks to avoid concretizing and keeping the work in its most schematic form possible. Of course, we can add, these two possibilities must constitute the two ends of a continuum with different intermediary possibilities selecting certain types of indeterminacies to be or not to be filled in, and selecting certain aspects to be actualized or not. Furthermore, the 'pooh-pooh' filling-in makes it necessary to distinguish two types of filling-in, the one thematic and with some degree of perceptual richness, the other one unthematic and thus not perceptually conscious, but rather consisting of the tacit co-representation of implied properties.

This leads to the most important set of issues: how many phenomena in the text have the schematic character admitting spots of indeterminacy – and, correlatively, what in the stratum of meaning provides the foundation of the holding-in-readiness of aspects? Ingarden himself is rather vague on these issues, as we have seen, and these aspects of his theory most decidedly requires a filling-in relating his doctrine to other results in semiotics and text theory.

Ingarden's theory, as often pinpointed, has its immediate background in Husserl's work – the foundation concept, the act theory and the (much-disputed) realism of *Logische Untersuchungen*, the perception theory of the *Ideen* I-II, the meaning theory surfacing in *Cartesianische Meditationen*, the quasi-mode of the writings published in *Hua* XXIII, etc. (cf. Chaps. 6–8, 14). The schematized aspect theory having its precursor in Husserl's doctrine of perception from *Ideen* I and the 'Sinnesschema' of *Ideen* II, places the schematized aspects very close to perceptual experience. But does this necessarily have to be the case? As discussed in Chap. 6, Husserl's theory of *kategoriale Anschaaung* implies that also categorial – schematic – objects possess their own mode of direct givenness. So do schemata not have aspects? We may address a circle by focusing on its periphery or on some select points on it, on its center, on a circle segment, on an inscribed polygon etc. – these are aspects of the circle in analogy to the different profiles or aspects of a perceived object. We may be able to grasp the entire circle from one such aspect, just like we see the complete object even from one apparent aspect only. The same goes even for abstract complexes not being of the geometrical kind which remains still rather close to perception.

Take for instance scripts or frames – Schank and Abelson's well-known restaurant example (1977) with a script connecting all the aspects of going out eating: selecting a place, ordering a table, going there, waiting to be shown a table, reading and deciding the menu, calling the waiter, ordering the meal, eating and drinking in a specific order, calling for the waiter to get the bill, paying, tipping, leaving – with side-scripts (going to the toilet, meeting friends and pulling tables together, leaving because the food was awful, etc.). This empirical gestalt configuration evidently has a general, schematic structure. The interesting thing in Ingardenian context is that such a schema permits *both* to function as provider of enormous amounts of spots of indeterminacies (ranging from the location or interior of the place over the other guests to menu details etc.) which may be filled-in in a more and less detailed manner, *and* to function as supporting aspects held in readiness – because very few words referring to such a script are sufficient to evoke the whole script in the mind of the reader (if the words 'table', 'menu', 'waiter' surface in adequate combination, the whole restaurant script may be actualized as *abstract, schematic* aspects held-in-readiness). It seems like Ingarden's focus on perception on the one hand and on word and sentence meanings on the other precludes him from seeing structures like this.

Ingarden's own – rather few – examples of filling-in and aspects held-in-readiness are thought-provoking as to which schemata actually function in them. Take, e.g., his discussion of the corridor walk in Thomas Mann's *Zauberberg* as a successful piece of continuous time representation – here, '... we not only find a manifold of successive, closely interconnected states-of-affairs representing an almost uninterrupted story, but, at the same time, closely interconnected aspects of corresponding objectivities are imposed on us. [...] ... we see almost continuously before our eyes the pertinent objects in appropriate aspects: first the room, then the corridor, etc. As we pass by, we see everything as it would appear sequentially in reality.' (283). What grants the unity of this series of aspects, of course, is not only their sequential ordering in the text (disorderly, expressionist glimpses which are discussed as counterexample, may occur equally ordered), but the mapping of this sequential ordering from the meaning onto the object stratum, kept together by the gestalt figure of source-path-goal. This 'kinaesthetic image schema' (Lakoff) here functions as a sort of abstract schema, implicitly present only, which forms part of the filling-in operation necessary to grant the continuous experience of the two patchwork sets of separated states-of-affairs and separated aspects. Another example displays a similar feature in the discussion of represented space which is '... as if pocked with gaps, which show up as, so to speak, spots of indeterminacy' (224). This conclusion is reached on the basis of the example of a room in a novel: '... a situation is represented as taking place in a given room and that there is no indication, even by a *single* word, that there is anything outside of this room. [...] If the actually represented space (within the room) does not end at the walls of the room, *it is only because it is the essence of space in general not to have any discontinuity.* It is only through this impossibility of spatial discontinuity that the space outside the room is corepresented: in turn, the space within the room is *co*represented; in turn, the space within the room consequently becomes a *segment* of space.' (223) But this conclusion implicitly relies on another gestalt: the schema which Lakoff nicknames 'container', consisting of an inside and an outside separated by a closed boundary. This schema is what makes possible the corepresentation.[439] But where should we seek the foundation of such schemata? Probably in the meaning level: the concept of room prototypically involves the container gestalt, just like the concept of walking along a corridor prototypically involves the source-path-goal gestalt.

The second base of Lakoffian cognitive semantics involves a ripe source of indeterminacy and filling-in possibilities as well: that of the basic level concepts of the psychologist Eleanor Rosch. A certain medium level in concept structure is taken to be basic, that involving cars, chairs, tables, dogs, running, walking, speaking, red, green, etc. The idea is that basic level concepts form the uppermost level where a common schematic image for the corresponding object, aspect or activity is possible, the uppermost level where a schematic motor program for interacting with the object, aspect, or activity is possible – which is why such concepts are easiest to learn and in most languages are associated with short, central word sounds. This opposes basic level concepts to more specified concepts on the one hand, adding further specifications to the overall concept (Chesterfield chairs, bulldogs, stepping) and more general concepts losing the common schematic picture and action schemata (furniture, pets, moving). This intermediate position – exactly like in the restaurant script example – makes basic level concepts good candidates *both* for holding large amounts of indeterminacy spots (which color, shape, age, etc. of the chair) *and* for holding aspects in readiness thanks to the detailed perception and action schemata associated with them ('he went among the furniture and sat down' permits us to imagine the typical bodily movement series and an average default chair associated with such an act, even if the latter has not been mentioned at all). The intermediate status of basic level concepts permits them to function 'both ways', so to speak, probably in different combinations in different contexts – both as general schemata and as providers of relatively specific imaginations of objects and actions.

The examples highlighted here thus span from transcultural gestalts like container and source-path-goal over basic level concepts with their mixture of crosscultural (body parts, animals, mothers, eating, sleeping, walking, etc.) and culturally central objects and practices (cars, driving, teepees, houses) and to highly culturally specific schemata like the restaurant script. To the former category must be added narrative regularities like the Greimasian narratological molecule of Hero, Witch, King, Dragon, and Princess (more technically, Subject, Helper, Destinator, Anti-Subject, Object) which also seems to be transcultural and functioning as a cohering factor in large chunks of text, interconnecting the states-of-affairs of many sentences. Also here, very few actualized aspects of this basic schema easily and almost automatically permits the reader to fill-in the whole structure of the narrative conflict, its many aspects and different possibilites of development. To the latter category must be added whole ideologies and *Weltanschuungen*, Islam, Christianity, Communism, Capitalism, etc. which form enormous, schematic complexes which may be actualized by very timid means (a sickle moon, a cross, the word 'proletarian', 'shareholder' etc.). Thus, the insight that the aim of filling-in of schematized aspects does not only involve sensuous schemata gives us an enormous amount of conceptual schematic structures held in readiness adding to the vitalization of the text during the reading process.

We shall not go further in this direction here, but a basic result seems to be that the somewhat crude duality between abstract and concrete in Ingarden's theory of concretization should probably be substituted by a continuous ladder of more and less abstract schemata (thus explaining the fact that many schemata may serve both as indeterminacy sources and as aspects-in-readiness sources).[440] The Austrian economist and philosopher Friedrich von Hayek who also was a philosopher of mind, has an interesting proposal in this context. His idea is that the whole of the mind's activity, all its variety and difference untold, has one common characteristics: categorization. The mind at all its levels involves categorization, of input on many levels of refinement, of output, of bodily dispositions, etc. – which implies that the contents of the mind, when they reach the threshold of consciousness, are always, at least to some extent, categorized. This he coins in the counterintuitive but highly interesting hypothesis of the 'primacy of the abstract' which we also discussed in the biosemiotics section.[441] Against the idea that perception is through-and-through concrete and that its possible abstract aspects or contents are added or distilled only by later, abstracting processings of the primary, concrete content, Hayek claims that what appears before the mind is invariably infused with abstract – that is, schematic – structure from the very beginning. Thus, ordinary perception is *already* ripe with abstract aspects (also a Peircean idea)– we need not perform any complicated, secondary, founded act in order to isolate these. People we pass by on the street, e.g., appear to us schematic, we do not notice their exact dress, facial expression, walking style, etc. unless now and then one of them are saliently singled out for further concretization. Rather, the degree of concreteness in ordinary experience is highly variable and dependent on ongoing shifts in attention distribution.

This implies, in our context, that the concretization operations in literary reading become easier to understand. If ordinary perception is *already* to a large extent abstract and involves constant changes in concretization, the distance so to speak becomes less drastic (but still large enough to require concretization and filling-in operations) than was the case if perception was 100 percent sensuous, particular, and detailed.[442]

But if these different schematic aspects of the text are indeed diagrammatical, in Peirce's sense of the word, then we should expect them to function as possible vehicles for diagram experiments. Indeed, in the previous chapter we developed the idea that, in reading or analyzing a text, the deductive reasoning phase implied more or less explicit cases of such experiments. In ordinary filling-in reading with its thematic focus on the level of represented objectivities, such experimenting is probably most evidently found in genres like detective novels, agent novels, thrillers, fairy-tales, fiction prose in general, where the experimenting with the possible outcomes of plot structures form a crucial part of the reading process – very often aided by the explicit diagram experimentation undertaken by a represented figure (a detective) which the reader may follow with awe or with the superiority of better insight provided by the narrator. This explicit, thematical case only forms, however, the tip of the iceberg: any case of deductive reasoning performed during the reading, thematically or unthematically, forms a diagram experiment in Peirce's sense. The relevance of such experiments adds further details to Ingarden's account for the filling-in procedure – it not only covers the dressing up schematic conceptual skeletons with imagined perceptual clothing, but it also covers a variety of logical inferences, thereby adding to the (quasi-) statements being explicitly represented in the literary work – on the basis of the schematic structures present in the work. Thus, the reader not only equips the work with much more sensuous illustrations than are explicitly present, he also develops the work logically, by thought experimenting from its schematic basis. This becomes strikingly evident in the art of suspense where the reader may be able to infer to the existence of some vague threat not being explicitly referred to in the text, or where informations present in the text allow for the drawing of a conclusion which is not (yet?) available for characters in the represented world with their delimited point of view. But still, these explicit examples are only salient representatives of a process going on at a much broader scale.

A good example of literary diagram experiment is provided by *metaphor* which has been widely discussed in literary theory and cognitive science of the recent decades. Both the metaphor theory of Lakoff and Johnson and the related blending theory of Turner and Fauconnier highlight the fact that in such textual phenomena as metaphors and blending, schematic objects (structures mapped from source to target space in the former, 'generic space' facilitating the integration of different inputs in a blend in the latter) play a central role. Metaphors thus also permit the duplicity of indeterminacy and holding-in-readiness – a source of indeterminacy relies in the fact that the text only rarely makes explicit the details of the mapping between source and target (and in some cases does not even make the target clear) – and a source for aspects held in readiness lies in the amount of semantic possibilities in the source concept which is *not* actualized in the given mapping, giving rise to the excess-of-meaning-experience so often associated with metaphor. As is evident from the theory of metaphors emerging from the cognitive semantics tradition, metaphor covers a continuum from basic, everyday expressions which most often pass unnoticed ('I see.' as metaphor for 'I understand.' due to the metaphor UNDERSTANDING IS SEEING), and to sophisticated, elaborated or combined or nested cases in literary works. While the former are processed automatically, the more complicated cases are less automatized. Already in rather simple cases, the understanding of metaphor requires thought experiments. Most metaphors, e.g., do not make explicit the range covered by the mapping involved, neither as regards the number of aspects of the source phenomenon to be mapped upon the target, nor as regards the possible generalization of the mapping to cover related phenomena. If my love is a rose, it goes without saying that the beauty and fragrance of the rose are among the aspects mapped, but what about the rose's verticality, its blooming during summer, or the fact that it thrives well on a diet of horse manure? And if the metaphor is accepted, is the possibility open that HUMAN BEINGS ARE FLOWERS in general, so that my love contrasts to other girls being petunias, tulips, thistles, and potatoes? Even a simple metaphor as that mapping rose upon girl may involve such thought experiments of diagrammatical nature, and much more complicated metaphor craft most often require elaborate thought experiment in order to be understood. The understanding of all sufficiently complicated schematic structures in the literary text, both on the level of themee, narrative structures, enunciation, style and rhetoric require diagram experiments. This highlights an aspect of the literary work of art which is underestimated in Ingarden's account: that of the literary work as a sort of thought-machine, requiring the reader to fill in inferences from the schematic structures in the text.[443]

The most important implication of seeing Ingardenian schemata as diagrams, however, is the Peircean conception of generality as continuity. Indeterminate spots are *unbestimmt* all right, but Iser's renaming of them to *Leerstellen* is a rather unlucky terminological choice. Such spots are *not* empty but remain filled with continuous generality involving the knowledge that variation experiments *could* be undertaken to give them further flesh. The Peircean-Hayekian idea that ordinary perception is perfused with generality makes it less strange that general spots of indeterminacy easily shade into proto-perceptual experience – there is no absolute distinction between perceptual and poor schematicity, rather, the two approach each other on a continuous scale. This is also why the filling-in of an indeterminacy spot in a diagram may be satisfied by *another*, more detailed diagram. We do not have to require filling-ins to be completely determinate; in many if not most cases a schematic filling-in is sufficient.

REALISM, TRUTH, AND 'IDEAS' IN THE LITERARY WORK

Ingarden's monumental work is, as mentioned, intended as a support for his position in the idealism-realism schism which he saw develop between himself and his master Husserl. His solution based on the analysis of literature as fictitious texts led him to assume three major object spheres: that of real, that of ideal, and that of purely intentional objects, the former two being autonomous, the latter not so. This, as we have seen, also made him develop his own brand of Husserl's act theory.

A counter-question must here be posted: literature now functions as the contrast case to acts aiming at real or ideal objects, as the phenomenon which thanks to its fictivity permits the discovery of the region of purely intentional objects – but does it not thereby make literature so to speak *more* fictive than necessary?[444] Where does it leave the common notions of literary 'realism', the idea of a 'truth' in the literary work, etc.? Ingarden closes his book by pondering over these issues. Of course, the very definition of the literary work of art by quasi-judgments precludes the possibility that the work should in any sense contain a true judgment in the proper sense (and the idea that this might be the case may have been more widespread when Ingarden wrote *Das literarische Kunstwerk*; nowadays it is an error rarely committed.) Still, there are three more loose senses of the word 'true' in which a literary work could be said to be true, as Ingarden claims. One is its function as reproduction (do the represented objectivities resemble certain real objectivities which they intend?), one refers to the work's 'objective consistency' (Do its represented objectivities cohere?), and one refers to the manifestation of the already discussed 'metaphysical qualities' (the latter are also Ingarden's interpretation of the commonplace of the 'idea' of the literary work of art).

These ideas require some comments. The first idea of the truth of the work, that of resembling reproduction, is placed by Ingarden as relevant only for historical novels, that is, for a very small part of literary works in general. But why reserve this notion of truth for the resemblance to a given empirical segment of the world? This truth may be generalized to covering also *types* of objectivities – does the work, e.g., satisfactorily recreate tendencies, structures, atmospheres of a given age? Such types range, in themselves, from clear to vague, and some of them may be close to Ingarden's metaphysical qualities, but it cannot possibly be the case that all attempts at describing central structures of a country during a period possess the existentialist insistence of 'metaphysical qualities'. This idea of truth is similar to the superficial idea of realism or iconicity in French structuralism and post-structuralism, e.g. in Greimas and Barthes who differ in many respects but whose accounts for 'reality effects' have much in common: it refers to the text's surface and its make-believe by means of the detail of typical reality-references, superimposed on top of more important layers of the text, be they structuralist semantics or streaming textual pleasure. Ingarden's account does not have the derogatory character of these French anti-realists, but still this kind of truth plays a (too) marginal role also in his account. To take a trivial example, what about key novels where main characters may be more or less reliably identified with really existing persons – this may not be a 'reality effect' only, but rather constitutes a central intention and issue of such works. Or take 'typical' realism as discussed in the Lukacsian tradition – novels aiming to representing central laws and regularities of the periods or societies depicted. Neither in this case such realism is a reality effect only, but a relation whose degree of truthfulness may be reasonably discussed – even if still, of course, on the typical level and not on that of more concrete empirical reality.

The second idea of literary truth, consistency, implies some problems not discussed by Ingarden. Taken at face value, consistency is an internal concept (cf. consistency truth theories in philosophy), and Ingarden's first description does indeed have this character: '... once represented objectivities are established by meaning contents of sentences as objects of a determined ontic type (e.g., as real objects and, in particular, as real psychic individuals), a consistency must be maintained in their further determination if they are to be constituted in the over-all course of the work as identical ...' (302). This internal consistency idea, however, is mixed up with what is, in fact, quite another idea: conformity to regional ontological structures: 'This consistency, however, can be maintained only if the content of represented objects is formed, at the very least, according to the a priori essential laws of a given ontic region.' (ibid.). Surely, the first idea also referred to 'objects of a determined ontic type', but earlier in the book Ingarden took care to note that it belongs to the privileges of the literary work to *mix* regional ontological domains to create improbable and impossible creatures. Such creatures, when first created, must also maintain an ontic stability – so they are permitted by the first, internalist consistency criterion, but prohibited by the latter, externalist consistency criterion. As a matter of fact, there is no contradiction here; the case is probably that Ingarden in his very short chapter confuses what is, in fact, two autonomous and equally valuable truth notions of the literary work, the former judging the coherence of the objectivity level represented, the latter judging the work's adherence to the a priori essence laws of a given regional ontology.[445] A further issue here deals with the question where to delimit this latter kind of literary truth – regional ontology – from the latter of the two 'effect of reality' truths above. Of course, typical expressions of a location, an age, etc. are empirical generalities only and thus differ from regional ontological essences (like person, mind, will, etc.). But the difference may be very difficult to tell – both in the single literary work which may easily mix up empirical and a priori issues, but also in philosophy and science where the precise amount of a priori regularities in single domains is far from mapped out. Probably the understanding of the different a priori structures of reality is as indefinite as that of empirical facts – cf. 'fallibilistic apriorism' discussed earlier. So even if the distinction between empirical and apriorical may be easily maintained in theory, the precise delimitation is blurred and different literary works may present their competing versions to be compared.

Finally, there is the 'noblest' among Ingarden's truth criteria, that of the 'metaphysical qualities'. This requires the mysterious manifestation of such qualities which – just like is the case with their real life appearance – has the unconditional form of revelation. Either they are there, or they are not there. This should not, however, prevent us from trying to indicate their status in the cases when they do in fact appear. In Ingarden's definition, they all pertain to the lived world of intersubjective human existence – and we may, again, ask for the precise delimitation of these qualities and regularities pertaining to the regional ontologies relevant for man. Husserl never developed in detail his doctrine of regional ontologies (apart from his crude distinction between those of physics, psychology and sociology, approximately), and even if a Stumpfo-Husserlian like Barry Smith's vision of a large series of more and less fine-grained regional ontologies remains provisional, realist phenomenological ontology must maintain that all empirical phenomena are governed by regional ontologies. This also goes for intersubjectivist, existential reality of the kind Ingarden speaks about when outlining his doctrine of 'metaphysical qualities'. Is the literary manifestation of such qualities a provisional grasping of them, until a proper regional ontology is developed for their understanding – or, the opposite way around, is literature perhaps a unique and appropriate way of grasping these core phenomena of human reality? Or, a compromise, are both ways parallel and equally valid ways of exploring human reality? Until further notice, I prefer to believe the latter.

To sum up, the discussion of truth and realism in literature leaves us with no less than five different notions[446]:

LITERARY PSEUDO-TRUTHS

(1) truth as empirical resemblance
(2) truth as typical resemblance
(3) truth as internal consistency
(4) truth as external (ontological) consistency
(5) truth as manifestation of metaphysical qualities

Even if an Ingardenian literary theory must be basically a theory of fiction, it thus may give rise to a rich palette of iconic ways in which literature interferes with reality.[447] The two former types refers to iconic relations to empirical reality; the third to an internal iconicity in the text (that figures and entities in the text remain sufficiently similar over the development of the text), the fourth pertains to iconicity in relation to regional ontologies, and the fifth iconicity in relation to metaphysical qualities as experienced in real life.

In the next chapter we shall look into how a subgenre constitutes itself on the basis on the fourth type, conformity to a material ontology – that of political science and, particularly, espionage.

CHAPTER 18

THE MAN WHO KNEW TOO MUCH

Espionage in Reality and Fiction: Regional Ontology and Iconicity

> The very fantasy of a spy's life, the loss of his own identity, his pursuit of pseudo-information through pseudo-relations, makes him a sort of hero of our time.
>
> Malcolm Muggeridge

SCHMITT AND SMITH

A basic form of iconicity in literature is the correspondence between basic conceptual schemata in literary semantics on the one hand and in factual treatments on the other – corresponding to the type 4) iconicity in the previous chapter. Thus, the semantics of a subject like espionage is not subject to arbitrary variation in literature, but is rather highly dependent on the regional ontology involving such phenomena.

Political science and historiography contain an enormous amount of concrete studies of famous espionage cases and agent operations, concerning the activities of both domestic and foreign services. Similarly, cases of this kind have caught public imagination to a huge extent with a whole literary genre – that of the spy and the agent novel – as a literary result. Just like its cognate the detective genre rises with Poe and Rue Morgue, the spy novel is born, albeit more gradually, with Kipling, Conrad, Ambler, Greene, Somerset Maugham, etc., to grow into one of the stable and comprehensive literary subgenres of the twentieth century.

It is a strange fact, however, that despite its firm grip around the imagination of the twentieth century, both in fact and fiction, espionage does not seem to have given rise to any significant amount of proper scientific treatment. No classic piece of writing entitled 'Vom Geheimdienste' by any Clausewitz exists in political science.[448] Despite the constant and delicate tension between the existence of secret services, necessary for the security of a democratic society, on the one hand, and the same democracy's basic principles about open administration, human rights, and equality, on the other, no tradition for deeper, theoretical understanding of this necessity and these tensions seems to exist. It is almost as if the natural secrecy of the subject is mirrored by a secrecy covering the principal reflection on it – whereas on the other hand both the factual and the fictitious coverings of single, concrete cases explode. The latter seems, in fact, to constitute a huge corpus of case-based reasoning governing the public – and maybe also the services' own – reasonings about the tasks, the constraints, and and regulations of the services.

I shall here attempt to outline the ontology of espionage, as a basis for the factual as well as the fictitious cases and for the possibility of iconicity holding between them. The clever reader will be quick to intervene: do I not confuse two separate problems? Is the description of the espionage novel not a piece of narratology dealing with genre literature – a task for literary studies – while the recurring structures of the object itself, espionage, is rather a task for political science and sociology? It is of course not possible to assume beforehand that these two tasks will be identical, but still it seems to me that a strong argumentation is at hand for the fact that they are intimately related. Not only because of the fact that all reflection of a subject marked 'secret' must keep on the distance of abduction from it, relying to some indefinite extent on the imagination and fantasies of the interpreter. But also because the relation between semantics and ontology for actual semiotics, as argued in this book, is rather different from what was assumed in the tradition running from structuralism to deconstrution and other post-structuralisms denying the possibility of iconicity in language and literature. The question of literary mimesis pertains, as we saw in the previous chapter, to several different levels: one is the possible similarity between aspects of textual expression and the subject treated (the figure poem as an example); another is the possible depicting value of a text in relation to certain empirical properties of reality (be they factual, as in journalism or historiography, be they more general like in the discussion of the possibility of literary realism to reveal insights about a given period, society or other issues). The iconicity at stake in the discussion in the following lies at an even more basic level: iconicity at the level of semantic structures used. I shall argue that the spy novel provides an example of this basic iconicity in so far as the very construction and understanding of a spy novel is only possible by the use of ontologically motivated concept structures similar to those incarnated in real life espionage cases.

We have already discussed Barry Smith's 'fallibilistic apriorism' extending the philosophical a priori realm to a long range of conceptual structures in the foundations of the single empirical sciences (Chap. 8). An implication of this idea is that a priori structures cover a far wider field than normally assumed; there is no reason to believe that formal ontology, common to all possible objects, is yet complete, and there is similarly no reason to assume that the single sciences' material or regional ontologies may not be investigated much more thoroughly than has been the case. The basis of each single science will contain, in its basic conceptual structures, a comprehensive network of interrelated terms of formal and regional ontology. It follows from this idea that works of fiction sharing the same subject as one of these sciences, will also share, to a considerable extent, one and the same basic conceptual structure.[449] Thus, Smith's approach entails that the semantics describing the content of a given domain will have iconic affinity to the ontology of the domain (even if many specific differences may of course prevail in the single case). This is the implication of one of Smith's slogans: 'putting the world back into semantics'.[450] So, we take the spy novel as an example of the fourth Ingardenian iconicity type – to conclude with the issue of how this type may indirectly activate iconicities of the fifth type, that of metaphysical qualities.

If we begin, naively, by taking a dictionary definition of a central concept for the agent novel like the term 'spy', we will find he is a person who 'illegally investigates (especially military) secrets'.[451] This definition refers to a whole range of implicit presuppositions belonging to an espionage script, an underlying highly structured diagrammatic scenario. Deprived of references to that scenario, the semantics of the word 'spy' would be ineffable. A spy investigates some subject secretly because of a certain danger or illegality in the investigation which, in turn, is determined by the fact that its subject is the business of some competing power, political or private, domestic or foreign. There is thus an a priori connection between the secrecy of the information and the relative illegality in which the spy indulges. The parenthesis of the dictionary definition implies that the spy typically has been sent out as an instrument to gather information by one power, militarily competing with another power possessing the secrets. Thus it is only in the light of this a priori, more general and more comprehensive, ongoing struggle that espionage becomes meaningful. The spy is, in essence, dependent on that schematic whole. Any fight sufficiently elaborated in time and space will always imply that knowledge about the opponent's next move adds to the probability for a positive outcome: this implies it is possible to try to anticipate that move and improve the efficiency of one's own next move. Or one may simulate such a move in order to seduce the opponent to open a flank giving a possibility for an even more efficient move. The agonistic structure of feints, simulated feints, etc. is implied here, as it is well known from mathematical game theory and instantiated in a long series of other fight or game types. The historiography of warfare is to a large extent based on the investigation of such structures of mutual deception strategies.[452] The raison d'être of the spy as collector of information lies in this scenario or script of struggle, and his role is to be a tool for one of the agonists of the battle waged.

Here we have isolated a minimal version of the regional ontology of espionage by looking at background presuppositions to a dictionary definition of the word 'spy'. A more systematic investigation might go the opposite way and try to develop the concepts of war, fight, game, or battle in order to distill espionage as one of the possible moments of fighting. A project of this kind is to be found in A.J. Greimas's narratology. Despite its apparent simplicity, this narratology remains one of the most sophisticated instruments to analyze narrative structures.[453] At a first glance, the 'narrative schema' of this theory is deceptively simple: a Destinator, defined as an actant impersonating central values, sends out an Operator Subject in order to solve a certain task. This subject is endowed with certain competences by a Helper during a first 'qualifying' trial; then follows the 'principal' trial where the Subject tries to defeat an Anti-Subject in order to take some Object in his possession. Back at the Destinator's, the Subject presents his results in a third and last, 'glorifying' trial and he receives – if the result is convincing – a Sanction judging the Subject's efforts. If the Subject wins this trial, he may receive a final Object as a reward or trophy. These three trials may, in specific cases, be realized in highly different ways, ranging from regular wars and to peaceful exchanges. A version of it clothed in fairy-tale garments makes the schema more intuitive: a King is threatened by a Dragon who has abducted the Princess, and he sends out a Hero to make up for it. The Hero must first gain a magical object or competence from some Sorcerer and he may now kill the Dragon and free the Princess. Back at the Court, the Hero displays the saved Princess and receives a reward, maybe the Princess and half of the Kingdom. If this diagram of imagination is so apparently simple, then it is probably due to its omnipresence in human imagination rather than to an inherent simplicity, not to talk about triviality. The schema contains a complexity generator due to the fact that every single phase of it refers to intersubjective relations with all the possible mirrorings, dialectics of recognition and possible misunderstandings involved. This has as a consequence that the schema may 'develop' in a huge bundle of different directions. The interaction between two actants which is in one version a raging battle may in other versions be a completely peaceful exchange – and, what is more, in each phase the teleological development mapped by the schema may go wrong. Maybe the Hero is too afraid to go to war; maybe the Sorcerer refuses to let go of his medicine copyright; maybe the Dragon actively tries to get rid of the awful Princess; maybe the King stubbornly sticks to both halves of his Kingdom; maybe there is a secret alliance between Dragon and King in order to fool the public, etc., and etc. As is evident, the schema is extremely plastic with respect to variations – at the same time as it has the stable character of being a prototype for the charting of socially integrable actions in general. As an addition to this powerful variability, the staging of narrative events in more or less artful enunciation may select single phases of the schema to emphasize and elaborate, while other phases are neglected. It may, moreover, display the events narrated, as seen from changing points-of-views of different actants, and, finally, it may recursively repeat the realization of it in different versions including the substitution of characters filling the actant roles and the embedding of local versions of the schema into more encompassing versions.

But the very question of social integration implied in the relation between Destinator – norm representative – and Subject – norm breaker – guides us on our way to the status of the spy in this schema. Of course, espionage may occur in each of the phases in the schema – in so far as the secret obtaining of secret information may be desirable in all intersubjective relationships. But because the Destinator incarnates socially stable values, the character of the Hero's task is decisive for the interest taken in the narrative in question. If the Hero's task follows ordinary procedures as governed by central administration, little remains to be told ('Once upon a time, there was King who should send a document to the council in one of his towns. He gave the task to one of his very best couriers, and the document did in fact reach its goal regularly. The courier received his contractual wage and lived happily ever after.'). A procedure of this kind is of course covered by the narrative schema's domain of modelization, but for a narrative to be interesting it is well known that it must contain some moment or other of norm break. This is, in fact, already implied in the very distinction between Destinator and Operator Subject: the frictionless action might as well be undertaken by the Destinator himself (if the Destinator in case is, e.g., central administration). The King might himself grab his good sword all at once and force it through the heart of the dragon. But he must have another actant do it, even one who receives occult, extraordinary, illegal abilities from some Sorcerer, that is, a person incarnating a competence transgressing what is usual and law bound and hence having powers and possibilities exceeding those of the King. The killing of the Dragon, moreover, most often takes place far from home – that is, far from the regular domain of laws and outside of public control. In this extraordinary competence in the Hero lies as a germ espionage, and more broadly, the secret agent, as an aspect of the Hero's deed. The Hero constitutes his own Special Task Force, and his deed is in itself to some extent a Covert Action. Now these features in the Hero actant do not distinguish the spy as opposed to e.g. the warrior, the detective and similar stereotypes derived from the same basic structure in the Hero.

Consequently, further *differentia specifica* must be found in order to grasp the difference between spy, detective, soldier, and the correlated fiction genres. We may as a first preliminary emphasize that the three of them share the Hero's character of being exceptional. The detective novel does not have the regularly working police officer as its hero, the war novel does not have the average, ordinary soldier as its hero. The detective novel favors precisely the private eye, and even more so, the deviant private eye who does not do his work 'by the book' but differs from the police in two respects: he does not, like they do, act correctly according to the rules, and, conversely, he is not involved in their muddle of corruption and mafia deals resulting from their rule breaks. Exactly because he does not act 'by the book' he may, paradoxically, act *by the spirit* and do the right thing. Even if we do, in fact, focus upon a regular police officer in the corps, we most often chose a deviant cop whose personal character and working methods transgress the average (model Colombo). Analogously, the modern war novel generally takes the point of view of a rebellious private, despising his superordinates at a comfortable distance to the front line, not obeying nor respecting their orders. Thus, this 'front pig', being an uncompromising survivor, may perform especially dangerous services. What distinguishes the spy – and the spy novel – from these stereotypes is that while the private eye and the front pig form individual cases of deviancy in the service of a higher cause (which they may serve so much more efficiently because of their disregard for rules), the spy's deviancy, on the contrary, is systematical. The very service for which he is working constitutes an anomaly in modern society.[454] The secret service is so to speak an institutionalized deviancy inside the state, a whole state organization characterized by *not* being forced to do things 'by the book'. As contemporary conflict researchers (like in Scandinavia Ole Wæver and Ola Tunander) have emphasized, we must turn to obscure political thinkers like Carl Schmitt in order to understand the specific character of these organizations. Schmitt began his classic of philosophy of state *Politische Theologie* from 1922 with the famous words: 'Sovereign is he who determines the state of emergency …'. In the context of Greimasian fairy tale logic, it is the Destinator who commands the state of emergency.[455] Ordinary law is valid only in so far a state of emergency is not declared – and the actant who decides whether the normal state prevails is of course endowed with the power of suspending it, to some (larger or lesser) degree depending on his own judgment only. Schmitt's cynical tradition turns Clausewitz upside down: the universalization of the schema of Friend and Foe makes politics a war continued with other means[456]. In such a tradition it will be a corollary that a preparedness or capacity outside ordinary legality must be kept, also during (apparent) peacetime. The state of emergency is always potentially present, and for this reason an organization is needed which is continuously able to judge which extralegal means are necessary to cope with occurring threats against the security of state.[457] Schmitt is, for this reason, the Cold War's and Terrorism's theoretician *avant la lettre*: any peace is according to him nothing but a cold war. In the Greimasian narrative schema the agent and the spy thus belong to a scenario in which the Destinator as a sovereign partially stops doing things by the book – and turns, instead, to the Schmittian book.

THE MAN WHO KNEW TOO MUCH – THE POSITIONAL CHARACTER OF THE SPY

This implies a series of distinguishing features in the spy as a potential aspect of the Hero – in contradistinction to the detective and soldier characters. In the most comprehensive and detailed text analysis which Greimas undertook – the book length Maupassant analysis *Maupassant*, the short story 'Deux Amis' analyzed has as its main theme precisely: espionage. During the Prussians' siege of Paris in 1871, two Parisian friends go fishing, and they receive a paper passport in order to cross the French lines into no-man's-land (which is a peaceful zone, there is still 40 years to World War I). After fishing, they are picked up by a Prussian patrouille who demands that they reveal the password they are supposed to possess in order to pass the French lines. They are unable to do so, of course, as the do not possess any password, and they are summarily executed. Greimas' detailed analysis finds that this killing represents the cruelty of power (especially Prussian power) as opposed to heroic citizens keeping a secret. The Danish semiotician Per Aage Brandt has, at this point, caught Greimas in a misinterpretation with crucial implications for the status of the spy. The two Parisians do *not* possess the password which the Prussians believe (they only have a paper passport), and they are unable to say what they do not know: they do not keep silent for heroic reasons. Correspondingly, the Prussian officer is not personally cruel, he just acts conforming to an ordinary logic of warfare.[458] The two of them *have,* in fact, seen the position of the German lines, and if they are allowed to get back behind the French lines, no Prussian may prevent them from informing the French defense. Even if the two fishermen are by no means spies, neither intentionally nor institutionally, they invariably *become* spies, functionally, because they are who they are where they are.[459] If you take a walk on a secret military area with your camera – we may recall certain Danish tourists arrested in Poland in the mid-eighties – then you *are* a spy, no matter whether the reason you do so may in fact be your innocent interest in a rare bird. In this light, the Prussian is not cruel, he just acts according to the *jus necessitatis* of warfare – exactly the same principle according to which secret services act during the cold war of peace. A classic of this species constitutes the Profumo affair, in which the British secretary of defense was forced to quit because he kept the same mistress, Christine Keeler – whether she took herself paid for her services or not – as a Russian intelligence officer, Jevgenij Ivanov. It is improbable that Keeler did in fact hand over sensitive information to the latter, but the simple fact of her position in the scenario was sufficient to release the scandal.[460]

This is of course the reason why it may be very important for the state to keep a file on persons with access to classified material. If they – who positionally are potential spies – should also decide to become spies *in actu*, then they must be made silent. They may be forced, for instance, to go out in the press and discredit themselves, maybe declare themselves insane, so all their sayings become polluted with ambiguity – and then they are maybe rewarded, in secret, with a pension that they would not have received under other circumstances. The specific methods of pressure are many, but the structure is stable – it is, as we know from a classic of the spy genre: it is impossible definitely to come in from the cold when you have first been out there. When first you have been a spy, then you continue being it, positionally, no matter what you may personally decide, because you now have the unavoidable property of knowing too much. This logic of position implies that the spy forms a radical example of impossibility of social reintegration. It is a well-known fact in fairy-tales that when the victorious Hero returns home with a Dragon's ear in one arm and a Princess in the other, a narrative problem may arise. Why should he be satisfied with a Kingdom and half of the Princess or whatever the King is prepared to offer – he, the Dragon slayer, who achieved what the King himself could not? Why shouldn't he take it all? The military coup as a structural possibility is inherent in this argument, just like revolutions, stabs-in-the-back and so on, and during peacetime the same logic seems to underlie the notoriously difficult reintegration of veterans after great wars. The extreme level of excitement and fear, the fact that every moment and every action concern life and death, the ultimate dependency upon the small *Männerbund* at the front and its unconditional friendship – all these experiences may make an ordinary civil life in peacetime seem like a dull superficial existence. It has often been remarked that the rocker organizations *Hell's Angels* and *Bandidos* were founded by American veterans from World War II and the Vietnam War, respectively, and the same goes for Nazism's triumph in the twenties and thirties which was only possible due to the support from enormous self-organized bands of First World War veterans in the SA and related *Freikorps*. Serbia in our time, marred by enormous mafia structures embedded in the state and led by former paramilitary troops from the wars in the 90s, forms an actual example of how difficult it may be to prevent the influence of such types on the state when first they are around.[461] The reintegration of the veterans is a psychological (and in large number cases a sociologial or political) problem which may be contained by different means – the reintegration of the spy an individual problem (and of course no large scale social problem), but then again so much more impossible. The spy may sing until he is dead, and hence he must be bound with pensions, threats, blackmail etc., because he can not leave the position of knowledge he now occupies. This structure is what, conversely, makes it possible for a spy to blackmail or punish his former organization if it does not treat him as expected. The British spy Leslie Nicholson was stationed in Prague in 1930 and spent 20 years there in the service of the SIS. When his wife became ill, he asked C, Sir Stewart Menzies (the 'M' of Fleming's Bond novels) for a loan which was refused. After his wife's death, Nicholson emigrated to the USA and took revenge on the SIS by publishing his *British Agent* there in 1964.[462] Peter Wright's *Spycatcher* from 1987 is a related example.

TWO SERVICE TYPES

The stable security structure of Post-war twentieth century features, in most countries, two organizations, foreign and domestic, and with connections to the military and the police, respectively. This structure has ancient roots (even if there was a tendency until the Second World War that services were founded ad hoc and cancelled in periods of peace[463]) and gives rise to a stable set of differences. Codes of honor based on mutual recognition is considered a military virtue and tend to have a certain influence on the former services, while the latter in its tendency mirrors the radicality of civil war as opposed to international warfare. Police-based services have as their object the state's own citizens (or domestic foreigners) conspiring against the security of the very state in which they live. Thus, they are aimed against traitors who are not seen as objects for the soldier's (potential, that is) gentleman-like behavior towards other soldiers only accidentally serving foreign powers, maybe being forced to do so by conscription. The French historian of religion Georges Dumézil once made an interesting observation in this respect when he discussed the relation of freedom to the second function (the military function) of Indo-European religion and ideology.[464] I translate the relevant passage from an interview:

Jacques-Alain Miller: 'Generally, as you analyze it, the second function displays a paradoxical aspect, because it effects the socialization of rather asocial features.'

Georges Dumézil: 'It is dangerous, but exactly for the reason that it does not respect laws, it may also happen that it may give rise to happy exceptions in those procedures where *summum jus summa injuria*.'

[cf. later in the interview: GD: 'The warrior is a creature who in all cases, not only sexuality, is always on the limit between the legal and the illegal, the ordinary and the exceptional.']

JAM: 'Thus you write that "the warrior keeps the features which takes him away from ordinary people and even puts him in an opposition to the social order which he has as his task to protect when necessary." '

GD: 'Ths possibility for opposition to the social order may appear for better and for worse. Deep down, it corresponds to the opposition between army and police. During the German occupation it was the opposition between Wehrmacht and Gestapo. It was much better to be involved with the former than with the latter. How could I forget the Mauss incident? He was saved because his flat had been claimed by the army ...'

JAM: 'But doesn't the army represent the military function here?'

GD: 'Yes. The army needed his apartment and its terrace at eighth floor, close to Porte d'Orléans, for anti-aircraft defense. One morning, I was at Mauss's place when a colonel, in a brusque but friendly manner, made him understand that the respite which he had been given had run out. Mauss negotiated and eventually got a new respite. Thanks to this, his library could be transported to the Musée de l'Homme and he himself could move into another place fifty meters from home in a flat required by the army.'

Jean-Claude Milner: 'That is Mars Tranquillus?'

GD: 'Let us say that is military honour.'

JCM: 'And the Gestapo?'

GD: 'They represented, unfortunately, the first function. Police has to do with the first function. The Rig Veda calls the stars "spies" for the sovereign god Varuna.'

Alain Grosrichard: 'That implies that deep down the descendants of the Germans still used trifunctionality during the war?'

GD: 'Let us not go that far. Let us just say that by coincidence, Gestapo's and the Party's relation to the army now and then mirrored the mythical depth of the relation between Varuna and Indra.'

JCM: 'You have written by several occasions that German culture has underlined two aspects of the second function: its violent side, the military gang, the *Männerbund*, as well as its autonomous side, freedom. But when I read you I have in the back of my head texts by nineteenth century historians claiming that the individual liberty was born in the forests of Germania. Is it possible, according to you, to find a sort of matrix in German law rather than Roman law, tied to the second function, which might serve as the structure of some sort of freedom?'

GD: 'A priori, it does not seem improbable. Let us think of the *thing*, the English and Scandinavian parliaments.'[465]

Apart from the fact that the right-wing royalist Dumézil here appropriates a Scandinavian-German myth about the origins of freedom which must have been felt like an insult on the Left Bank, the distinction made is interesting. The warrior makes possible freedom, honor, mutual recognition and has his place on the limit between law and transgression; the high priest and judge – and their spies – of the first function seem highly elevated over that very limit. The spy and the police are connected to the somber first function which, unlike the second, has nothing to do with freedom, autonomy, and honor. We glimpse the ravens of the first function sovereign deity Odin, these scouting scavengers, as an archetype of the spy – and the crafty Odin versus the brave Thor as an adequate Nordic icon of the Indian Varuna/Indra distinction. According to this distinction, the first function's police is thus potentially less concerned about rules and rights and hence less 'democratic' than is the second function's army[466] – and so the intelligence services of the two organizations, although both tend towards the first function side of the distinction so that army intelligence rather forms a sort of intermediate compromise between the constraints of the two functions. The first function, however, is superordinate to the second; it is precisely sovereign, also in a Schmittian sense of the word, because it is a task of the first function to determine whether ordinary law prevails. Prisoners of war are respected due to conventions and are returned to the enemy after the war; foreign spies are relegated, while the country's own undermining forces are classified as traitors and even criminals of an especially malignant type – this indicates that Dumézil's distinction remains to some extent valid in our days foreign and domestic services and their different treatment of their opponents.

The foreign services meet as their opponents organizations, similarly organized and equipped, from enemy or neutral (or even friendly) states. This implies a mirror structure we recognize from many spy novels, and it entails defection as a constant possibility. For the double, triple or nth level agent it is a possible way out when the earth is burning beneath him and the threat of exposure comes close; for the agent in general a way out, also to escape from other possible, maybe personal problems. The capital you may use to buy defection is, of course, inside knowledge which will subsequently be paid off in long-stretched debriefing sessions. The defector will now find himself at the mercy of the receiver country and an obvious possibility is the emergence of a new relation of mistrust given by the fact that the defector's interest is to feint more knowledge than he actually has and to delay the disclosure of it until he has gained maximal advantages from it. Domestic services most clearly representing Dumézil's first function are only part of this mirroring by their involvement in counter-espionage, while its other measures taken against the state's own citizens rarely face a similarly organized resistance. In return, the interior service in many modern states is likely to suffer from a structural paranoia due to its status as subject to a controlling Destinator in the form of public, parliamentary control.[467] It may seem natural for this service to act as an autonomous instance – also in a stronger sense than indicated by the natural Weberian tendency of all bureaucracies towards secrecy. Thus, it may seem a matter of course for it to extend its interests also to powers or persons which may not be a threat to security of the state but are merely threats to the service's interests, that is, politicians or writers with critical or even merely controlling intentions related to the services. A continuum thus stretches from security of state and to security of the service, and it is hard to exclude the possibility that a service may in case of crisis chose the latter rather than the former. The military coup is, by a homologous structure, the foreign service's corresponding possibility; the domestic service's possible unfaithfulness is bound to follow less conspicuous ways – for an unverified example, take the recurring rumors about right wing circles in the

Stockholm secret police *SÄPO* and their support to the Palme assassination.[468] According to Seymour Hersh's recent book on Kennedy's presidential period, J. Edgar Hoover was able to guarantee his continued leadership of the FBI under the newly elected president (who disliked him) by maintaining huge dossiers involving sensitive information on Kennedy, including his first and blacked-out marriage in the forties.[469]

Generally, democratic control with such organizations is by nature a delicate issue. Control commissions must keep silent, even regarding the controllers' own political parties, and on the other hand, how can a commission make sure it has received access to all relevant information from the services? This tension has a principal a priori character, in so far as total public control with such organizations would severely limit or even reduce their possibility of action – it is a given thing that such organizations must, for the sake of efficiency, be given a certain margin in which to operate, both as regards secrecy and as regards violation of law for the sake of security – even if this fact makes the organizations constantly vulnerable to potential public scandals. The old saying, attributed to Lenin: *trust is good; control is better*, cannot be applied here. The problem about control of the controller leads, of course, to an infinite regress which is only doubled by the necessary secrecy in the control of secrecies. Control must, sooner or later, at some delicate level, meet a limit, beyond which only trust remains.[470]

TWO INSECURITIES

A further a priori necessity in the spy and agent characters is the particular recruitment circumstances. No matter how upright, well-educated, and clear-minded the leading figures in an intelligence organization are – and they must be, if any – they are in no position to impose the same requirements on all their subordinates. To undertake the dirty work, they will have to use occurring characters of different kinds. The Greimas asymmetry between Destinator and Hero thus multiplies internally within the organization: the director of secret actions must make use of concrete tools operating in that part of reality which must be kept under surveillance, influenced and manipulated – and for secrecy reasons it is obvious that you can not arbitrarily plant anybody anywhere. You must, to a large extent, use persons who by coincidence have a character, a past, or a position making it possible for them to fit unseen into the environment in question. And this implies that you cannot afford to be too fastidious: this is not tasks which it is possible to educate people to fulfill, except for – exactly – exceptional cases. This does not imply unfaithfulness as a necessity but as an always threatening limit possibility: this personnel outside of perfect control consists to some extent of misfits, persons of a peculiar psychology, persons who for odd reasons feel attracted to intelligence work, people easy to convince to betray their employers, ideological fanatics, people who feel drawn by sinister affairs, people who undertake such duties of bitter necessity, people who are easy to threaten to do such tasks – even if you may ever so much hope that their main motivation be idealist. As in so many aspects of intersubjectivity, these actors' motivations are hard to determine: idealism, loyalty, excitement, desire for recognition, money, threats, brute force... the manifold of possible motivations implies that the superordinate person will have a tendency to make sure that he, just in case, has access to the latter means of influence. This insecurity implies that an elementary relation of mistrust inside the organization is obvious – which is proved by the many cases of important information that was not taken seriously. Dusan Popov informed the American army about the Japanese Pearl Harbor plan but he was not believed; the Soviet Western intelligence was well-informed about Operation Barbarossa, but they were in no position to convince Stalin; CIA was under severe political pressure to privilege information about Iraqi weapons of mass destruction and suppress information claiming the opposite... this insecurity spreads like a fog in the relation between the intelligence organization and its own informants and henchmen on all levels. An obvious danger in this fuzzy subworld is that the organization believes what it wants to believe, and it must face the paradox that the more information a message contains – that is, the more surprising it is – the less probable is it that the message will be believed.

This top-down insecurity is doubled, however, by a parallel and even more decisive bottom-up insecurity. For security reasons, the single agent must of course know as little as possible about the general plan of which he forms a part – not to speak about other parallel operations – but this necessary 'compartmentalization' of information and tasks implies a fundamental insecurity about the very character of the operation as seen from below.[471] Not only is the individual, like in all struggles, at a feint's distance from the enemy and his intentions; this basic indecidability is doubled, for the spy's point of view, by a parallel insecurity as to the exact intentions of his own side. This structure has its most prominent result the heavy weapon of counter-espionage, the double agent, who acts as if he belongs to one side while employed by the other (probably, maybe his sympathies are changing...). You will never know, as a spy, if your spymaster or leading officer is miserly with information because his deepest sympathy lies elsewhere – cf. the classic uncoverings of the third, fourth, fifth man of the Cambridge Five, all of whose existence was known long before a name could be singled out. This structure entails that a fundamental insecurity spreads in the whole spy world, pinpointed in Len Deighton's description of how Bernard Samson's own wife Fiona all of a sudden disappears as the enemy's most treasured double agent. This insecurity has several sources (apart from the enemy's natural attempts at spreading fog): the insecurity whether the mission you are on has a real purpose or if it is rather a deception operation destined to fool the enemy while the really important operation takes place elsewhere; the insecurity whether the mission you are on is in fact planned by double agents in your own organization; the insecurity whether your own organization does in fact attempt to satisfy the goals it presumes and not completely other purposes.

In extreme cases an agent may, as a 'useful idiot', function as a tool for an organization without even knowing it himself. And such matters may not even be settled by archive files – because a spymaster has his own interests, in turn, in relation to *his* superordinates on a higher level. A well-equipped archive with files on alleged 'agents' and 'spies' may keep him safe, even if the persons mentioned are to a large extent not at all spies but only innocent people to whom he maintains loose contacts.

In the secret organization, this very secrecy principle has an ambiguous character which adds to the spreading of fog. The basic motivation for secrecy is naturally immanently given: the enemy must not know what we know. But to this, a procedural secrecy is quickly added: the enemy must not know the illegal procedures undertaken in order to gain information etc. – this becomes in itself a potential conflict cause. And this problem doubles once more internally in democratic societies: the public must not know (too much) about the types of method used because this may delegitimate democracy's own laws and ideals. These constraints have led to a violent growth in the use of the three classic grades of secrecy: confidential, secret, and top secret. In the American context, this has recently been investigated by senator Moynihan (1998) finding that the extent of secrecy is now so all-encompassing that it forms a threat to the very efficiency of the services, and, in the last resort, to the security of state. Secrecy is naturally a basic problem in an open society, but in addition to that comes the fact that secrecy may blind the intelligence organizations for important real-world facts. Moynihan predicted the fall of the Soviet Union as early as in the late seventies, and he wonders why the CIA did not have the slightest idea of what was to come, even immediately before the breakdown – in spite of the obvious crisis in Soviet economy and the international decline of Marxism as an ideology.[472] Too much secrecy not only entails that the organization may loose grasp of its own informations; it may, furthermore, lead to the widespread misunderstanding that just because something is marked Top Secret it is *eo ipso* true. But even worse, Moynihan argues: the enormous increase in secrecy has given the American services a reputation as state monsters turned against the population, and a frightening statistics proves that around three fourths of the American population believe in conspiracy theories involving the services, among them the Kennedy and Martin Luther King assassinations, the lore about the secret military shootdown of UFOs and autopsy of extraterrestrials at Roswell in the fifties – and much more. The extent of secrecy thus may threaten to destabilize the very relation between state and population – and Moynihan proposes a radical intervention. Obligatory discharge of secrecy marked material after 10 years (of course with the possibility of withholding especially sensitive cases). In all cases, his diagnosis is thought-provoking: the very act of secrecy may contribute to inefficiency as well as to discrediting of the organization using it.

It is thus a part of the nature of espionage that a potential insecurity with several structural sources spreads at many levels. This should not, of course, hide the fact that most of the everyday work in such an organization is probably completely undramatic and is concerned with information taken from official or other public sources. Very often 75 percent is mentioned as an estimate of the part of the organizations' work which remains completely untouched by such insecurities. But even if the insecurities do not have to be part of one and every operation, they are constantly present as a potential limit condition. Moreover, they are most often involved in sufficiently complicated, spectacular, and embarrassing cases, which is why they play a main role in the spy literature – cf. Muggeridge the spy's quote at the beginning of this chapter.

LITERATURE AND AGENTS

These fundamental insecurities distinguish the spy from the detective and the front pig. The detective may be ever so insecure about who and what may be trusted in the world of crime and police, but his own common sense is unshaken, even if he is disillusioned about everything else; his own mission, be it with or without success, is basically out of the reach of doubt. The front pig may be ever so let down and deserted by superordinates and under attack from enemies, left behind in the most meaningless and disillusioned battle on Earth – but his own and his front pals' fundamental fight for survival provides a firm ground of reality not to be shaken. Before we go on to investigate the spy novel's relation to these a priori structures in the ontology of espionage, it is worth mentioning that exactly the insecurities mentioned give writings about espionage a rather particular relation to categories like fact and fiction. Even the memoir writing by indubitable spies is ripe with paragraphs to which they themselves are the only possible or only actual sources. This implies that they may potentially take themselves all possible freedoms when describing the events in question, just like they may owe different persons and organizations to cut things out or color the narration in various ways. The insecurity moment in the very object thus implies that a potential fictivity sticks to even the most well documented spy accounts. It is very few other factual domains which could give rise to volumes like the *Faber Book of Espionage*[473] which as matter of course mixes up excerpts from spy novels with ditto memoirs. Ian Fleming side by side with Kim Philby, Graham Greene with George Blake, Somerset Maugham with 'Dusko' Popov. This intricate relation between espionage and literature is also emphasized by the fact that very many top agent novel authors possess a first hand knowledge about the business. This includes Fleming, le Carré, Somerset Maugham, Greene – who even worked together with Philby – which is why their works of fiction might be suspected (and are in fact so suspected!) for, to larger or lesser extents, to be key novels. Is James Bond a fiction copy of Popov (minus his hump), is Leslie Nicholson the model for Greene's *Nobody to Blame*, etc. . . .?). In the same vein, anecdotes flourish about the really existing organizations having lent inspirations from spy novels or their authors; cf. the idea that Fleming should be the source of CIA's plans of killing Fidel Castro by a cigar explosion or infecting him with barber's itch so he would lose his charismatic beard and, with it, political power. Thus, there is a fluid borderline between fact and fiction because of the fact that fiction is a core part of the ontology of espionage. This does not imply that a unanimous reality of espionage does not exist – it just entails that we have no methodologically granted access to that reality.[474] Thus, there is a structural connection between literature and espionage. The author shares central features with the spy in so far he is a (partly) disinterested observer on the margin of the society in which he lives – but in addition to this structural analogy, there seems to be an empirically well-founded correlation between writers and intellectuals on the one hand and spies on the other. Already the playwright Christopher Marlowe performed counter-espionage for Queen Elizabeth I and her chief of intelligence Walsingham against Mary Queen of Scots and paid with his life for it. Geoffrey Chaucer is believed to have spied for John of Gaunt, and the famous French eighteenth century spy, the Knight of d'Eon (often disguised as Miss Lia of Bermont) was also the author of a treatise on economics. The first intelligence service in England around the Duke of Marlborough involved Daniel Defoe who later became the first leader of organized intelligence in England and thus, in a certain sense, one of the founders of *Secret Service*. In addition to many deeds as active spy, e.g. against the Scots, he even wrote one of the first papers on intelligence 'A Scheme for General Intelligence' (1704).[475] 'Intelligence is the soul of public business,' so Defoe, who continues to define counter-espionage: 'For as intelligence is the most useful to us, so keeping our enemies from intelligence among us is as valuable a head.', just like he recommends the organization of archives with files on all potential problem sources. Already Defoe used his literary work as an alibi during political information gathering, and he thus forms a prototypical example of a practical connection between the author's and the intellectual's free, philandering lifestyle and their potential use for intelligence purposes.

'EXISTENTIAL CORRELATE' AND ENUNCIATION

Both the detective, the agent, and the front pig novels are meaningless without some version or degree of a first person narrator perspective – without which the elementary suspense of these genres is difficult to maintain. Omniscient narrators, especially with respect to the time aspect, but also with regard to the inner life of many persons, would spoil these effects, apart from the fact that they would seem unnatural with their unrealistically, unavailable amount of knowledge. But the spy novel seems even more tied to the first person perspective than the other two, in regard to time, space, and persons, because the first person perspective is a prototypical point of view for the experience of radical insecurity.

This leads to the question of what could be called the 'existential correlate' of these genres, that is, their iconic relation to other domains on a higher, secondary level of iconicity, connecting to the fifth iconicity type of Chap. 17, that of the 'metaphysical qualities'. We should of course not suppose that the legitimacy of these genres lies only in their ability to allegorize basic existential experiences for the reader. There is a huge amount of direct information about espionage ontology as well as empirical facts of real milieus and experiences in them, and espionage is in itself, moreover, a complicated facet of existence – but all the same it seems to call for an explanation that these genres possess the mass appeal which have made them huge popular genres of the twentieth century. It seems to be connected to the fact that these related genres make possible a bouquet of rather different allegories in relation to other domains of life. The detective genre's affinity to cool intellectual games, solution of enigmas, intelligence tests, a heart of gold behind the tough appearances, lonely cinema noir rainy day melancholy, etc. probably forms the most well-known of these male cocktails. The front pig genre rather has a connection to fundamental feelings of misfit, hatred towards superordinates and ordinary life, violent reaction, radical and unanimous choosing side, bodily primitivism, and the dream of *Männerbünde*, the blending of blood and unconditional male friendship. The spy genre, on the other hand, lies on a continuum from idealism over the violation of idealism and to charades, loss of identity, fundamental lack of orientation and insecurity where any firm supposition about reality may vanish and initiate a foggy *Nebenwelt* in which a dark and somber worldwide destiny develops unpredictably. The spy may despair, but his loneliness is not the outsider's like the detective's, it is rather the loneliness of being tied to an irreversible position in a structural paranoia where any figure like in a puzzle picture may all of a sudden change into its opposite. The connection to politics is of course direct and in no way allegorical, but in addition to that, these structures seem to give the spy genre a special relationship to love and religion, maybe even the more dark and despairing aspects of the two. Love, jealousy, sex, and so on play marginal roles for the detective and the front pig who may deal with these matters in a unashamed Hemingwayian toughness; for Marlowe or Kelly's Heroes women are interesting staffage but no intrinsic issue – but these issues are evidently generic in spy literature. Already in the objectivity itself, there is a connection, cf. the classic features of female spies, both as *honeytraps*, patiently waiting for the appearance of the classical *pillow talk* (while maybe the seances are filmed or in other ways documented for use in blackmail).[476] Here, a common sense insecurity as to the continuum between sex and love is mirrored in a continuum between sex and blackmail. But in addition to these structures in espionage itself, the stable occurrence of these themes in spy literature is probably motivated in the structural analogies in the respective domains which make them obvious to use as allegories for each other. To many literary spies, the mysteries of love seem to be realities into which you may flee when the insecurity on the first level becomes unbearable – just to discover that a structurally analogous insecurity repeats itself at the second level.

A similar analogy of structure which may be a reason behind the popularity of the genre, is theology. We have already remarked upon Carl Schmitt's idea of the theological genealogy of modern political science concepts. Theology becomes – via the deism of Enlightenment – constitutional law; the priest becomes the lawyer; God becomes the sovereign; epiphany becomes the state of emergency. You may continue the parallels yourself: the religious community and the heathens become friend and foe, respectively; atheism and doubt become the ideologies of the bourgeoisie (the 'discussing' class trying to evade decision). Just like faith makes only sense for a believer, thus politics requires, according to Schmitt, 'existentielle Teilhaben'. It is not necessary to join Schmitt in his fascist conclusions to these analogies in order to see the spy novel as intensely occupied by a theology of the state[477]. Democracy, to Schmitt, was a naïve belief in the possibility of the definitive extermination of sovereignty; for a more cool point of view, democracy is rather a strong – if not the only – means to contain and control a sovereign position which may not be exterminated, and among the chief political virtues of democracy is precisely the fact that you need not be existentially aroused by it in order to claim your rights as a political citizen. But the position of sovereignty in democracies is precisely located in the secret services and the (most often, small) central parts of the political elite controlling them. In and around the intelligence organizations, all the theological paradoxes repeat themselves. This goes for the political science understanding of them, but also for the participants: the continuum in espionage from existential bottomlessness and to idealist confession mirrors the continuum from doubt to faith, and just like the ways of the Lord are past understanding, even for the believer, so exactly the same holds for the ways of the State, even for the most devoted spy. He becomes a mystic of the state, be it real or dreamt-up, serving an enigmatic entity which by its very nature never can be met with face to face, which he may only meet in his own doubtful deeds where any victory is provisional, open to dispute and maybe even a defeat in disguise. In theology, the spy novel thus finds another 'existential correlate' – and with it all the passions, the rare epiphanies, and the dark-nights-of-the-soul – both in dogmatic theology and popular belief versions.

But here, the espionage novel adds to these existential passions a cool and comfortable objectivation in so far as they are here played out in full intersubjective orchestration. In doing so, the spy genre may stage these existential and theological structures without the first person perspective leading to orgies of expressive psychologisms. Most often, the first person perspective is – in spite of its status as point-of-view – minimally described, exactly because the objective scenario of the plot stages the existential figures. This allows for a cool and objective rendering of structures which in other genres may be given rather juicy and self-indulgent psychological descriptions.[478] This force of the spy novel may be that it orchestrates the passion of the state at the same time as it provides an objective iconic tool to grasp certain Ingardenian 'metaphysical qualities', the bottomlessness of love, existence, faith, and doubt: a stable instrument to understand a set of basic insecurities.

ICONICITY IN ESPIONAGE REPRESENTATION

To conclude, a basic condition of possibility for the spy novel lies in its iconic use of basic a priori structures of espionage in an Ingarden iconicity type 4). A basic outline of this schema can be found in the presuppositions of any definition of the word spy. A further analysis demonstrates a series of structural corollaries to this spy definition:

1. The spy as a special moment of the narratological hero (as opposed to the related characters, the detective and the front pig)
2. The positional character of the spy – the possession of secret knowledge as determination independent of any espionage intention or affiliation in the person in question

3. A tendential structural difference between foreign and domestic services
4. Two types of basic insecurities in any espionage hierarchy: one top-down insecurity eroding the superordinates' trust in the subordinates. And one bottom-up insecurity inflicted by the 'compartmentalization' of secret services, eroding the spy's trust in the organization employing him.
5. The secrecy and insecurities of espionage makes fiction a virtual aspect of every factual writing about it.

These basic ontological features of espionage are iconically reproduced in the spy novel genre and contributes to its very definition as such.

The fifth property, moreover, implies a particular, intrinsic relation – and iconicity – between the role of authors and the role of spies.

Finally, the clarification of these basic diagrammatical properties of espionage makes possible a hypothesis about a second-order iconicity holding between the espionage novel and the metaphysical qualities of other discourses, namely those of love and of theology. These two fields structurally share the basic insecurities of espionage which is why it may be used iconically to address, more or less directly, and more or less allegorically, central problems of love and religion.

Thus, iconicity is at stake in at least two different aspects. Basically, an iconicity between espionage as such and the novels about it is made possible by shared diagrammatical semantics. On this basis, other important iconic relations become possible, namely those between spies and authors and those between espionage on the one hand and love and religion on the other.

This conclusion forms an empirical case against two ideas of the relation between iconicity and literature. One is the skepticist idea that iconicity should play no role at all in literature and that, consequently, it should be possible to describe literary issues with literary theoretical concepts only. Against this, it may be argued that the very existence of stable genres – as for instance the spy novel – point to iconical, realist foundations outside of literature proper. Another is the idea that iconicity in literature should concern only the relation between expression and content (like figure poems, basically). Against this, it may be argued that a more basic iconicity concerns also the relation between meaning and reference, in this case between a literary subgenre and a schematic a priori structure defining its domain of reference.

Our attempt at charting the regional ontology of espionage took its point of departure in the necessary presuppositions to a dictionary definition of 'spy'. Our further development of this ontological structure involved theoretical and empirical observations and investigations from a host of rather different sources. This points to the special conditions of charting regional ontological structure. Even if a priori, such structures may not be fruitfully investigated by pure armchair speculation only. Even if basically diagrammatical, such knowledge must base itself on the analyses of the empirically presented structure of the field, pendling back and forth between a purely diagrammatical grasp of regional necessity regularities on the one hand and empirical concepts produced in the ongoing scientific and other research in the field.

PERSPECTIVE

In the course of this book, we have argued for the actuality of two traditions: that of Peircean pragmatism and Husserlian phenomenology. We have even argued that they share a considerable amount of valuable thought - particularly involving the Peircean notion of diagrammatical reasoning and the Husserlian conception of the synthetic a priori of material ontologies.

Roughly, these two ideas connect in two ways. Diagrammatical reasoning is a major road to the understanding of ontological structure – and diagrams, in turn, form a core notion in the material ontology of semiotics.

These ideas are developed in two steps – one purely philosophical, and one intervening in the material regions of biology, pictures, and literature.

The development of this meeting between Peirce and Husserl is thus only to a limited extent seen as an issue of history of ideas. Rather, it is taken to be an intervention in the actual discussions of phenomenology, ontology, and semiotics, of biosemiotics, pictures, and literature – and this is why the discussions of Peirce and Husserl have – all through the book – been embedded in a host of actual issues and discussions with other philosophers, scientists, and critics.

The resulting *diagrammatology* is intended to contributing to the survival of the enlightened idea of the study of the humanities as a rational, interdisciplinary endeavor – during a Dark Age at the humanist faculties strangely wedding constructivist irrationalism to glib marketing and political correctness.

APPENDIX: PEIRCE'S CONCEPTION OF CONTINUITY BETWEEN MATHEMATICS AND PHILOSOPHY

> Mathematical reasoning consists in thnking how things already remarked may be conceived as making a part of a hitherto unremarked system, especially by means of the introduction of the hypothesis of continuity where no continuity hitherto had been thought of...
>
> Peirce ('Spinoza's Ethic', NATION III, 85)

When René Thom claims, in the preface to his *Ésquisse d'une sémiophysique* (1988), to be the first philosopher of the continuum since Aristotle, then it is correct only to the extent that there are not indeed very many philosophers who have been interested in that concept – unlike the heavy interest in continuity in mathematics. But Thom's claim, in any case, not only overlooks figures like Leibniz, Kant, Veronese, and Weyl, but most conspicuously Peirce's effort in the field, especially in the mature versions of his thought in the years around 1900. The concept of continuity is central to the whole of Peirce's thought. 'Of all conceptions Continuity is by far the most difficult for Philosophy to handle' (R&LOT 242), writes Peirce in 1898, and consequently continuity is 'the master key which adepts tell us unlocks the arcana of philosophy,' (untitled manuscript, 1897, 1.163), in short it is '... the most difficult, the most important, the most worth study of all philosophical ideas.' ('The Categories', undated, NEM IV, 310).

It plays, in particular, a main part in Peirce's metaphysics and consequently – following Peirce's Kantian idea that metaphysics mirrors logic[479] – in Peirce's comprehensive logic. This is why it also becomes of seminal importance for what connects logic and metaphysics in Peirce: his conceptual realism. Peirce's 'extreme realism' is built on the concept of continuity. It is regrettable, thus, that excellent discussions of his realism – like Boler or Haas – only marginally involves continuity; this might be for editorial reasons, though, since some of the decisive texts have only been published since their books. Finally, the concept of continuity provides the basis – often overlooked – for Peirce's more well-known semiotics, in particular in relation to the central concept of the icon.

Continuity as a metaphysical concept, however, is intimately tied to the mathematical presentation of it in Peirce, following his architectonic system for the sciences, according to which mathematics is the first science.[480] Peirce was one of the first philosophers to see the problems in a consistent mathematical treatment of infinity and continuity[481]; according to himself his very first definitions in the field go back to 1867, but it became central from at least 1880 (Peirce's strongly mathematical period, when he held a position at the Johns Hopkins University from

1879–84), that is, before he could have known neither the works of Dedekind nor Cantor (*Was sind und was sollen die Zahlen* appears in 1888[482], and Cantor's papers only gain recognition among international mathematicians during the eighties[483]). Potter and Shields (1977) refer to Peirce's 'pre-Cantorian' period up until around 1884 when he held the Kantian idea that continuity could be defined as infinte divisibility. During the 80s, Peirce became familiar with the writings of Cantor whom he greatly admired. Not without certain aberrations and criticisms, he basically shares Cantor's definition of continuity up until the mid-90s. After around 1895–96, Peirce gradually develops his own version of set theory with the aim of providing a continuity description on that basis, and so his main papers on the issue fall in the period from 1897 to 1908. The continuum definition now revolves around the idea that in continuity, all parts have parts of the same kind. Potter and Shields refers to this period as 'Kantistic', maybe a little wildleading as Peirce's earlier, pre-Cantorian idea of continuity as infinite divisibility was also a Kantian idea. Potter and Shields outline a last period from 1908 to Peirce's death in which he takes the continuity of *time* to be the measure stick of all continuity, thus prolonging an older tendency to describe continuity as no collection at all.[484] Detailed treatments of the problem of continuity thus only appear in the decades around the turn of the century; main points in his reflection on the issue are to be found in 'The Law of Mind' (1892, EPI, 312ff, 6.102ff), 'The Logic of Quantity' (1893, 4.85), Robin-numbers 14 and 28 (christened 'On Quantity' and 'Multitude and Continuity' by Eisele) from 1895 and 1897, 'Fallibilism, Continuity, and Evolution' (c. 1897, 1.141) the lecture series known as 'Reasoning and the Logic of Things' (1898, R&LOT), the 'Lowell Lectures' (1903, partly in CP 4.510, partly in NEM), the papers on existential graphs (1903, 4.418), 'The Bedrock beneath Pragmaticism' (1906, parts in CP (4.553n2; 4.561n1; 6.174–176)),'Some Amazing Mazes' (1908, 4.585), and 'A Sketch of Logical Critics' (1911, EPII, 451ff, partly 6.177) – and recurring reflections on the continuum are to be found in much of his mature work in this period.

THE LAW OF MIND

Let us begin by presenting the whole complex of problems which Peirce connects to the concept of continuity, as it appears in his 'Cantorian' period in his characteristic excursive style, in which he in a few lines moves from mathematics to ontology to psychology to religion. A critical presentation of 'The Law of Mind' (1892) which forms Peirce's first large discussion of his metaphysics of continuity, his so-called 'synechism', will illustrate the range of problems he expects it to solve.

'The Law of Mind' is Peirce's first major discussion of the problem. In the preceding article in the series written for *The Monist*, of which 'The Law of Mind' is the third, Peirce proposed the concept of 'tychism' as referring to a doctrine of ontological absolute chance. He now adds the idea that this concept must give rise to an evolutionary cosmology 'in which all the regularities of nature and of mind are regarded as products of growth, and to a Schelling-fashioned idealism which holds matter to be mere specialized and partially deadened mind' (6.102).[485] This programmatic monism is now taken as the basis for the validity of the law of mind – briefly, the tendency of ideas to spread – holding for mind in a more narrow sense as well as for matter, corresponding to Peirce's contention that the ideas of Thirdness constitute a 'thought-like' reality without reference to any thinking human subject. Peirce refers to the fact that many years earlier (he thinks of 'Questions Concerning Certain Faculties Claimed for Man' and 'Some Consequenses of Four Incapacities', both from 1868) he had tried to develop this doctrine, albeit by then being too blinded by nominalist prejudices. What Peirce refers to in these early papers is partly the problem of epistemology having to begin from something 'first' which he rejected on the basis of an Achilleus-and-the-Tortoise argument from continuity (EPI, 26; 5.263) claiming that science is always-already in the process of development. And partly the idea that science continually approaches reality due to the ongoing effort of the community of researchers (EPI, 52; 5.311).[486] In these early papers, his reflections referred to continuity in the development of knowledge only, hence the later rejection of the ideas as nominalist. In 'The Law of Mind', on the contrary, the idea is revised so as to refer to ideas understood as fully objective entities: the very Law of that title refers to the fact that 'ideas tend to spread continuously and to affect certain others which stand to them in a peculiar relation of affectibility. In spreading they lose intensity, and especially the power of affecting others, but gain generality and become welded with other ideas.' (EPI, 313; 6.104) Thus the strongly idealist formulation of the Law, and the rest of the paper consists of comments to this. Peirce's introductory argumentation of the objecitivist continuity of ideas rests on an analysis of *memory*.[487] How is it possible at all for past ideas to be present in actuality? If not by some kind of vicar or representative (which we might never remain sure to understand), then only by 'direct perception'. The problem has a certain analogy to the old Husserl's famous discussions in 'Ursprung der Geometrie': how is it possible to secure the stable heritage of knowledge in a represented form? – only by the maintenance of that original evidence which gave rise to its symbolization in the first place.[488] In Peirce, the answer is '... by direct perception.' (EPI, 314; 6.109). This is why an idea, if present now, can not be complete past at all; the present now must be connected with the relevant parts of the past by a series of infinitesimal causal steps – by continuity. The fact that ideas may thus possess a constancy during time becomes an argument in Peirce for a continuous conception of time, an idea which consequently cannot tolerate the idea of a time line dissolved into a powder of separate points. Consciousness must thus – monism here allowing Peirce to pass directly from experienced time to consciousness – cover an extended interval of time which can not, on the other hand, be a finite interval (in that case time would not pass), but precisely an infinitesimal interval. This is a decisive step in Peirce's argumentation which explains his metaphysical sticking to the formulation of calculus in infinitesimal terms in his philosophy of mathematics (as against Weierstrass's version using limits, which had become the canonical version in Peirce's time and well until the 1960s and the appearance of non-standard analysis).

To Peirce, infinitesimals are not only mere conventions for calculating; they possess real existence, and it is they that grant the continuous connectedness of time as well as of consciousness. The fact that the *object* of consciousness must also be continuous, Peirce now tries to argue with a (rather weak) argument of Kantian flavour: '... in this infinitesimal interval, not only is consciousness continuous in a subjective sense, that is, considered as a subject or substance having the attribute of duration, but also, because it is immediate consciousness, its object is *ipso facto* continuous.' (EPI, 315; 6.111) This subjectivist – and hence potentially nominalist – argument is probably part of the reason why Peirce later sees 'The Law of Mind' as a mere step in his development of a continuity theory. Time consists of *moments* possessing an extension, not of *instants* being mere points. Instants may appear only relative to moments, not as autonomous entities in themselves. In this claim for an ontologial argument for the existence of infinitesimals, Peirce may be compared to the non-standard analysis of the recent decades (A. Robinson) which reintroduced infinitesimals, even if not for ontological reasons, and without any direct practical implications to the extent that 'non-standard analysis' is consistent with the standard theory using limits (cf. below).[489] Peirce, on the contrary, has a metaphysical background for preferring infinitesimals – his logic of relations: 'The illumination of a subject by a strict notation for the logic of relatives had shown me clearly and evidently that the idea of an infinitesimal involves no contradiction, before I became acquainted with the writings of Dr. Georg Cantor [...] in which the same view is defended with extraordinary genius and penetrating logic.' (EPI, 316; 6.113). Peirce here overlooks the fact that Cantor in no way endorsed infinitesimals, quite on the contrary. It allows him, however, to introduce his old distinction (from 'The Logic of Number', 1881) between finite and infinite sets, defined by the fact that only finite sets are subject to de Morgan's 'syllogism of transposed quantities'. This idea is equivalent to Dedekind's and Cantor's definition of infinite sets as sets with subsets having the same size as the sets themselves,[490] and as they it takes as its basis the Bolzano measuring of the relative size of sets the one-to-one pairing of their elements. In Peirce's gay example, reference is made to Balzac's *Physiologie du mariage*, wherein it is claimed that any young Frenchman boasts of having seduced a French woman. As a woman in this use of the word may be seduced only once, and as there are as many French men as women, it follows that no French woman escapes seduction. But this argument only holds provided the set in question is finite. If the French population is growing (and why shouldn't it, considering the amount of seductions), the conclusion is not necessarily true. In this case, the example corresponds to Cantor's well-known proof – which Peirce refers (EPI, 316; 6.115) – that there are as many even numbers as there are whole numbers, because to any whole number, an even may be found, and vice versa. But still they remain a subset of the whole numbers.[491]

Now follows Peirce's first attempt at classifying transfinite sets. He claims that 'of infinite collections there are but two grades of magnitude, the *endless* and the *innumerable*' (EPI, 317; 6.116). This is obviously wrong, as Peirce later acknowledged, but let us follow his argument. The merely endless sets are countable, just like the finite sets, but even larger sets exist which are not countable, because they are not subject to what Peirce calls the Fermat inference. Peirce gives an unnecessarily complicated algebraic example; the idea is simple and equivalent to mathematical induction (which Peirce for terminological reasons finds 'improperly'). The Fermat inference consists in the possibility of proving a theorem valid for all n, if it is possible to prove it for $n = 1$ and then prove that if it is valid for n, then also for $n + 1$. This conclusion requires, of course, that the set in question – as may the whole numbers – be arranged in a list, in a countable set. And that is not the case for all sets, as proved by Cantor: the irrationals may not be so arranged, because the distinction between them requires an infinite decimal expansion: 'Now if they cannot be exactly expressed and discriminated, clearly they cannot be ranged in a linear series.' (EPI, 319; 6.117). Peirce does not refer Cantor's proof, the famous diagonal proof, later so crucial to computer science, but the very wording of his own argument is interesting: he interprets uncountability as implying that the single elements of such a set may not be clearly distinguished as individuals, and this points forward to his own ontological interpretation of the concept of continuity as a concept of a hyper-set in which the single elements totally lose identity. In contradistinction to the bipartition of infinite sets, Peirce immediately (EPI, 319–20; 6.118) introduces a further distinction of the uncountables in two. First, he claims with Cantor that a line contains exactly as many points as a 2-D plane or a 3-D body.[492] The coordinates of the single points may be compressed into one single decimal expansion, no matter how many dimensions (and correlatively, coordinates) the entity in question may have: place the first coordinate's first decimal on the first place of the number; the second coordinate's first decimal on the second pace; the third coordinate's first decimal on the third place, and so on. This, of course, becomes impossible if the number of dimensions is infinite; a set of infinite sets of numbers which may each of them vary uncountably must transgress the simple uncountable set, and Peirce consequently calls it endlessly infinite: 'The single individuals of such a collection could not, however, be designated, even approximately, so that this is indeed a magnitude concerning which it would be possible to reason only in the most general way, if at all.' (EPI, 319; 6.118). Again, Peirce forwards the idea that the single, distinct individual may become blurred in very large sets, without his argument being particularly clear: it is not evident why it should not be possible to reason about single individuals in such a set, and later Peirce must admit that the elements of even uncountable sets are distinct so that any idea of the anticipation of continuity in transfinite sets must be argued in other ways. The fact that Peirce has not yet grasped the rules of calculation for transfinite numbers is demonstrated by his idea that the product of two uncountable numbers should be larger than each (as e.g. the set of possible pairs composed by one number from each set cf. note 16). It will later occur that a given transfinite entity must appear as exponent in order to let a larger transfinite set appear: $\text{Aleph-n} < 2^{\text{Aleph-n}}$.

These deliberations, however, take Peirce to the decisive question: what *is* continuity? Here he takes his basis in one version of Kant's definition – infinite divisibility – which quickly, and following Cantor, must be dismissed: the rationals are infinitely divisible but not continuous, because between two given rational numbers uncountably many irrational numbers may be found.[493] Cantor's definition is, instead, that a continuous series is '*concatenated* and *perfect*. By a concatenated series, he means such a one that if any two points are given in it, and any finite distance, however small, it is possible to proceed from the first point to the second through a succession of points of the series each at a distance, from the preceding one, less than a given distance. This is true of the series of rational fractions ranged in order of their magnitude. By a perfect series, he means one which contains every point such that there is no distance so small that this point has not an infinity of points of the series within that distance of it. This is true of the series of numbers between 0 and 1 capable of being expressed by decimals in which only the digits 0 and 1 occur.' (EPI, 320; 6.120).

Peirce here refers Cantor 1879–84, 194 where the corresponding concepts are 'perfekt' and 'zusammenhängende', and the continuum consequently 'perfekt-zusammenhängende'.[494] 'Perfekt' is here defined as being the same as non-reducible, that is, not reducible by 'Ableitungen', which successively isolate the limit points of a set,[495] while 'zusammenhängende' in our days' terminology will be dense (in itself dense).

Even if Peirce extensionally agrees with this definition (in so far as he finds it includes continua and excludes non-continua), he is not satisfied for intensional reasons: the definition is metric (which is only true in a relative sense: 'Es handelt sich also um eine 'metrische' Eigenschaft des Kontinuums' (194); Cantor here refers to the definition of Zusammenhang by distances less than a certain neighborhood distance epsilon) – which is not the case for the distinction between continuous and discontinuous. The fact that a perfect series is defined by containing every point of a certain description is regarded by Peirce as a definition by negation (because points not caught by this description are excluded), while no positive idea is given about 'what all the points are'. This cricitism seems to presuppose Peirce's own idea of the continuum which excludes that it may be described as a set of points of any properties whatsoever, but requires a description on its own level, prerequisite to any division of it into points; without the presupposition of this idea, Peirce's criticism seems ill-placed. Finally, he attacks Cantor for giving no distinct definition of the concept of continuity, because his definition falls in two parts, not yielding any intuition of what the continuum as an entity really is. As mentioned, Cantor believed at this point that the continuum hypothesis could be proved within set theory, so that the continuum was equal to the power of the arithmetical line which was again equal to the first transfinite cardinal after the countables, and he consequently viewed the definition given as holding for the geometrical line. Peirce in a certain sense agreed in the continuum hypothesis, in another sense he did not – and none of them knew, of course, it was later to be proved indecidable in set theory (Cantor's version as it was later formalized by Zermelo and Fraenkel).

Peirce now undertakes to provide an alternative definition based on Kant's admittedly incomplete suggestion. How are the 'holes' in Kant's definition (infinite divisibility corresponding to density) to be 'mended'? 'What is required (...) is to state in non-metrical terms that if a series of points up to a limit is included in a continuum the limit is included' – from the argument that given a 'hole' in a Kantian series, at most one of the end points may belong to the set (if both belonged to the set, then there would be, according to Kant's definition, at least a further point between them belonging to the set, cf. the Dedekind cut). So if we can require both points to belong to the series, the hole is 'mended'. Peirce notes that this seem to be the property Aristotle saw in defining the continuum as something whose parts has a common limit (Peirce refers to Physica 227a, Metaphysica 1069a[496]). This definition Peirce baptizes 'Aristotelicity', to be added to the 'Kanticity' of infinite divisibility: 'The property of Aristotelicity may be roughly stated thus: a continuum contains the end point belonging to every endless series of points which it contains. An obvious corollary is that every continuum contains its limits. But in using this principle it is necessary to observe that a series may be continuous except in this, that it omits one or both of its limits.' (EPI, 321; 6.123).[497]

Frankly, this definition of continuity seems no less 'double' than Cantor's. Partly, it requires density ('Kanticity'), partly it requires that any delimited part must contain its limit ('Aristotelicity', corresponding to Cantor's 'perfection'[498]), moreover, it corresponds well to the standard definition today. Peirce now connects it to the existence of infinitesimals: the word 'infinitesimal' simply is Latin for the ordinal number the 'infinitieth' and it refers to the fact that incommensurable numbers need an infinitieth number in their decimal expansion in order to be fully described. This is why continuity makes possible infinitesimal entities, Peirce claims in an argument which must presuppose the continuum hypothesis in Cantor's version. His criticism of Cantor here – as Marco Panza puts it (Salanskis (ed.) 27) – seems to amount to the fact that Cantor wants to give a definition in order to constitute the object, while Peirce rather conceives of the task as to satisfactorily describe something already given beforehand, cf. Dipert's (1997) characterization of Peirce's disagreement with Dedekind as pertaining to intensional vs. extensional description of sets. Rather than a mathematical difference, it is a disagreement as to the requirements for the very character of a description.

After this mathematical interlude, Peirce returns to the philosophical use of the concept of the continuum. First a problem of perception: which color should be ascribed to the borderline between to colored surfaces? As the color of a point, seen from a continuist point of view, must be identified as the color in a certain neighborhood around the point, we may only say that the color of the borderline is half of each of the two adjacent colors. The present now must be another case of such a limit surface, consisting half of past, half of future – corresponding to the definition of velocity in the differential calculus as the middle value of speed in an infinitesimal interval. These ideas painlessly are generalized to consciousness: 'Just so my immediate feeling is my feeling through an infinitesimal duration containing the present instant.' (EPI, 322; 6.126)

These quick analyses of time and consciousness now hold decisive implications for the Law of Mind, for the distinction between past and future makes the Law of Mind – unlike physical laws of force – irreversible. The present now is influenced by the past, not by the future. From this truism[499], Peirce derives radical consequences: the continuity between past and the present implies that '... every state of feeling is affectible by every earlier state'. In this version, the theory seems rather radical, because it does not delimit the influence to a local neighborhood around the single 'state of feeling' – if it is not to be read in a fashion making consciousness itself identical to that local space. That seems not to be the case, though, if we are to believe the following definition of an original continuum of feelings towards which human consciousness only possesses a highly restricted window: 'The development of the human mind has practically extinguished all feelings, except a few sporadic kinds, sounds, colors, smells, warmth, etc., which now appear to be disconnected and disparate.' This idea is connected to Peirce's cosmology, which is for the first time developed in this series of lectures – according to which the beginning of evolution was constituted by an infinite space of possibilities, from which evolution gradually selects some to realize. The realized feelings are in themselves continuous (albeit in less dimensions than the original space of possibilities) so they appear temporally articulated in infinitesimals. Correspondingly, they have a continuous spatial extension, which takes Peirce into his strange doctrine of slime, of protoplasma which is seen as the physiological continuum permitting emotion to enter into motion. This spatial continuum has wide implications in Peirce: 'Since space is continuous, it follows that there must be an immediate community of feeling between parts of mind infinitesimally near together. Without this, I believe it would have been impossible for minds external to one another ever to become coördinated, and equally impossible for any coördination to be established in the action of the nerve-matter of one brain.' The continuity of space and the continuity of feeling are taken to solve all at once both the problem of the existence of other minds and of the mind-matter relation. Both these problems' character of the transgression of boundaries is taken to rely on the lack of insight in a monistic continuity underlying the apparent limits.[500]

Continuity, thanks to monism, immediately becomes the question of the continuity of ideas. Ideas are defined by three aspects, after Peirce's three metaphysical categories: their quality as feelings; the energy with which they affect other ideas[501]; their tendency to imply other ideas. Energy is supposed to diminish, as the idea spreads. It seems like Peirce here argues in some kind of analogy with the decrease of light intensity proportional to the squared distance. Its quality, unlike its intensity, is unchanged (a weak recollection of a strong red colour is not a weak red...). The continuity of feeling of course implies that 'A finite interval of time generally contains an innumerable series of feelings; and when these become welded together in association, the result is a general idea. For we have just seen how by continuous spreading an idea becomes generalized.' Now we approach the very exegesis of the Law of Mind, and there are several interesting points to comment here. If we accept the premiss that any finite time interval contains an uncountable number of feelings, then there is still some distance to the conclusion that these should be able to synthesize into one general idea. It is obviously only possible in some cases, namely when they are 'welded together in association'. Here Peirce's doctrine of ideas all of a sudden becomes rather brutally psychological, and even in a rather primitive associationism, not unlike the British empiricists: if anything may be associated with anything, then the Mind may entertain any 'general idea' it might wish. Consequently, Peirce must describe this general idea as 'vague' due to its unlimited character, still, on the other hand, it is claimed to have the inner property of a 'living feeling' with strong anti-nominalist implications. Ideas are seemingly essentially related, 'we can directly perceive the one gradually modified and shaping itself into the other. Nor can there any longer be any difficulty about one idea resembling another, when we can pass along the continuous field of quality from one to the other and back again to the point which we had marked.' (EPI, 326; 6.139). Just like the case in the other great phenomenologist Husserl, a refined and continuous version of the psychology of association pays a central role in the description of the synthezicing activity of the mind in Peirce.[502] As the synthesis is here described, its object is not merely ordinary associations of ideas potentially being able to synthesize anything, but specifically *iconic* associations, connections by similarity, objectively existing in the multidimensional quality space. Thus, it is a variety of eidetic variation in this quality space which grants the real connections of ideas in Peirce, making possible, in turn, their association for the mind as well as in the world. After an interlude about the intensity of ideas, Peirce specifies this connection: 'We can now see what the affection of one idea by another consists in. It is that the affected idea is attached as a logical predicate to the affecting idea as subject. So when a feeling emerges into immediate consciousness, it always appears as a modification of a more or less general object already in the mind.' (EPI, 326; 6.142). This points towards Peirce's theories of abduction and his distinction between abduction, induction, abstraction, and prescission in the field which were earlier most often covered by simplicist empiricist-psychological abstraction theories (see Chap. 11). Ideas may not at all be connected without continuity (6.143). It may happen, however, in a universe suffused by chance – cf. Peirce's 'tychism' – that ideas are associated which may not form a general idea. But association does not follow mere coincidence, generally, there is a Darwinian tendency, by selection, presumably, in ideas to gather in more general ideas. The scientific quest for truth in the community of researchers, the development of civilisation in general, and behind them biological evolution, lie, implicitly, behind this argument. This 'mental law' conforms to the forms of logic, however, deduction corresponds to the fact that a general idea releases a reaction (like when the frog's dissected legs react when pinched), in induction, a habit is constructed which is present as a general idea in a series of single feelings, and in abduction (here: hypothesis), the general idea is called for by a single feeling (the logical inference drawn from a single observation to the fact that the entity in question may belong to a general category making its behavior understandable). Only deduction is necessary, while the two other inference forms are probable and possible, respectively (in this context both are called probable, EPI, 329; 6.147), mental action is not necessary and invariable, quite on the contrary it is insecure, plastic, and vital. This argument draws the consequence of monism: the Law of Mind is claimed valid not only for human thought but also for the evolution of the world even down to details like the claim that the mind's tendency to follow logical forms is also shared by the movement processes in the world: this is what constitutes Peirce's famous and extreme 'logical realism'.

This now permits Peirce to restate the Law and its implications: 1) from a nominalist point of view ideas may not be similar to each other nor influence each other, 2) momentaneous feeling flows together into one continuum of feeling, gaining generality – in this conceptual realism, similarity, connections between feelings and the world cease to constitute problems, 3) because of this, general feelings are no longer mere words, but *even more* than the feelings in which they incarnate (cf. 'extreme realism'), 4) the highest law – which Peirce does not abstain from calling 'heavenly and living harmony' – does not require the single feelings to resign from their individuality, but merely to influence one another. The potentially totalitarian implications of this somewhat Hegelian-Schellingian idealism is sought avoided by this turn: the Law does not require the single ideas to do something *determined*, they are just urged to self-organize without the result of this process being dictated beforehand, 5) we are thus unable, at our present level of knowledge, to say to which extent this evolution is governed.[503] (EPI, 330; 6.150ff).

With this theory, Peirce declares, in yet a surprising turn, we are close to an explanation of personality. Personality is also a general idea which may not be grasped in one glance: 'It has to be lived in time: nor can any finite time embrace it in all its fullness.' Our finitude as human beings here dampens realism: we are not able to grasp sufficiently complex general ideas in finite time: personalities and other such generalities belonging to the realm of continuity.[504]

To sum up, Peirce's metaphysical doctrine of continuity, this 'synechism', entails logical realism, objective idealism and tychism as well as a thoroughgoing evolutionism.

As is evident from this presentation, Peirce's mathematical speculations on continuity is intimately related to his phenomenology (Firstness and Thirdness, characterized by potential and actual continuity, respectively), his philosophy of mind (the doctrine of motivated associationism), epistemology (the tight analogy between evolutionary processes in the world and logical inference processes in the mind), ontology and metaphysics (the continuous character of being, the continuous reality of universals), not to speak about religious speculations (a continuous and therefore (!?) personal God). Even if the motivation for Peirce's strong interest in the continuum is obviously extra-mathematical and related to the need for an ontological 'glue' to make the different parts of his architectonic stay together, Peirce maintains that the possible mathematical definition of the term is necessary for its use elsewhere. This is implied by his conception of mathematics (inherited from his father, the mathematician Benjamin Peirce) as the science which draws necessary conclusions from hypotheses. Even if mathematics is in itself hypothetical through and through (and thus not in itself true or false, it is true only as a corpus of if-then propositions which may have mutually contradictory implications, so as e.g. the different non-Euclidean geometries), it is the reasonings of mathematics to which the more empirical sciences have to look in order to find formal structures. This is why the mathematical determination of the continuum is necessary for Peirce in order to develop the concept in its ontological and empirical applications. The 'Law of Mind' is often referred to as insufficient (and rightly so) in later Peirce texts, among other things exactly with reference to its conception of continuity. Its chaotic mixture of subjects, though, form a good introduction to grasp the wide span of issues continuity is supposed to explain in the mature Peirce's thought in the years around the turn of the century.

CONTINUITY BEYOND THE TRANSFINITES

The presentation of the purely mathematical aspects of the concept of continuity is – as so often the case in Peirce – spread out in a long series of more or less finished works from the period around the turn of the century. One of the most thorough versions is to be found in Robin nr. 28, published in NEM III with the title 'Multitude and Continuity' (dated by Robin as 1897?). The title is well chosen, for the goal of this cautious presentation is to situate the concept of continuity in relation to the doctrine, inspired by Cantor, of transfinite sets and their size, 'multitude' (Peirce's translation of Cantor's 'Mächtigkeit', our days' 'power'). It might surprise that Peirce choses to treat Cantor's development of the transfinites so detailed as is the case: his main point of view remains, as already in 'The Law of Mind' that the continuum is a primitive phenomenon which may not be derived from simpler phenomena – and yet he devotes great energy to the reconstruction of his own version of an analytical, Cantor-style theory whose aim is to build continuity 'from below' by means of sets of points. We shall return to this seeming paradox.

Peirce had developed his own terminology since the beginning of the eighties[505] which calls for a presentation. 'Multitude' is, as mentioned, his term for Cantor's 'Mächtigkeit', measuring the size of sets from one-to-one mappings; Peirce often uses the term as synonymous with a set equipped with such a measure. 'Collection' is his translation of Cantor's 'Menge' (today: 'set') even if the two terms do not coincide in all respects. 'Enumerable' is his term for the power of sets with a finite number of elements. 'Innumerable' refers to the opposite property and thus corresponds to Cantor's concept of 'transfinite'. This field may, of course, be subdivided into several categories, of which the smallest is called 'denumerable', corresponding to Cantor's 'abzählbar' (today 'countable'), referring to the power of the natural numbers and related sets. Sets transgressing this size are 'abnumeral' (today 'uncountable'), – referring e.g. to the power of the reals. This power constitutes – Peirce here more or less tacitly following the continuum hypothesis (CH) – the first abnumeral number. Cantor believed to have given a proof of the CH, and general opinion of the period held it to be true and possible to prove; as mentioned, it was only in 1964 finally proved indecidable with respect to ZFC set theory. Peirce's overall stance here thus does not differ from that of the period.[506] With respect to the use of the concept of 'abnumeral', Peirce sometimes lets it include the 'denumerable' so that countable sets are seen as the zeroth abnumeral number. Later texts often substitute

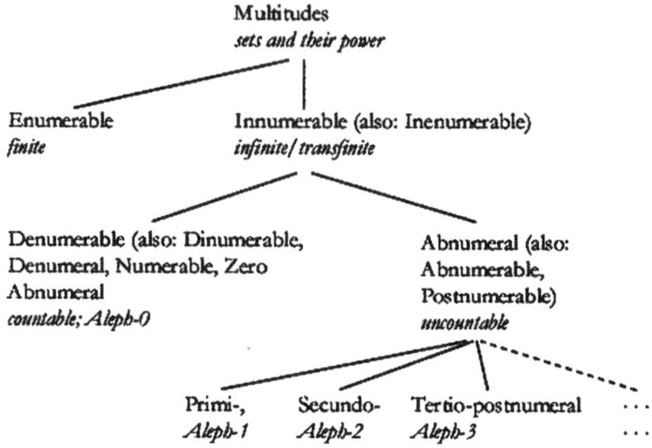

Figure 55.

'-postnumeral' for 'abnumeral', the former with the addition of a prefix addressing which uncountable set is referred to ('primi-', 'secundo-', 'tertiopostnumeral', etc.), just like 'denumerable' is later changed into 'denumeral' (because only one class of countables exist), and 'abnumeral' conversely into 'abnumerable' (because several such classes exist). We may sum up Peirce's concepts in the diagram above.

We said above that Peirce in this text – as in general – supports the continuum hypothesis (CH) as it is usually presented: Aleph-1 $= 2^{\text{Aleph}-0}$. This is, however, a truth with some modifications. Peirce (most often) agrees fully in the equation as given here: that the first cardinal number after the countables is equal to two elevated to the power of the countables. In this he agrees for several reasons. First, he agrees with Cantor in Cantor's theorem proving that the set of subsets (today: the power set) of a set with *a* members is given by 2^a, and that this number is always larger than *a*. This proof Peirce repeats at several occasions – and he even thinks he preceded Cantor in proving it. Furthermore, he agrees in the non-trivial content of CH that no further powers exist between the power of the natural numbers and the power of the reals: 'Let us now ask, what is the smallest multitude which exceeds a denumerable multitude? That there is a multitude which exceeds a denumerable multitude has already been shown. Namely, the possible combinations of whole numbers is such a multitude. The possible combinations of whole numbers each of which consists of an enumerable collection of whole numbers is merely denumerable. But the possible combinations of denumerable whole numbers form a collection which exceeds a denumerable collection. I call this multitude the *first [abnumeral multitude].*' (NEM III, 83–84).[507] The opposite idea would be that even if the reals can be proved (Cantor's diagonal proof) to have greater power than the natural numbers, it is not thereby proved that a certain specified subset of the reals might not be greater than the natural numbers and smaller than the whole of the reals so that $2^{Aleph-0}$ would equal Aleph-2 or even a higher Aleph; this is what Cohen's forcing technique in the 1960s proved indecidable. Peirce even agrees with

Cantor in what was later called the generalized continuum hypothesis (GCH): the idea that this relation not only holds between the Alephs 0 and 1, but between each successive pair of Alephs: Aleph-n + 1 = $2^{Aleph-n}$, so that the transfinite numbers form a well-ordered set of Alephs where the next Aleph in line has the same power as the set of subsets of the former Aleph which again equals the set of ordinal numbers having the former Aleph as its power. When Peirce does not, in another sense, agree with CH at all, it refers to a prerequisite so taken for granted that it does not at all appear in the canonical formulation of it. It is the idea that Aleph-1, the power of the reals, is supposed to be identical to the power of the arithmetical line, in turn identical to the power of the geometrical continuum (which is apparent only in the name of the hypothesis). This intuition which dates back at least to Descartes' analytical geometry identifying arithmetical numbers and geometrical line, so that for any point on the line there is a number, and vice versa, seems so evident that it need not be mentioned. In a Peircean light, however, CH as we usually understand it, should rather be stated as Aleph-1 = $2^{Aleph-0} = c$, where c refers to the power of the geometrical line. It is the last part of this equation which makes Peirce strongly disagree, and it is his alternative idea of c which the text in question sets out to prove. This idea is already inherent in the special way Peirce conceives the relation between infinite entities. Let us take a look at his 'proof' (which is of course not valid) of CH, after he has just articulated it in the quote above:

That there is no multitude greater than the denumerable multitude and less than the first abnumeral, I argue as follows: the demonstration does not seem to be perfect [sic]; but I think it is only because I do not succeed in stating it quite right. The addition, or aggregation, of denumerable collections is entirely without effect upon the multitude, even if the multitude of collections aggregated is denumerable. The multiplication of denumerable collections, that is the formation of all possible compounds each being a set composed of an individual out of each collection, is also without effect, if the multitude of denumerable factors is enumerable. But if the multitude of factors is denumerable, the result is the first abnumeral. Now since there is no multitude greater than every enumerable multitude and less than the denumerable multitude, it follows that there is no multitude greater than the product of every enumerable collection of denumerable factors and less than the products of a denumerable collection of denumerable factors which is the first abnumeral multitude. (84)

This sounds immediately seductive: Peirce lists the types of calculation – addition and multiplication – which are without effect on the size of a countable set in so far as they are used a finite number of times. Is multiplication used, on the other hand, a countably infinite number of times (corresponding to $2^{Aleph-0}$) you will get a larger set consisting of possible combinations, that is, possible subsets of the countable set. This implies, Peirce infers, that when you pass from a finite to a countably infinite set of operations, then you go from a countable set to an uncountable set as a result. But if this should count as a proof, then the burden is passed a step down in the Aleph hierarchy, the premiss being that there is no transfinite entity less than Aleph-0, that is, no infinite large numbers exist which are smaller than the set of the natural numbers. But it is, in fact, a possible implication of Peirce's point of view that such numbers exist.[508] But even apart from such Peircean peculiarities in the proof, it does not hold. It might be possible to define a set with power intermediate between the naturals and the reals with *other* means than multiplication. This is not excluded by the argumentation.

Peirce now goes on to consider uncountable sets and he observes, with Cantor, that just like it was the case with Aleph-0, the product of a finite set of uncountable factors is still only uncountable (here, Peirce has learnt things since 'Law of Mind'): if every factor of n factors is given in decimal expansion, then we may make one combined decimal expansion so that the first factor of the first number is given the first place the first factor of the second number the second place... the second factor of the first number in place n + 1, the second factor of the second number in place n + 2, etc. This implies that the product of a finite set of uncountable numbers may be described in one single decimal expansion and thus be contained in Aleph-1. But again the picture changes – according to Peirce – by an infinite (but countable) amount of factors, taking us into Aleph-2. Peirce's argument, thus: $\text{Aleph-1}^{\text{Aleph-0}} = \text{Aleph-2}$. But this is not the case, given his premisses; the result remains caught up in Aleph-1. The exponent must be uncountable in order for a shift to take place: Peirce here argues against GCH – or he has rather misunderstood it. About Aleph-2 Peirce goes on:'Mathematics offers no example of such a multitude. Mathematics has no occasion to consider multitudes as great as this' (85).[509] In the same way, Peirce constructs Aleph-3 and then he continues:

> And so the multitudes succeed one another indefinitely; and the constituent individuals of collections of those multitudes are distinct from one another.
>
> Thus, the whole series of multitude, so far as yet made out, begins with the multitude of a non-existent collection, or zero, and then comes the multitude of a single object, and then the multitude of 2, and so on increasing by one without end. After these multitudes comes the denumerable multitude which may be called the zero abnumeral multitude, then the first abnumeral multitude, then the second abnumeral, and so on increasing in order by one without end. All these multitudes thus form two denumerable series, and consequently there is only a denumerable multitude of different possible multitudes, so far as yet made out. (85–86)

All this might appear to be a (partially unsuccessful) attempt of an relatively orthodox presentation of Cantor's set theory; yet there is a strange emphasis on the role played by the countable sets in Peirce's account. In all cases it is the transition from a finite to a countably infinite set of operations which is supposed to give power change as its result. This is of course not allowed by GCH, according to which Aleph-x + 1 is achieved by elevating two to Aleph-xth power (and not only to the Aleph-0th power, as Peirce seems to assume). Peirce furthermore seems to have the idea that the whole series af Aleph powers is itself countable – opposed to Cantorian set theory in which Alephs may get ever larger transfinite ordinal numbers as their indices. Despite the generativity in the set of transfinite sets, it appears to Peirce to be somehow 'tamed' by the countable infinity, by the natural numbers.

After these preparations we now reach Peirce's own continuum hypothesis, presenting his idea of the relation of the geometrical line to transfinite sets (PCH, we might call it). It is given first in the form of a hypothesis:

Let us now suppose that there is a collection of distinct objects of each of those multitudes. Then, taking any one of those collections, no matter what, there is, among the whole collection of those collections, a denumerable multitude of collections each of which is greater than the collection chosen. Let us then throw all the distinct individuals of those collections so as to form an aggregate collection. This aggregate collection is greater than any of the single collections; for it has, as we have just seen, a denumerable collection of parts greater than any one of those collections. I shall call it a *supermultitudinous* collection. (86)

Peirce imagines the construction of a set constituted by the sum of the elements in a sum of sets composed by one set from each power class. In doing so, he of course unknowingly uses what was later identified as the Axiom of Choice (AC) and proved indecidable – it allows us to chose one element from each of an infinte set of sets. This union of sets of all powers U(Aleph-n), n $\rightarrow \infty$, Peirce now claims constitutes a 'supermultitudinous' collection *S*, larger than any cardinal. It is of course true, as Peirce claims, that such a set will be larger than any single Aleph-n, because it contains the elements of a set with the size Aleph-n + 1, and he even adds an equivalent to CH on this higher level: 'It seems to be sufficiently evident that there is no collection at once greater than every abnumeral collection and less than such a supermultitudinous collection.' (86), a claim which may be proved if (in addition to Peirce's other, fallacious premisses), the simple CH is assumed to be valid.

But why is it not the case that this set is merely identical to Aleph-ω_0, as it would be the case in orthodox set theory? First of all we must admit that Peirce differs from the later standard ZFC set theory by not assuming the so-called 'Axiom of Replacement' saying that for every set it holds that if the elements are replaced by something else (e.g. other sets), the result will still be a set. By this axiom, ZFC goes beyond the countable set of cardinal numbers because we may substitute for every number of the natural numbers the cardinal having the number in question as index and then take the union of the sets in this set – this set will have a cardinality exceeding any cardinal with a countable index. Peirce does in fact construct such a set, but he now argues that the decisive interest in such a set lies in the fact that it may be proved that this set *S* has properties differing from all other transfinite sets, making it the equivalent of the very power of the continuum. The kernel of this argumentation is the following 'proof' which just as little as his 'proof' for CH is valid. Still, it expresses nicely which metaphysical requirements a Peircean continuum definition must meet:

The collection of possible ways of distributing the individuals of a supermultitudinous collection, *S*, into two abodes is no greater than that supermultitudinous collection, *S*, itself. For denoting by γ the denumerable multitude, the abnumeral multitudes are [...] Exp γ, $(\text{Exp})^2\gamma$, $(\text{Exp})^3\gamma$, etc.[510]; and the magnitude of the supermultitudinous collection is the limit of this series. It is, in short, the result of a denumerable succession of exponential operations upon the denumerable multitude. But the magnitude of the collection of possible ways of distributing the individuals of a collection into two abodes is simply the result of an exponential operation upon the magnitude of the collection itself. *Hence the magnitude of the ways of distributing the individuals of a supermultitudinous collection into two abodes is obtained by adding one more to the collection of exponential operations successively performed upon the denumerable multitude.* But this collection of operations being denumerable, the addition of one operation to it does not increase the magnitude. Hence, the collection of possible ways of distributing the individuals of a supermultitudinous collection into two abodes equals that collection itself.

But we have already seen that the multitude of ways of distributing the individuals of a collection of distinct individuals into two abodes exceeds the collection itself. *Hence, it follows that a supermultitudinous collection is so great that its individuals are no longer distinct from one another.* (86–87, our italics)

Peirce here argues that the set S is larger than any transfinite number and consequently makes its single elements lose individuality. In the above quote we have emphasized the two decisive steps in the argumentation. In the period immediately before the first step we find the correct claim that the power of a set of subsets of a set is an exponential function of the power of the set. But after this claim, the argumentation fails when Peirce claims that the power of S should be obtained by adding 'one more' to the series of exponential operations quoted. What is that supposed to mean? In the series 'Exp γ, $(\mathrm{Exp})^2\gamma$, $(\mathrm{Exp})^3\gamma$, etc.' it is already presupposed that it goes towards infinity; hence S will not be obtained by the 'one more'. It would rather require a countably infinite ordinal ω_0 as index for the set, as it would be the case in standard Cantorian set theory where the series of Aleph-n's is followed by a series of Aleph-ω_n's, and so on, with no largest transfinite set showing up. By this fallacious argumentation Peirce 'proves' that S has the same power as the set of its own subsets: $S = 2^S$, at odds with Cantor's theorem $x < 2^x$. This contradiction Peirce attempts to solve – in the second step – by claiming that this inconsistency implies that S can not at all be a set (for which $x < 2^x$ must hold). Which kind of an entity, then, is S? S ceases to have distinct elements, because $x < 2^x$ describes the power relation between a set and the set of possibilities for combining its distinct elements into two sets (corresponding to its power set). 'It follows', Pmirce bravely claims, but why it should be exactly the distinctness of this definition that must be given up is not clear. It would be equally possible to conclude that S does not exist; that S is inconsistent; that S is neither a set nor a collection or class; that it is impossible to take out subsets from S... etc.[511]

No matter which construction Peirce makes of S, though, it is evident that it suffers from exactly the same inconsistency as does Cantor's 'set of all numbers' Ω which cannot include its own ordinal number, because the inclusion of that will make its own power larger. Cantor saw this already in the 90s, and if Peirce realized this he might have a sounder version of his idea of S: Cantor's inconsistent set of all numbers might be interpreted as referring to the fact that the continuum as constructed by points is necessarily inconsistent.

Thus, even if Peirce's 'proof' of the existence of a superpower set is thus far from convincing, it demonstrates how he imagines the continuum conceived of as something that transgresses any possible transfinite number referring to sets of distinct elements. An obvious question is why Peirce is so concerned to grasp the continuum in relation to the series of cardinal numbers when, at the same time, he conceives of it as a 'superpower' beyond all of them. The ensuing presentation answers that question. He is forced to ask the question whether S really 'stops' the transfinite series, that is, whether all supermultitudinous classes will now have the same size ('magnitude'; Peirce must of course give up the notion of 'multitude' here). He begins by considering a subset of *S*, namely finite decimal expansions, of which he may claim that 'all individuals which are marked each by an enumerable collection of characters form a denumerable collection of individuals.' (87). The crucial feature of this claim is that the set of finite combinations is not in itself finite, but countably infinite; this simple implication of the calculation of a power set has an interesting implication to Peirce: 'Any one of them or any collection of them may be taken away without affecting the remainder: but yet there are already symptoms of incipient cohesiveness in them, a premonition of continuity. Remember that the multitude of grades of enumerable multitude is denumerable; and the multitude of all the grades of multitude is no greater.' Time and time again, Peirce seeks an anticipation at lower levels of the transfinite hierarchy of the cohesion of the continuum and the vanishing of the distinctness of its elements; here his focus upon the countable infinity recurs. Peirce introduces the term 'arithm' to signify the set of size levels *below* a certain set, so that the arithm for 0 is 0, for 1 is 1, for 2 is 2, and generally is identical with the number itself for all finite numbers. But as there are only – to Peirce – countably many transfinite numbers, then it must be the case that, '...the arithm of all higher multitudes is the same. It is the denumerable multitude, which may be called *infinity*, ∞' (88) (which would not be the case in ordinary set theory where PCH is not assumed). Peirce interprets this as follows: '...this unity of the arithms of all the higher multitudes is the first embryon of continuity' (88) It is difficult to appreciate what exactly are the implications of this claim, but Peirce makes the following explicitation:

> The entire collection of those individuals that are marked by enumerable series of characters, is *denumerable*. But as soon as the denumerable multitude of interpolations has been made, the collections of characters attached to the last inserted individuals are denumerable, and the collection of individuals here is of the first abnumeral multitude. Stop it at any point you please, however early, if all the individuals marked by enumerable series of characters are there, there is also along with them a first abnumeral multitude of individuals marked by denumerable series of characters. *They stick together.* (88)

The quote refers to the imagination of a procedure where you are on the way to fill the set of finite numbers to the top, an infinite operation because of the infinity of the set. Peirce then imagines – contrary to intuitionism – the termination of this infinite series of procedures which is only marked by the fact that one all of a sudden must use infinitely many figures in order to give the next numbers, the first irrationals, so to speak, so that you are already in the process of construing the next, uncountable, set, the reals. You can not fulfill the finishing of the countable set without already implying its uncountable successor, this is Peirce's decisive argument making the bottom-up construction of transfinite sets *presuppose* the existence of higher-level sets. It is in this way, allegedly, that the continuum is anticipated in these sets' 'sticking together': we may not terminate the construction of Aleph-0 without already beginning on construing Aleph-1. By introducing the operational point of view, Peirce in one sense anticipates constructivism, even if with a totally different aim: constructivists will claim that a process involving an infinity of operations may not be ascribed any reality; Peirce will quite on the contrary claim that when you not merely define your way ahead formalistically, but imagine how the operations are to take place, then it is indecidable beforehand 'when' the construction of a transfinite set has terminated. You will first 'discover' this when elements of a higher set 'come along'. You cannot have Aleph-0 without being already fully involved with Aleph-1. The quotation marks above indicate that the apparent psychology of this presentation should only be taken for heuristics: the anti-psychologist Peirce will not in his interest in the character of the construction operation refer to psychology, but only raise the question as to the criterion for the conceivable, succesful termination of the procedure. The fact that there are no other conceivable criteria than the appearance of higher elements from a set of the next power then becomes the crucial argument: '... and before there can be any past after the denumerable collection is all there, a first abnumeral collection must already be there.' (92) Peirce repeats this point in different guises over several pages without any substantially new arguments occurring. It seems problematic to assume that the *whole* of Aleph-1 is supposed to appear in the completion of Aleph-0; it must be sufficient if one single real 'showed up'; in which sense would this one number presuppose the existence of whole sets of Aleph-1 power?

The argumentation might seem made with a guided hand. Elsewhere, Peirce may suggest that it is of no importance where in the Aleph series the strange transition to S takes place: 'It is, however, of secondary importance at what particular place in the series of multitudes this phenomenon occurs. It must occur somewhere...' (NEM III, 121 n1). While continuity thus seems anticipated in arithmetics, it is the case, conversely, that numbers as distinct entities may only express discrete order, never size ('magnitude'): numbers 'cannot possibly express continuity' (93). On the contrary it will be the case that between a limit (e.g. π) and a series converging onto it, any magnitude whatsoever may be interpolated. This leads back to S, for may this 'set' be so interpolated? Here, the existence of S is again defended with a fragile argument. Peirce talks about the possible interpolation of the different 'postnumeral quantities', that is, uncountable sets: 'But if we suppose that *all* such orders of systems of quantities have been inserted, there is no longer any room for inserting any more. For to do so we must select some quantity to be thus isolated in our representation. Now whatever one we take, there will always be quantities of higher orders filling up the spaces on the two sides.' (95) Here, the argument for the claim that $c = S$ comes from the infinitesimal side, form the possibility of 'filling up' the geometrical line. Here we hear the surprising claim that if we have filled up the line with all Aleph-n's, then there is room for nothing more, and thus the whole line will be complete. But there is no precise definition of this 'filling up', and the argument is not valid. It would be possible to fabricate a completely parallel argument (satisfying constructivists!) against the existence of the reals: take a real number, it must lie between two fractions, but no matter which divisor is chosen for the fractions, there will be fractions with higher divisors 'filling up'. But what is the proof that there is no more space left?

It seems like Peirce himself realizes the fragility of his argument: the text concludes symptomatically with a whole bunch of further arguments. The first is a *Gedankenexperiment* with knifes cutting the line into parts, a surprising picture in so far as it runs counter to the whole Cantor-like tendency in the argument up till now, having attempted to synthesize the line from a combination of elements: 'Across a line a collection of blades may come down simultaneously, and so long as the collection of blades is not as great that they merge into one another, owing to their supermultitude, they will cut the line up into as great a collection of pieces each of which will be a line, – just as completely a line as was the whole. This I say is the intuitional idea of a line with which *the synthetic geometer* really works, his virtual hypothesis whether he recognizes it or not; (...) cast aside all analytical theories about lines, and looking at the matter from a synthetical point of view to make the mental experiment and say whether it is not true that the line refuses to be cut up into points by any discrete multitude of knifes, however great.' (96, our italics). This would imply that if you – the synthetic geometer – should throw away all the analytical arguments, then this should also include the vacillating proofs of the preceding pages trying to construct the line from the 'merging' and 'welding' of distinct entities. This you might do without if you accept the continuum as a geometrically intuitive primitive object, on which points are mere different classes of intersections. The number of points into which the line presumably may be cut, is of course also taken to be *S*, of which Peirce gives no less than two blitz proofs:

> The possible lines into which any line may be cut at one chop exceed any discrete multitude. Now the points on a line form a collection at least as great as the collection of the possible lines into which it can be chopped. Hence, the points of a line are supermultitudinous.
>
> The other proof is this:
>
> A line consists wholly of points; but in order to chop a line into points, the two ends of each piece must unite; and to do that without shrinkage they must merge into one another. Hence the collection of blades of the chopper must be so great that its constituent blades are no longer distinct. (97)

It is not difficult to complete this 'proof' which just like the first one rests on the intuition of the line as composed of a set of small lines any of which is larger than points. The premiss is thus that the synthetic line contains pieces of line in the number *S*, not so far from what should be proved, and you need only the further premiss that points are smaller than lines in order for the proof to be finished. The first premiss, however, is little less problematic than is the conclusion.

It is interesting that Peirce – after all these proof attempts – return to his definition of mathematics as the science which draws necessary conclusions from hypotheses. It must be this idea which lies behind his claim that: 'This I declare to be the synthetic geometer's hypothesis of the relation of a line to its points. *But it does not affect my argument if it be not so. It is sufficient for my main purpose that it is a perfectly consistent hypothesis*. For all I am trying to do is to elucidate the conception of a supermultitudinous collection and show that it involves no contradiction.' (97, our italics) The whole of the argument is so to speak fictitious, what Peirce wants to prove is only that a certain concept – *S* – is consistent – not that it exists in any more ontological sense. Mathematical existence in this claim seems formalistically to be reduced to consistency. After having regarded and rejected the possibility that *S*'s 'cohesiveness' could be due to a certain tricky arrangement of its elements only, Peirce admits, before his final conclusion, that 'There is, I confess, a paradoxical aspect in the proposition that a collection may be so great that its individuals lose their separate identities. But the key of that paradox will probably ultimately be discovered to lie in *some unnoticed condition* in the general hypothesis of a collection which requires the mergency of individuals.' (100, our italics)[512]

But the proof that PCH is consistent in relation to set theory (if we assumed it could in fact be proved) is not the same as the claim that it is implied by set theory. 'Half' of Cohen's proof of CH's indecidability in relation to ZFC set theory consists of Gödel's proof of its consistency in relation to ZFC set theory. Peirce consequently only believes he has proved that PCH is not inconsistent with set theory, and consequently he must await the appearance of a new axiom, 'some unnoticed condition' in the definition of the concept of set may finally connect the two theories – not unlike Gödel when he, 50 years later, hoped for a solution to the CH problem by the assumption of new, abstract, and complicated axioms concerning the concept of set which might be obtained by phenomenological, non-formal means.

If we look at Peirce's argumentation for *S* from a distance, there are several paradoxes to be solved. We already mentioned one: Peirce's strange interest in the construction of the transfinite hierarchy of sets, given his explicit insistence on the intuition of the synthetic line before any divisions – a line which it should not, then, be necessary to create by the 'merging' of discrete points.

Another strange issue is the fact that given Peirce's idea of mathematics as the pure investigation of the implications of hypotheses, it seems peculiar he tries to found the decisive concept in his metaphysics on a mathematical argument. The answer to this question can only be that Peirce, in metaphysics and logic, regards mathematics as a sort of aide, a supply of deductively valid diagrams whose application in less hypothetical cases mathematics assumes no responsibility for. But in that case, Peirce's argument is metaphysical and not mathematical, and all the mathematical argumentation only goes to show that the idea of a superpower continuum is not inconsistent: the mathematical argument, were it valid, does not *prove* *S*'s existence, but only its *possible* existence – the further verification of the hypothesis is no mathematical issue but a case for phenomenology, logic, semiotics, and metaphysics.

A third reason for puzzlement is the combination of similarities and differences in relation to prominent positions in the philosophy of mathematics of the twentieth century The idea of mathematics as the investigation of hypotheses approaches Peirce to Hilbertian formalism; just like his unproblematic acceptance of objects requiring infinities of operations to be constructed (which is no problem for the formalist as long as the calculus in question may be stated in a finitist metalanguage) – but on the other hand, Peirce displays certain affinities with intuitionism, partly in the synthetic aspects of his arguments for *S*, partly in his idea that the generativity of transfinite cardinals must somewhere be stopped. Peirce will not be hindered in undertaking innumerably many operations, but he can not subscribe to the idea of Aleph-ω many operations.

A fourth reason, finally, is the tension between his insistence that the continuum exceeds any attempt at grasping it analytically with discrete means – and on the other hand his claim that sufficienty large discrete sets seem to be inhabited by germlike tendencies towards continuity, the idea that one cardinal in some sense implies the next one – in general the idea that the transition between discontinuity and continuity must, in itself, be continuous. We shall return to these points.

In a lecture series from 1898 which has recently been published in its entirety under the title *Reasoning and the Logic of Things* (RLOT) Peirce again takes up the definition of the continuum, now connected to his claim for the reality of infinitesimals and the most elaborated version of his continuum metaphysics. Let us discuss Peirce's metaphysics of infinitesimals and the publishers' Ken Ketner and Hilary Putnam's reconstruction of Peirce's position.

INFINITESIMALS AND THE STATUS OF MATHEMATICS

In the third of the 1898 lectures, Peirce introduces his logic of relations, to which we return below.[513] We shall not here discuss Peirce's diagrammatical notation techniques for logic, but only take note of those ideas implying continuity and infinitesimals. Peirce implicitly takes his point of departure in the traditional predicate judgment 'S is P' which is generalized to the extent that more than one subject may appear in it. This is well known from everyday language, as for instance in constructions like 'X sells Y to Z for the price of Æ'. Peirce then determines the predicate as that structure in the construction which is left when all subjects are replaced with empty slots. In this way the predicate outlines a relation between a plurality of subjects, and predicates may take potentially infinitely many empty slots to be filled with subjects. By his famous theorem of reduction, Peirce claims that all predicates with a valency larger than three may be reduced to a combination of predicates each of which are no more than trivalent. Trivalent predicates exist, on the other hand, which may not be further decomposed into bivalent predicates. The predicate in the logic of relations thus implies a more extensive generality (and, consequently, possibility for comparison) than does the classical predicate judgment. Now subjects may be related not only in terms of satisfying (or not) the same predicate, but also in terms of satisfying one predicate in different slots; appear in analogous slots in different predicates, etc. Similarity, in this respect, is just a simple relation among many, and the concept of a class of objects defined by a predicate is substituted with the concept of a 'system' defined with the interrelations between its objects (156).

Second, it is crucial to note that to Peirce this relational logic makes possible a tripartion of the types of signs necessary to describe these relations. First *verbs* describing predicates (to which he counts adjectives, common nouns, etc.), second *demonstrative pronouns* (including proper names), referring to the subjects involved in the expression, and finally logical operators (negation and conjunctions, etc., see Chap. 1). These three classes of signs now provide the meaning, reference, and inference, respectively, of the proposition.[514] (155).

The crucial issue here with respect to the continuum is that the classes (Peirce uses the term even if he has just argued it ought to be replaced with 'system') of objects referred to by propositions of the logic of relations very often possess generality (predicates are general and subjects may be). That this generality can only be defined by continuity forms a cornerstone of his semantic realism.

Now he runs through, in high speed, his 'proof' for the supermultitudinous set S, and then he divulges in new deliberations on the nature of the continuity of a line. Take, he says, a circle, it is a set of points, because a particle which in every moment contains a point, moves along the periphery and draws a line which consists only of points in which the particle has been. This appears to be a traditional description of the matter, but Peirce, importantly, adds that no point on this line has any distinct identity; it is only when we single out a point on the line that it becomes distinct from the others – because a discontinuity is then created. If we now cut the line in this point, *then the marked point has become two points*. This claim, of course, is contrary to the concept of the Dedekind cut which defines a real number (and hence, in the classical interpretation of the arithmetical line, a point on this line) by dividing the line in a closed and an open section (that is, one part containing its limit point and another not so):

It has become two points. And if those two ends were joined together so as to show the place, – they would become one single point. But if the junction ceased to have any distinguishing character, that is any discontinuity, there would not be any distinct point there. If *we* could not distinguish the junction it would not appear distinct. But the line is a mere conception. It is nothing but that which we can show; and therefore it follows that if there were no discontinuity there would *be* no distinct point there...(160).

The two points are two as soon as we can distinguish them. What is at stake here is of course not mere psychological distinction skill, the crucial thing is what can be demonstrated with logical means. One should not be fooled, however, by the apparent simplicity of the criterion; even if the line is only our 'conception', the evidence it presents to us is not complete – cf. also Peirce's Gödelian idea that new axioms might settle the continuum hypothesis.

Even more counterintuitive to the present conception of the line is the following claim: one can remove one point from the end of the line segment, but still an end point will remain – as against standard arithmetic, according to which we are left with an open interval with no limit point included in the line if we remove the limit point from a closed interval. As is evident, Peirce's conception of the geometrical line is very different from the arithmetical line:

'The end of a line might burst into any discrete multitude of points whatever, and they would all have been one point before the explosion. Points might fly off, in multitude and order like all the real irrational quantities from 0 to 1; and they might all have had that order of succession in the line and yet all have been at one point.'(160)

Points in an amount corresponding to the real numbers may be removed in an ordered series from one single point on the line. As is evident, this conception of the line also implies a completely different conception of the point which is close to the Aristotelian idea of the line being primitive and the point being the addition of a discontinuity to the line. It goes even further than Aristotle's idea which does not contain any idea of a point being able to contain other points, let alone an ordered series of points.

The editors of *Reasoning and the Logic of Things*, Ketner and Putnam, have added to their preface an attempt to construe a consistent hypothesis from Peirce's doctrine of point and line. They observe that none less than Gödel remarked that it seems intuitively correct (as against the Dedekind cut) to assume that if the geometrical line is cut into two parts, the result will be two symmetrical pieces (personal communication to Putnam, 38n). The division of the line should not be conceived of as the mapping of sets of points (because the line is no set of points), but quite on the contrary as an irreducibly geometrical transformation, and they propose that Peirce's point of view may be expressed in analogy with non-standard analysis which, like him, reintroduces infinitesimals in analysis. Thus, when Peirce claims that the infinite series of points 'are' one single point, then this is no identity claim, but rather a *mereological* use of 'to be' so that it should be read as saying that the point 'consists of' of many points, just like in the proposition 'India and Pakistan were one country in 1940'. Hence, Peirce's conception of the point implies that the points may have parts – and transfinite reasoning, of course, parts may have the same power as the whole they partake in.

Formalized in non-standard analysis, they continue:

> ...we say that two points P and Q whose distance is infinitesimal are 'identical modulo the infinitesimals', and we symbolize this by using a wiggly equals sign: $P \approx Q$. If P is a point, the collection of all points Q such that $P \approx Q$ is called the *monad* of P.(44)

It can now be shown that every point must lie within the monad of a standard point, and they propose that Peirce's idea can be expressed in the following way: his concept of 'point' corresponds to the concept of 'monad', and parts of points corresponds to normal points. Then it becomes meaningful to say that every point (monad) contains a set of parts (points) with the power of the reals. Given Peirce's identification of the power of the geometrical line with the supermultitudinous set S, one may infer that the set of non-standard points on the line (exceeding Aleph-1) must be vast; in Peirce's construction of this phenomenon it does not, however, give meaning to distinguish standard and non-standard points, respectively, because Peirce does not share the ambition of non-standard analysis to correspond to standard analysis. So is it, from this point of view, possible to 'remove' a monad from the end of the line? Ketner and Putnam plausibly argue that this point of view must be constructivist: by removing the monad we create a new monad of possible points; what is removed is a 'notion of a possibility', and it would give no meaning to say that we removed the whole set of points which could ever be construed in the monad of the original point (53). We may add to Ketner and Putnam's discussion that Peirce has already elsewhere, in his analysis of the continuity of time (which we have already touched upon), proposed a distinction analogous to that between monad and point, namely *instant/moment*. The latter has an infinitesimal extension and hence actual reality as opposed to the former which are mere, punctual aspects of the latter and which may be actualized or not.

Construed in this way, Peirce's position with respect to infinitesimals is consistent, and Ketner and Putnam may propose that Peirce's use of the concepts of potentiality and actuality in relation to the possible and actualized points takes him close to Putnam's own 'modal logic view of set theory'[515] (50): sets are not mathematical objects in the sense that we, e.g., have the set of all real numbers[516] before us, the sets are constructions which implies no more and no less than given a certain set of operations we *may* construct them. In this respect – they continue – Peirce comes close to e.g. Brouwer's intuitionism with the crucial difference that Brouwer would let the finitude of the human mind prohibit infinite series of operations (which implies, in turn, the sacrificing of well-known mathematical objects as for instance the real numbers), while Peirce, given the free domain of potentiality on the one hand and the doctrine of 'real possibilities' on the other, did not erect prohibition signs of that kind. Given the potentiality of concepts, Peirce does not refrain from assuming a set transgressing the power of any transfinite set, his S – but this has the epistemological reason that intuition and perception always already contains this set. The faculty of intuition thus radically exceeds what any potential, transfinite series of operations would ever be able to accomplish.

These vast sets of points *are* not, actually, on the line which is not, in turn, composed of them. As an Aristotelian, Peirce may indeed say that the points on the line *potentially* are there but that they are only actualized when we perform some operation making them distinct. It is this Aristotelian train of thought which forms the base for Peirce's constant link between the continuum and the concept of potentiality – and it is *this* metaphysical intution which provides the motor in Peirce's various attempts to chart the domain of the transfinites: multitudes of ever so great a power are distinct and hence potentially actualized while the realm of the merely possible always remain greater than that of the actual. This is why Peirce seeks a mathematical foundation for the point of view that the continuum exceeds any transfinite power.[517] In this way, Peirce founds his doctrine on a modal triad: the potential, the actual, and the real. Already the whole numbers constitute a set which we can only potentially construe completely but which in principle would be actualisable as infinite procedures can be completed. Beyond the actual lies, consequently, a realm which is not only potentially actualizable but, as Peirce will put it, realizable: evolution does in fact realize continua in the shape of 'real possibilities', as universals with real existence. Beyond that, on the other side, remains the continuum in itself, which may never be actualized in toto. Ketner and Putnam express this idea in terms of possible-world semantics: '...one can never construct *all possible* points on the line, because there is no possible world in which there are actually Ω [their Cantorian name for the size of S, fs] 'distinct individuals.': '(...) *What answers to our conception of a continuum is a possibility of repeated division which can never be exhausted in any possible world, not even in a possible world in which one can complete abnumerably infinite processes.*' Also here, the two of them continue the ideas of Putnam 1967: '... it is not possible for all *possible* sets to exist in any one world!' (311)

THE RETURN OF NON-COMPOSITIONALITY – A PEIRCEAN CONTINUUM HYPOTHESIS AS METAPHYSICAL POSSIBILITY?

Putnam and Ketner's Peirce interpretation forms part of the tendency to question the Cantor-Dedekind interpretation of the reals which has been state of the art for most of the twentieth century set theory and arithmetics. The force of that position (**R** as an uncountable, connex set (which happens, that is true, to make discontinuous real functions much more probable than continuous functions!)) has been to unite arithmetics and spatiality, so that different geometrical and topological spaces can be conceived as having $\mathbf{R}^n$ as their base. Thus cardinality and topology are unified (because uncountability and connexity are identified), logic and mathematics are unified (Salanskis 194). The continuum hypothesis is undecidable, that is true, but this inconvenience has been neglegted and has had few consequences in practice. The duality within formalism of pragmatism (we assume CH when needed) on the one hand, and scepticism (in reality we know nothing about continuity to the extent that it exceeds formalisations) on the other now gives way to a new obsession with what Salanskis calls 'the return of non-compositionality'. This has been evident, inter alia, in the volume *Le labyrinthe du continu* which he edited (1992) with the participation of a series of mathematicians and philosophers of mathematics. Non-standard analysis with the return of infinitesimals, non-Cantorian set theory, renewed interest in Gödelian Platonism, the search for new axioms in set theory – all these tendencies might point in the direction of Peirce's metaphysics of the continuum, and in this section we shal briefly consider a bundle of these actual deliberations on the status of the continuum.

The possible consistency of Peirce's construction – despite all the blatant errors in his attempts at appropriation of Cantorian set theory – lies first and foremost in Cohen's proof of the autonomy of the continuum hypothesis. By the so-called 'forcing' technique this proof shows that within standard ZFC set theory, the continuum may be ascribed any cardinality you wish without this leading to inconsistency.[518] It could be added that Peirce's idea of *S* has an interesting equivalent in Cantor's concept of the set of all ordinals Ω which he as mentioned realized was an inconsistent concept, an 'absolutely infinite set'. The proof of the inconsistency of this set seems to be vaguely foreboded by Peirce in his 'proof' for the existence of *S* – the fact that the sum of the numbers preceding a given number equals that number gives the consequence that 'the set of all numbers' is inconsistent. Peirce commits the error of assuming the countable series of transfinite cardinals to constitute a superior limit to the series of transfinites, but apart from that his argumentation takes the same course as the argument that 'the set of all numbers' is inconsistent. We can add that Peirce certainly would not oppose the idea that *S* is inconsistent. In his trivalent logic this could be interpreted as saying that claims involving elements of *S* have the third, undecided truth value between truth and falsity – anticipating intuitionist logic. Precisely the infinitesimal parts of the line are ascribed such a status by Peirce: he claims that the law of excluded middle (implied, of course, by a two-value logic) does not hold for the domain of the possible and particularly not for these points which are consequently 'vague'. On the other hand, if falsity is taken to be the only source of inconsistency in logic, *S* would yet be consistent in this (admittedly) more broad use of the word.

All the time, certain voices from the past kept protesting against the scepticist triumph of formalism in the twentieth century. On the one hand, the hyper-scepticism of intuitionism connected to (psychological, it must be admitted) references to intuition (Weyl, Brouwer), on the other hand, Gödel and his updating of Platonism. Gödel kept insisting that the continuum had existence as an autonomous object to be investigated. It just needed a more complicated and competent version of set theory, obtained by the addition of new axioms, in order to be decidable. Gödel claimed – and rightfully so – that formalism tends to overlook the fact that there must always be extra-formal motivations for the assumption of new axioms – and he claimed, moreover, that we have no reason to expect such motivation to be particularly simple in every case. Against the formalist main current Gödel thus interpreted his own famous 1931 proof, not as an indication that you had to stay within formalisms (because it proved that every sufficiently complex formalism makes possible true propositions which cannot be proved), but quite on the contrary as an anti-formalist insight that formalisms are motivated from their outside. Thus, a compuational theory of consciousness was impossible and the main task ought to be the search for new and more ambitious axioms by an intuitive, phenomenological procedure, not a formal one. The older Gödel for the same reason took a strong interest in Husserl and assumed that a 'clarification of meaning which does not proceed by logical definitions' could take things further, for instance by a phenomenological analysis of the concept of 'set'[519]. Gödel shifted between believing and disbelieving CH, but his Platonism remains a corrective to the extreme formalism which will renounce any intuition at all (or better: *Anschauung*) access to mathematics except the tiny subset of intuition needed to read finitist calculi of symbols, a subset which is assumed to be fully controllable.[520] Apart from this, it remains undecidable (if we are to take formalism at face value) what a formal calculus is in fact 'about'. Opposed to this, Gödel's point of view is an ongoing questioning of formalism, to the extent that intuition – in this case the primitive intuition of the continuum – is taken to be irreducible and the formalisms are seen as vehicles for knowledge about intuitions, not for the extermination of them.

In the last decades, Platonism and its claim that CH must be decidable because mathematics is taken to have real, autonomous entities as its objects, have had a renaissance, e.g. in Penelope Maddy's philosophy of mathematics (Maddy 1990 and 1997) claiming a so-called physicalist Platonism. It takes its point of departure in physicalism in much of recent American epistemology, as it is expressed e.g. in the classical Quine-Putnam position where the applicability of mathematics in physics is the main argument for a kind of Platonism. This Platonism claims, at least, that mathematics is not only a mental construct of ours nor empty logical formalisms (even if Quine still clings to conventionalism). As Maddy rightly observes, this position does not account for the validity of non-applied mathematics, neither for the classical issue of the apparent evidence in simple mathematical intuitions. Yet, she agrees in the demand for naturalization of mathematics, taken in the sense that mathematics must be seen as a science on a par with other empirical sciences[521], so that a physical explication is needed to account for the mathematician's knowledge about his objects. This idea gained authority due to a classical paper by Benacerraf ('Mathematical Truth', 1973, reprinted in Benacerraf and Putnam 1982) which takes as its main argument against Platonism that it does not account for how it should be possible to achieve knowledge about timeless, abstract entities without any 'causal power'. This idea is influenced by the 'causal reference'-semantics of its time claiming that reference in general would be explainable by causal influence of the object on the utterance. This somewhat simplistic idea is taken up by Maddy in her idea that sets may be perceived directly. In a certain sense, this is of course correct as a Gestalt Theory claim, and in this sense it is important because it can be argued that many sets have an objective, empirical existence. But its naïvety relies in the fact that this claim is identified with the requirement that the objects in question have a physical existence only, that is, a purely material existence. Another problem is that this idea holds only for finite sets – nobody ever saw a transfinite set, so taken at face value, such an idea might lead to strongly finitist ideas which Maddy does not embrace. A more cautions expression might be to say that the schematic aspect of perception which allows us 'directly' to see shape, contour, number, sets, etc. of everyday perceptual objects is the same ability which is active in the pure intuition of mathematics.[522]

But let us leave the physicalist part of Maddy's Platonism in order to discuss her insistence that realism demands of CH that it be decided.[523] Her point of departure is Cohen's 1966 proof that CH is undecidable in ZFC. This proof basically assumes a tension between the two methods for obtaining higher cardinals (by the power of ever larger sets of ordinals; by the power sets of the preceding cardinals, respectively). Cohen finds an incurable opposition between these two methods, because exponentiality in the power set construction is always 'stronger' than the generation through ordinals. Thus, Aleph-1 must be less than $2^{\text{Aleph-0}}$ – and CH consequently false:

> A point of view which the author feels may eventually come to be accepted is that CH is *obviously* false... Aleph-1 is the set of countable ordinals and this is merely a special and the simplest way of generating a higher cardinal. The set [of subsets of the natural numbers] is, in contrast, generated by a totally new and more powerful principle, namely the Power Set Axiom. It is unreasonable to expect that any description of a larger cardinal which attempts to build up that cardinal from ideas deriving from the Replacement Axiom can even reach [a set of size $2^{\text{Aleph-0}}$]. Thus, [$2^{\text{Aleph-0}}$] is greater than Aleph-n, Aleph-ω, Aleph-α, where α = Aleph-ω, etc. This point of view regards [the power set of the set of natural numbers] as an incredibly rich set given to us by one bold new axiom, which can never be approached by any piecemeal process of construction. (Cohen (1966), 151; quoted from Maddy 131)

– a theory which just like many in our days considers the continuum large as compared to Aleph-1 (unlike Gödel who most of his life did not believe in CH neither, but tended to identify the continuum with Aleph-2). In this light, the assumption of CH would be an unrealistic delimitation of the continuum, but the matter may, as Maddy says, just as well be seen the other way around: that CH ascribes to Aleph-1 an unrealistically large size. Cohen's proof, however, paved the ways for the attempts which have since then been made to construe a new and better version of CH by reparing Cantor's original version or by stating a version making the continuum stronger than Aleph-1.

The former road takes its departure in Gödel's first half of the indecidability proof (the proof for CHs consistency in ZFC) which is built on a version of the classical, iterative set theory with its axioms – but delimited in so far as only those subsets are allowed which may be explicitly defined by predicative formulae, that is, expressions which refer to sets already constructed. This so-called constructible universe is called L, and Gödel's proof thus adds a constructibility axiom to ZFC which identifies the whole universe of set theory with the set of merely constructible sets L, so that $V = L$. Gödel's proof now demonstrates that no contraditions may be derived from $ZFC + V = L$ which were not already in ZFC. But as it may be proved that CH follows from $ZFC + V = L$, CH can not be disproved by ZFC. $ZFC + V = L$ is thus a straw man built with this one intention – but could it not be converted into a positive idea so that $V = L$ is taken as a new axiom which makes not only CH and GCH but also AC (so all sets may be well-ordered) true? This solution has tempted many which find it an attractive and even 'natural' solution, because constructibility may be seen as a consequence of the very concept of 'set' (Keith Devlin). In this train of thought, the set concept is interpreted in a way appealing to Peirce: intensionally (as an explicitation of certain predicates referring to properties which, in turn, determine the elements) instead of extensionally as defined by its elements. This solution, however, is untenable, also for Maddy, because it goes against the fundamental, combinatorial assumption of set theory by simply prohibiting the existence of certain subsets.[524]

The other main road leading from Cohen 1966 receives Maddy's support: to see the continuum as stronger than Aleph-1. This strategy must thus seek new axioms supporting such an idea, partly the different large cardinal axioms and so-called 'dependency axioms'. Already Gödel rested his head on such axioms, and the first of them, the axiom of inaccessible cardinals (going back to Zermelo as well as Sierpinski and Tarski around 1930), is built on the simple intuition that all sets which may be built from the other axioms in themselves constitute a set, on which those very same procedures may in turn be used (Maddy 135) – by which procedure otherwise inaccessible cardinals are reached (because of the infinity axiom is used on already infinite sets, so to speak). This operation may now be repeated, which, in turn, gives as a result the axioms of 'measurable cardinals' and 'supercompact cardinals'. The former assumes that there are inaccessibly many cardinals between the inaccessible cardinals and the first measurable cardinal, and was proposed already by Ulam in 1930 and gaining support because of its promising implications elsewhere in mathematics.

Unfortunately, it was quickly proved in 1967 (Solovay and Levy) that none of these attempts decides CH. Axioms of 'very large large cardinals' led to the introduction of so-called supercompact cardinals (there are measurably many measurable cardinals below the first supercompact (SC) cardinal) which proves projective determination, and ZFC+SC were thus seen by many in the 1990s as an attractive maximalist alternative to the minimalist ZFC+V = L solution. But the status of CH has not yet been settled by this development:

> If [CH] is true in L, a minimal environment, perhaps it is false in a maximized world of large cardinals. There is some evidence in that direction [...] But in Cantor's favoured form - Aleph-1 = $2^{\text{Aleph-0}}$ - it remains false; the continuum isn't of size Aleph-alpha for any alpha because it can't be well-ordered.(142–43)

No unanimous consequence for CH follows, and different positions among the adherents of ZFC + SC seem to exist – this, in turn, because they do not agree whether CH or its negation is the most restrictive claim (143). Even if no solution thus is at hand, Maddy calls on a realist basis for research to continue, and she does not hide that her sympathy lies with ZFC + SC. The strange thing is that this preference in no clear way is connected to her physicalist Platonism.

Maddy is concerned with the question whether arguments for the different axioms have inner or outer character, respectively, referring to whether the arguments are related to the plausibility of axiom itself or whether they relate to intertheoretical effects in relation to other parts of mathematics. She concludes that the arguments for very grand cardinals do not fit easily into one of those groups, but rather follows certain 'rules of thumb' which are – just like Gödel imagined, in fact – is connected to basic intuitions about what the concept of set entails. These are intuitions like the idea that the universe of sets continues to as large cardinals as possible or the idea that the hierarchy of cardinals must contain subsets which looks alike – a sort of generalisation of the axiom of replacement. The reason why Maddy refuses to give these rules of thumb an 'inner' status goes as follows: '... because I think they extend beyond anything that could plausibly be traced to an underlying perceptual, neurological foundation.' (141)

But this already goes for **N** taken as an actual infinity?! Maddy admits that the idea that sets are formed from already existing elements and that they are formed by combination without necessitating any explicit rules, does have an intuitive basis, while the continuation of the hierarchy to infinity (cf. already the axiom of infinity) and the continuation of it as far as possible, is based on the 'methodology of set theory' rather than any simple intuition (ibid.). But is this methodology without intuitive basis? It seems as if Maddy's concept of intuition is constrained by her physicalism to a small, strongly finitist inventory. It is indeed correct that no sets larger than finite or, at most countably infinte sets, seem to have any probable existence in the physical world. This gives rise to further questions: a really stubborn physicalist would probably not find questions far upwards in higher cardinals interesting (cf. 'materialist' Soviet mathematics refusing the 'idealism' of set theory), because they really only pertain to non-applied math? And, as Maddy herself touches upon: the number of neurons and their interconnections in the brain is vast but still finite: how could they (presupposing a computational theory of consciousness which Maddy does not discuss further) give rise to representations of the infinite? But there is no need to go so far as to the strong (even if yet undecidable) arguments for the existence of non-computable processes in the physics of the brain (like Roger Penrose) to find Maddy's version of basic mathematical intuition unsatisfactory. For it *is* indeed the case that a simple intuition of the continuum exists which does not, it must be admitted, imply the assumption of actual infinity but only refers to the continuum as a primitive phenomenon characterized by homogeneity between its parts, lack of distinction between parts, etc. – to a large extent corresponding to the Aristotelian notion of the continuum as potentiality. The two Aristotelians Peirce and Thom refer, both of them, to the idea that the givenness of continuity for intuition in some sense grants its reality. Thom: the idea of the continuum can not be an illusion – it is intersubjectively verifiable and it is even subject to commercial exploitation, as in cinema where we without any problems interpret the presentation of 24 discontinuous pictures per second as the continuity of time: 'thus, it is an "objective" scientific fact and must, as such, be "objectively" explained. If you do not have an immanentist theory of the continuum in our neurophysiology, how will you explain to me the origin of the illusion!'[525] Essentially the same argument may be found in Peirce who argues that continuity – together with thirdness and argumentation – is not the privilege of the mind, because it can be found already in nature: 'Whatever unanalyzable element *sui generis seems* to be in nature, although it be not really where it seems to be, yet must really be [in] nature somewhere, since nothing else could have produced even the false appearance of such an element *sui generis*. (...)... the very fact that there seems to be Thirdness in the world, even though it be not where it seems to be, proves that real Thirdness must somewhere be. If the continuity of our inward and outward sense be not real, still it proves that continuity there really is, for how else should sense have the power of creating it?' ('Detached ideas continued', 1898, NEM IV, 344) The Peirce-Thom argument has a strong thrust against physicalists whether they be nominalists, or, like Maddy, realists, because the core of the argument is naturalist: basic existing ideas which can not be reduced to other ideas (such as may unicorns, etc.) can not be naturalistically dismissed as illusions. They must possess birthright in nature itself, for otherwise the naturalisation of epistemology can not be pulled through... One can add, a propos set theory, that it is the primitive idea of the continuum which gives rise to the rarely discussed premiss in CH that $c = \mathbf{R}$ – that the real numbers (whichever vast cardinality they may possess) do in fact exhaust the geometrical line – and hence, in turn, gives rise to the whole search operation for grand grand grand cardinals, or, the other way around, a constructivist constraint on the real numbers. From a Peircean point of view, the search for very large cardinals could indirectly appear as a proof for continuity as a primitive intuition, because the larger cardinals are being invoked in attempts at making CH decidable, the more probable the inexhaustibility of the continuum seems to be.

But Peirce's rejection of identifying the continuum with the real numbers makes him, on the other hand, immune to the possibility of CH being decided by some new grand cardinal axiom: even the eventual identification of **R** with some mighty cardinal will, from his point of view, never exhaust the intuitive continuum.

LACK OF DISTINCTION

A further mathematical development which might support a consistent interpretation of Peirce's concept of the single elements' lack of distinctness in uncountable sets is Gregory Chaitin's algorithmic information theory and its 'random numbers' which follow from an extension of Gödel's theorem. Chaitin defines a random number as a number which can not be expressed in any shorter way than its own decimal expansion. Many numbers may be written shorter than that, the integers, the rational numbers (being expressible as fractions), but also many irrational reals, π, e – in short, all real numbers which may be expressed by some finite algorithm. The interesting issue in Chaitin's theory is that it can be proved that the random numbers (unlike the non-random numbers) have the same power as the reals – so that 'almost all' numbers on the arithmetic line are random. On the other hand, it can be proved that it cannot be proved about one particular number that it is in fact random. It is impossible, given a candidate random number, to exclude the possibility that some subtle order in its decimal expansion might once be found, thereby making the decimal expansion reducible to some shorter expression. This implies that random numbers are uncomputable, they may not appear as the output of a computer (not only because of the infinte amount of time printing would require) – they can not be determined distinctly. In this interpretation, the random numbers as a vast subset of the reals seems to satisfy Peirce's demand for non-distinctness, vagueness: it is impossible to pinpoint any single random number even if they are overwhelmingly there.

Chaitin does not involve the continuum question, but a related argument is presented by Harthong (cf. Salanskis, 207) who makes a distinction between simple, programmable integers on the one hand and complicated, aleatoric integers on the other hand. For sufficiently large natural numbers, the simple numbers are surrounded by complicated numbers – just like standard reals are surrounded by non-standard reals – and Harthong consequently proposes a metaphorical interpretation of the continuum: it is like the integers viewed at a far distance: '**Z** vu de loin'! Thus the question of the continuum is taken by him to be an irreducible *scalar problem*: the same phenomenon may, in different blow-ups, be continuous and discrete, respectively. In a philosophical bent, Harthong proposes that the duality continuous/discontinuous (just like René Thoms idea of 'apories fondatrices' for different sciences) constitutes a veritable antinomy for pure reason in the tradition of Kant's well-known four antinomies from *Kritik der reinen Vernunft*[526], and that it for that reason constitutes a question which remains necessarily unsolvable.

Salanskis sums up in his synthesizing paper which touches upon many of the positions outlined here[527] when he depicts the actual understanding of the continuum as a necessary surplus exceeding any possible description of it: '... what thus appears in this set of works of non standardists continuing the Borelian tradition, is that the continuum is some sort of 'effect' which results from an *excess*. If the approach from the point of view of 'de-setting' the destiny of the Cantor-Dedekind continuum emphasizes *incommensurability*, what is won on the other hand is the attention towards the constructive and its place in the world of mathematical entities and it makes us confront the philosophical theme of *excess*. The link between those two terms seems clair: the excess engenders incommensurability, and the distribution of incommensurability on the line adds to it the constitutive *quality* of the continuum...'[528] This excess appears, according to Salanskis, rather as an *indetermination within* the finite (rather than the Gödelian idea, as a *determination of* the finite calculus). This difference points towards Petitot's reformulation of Gödel's position.

René Thom repeats, in the Salanskis volume, his conviction about the irreducibility of the continuum, from a simple argument. The continuum is ontologically prior to discontinuities, measured by a simple criterion: X is ontologically prior to Y, if X naturally takes Y as a predicate, but not vice versa. (This hypothesis is interesting, by the way, to compare to Peirce's ontological observations as to the subject-predicate distinction.) By the application of this criterion he reaches the observation that the continuum may easily be modfied discontinually – which in some sense corresponds to taking it as a predicate ('a broken continuum') while the other way around is impossible: a discontinuous entity or set of entities can not take a continuous predicate without becoming in some way continuous itself (138). We may imagine something like 'merging points' as an example.The sets of set theory are like flour sacks (without sacks, because they are the result of the mind) in which every flour particle is endowed with a label, Thom sarcastically says. Instead he refers to an Aristotelian space characterised by total qualitative homogeneity where 'rien n'y est identifiable' (141) as a paradigm for the continuum. The points, then, could be derived only secondarily by taking Aristotelian *chora*, places, and letting them shrink to zero size. Generally, discontinuities are introduced in the continuum by *cuts*. Even non-standard analysis is judged too rigid in its insistence to stay within the confines of set theory where the continuum is supposed to be the result of an arithmetical derivation process. Thom hints at the possibility that richer classes of irrationals may be defined and ends up with a promising Gödelian-style hint: maybe the concept of point has not yet revealed all its secrecies for us. We are tempted to believe that Thom has up his sleeve a more or less vague idea that the point in some yet unknown way will appear able to demonstrate its continuous affinities. We shall return to Thom's deliberations on the continuum and the subject-predicate calculus.

The most detailed analysis of the continuum in Salanskis' volume is undertaken by Thom's disciple Jean Petitot[529] who, at the same time, makes a plaidoyer for a Kantian-transcendental rehabilitation of Gödel's Platonism. Petitot's point of departure is the observation of a strange duplicity, a so-called 'bimodality' regarding the status of the continuum in mathematics and physics, respectively, as critically construed objectivities. While the continuum in mathematics has the status of 'meaning', as a thing-in-itself and consequently is endowed with the inaccessibility of the *noumenon* as a mere regulative idea, the continuum in physics has the status of an unexplained empirical phenomenon appearing in the sensuous intuition – logically anterior to the constitution of a physical objectivity by the transcendental-aesthetical determination of the status of form and the development of physical concepts therein by means of schematisms. Here, the *Ding an sich* of physics on the one hand becomes the 'inner substance' of matter, and, on the other, it refers to forms more complicated than those which mechanical physical objectivity governs: phenomena of self-organisation and related morphological phenomena. Petitot's idea now is that this constitutive duality between mathematical inaccessibility and physical givenness is what gives rise to the constantly diverging interpretations of the continuum. He thus separates 4 interpretations:

1. (Neo-)Aristotelian ideas of the continuum as primitive phenomenological intuition, characterized by the merging of parts and by the merely potential infinite point sets which may, in turn, be realised herein. This position evidently encompasses Thom and Peirce, and, in addition to them, Brentano, Stumpf, Husserl, Weyl, Poincaré, Gestalt Theory – they all chose different ways of interpreting this basic status of the continuum. Brentano, Poincaré, and the gestaltists have a tendendy to psychologize it (and thus make of it a sort of necessary illusion), while Peirce, Husserl, Weyl accordingly should phenomenologize it. Naturally it is correct that the latter will refuse a psychologisation of the continuum, and Peirce and Thom can, as we saw above, directly use basically the same argument in this refusal[530], but the question is whether it is correct to make Peirce a phenomenologist as opposed to Thom. As we have seen, the continuum has a basic ontological status in Peirce, and Thom's and his positions rather seem very close. Husserl of course takes the intuitive plurality af continuous morphologies – 'inexakte Essensen' – to be irreducible, but also as a fact distinct from scientific, purely formal symbolization so that there is no direct road from the former to the latter.
2. A second position is the continuum as a case of actual infinity given in intuition; this includes Kant, Neo-Kantians, Veronese and Salanskis.
3. A third interpretation is the positivist-eliminativist in Carnap and nowadays Hartry Field who conceive of the continuum as a purely physical phenomenon.
4. A fourth interpretation is Gödel's who as already mentioned regards it as possible to grasp the continuum – including its inexhaustibility and potentiality – by set theoretical tools, if they are enlarged with suitable stronger axioms. To this, it must be added, despite Petitot's placing of Gödel in 'the opposite end of the spectrum' (246) that he, as already mentioned, stick to a phenomenologically primitive intuition as a prerequisite to the formalizations of set theory. Gödel's position has, of course, appeared like a constant thorn in the flesh on Hilbertians taking further Hilbert's program (e.g. Feferman). Feferman has approached the problem via his idea of a 'reverse mathematics' (250) which, given a theorem, seeks which axioms will be sufficient to prove

that theorem – an abductive method which Petitot praises as 'deflationist' and highly anti-Platonist. Thus is indeed the intention behind Feferman's argument, but the question remains whether this procedure is necessarily anti-Platonist. Gödel's own idea (based on CH as a theorem from which to search which axioms may prove or disprove it) has exactly the same reverse methodology, even if, admittedly, other expectations. We can not here go into Feferman's technical fight with Gödel: the decisive thing is that he sticks to the idea that the power set construction ($P(x)=2^x$) is well-defined for all sets – whereas Feferman claims that this Platonism is medieval and untenable, because CH is essentially undecidable and thus will remain unsolved forever.

Petitot's solution is here – as so often – to seek a middle way between the two extremes: he wants to keep Gödel's Platonism equipped with a criticist, transcendental basis which will, on the other hand, agree with Feferman in the principal indecidability of CH (not only in relation to ZFC). Petitot here argues on the basis of a discussion (235) of the status of the concept of mathematical existence. Either this concept is formalistically interpreted as referring solely to true expressions involving existential quantifiers – or it is interpreted as referring to something extra-mathematical (that is, something outside of the symbolic calculus). But the former case is unsatisfactory because it is possible to reduce existential quantifiers by means of alternative logics and the introduction of sufficiently many ideal elements (and also, supposedly, also because this idea suffers from the constant Hilbertian problem that a consequent 'formale Redeweise' exterminates any appeal to intuition). The latter possibility is also unsatisfactory, however, because it gives the continuum the status of an object where instead it must be maintained as a Kantian 'pure intuition', that is, determined by mathematical idealities. This interpretation contributes a classically Kantian duplicity: the continuum may be determined as objective reality – at the same time as keeping its character of being a 'horizon-object' escaping any complete determination. This Kantian have-your-cake-and-eat-it-too is made possible in Petitot by the interpretation of mathematical existence as derivative from a transcendental constitution of objectivity: 'The transcendental Platonism is 'negative' and not 'positive' like the ontologically vulgar Platonism. It permits to *invert* the philosophical claims of anti-Platonism *and to transform the phenomena of incompleteness and indecidability into arguments supporting Platonism.*'[531] These apparently depressing results should, quite on the contrary, be taken as the very signs that objectivity is no ontology; they are so to speak in themselves Platonist properties, not of the object, but of the determination of the object.

Petitot runs through the series of different proposals for additional axioms which we have already mentioned, e.g. the axiom of measurable cardinals (building on the intuition – a sort of set theory analogue to the cosmological principle in astronomy – that the continuum must be 'equally dense' on all scales. Just like the domain of the transfinites is inaccessible from finite side and requires the axiom of infinity, there must be greater numbers inaccessible from the transfinites, etc.). An extension of these axioms lead to the axiom of determinacy which prohibits that certain complicated sets may be 'forced' into being interpreted as projective[532]. The toughest version is the axiom of complete determination (AD) which claims that any subset of **R** is determined which makes CH false because **R** may now be ascribed very large cardinalities (even if there still will be no transfinite number between Aleph-0 and $2^{\text{Aleph}-0}$), because **R** can no longer be well-ordered (AD excludes AC). Other, weaker versions of AD only applies this demand on projective sets and allows the well-known 'pathological' sets associated with AC to unfold in the non-determined parts of the set universe. This construction, according to Petitot, contains a sufficient realisation of our intuitive idea of the continuum (258). This is what made Gödel insist that ZFC does not contain a satisfying description of the objects of mathematics which transgress the abilities of formal systems – and to insist on the objectivity of mathematics, our intuitive access to it prior to formalisms, a non-computational theory of consciousness, and a philosophy critical against pure conventionalism, scepticism, syntax without semantics, antiplatonism, etc. These Gödelian standards are now reinterpreted by Petitot in a criticist way as referring to intuition, transcendental aesthetics and the synthetic a priori judgments as a charting of it – that is, as referring to the fact that the intuition of the continuum might be ineradicable but is, on the other hand, inexhaustible by formal descriptions.

At this point, Petitot's new foundation for Platonism seems to support a renaissance for Peirce's intuitions of the continuum as being at the same time primitive and inexhaustible. Yet, Petitot's position involves some open issues. In his presentation of the 'bimodality' of the continuum it seems strange – also given his emphasis on intuition in relation to Gödel – that the mathematical intuition of the continuum is relegated to thing-in-itself while the form of manifestation of mathematical phenomena is a pure, Hilbertian 'littéralité symbolique', and the form of objectivity of mathematics correlatively purely symbolic. It seems as if Petitot in a Hilbertian craze relegates intuition out of mathematics as such, which symptomatically gives rise to the fact that no schematisms to production of synthetic a priori judgments are needed here: mathematics is through and through analytic. Even if it might indeed be plausible to relegate the continuum as object to the *Ding-an-sich* position, it hardly seems reasonable that its 'density', 'homogeneity', 'mergedness', 'potentiality' and so on are not represented as primitive phenomena for pure intuition. Maybe this development in Petitot's thought is connected to the conclusion of the paper where he – and plausibly so – supports Roger Penrose's claim that the non-computationality of consciousness (Gödel) for a naturalist viewpoint must necessarily lead to the claim that parts of physics are non-computational. This deep idea – and the research program connected to it regarding a non-computational physics of the mind, maybe even implying a Penrosean revolution in quantum mechanics – leads Petitot to the strange claim that 'Obviously, the riddle of pure intuition remains unsolved, that is, of how we are affected by the outside. This problem is not first of all mathematical, but *physical and cognitive*.'[533] Penrose's argument certainly leads to the assumption of a physical and empirical foundation of consciousness – but this refers to *empirical* consciousness, to which pure intuition could hardly be taken to belong. Does Petitot here commit an advanced version of the classical mistake of identifying the empirical and the transcendental subject to the disfavor of the latter?

In the transcendental realm seems to remain only Hilbertian finitism, while all the questions regarding intuition may be thrown to the lions of empirical cognitive research. Even if it may be correct, as Petitot adds, that Kant vaguely anticipated such points of view in his *Opus postumum*, when his idea of a 'phenomenon of the phenomenon' in that work refers to a non-mechanical and non-atomistic physics, then this as yet remains a speculative idea and no empirical physical result. Should one really take the consequence of this unhappy partition between a Hilbertian finitist metamathematics referring only to computational symbol strings on the one hand and an vague ungraspable 'meaning' open to empirical cognitive science only, then it might amount to something like a resurrection of parts of that 'transcendental Jalta' which Petitot has himself bravely spent much of his carreer demolishing.[534] Petitot's hasty naturalisation of pure intuition probably must be seen as an insistence that it is, naturally, natural and indeed highly interesting to investigate the way any single one of the Kantian faculties are implemented in empirical subjects – without thereby forgetting that such an ambitious piece of psychology will still remain a special science referring to a regional objectivity, in turn to be grasped by transcendental tools (among them the ineradicable pure intuition itself with its inventarium of Hilbertian finitist tools as well as primitive, pre-combinatorial and pre-constructive intuitions about the continuum.)

Still, Petitot's delicate distinction between the continuum as thing-in-itself on the one hand and mathematical formalizations on the other (whether they involve Hilbertian symbolisms only or richer representations) forms a strong parallel to Peirce's version of an Aristotelian continuum. Peirce's interpretation of the continuum in terms of possibilities (firstness) and 'real possibilities' (thirdness) indeed places it at a distance from any finite actualization whatsoever.

To sum up, we may say that various actual mathematical and philosophical discussions of the continuum support the reactualization of a version of Peirce's continuum doctrine. The set theoretical definitions of supermultitudinous set *S* and the search for its exact place in the hierarchy of cardinals must given up, while the very idea of the continuum as a basic intuition equipped with a potentiality exceeding any formal attempt at actualization is preserved. Moreover, the actual initiatives seem to take further Peirce's intution that even if set theory may not exhaust the continuum, then it may indirectly depict it, to the extent that different formalisation attempts in some way or other display the 'indistinctness' which is a crucial property of that intuition. This goes for e.g. the indecidability of generic subsets of reals even in **R** (Chaitin, Harthong), it goes for the 'halo' of points in non-standard analysis (Robinson, Harthong), it goes for the different categories of 'inaccessibility' regarding the axioms of 'large cardinals', it also goes for the 'ungovernable' sets which the axioms of partial determination allow for (and which also Chaitin refers to when he points to the fact that there is no winning strategy in the even-uneven game), it goes for the determination of the continuum as 'excess', referring to the fundamental status of incommensurability in all reasoning about the continuum (Salanskis), it goes for the arguments for the ontological anteriority of continuity in relation to discontinuities, connected to an ontological interpretation of any predicate calculus (Thom), and it finally goes for the criticist relativisation of Gödel's Platonism which makes CH's apparent indecidability to an argument *for* a Platonist idea of the continuum rather than the opposite (Petitot).

PEIRCE AS PHILOSOPHER OF MATHEMATICS

Given that Peirce's position as reconstructed by Putnam and interpreted within the framework of his own (partly) intuitionist logic is consistent, we may ask the anachronistic question: to which position in the philosophy of mathematics would it belong? We have already mentioned Kurt Gödel whose controversial philosophy of mathematics is apt to put Peirce's position into profile. The implication of Gödel's 1931 proof – proving the incompleteness of any logical system which contains finite arithmetic – has very often been taken to constitute the definitive farewell to Platonism – the idea that the objects of mathematics have some sort of independent existence which our mathematical operations serve to reveal – with the further formalist implication that mathematics must now rely upon the tentative experiment with different axiom sets which remain our constructions, without any claim to referring to some ideal world.[535] Against this consequence, Gödel took his incompleteness proof to show that *formal* methods might never make mathematics complete, and that this fact must force mathematics to seek refuge in non-formal, phenomenological methods able to clarify the abstract content of our intuitive assumptions and thereby lead to new axioms which might make some of the indecidable axioms (AC, CH, etc.) decidable. Thus he tended to regard AC as true and CH false (so that $2^{\text{Aleph-0th}} = \text{Aleph-2}$ – even if he changed to the opposite viewpoint late in life.)

Gödel is thus in opposition to formalism and Hilbertian finitism, claiming that proofs may be obtained by symbolic calculi alone and let intuition about mathematical objects remain on the metamathematical level where axioms are selected and proofs interpreted – and he is also, on the other hand, in opposition to intuitionism with its very tough interpretation of human finitude tending towards psychologism and, in many cases, towards constructivism prohibiting infinite operations and the principle of excluded middle.

Many aspects of Peirce's philosophy of mathematics seem to make him quite the opposite of Gödel: Peirce's constructivism implies that he does not regard the objects given in mathematical intuition (the line, for instance) as being exhaustible by operations, so that the actually construed always exists within a modal horizon of possibilities. His insistence, moreover, close to formalism, that we are free to choose widely different procedures, each of them consistent, but not mutually consistent, within this field of possibility (cf. his obsession with non-Euclidean geometries) which might seem to give him a MacLanean affinity. Still, we find other respects in which Peirce may express himself as a through-and-through mathematical Platonist, just like he, in analogy to Gödel, may expect CH to be settled by future axioms – all in all the idea that he sees the issue of the reality of mathematics as a question which may be settled with metaphysical means. How is this strange knot to be untied? Let us consider one of the cases where Peirce declares himself Platonist. He observes an ellipsis drawn on a blackboard as a continuous figure, 'inheriting' its continuity form the plane on which it is drawn. Its continuity is actualised due to its potentially infinite set of actual tangents governed by that habit which constitutes the ellipsis. Then he claims:

> Many such reacting systems may spring up in the original continuum; and each of *these* may itself act as first line from which a larger system may be built in which it in turn will merge its individuality.
>
> At the same time all this, be it remembered, is not of the order of the existing universe, but is merely a Platonic world, of which we are, therefore, to conceive that there are many, both coördinated and subordinated to one another; until finally out of one of these Platonic worlds is differentiated the particular actual universe of existence in which we happen to be. (R&LOT 1898, 263)

Here, we are close to Peirce's cosmology where the undifferentiated continuum appears as a sort of ever-creative Ur-cosmos, the beginning of everything. Here we find a surprising chain of cosmic development:

Continuum → many Platonic worlds → one Platonic world → the particular actual universe

We will not here discuss this extravagant onto-cosmological claim as a proposition referring to an real genetic developing process, but restrict us to discussing its implication in philosophy of mathematics, as a series of presupposition relations. Mathematics, to Peirce, is no science of actuality, its domain is the iconic field, formally apprehended, its point of departure hypotheses, and its task is to draw necessary conclusions from them. No formalist would disagree. Peirce and the formalist split, however, when Peirce's idea of 'many Platonic worlds' is taken to imply that *all competing consistent axiom system have a Platonic real character*, they are not only symbolic nor human constructs from which we can demand no further reality. Quite on the contrary, different axiom systems so to speak map a series of different, possible Platonic worlds, and the role of mathematics is to cover this field of possibility. 'The particular actual universe' as an object must of course be covered by different empirical special scientists – what will happen, then, with 'the one Platonic world'? It is investigated by applied mathematics, not in the special sciences, however, but in metaphysics. The import of mathematics in various special sciences guarantees the validity of their deductive schemata and diagrams and thus their generality beyond the always finite field of particular empirical observations. It is those very empirical observations, on the other hand, which investigates which one combination of the possible Platonic worlds we actually live in. This is probably why Peirce may seek to let metaphysical considerations decide CH: mathematics may supply many different consistent systems, they are merely hypothetical – what is decisive is what metaphysics, in the Kantian way built on logic, as a real science will chose in the mathematical candy store.[536] As against Gödel, Peirce would (probably) support formalists and MacLaneists: there are several different axiom systems for set theory, and we can not from consistency alone decide which of them is mathematically correct – being consistent they are equally correct. But against formalists, on the other hand, Peirce would say: *all* these different systems have equal objective Platonic reality – they are true investigations in the domain of the possible and thus not only (MacLane) 'correct'. As regards the role of intuition, Peirce would agree with Gödel and intutionists, each of which will claim intuition's role to be irreducible, and against the formalist tendency to strong reduction or even elimination of intuition. Peirce would also agree in the intuitionists' skepticism towards the law of the excluded middle, and he even counts as an important forerunner to their doctrine in his characterization of firstness and thirdness, as not obeying the law of contradiction and the law of excluded middle, respectively. But against (at least) the constructivist aspect of intuitionism, Peirce will argue against the idea that human finitude must constrain us to prohibit infinite operations[537] – here he would rather agree with Gödel that intuition is nothing simply given but something which must be investigated further with phenomenological means – diagrammatical reasoning – and potentially with new formalist results as its consequence. Even if we are finite beings, our finitude is indefinite, because its exact limits are not given and will only be clearer from ongoing research.

This will also further clarify Peirce's idea of the continuum. Its basic motivation is metaphysical; it comes from the generality of universals in the logic of relations. It comes from the basic fact that predicates do not possess any directives for which existing objects they hold for. Thus, they are potentially valid for an inexhaustible continuum of possible subjects. This is why any extensional semantics trying to define the property expressed by the predicate by the existing subjects possessing the property, is impossible, because these subjects may not be listed, not even in any possible world. This Aristotelian universalism whose special Scotist background in Peirce we covered in Chap. 2 becomes the motivation for his metaphysical continuism, and mathematics becomes the formal apparatus to investigate it. Peirce's realism with respect to universals is thus the decisive basis for his continuity doctrine: universals are would-be's, and such phenomena may be described only as general continua of possible events. The competing Platonic worlds which are mapped by mathematics grow out of an original continuity which eo ipso *is given before even the mathematics supermarket of competing sets of hypotheses*. Mathematics thus *must* be continuist; even if it possesses its full rights, even duty, to investigate discontinuous hypotheses, then it is a metaphysical premiss that they may only be understood on the background of a more encompassing continuity. This is due to metaphysics – which given its systematic place in Peirce's encyclopaedia is restricted to deal with general properties of *this* world (and thus much more constrained than mathematics). But it is also due to that much more encompassing metaphysics which is not explicitly called by that name in Peirce, but which appears in the very shaping of Peirce's encyclopaedic architectonics, and in which mathematics is first science, dealing with a continuum of pure possibilities.[538]

NOTES

[1] References to Peirce and Husserl are in much-used cases abbreviated (like EPII or LU) – key to abbreviations is in the bibliography. As to Peirce's work, references are to the *Writings* and *Essential Peirce* when possible. If the same texts appear also in the older *Collected Papers*, references to that edition are given immediately afterwards. In many cases, though, reference is possible only to *Collected Papers*. In both cases, references to that work follow normal procedure: volume number followed by paragraph number (like 3.445).

[2] Lakoff and Nuñez (2003) has a similar point in investigating the metaphors and blendings used to obtain central mathematical results – their book considerably adds to our understanding of how embodied beings may get in touch with mathematical truths by means of the abilities charted in the cognitive semantics tradition. I do not agree, though, in their jump from this epistemological investigation and to reductionist ontological claims that mathematical objects simply *are* nothing but the result of such mapping processes (rather than independent objectivities *investigated* by such processes). This seems like a confusion of *Genesis* and *Geltung*, origin and validity, and thus reawakens that psychologism which both Peirce and Husserl were out to bury.

[3] The art historian W.J.T. Mitchell has, on a couple of occasions, used the word 'diagrammatology' in a sense not wholly unlike my own, based on the plausible idea that our access to literary (and other) form takes place by means of ' "sensible" and "spatial" constructs'. Mitchell first used it in a small debate article (W.J.T. Mitchell 1981): 'Diagrammatology', in *Critical Inquiry*, Spring 1981, Vol. 7, nr. 3, 622–33, responding to Leon Surette's critique of his own paper 'Spatial Form in Literature') – here, he poses the question: 'If we cannot get at form except through the mediation of things like diagrams, do we not then need something like a diagrammatology, a systematic study of the way that relationships among elements are represented and interpreted by graphic constructions' (623). More recently, he has taken up the notion again in an internet interview (O.N.C.Wang 1997) 'The Last Formalist, or W.J.T. Mitchell as Romantic Dinosaur', Interview with Mitchell by O.N.C. Wang, in *Romantic Circles Praxis Series,* Series Editors: Orrin N. C. Wang and John Morillo (http://www.rc.umd.edu/praxis/mitchell/mitch-about.html) which does little more than refer back to the earlier, 'strange, little article'.

[4] With diagrammatology in general as its subject, this book does not contain analyses of particular diagrammatic representation systems – however important this subject indeed is. Peirce's general doctrine of diagrams developed in the years after the turn of the century is motivated, to a large extent, in his parallel development of existential graphs. The results of these developments, Alpha-, Beta-, and Gamma-graphs for logic representation are thus discussed here only in the context of constituting an important part of the specific material from which Peirce's general diagram doctrine is developed.

[5] Quote from the presentation of a planned work with the title *The Principles of Philosophy: or, Logic, Physics, and Psychics, considered as a unity, in the Light of the Nineteenth Century*, c. 1893; the CP bibliography Vol. 8, 283.

[6] I originally planned to open this book with a thorough chapter connecting Peirce's mathematical attempts at understanding the continuum with his scattered outlines of a continuity metaphysics. It was, however, judged too harsh as a starting point which might dissuade too many otherwise interested readers from continuing – and to some extent also wildleading in relation to the overall diagrammatological aim of the book. A more detailed discussion of these issues may now be found in this book's Appendix – the curious reader is encouraged to continue there. Further discussion of Peirce's continuity concept may be found in Parker 1998.

[7] For an actual elaboration on Peircean knowledge in the limit, see Hendricks 2001.

[8] Yet, one time Peirce might refer to Poncelet when talking about his continuity principle – namely in a prospect for a book never written about nineteenth century ideas (in the bibliography of CP, Vol. 8, 283).

These ideas embody the principle of continuity, and among them is listed 'modern mathematics and its imaginaries'. This might refer to Poncelet whose generalizing interpretation of geometrical figurae allowed them to refer not only to possibilities present but also possibilities absent (thus, the figure of two crossing lines might refer to all such different crossings, but also to the case of the two lines being parallel, that is, crossing at a point in infinity, cf. Greaves 2002, 46–47). This is, in fact, Poncelet's continuity principle – that what holds for one figure, also holds for continuous deformations of it – and it seems likely that Peirce's principle is influenced by him, the more so because of the evident relevance of such general figure interpretation for Peirce's diagram doctrine.

[9] Peirce's conception of the system of sciences is in constant development. Here, we stick to the relatively elaborated versions which he crafts after the turn of the century, especially the 1902 Carnegie application (MS CSP L75, i NEM IV and LCS, Joseph Ransdell's synthesis of L75), 'A Detailed Classification of the Sciences' from 'Minute Logic', 1902 (EPII, 115ff; 1.203ff) and 'Syllabus' from 1903 (EPII, 258ff; 1. 180ff).

[10] Peirce takes the term 'Grammatica speculativa' after a treatize ascribed to Duns Scotus (properly written by Thomas of Erfurt which Peirce was then in no position to know), meaning 'mirroring grammar', that is, a grammar mirroring metaphysics. Hence its connection to epistemology: 'Kant, who first raised these questions to prominence, called this doctrine *transcendentale Elementarlehre*, and made it a large part of his *Critic of the Pure Reason.* But the *Grammatica Speculativa* of Scotus is an earlier and interesting attempt. The common German word is *Erkenntnistheorie*, sometimes translated Epistemology.' ('Minute Logic', 1902, 2.206) Thus, in Peirce semiotics and epistemology are generally two sides of the same coin.

[11] As Thirdnesses, teleologies are naturally general, vague, variable and bundle-creating ('Minute Logic', 1902, EPII, 118ff; 1.206–7), and we must not expect sharp, but only provisional, fallible (but gradually improving) demarcations between the sciences.

[12] Cf. Parker 1992, 67.

[13] Short (2004) argues that the sign theory of the early Peirce had the consequence of making of the concrete object a bundle of qualities only (much like the trope-bundles in present discussions), necessitating him in 1885 to add the Scotist idea of 'haecceitas', 'thisness' and the corresponding sign type of indices, in order to grasp actual existence without giving up the idea of the particular as not fully determined.

[14] In Peirce's theory of perception, perceptual judgments form a borderline case of judgments because they, as a rule, and unlike other judgments, are beyond conscious control. They judge on preceding sensory 'percepts' to which we have no direct access but which must be postulated as the matter of perceptual judgments – but the generality revealed in perceptual judgments must inhere already in these percepts even if they are not propositional in structure.

[15] Even if the phenomenological categories are logically anterior to logic, they have, in Peirce, been discovered through logic, by generalizing abstraction from logical relations (in the three categories, it is easy to recognize the properties of predicate (1), subject (2), and copula (3)). But here Peirce separates what the theory of science of our days would call context of discovery from context of justification.

[16] 'The acquiring [of] a habit is nothing but an objective generalization taking place in time. It is the fundamental logical law in course of realization. When I call it objective, I do not mean to say that there really is any difference between the objective and the subjective, except that the subjective is less developed and as yet less generalized. It is only a false word which I insert because after all we cannot make ourselves understood if we merely say what we mean.' ('Abstract of 8 lectures', undated, NEM IV, 140)

[17] A tautology or a petitio principii, it might be critically argued?

[18] The quotes are from 'Abstracts of 8 lectures', Robin-nr. 942 which is undated. Clearly, it is a rather late text, in any case later than 'Law of Mind'(1892), and probably also later than the 'Reasoning and the Logic of Things' series (1898).

[19] This is, in fact, an interpretation of the Kantian doctrine of human finitude: we are finite beings because we do not possess intellectual intuition. Being bound to sensory experience, we must organize this in different ways. This finitude is here seen as a result of the fact that sensory experience is limited to

certain contingently constrained senses which defines what we may imagine in intuition. As a contrast, we might imagine a god who has the ability of sensing in the whole continuum of qualities.

[20] This idea, due to the attempted definition of Firstness as that whose constituents are independent of anything else, evidently runs counter to Peirce's insistence on non-distinctiveness as basic property in continua. It seems like Peirce's determination of Firstness qualities as *vague* is a compromise between these two ideas.

[21] In Peirces evolutionism, he even ventures the hypothesis that Thirdness is growing as a tendency towards the appearance of still more correlations between still more Secondness events. Thus the relation between the three categories as presented here may be given two different interpretations. One is structural, picturing the atemporal relation between the three. Another is temporal, supposing that Firstness is also first in a temporal sense, so that the growth of Secondness and Thirdness are cosmological processes empirically taking place. Here, we stick to the former interpretation; it seems uncertain whether evolutionism prospers from the attempt at rendering it in these categories. Cf. the discussion in the biosemiotics section.

[22] The difference between ordinary modern logical use of PC and PEM, referring to propositional logic, and Peirce's use, referring to contradictory predicates within propositions, has only recently been clearly outlined in Lane (1997).

[23] The rain examples pertaining to both PC and PEM are from MS CSP 678, quoted from Lane (1999), 289.

[24] Modern modal logic regarding modal propositions also has necessity not obeying standard PEM and possibility not obeying standard PC. If N is the necessity operator, then PC holds: *Not both Np and N non-p* – but not PEM: *Either Np or N non-p*. And, conversely, if M is the possibility operator, then PEM holds: *Either Mp or M non-p* but not PC: *Not both(Mp and M non-p*. (Øhrstrøm and Hasle 1995, 144-145). Modal propositional logic thus give parallel results to Peirce where modality is expressed inside the S-P proposition in first order predicate logic.

[25] The semiotic description is adapted from "Pragmaticism, Prag. [4]", c. 1905; 5.505.

[26] Peirce's introduction of a three-value logic in 1909 has no direct connection to the analysis of generals by means of non-appliance of PEM; it rather points to discontinuous limit cases in continua (like the status of the borderline between two coloured areas, etc.), calling for a third, limit truth-value where PEM does apply but is false. These issues have often been confused, but is admirably sorted out by Lane (1999).

[27] Noble (1989) argues against Murphey's (1961) idea that it was Peirce's new conception of the continuum which gave rise to the doctrine of real possibilities, claiming that it was rather real possibilities which – after the abandonment of Cantorian continuity – gave rise to Peirce's mature idea of the continuum as inexhaustible (170). Our contention that Peirce's continuum concept is metaphysically rather than mathematically motivated seems to support Noble's account. We return to 'real possibilities' in Chap. 2.

[28] Or, more properly, the inside of the cut appears in the mode of 'If'. Only if the cut contains a further cut, thus constituting a 'Scroll', the contents of the cut are denied.

[29] We shall not go further into Peirce's three systems of logic graphs here, just notice that Peirce's work on them undoubtedly formed part of his motivation for a general diagrammatology. For the time being, a fertile research into the possibilities of the Graphs is developing, cf. Barwise and Etchemendy 1995; Øhrstrøm 2001; Pietarinen 2006; Shin 1994, 2000; Sowa 1997, 1999, etc. In Stjernfelt 2006, I argue that the Graphs urged Peirce to add a stronger criterion of iconicity to his operational iconicity discussed in the present book, here called 'optimal iconicity' – with the motivation of shaping the graphs to give as fitting a description of logical reality as possible.

[30] Cf. Stjernfelt 2006.

[31] In the letters to Lady Welby, Peirce explains that he often uses the human mind as an illustration when introducing his sign concept, but that this is only for the sake of giving a clear example - this reference being a 'sop to Cerberus', as he puts it (Dec 23, 1908, EP II, 478). This implies a theory of mind in which consciousness is not a necessary ingredient, because the mere existence of purposive action is sufficient to grant the presence of mind in this broad interpretation: 'I hold that purpose, or rather, final causation, of which purpose is the conscious modification, is the essential subject of

psychologists' own studies; and that consciousness is a special, and not a universal, accompaniment of mind.' (From 'Minute Logic', 1902, 7.366 – for a discussion, see Skagestad 2004, 247f). It must be added that Peirce is often very unclear as to the extension of this mind concept. Does it cover all of Thirdness, so that all regularity in the world, every abstract object which is not a mere *ens rationis*, has a mind-like character – or do purposive, mental phenomena constitute special subclasses of Thirdness? In the more extreme versions of his cosmology, Peirce without further notice presupposes the former; we shall stay content with assuming the latter.

[32] Of course, it is an open issue whether this general definition is not too general. Does it add anything to our knowledge of a physical process to describe it in sign terminology? I tend to say there is not, cf. below. Even if a fact shares the same structure as a proposition, as Peirce says, it *is* no proposition but rather the *fundamentum in re* of a proposition.

[33] We might imagine, as examples, that an animal species can be conceived of as the general conclusion to an evolutionary argument, or, to take a less teleological example: the solar system will be the conclusion to the premises constituted by mechanical laws and a set of initial conditions. In his later years, Peirce is rather definitive on the existence of natural arguments, even to the extent of claims like '... the universe is a vast representamen, a great symbol of God's purpose, working out its conclusions in living realities. Now every symbol must have, organically attached to it, its Indices of Reactions and its Icons of Qualities; and such part as these reactions and these qualities play in an argument that, they of course, play in the universe – that Universe being precisely an argument.' (Lectures on Pragmatism, 1903, EP II, 193–4, 5.119)

[34] Peirce often characterizes icons and indices as 'degenerate' symbols. He does not use that term about simpler symbols (propositions, rhemes) as compared to arguments, but still he has an analogous idea: 'If we erase from the argument every trace of monstration of its special purpose, it becomes a proposition; usually a copulate proposition, composed of several members whose mode of conjunction is of the kind expressed by 'and'; which the grammarians call a 'copulative conjunction'. If from a propositional symbol we erase one or more of the parts which separately denote its objects, the remainder is what is called a *rhema*...' ('New Elements', 1904, NEM IV, 244). Here, both proposition and rheme are seen as less complete arguments.

[35] The three basic trichotomies pertaining to sign, object, and interpretant, respectively, may not be combined freely. A subtype of a higher triad may not be combined with lower subtypes of lower tirads (as all symbols are legisigns, e.g. symbols do not combine with quail- or sinsigns). Combination of the nine sign aspects recorded in the trichotomies thus only yield ten combined sign types, presented by Peirce in this triangle ('Syllabus', EPII. 296; 2.264) with the simplest sign upper left and the most complex upper right:

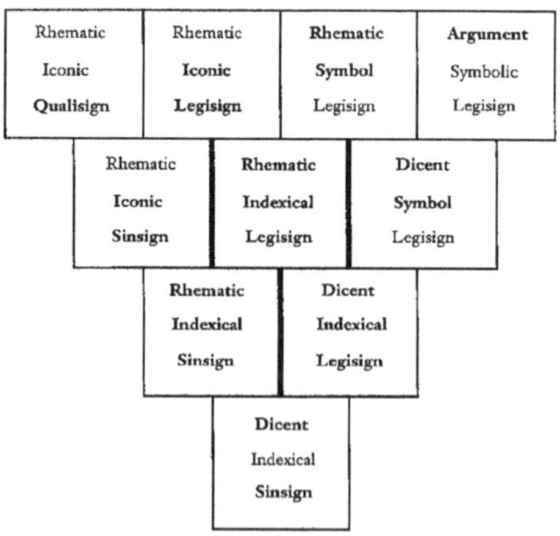

Figure 7.

It is rather rare, however, that Peirce (nor we) need to refer to the precise type of single signs – primarily because signs rarely occur as isolated specimens but most often appear as parts or aspects of an ongoing reasoning process where it is sufficient to highlight the sign aspect actually scrutinized.

[36] These trichotomies are the basic three among the ten trichotomies which Peirce outlined in the Lady Welby letters during the last decade of his life. Most of the others are only sketchily outlined, but some of them merit more study, thus the trichotomy pertaining to the nature of the dynamic objects of signs, distinguishing 'Abstractives, Concretives, and Collectives', that is, roughly, universals, particulars, and sets.

[37] Some authors maintain that there is a decisive difference between the triads of qualisign-sinsign-legisign and tone-token-type, respectively. This is not the case. Both are defined, as the only triad, with respect to subtypes of the sign itself. 'As it is in itself, a sign is either of the nature of an appearance, when I call it a qualisign; or secondly, it is an individual object or event, when I call it a sinsign (the syllable sin being the first syllable of semel, simul, singular, etc.); or thirdly, it is of the nature of a general type, when I call it a legisign.' (Letter to Lady Welby Oct 12, 1904, 8.334 – the same distinction as used in the 'Syllabus' quote given here). Two years later in the 'Prolegomenta to an Apology for Pragmaticism': 'Only one division is concerned with the nature of the Sign itself, and this I now proceed to state.' – and then follows the tone-token-type distinction (1906, 4.536–7). Also the CP editors in a note claim the identity of this triad with the former. Two more years pass, and in another letter to Lady Welby, Peirce takes up a terminological discussion. Here, he prefers the triad potisign-actisign-famisign for the same distinction, even if toying with maintaining tone-token-type with 'mark' substituted for 'tone' (EPII, 488; 8.363–64). In all cases, though, the distinction pertains to the merely potential quality of a sign as opposed to a singular, active sign as opposed to a lawlike sign able to exist in identical tokens. These tokens of a type or sinsigns of a legisign are tokens/sinsigns of a 'special type' because governed by a legisign (not all tokens/sinsigns are so governed), and the term proposed for them is 'replicas' or 'instances'.

[38] Here, Peirce's special conception of 'set' as intensionally defined as a unity is important, cf. his argument with Dedekind on this issue. If sets are defined extensionally, then the unity of a set of sinsigns becomes a fragile notion, and types, important to grant the identity of the same sign from one occurrence to the next, become problematic. As Randall Dipert remarks, Peirce's refusal to accept Dedekind's notion of 'echter Theil' (proper subset) can be seen as an insistence that sets be defined intensionally and in a certain sense constructively (Dipert 1997, 62ff). If we say P is a proper subset of W, the crucial issue is 'How do we know a set *is* a proper subset of an infinite set? [...] some property or relation must hold on some of the members of the collection, W, but not others (i.e., those in P), and we must know this. In other words, we are back, if we choose Dedekind's method, of requiring certain kinds of knowledge about the internal structure of the original infinite set W.' (62–63). In order for an infinite set to be treated as an individual, it must be somehow intensionally defined, for otherwise we should be able to list all its elements. Thus, Peirce's realism and intensionalism make him insist on a constructivist idea of sets based on the reality of rules defining such sets rather than on the actuality of each of the elements.

[39] René Thom has exactly the same argument: to say that a sign is a mere 'flatus vocis' begs the question, for how could a mere blow of air be absolutely identical from one occasion to the next without assuming realism of that sign expression and thus realism about universals?

[40] Douglas Hofstadter has made a strong argument for the complicated inner structures of apparently very simple types, such as for instance letterforms. The possible token instantiations of the letter type A is infinite and even subject to constant innovation in designer typefaces, even if the A 'tone' for a superficial glance seems extremely simple. His conclusion is that attempts to make explicit all possible A's by an algorithm are subject to incompleteness constraints formally analogous to Gödel's 1931 proof (in Gödel 1986). This thus forms a strong argument against semantic extensionalisms without resting on psychological arguments (Hofstadter 1986, see also Stjernfelt. 1992c).

[41] The somewhat strange notion of 'dynamical' object in Peirce refers to a distinction between the 'immediate' object – the object as the sign presents it – on the one hand, and the object as it is in itself, regardless of the sign, on the other. The latter is the 'dynamical' object.

[42] At a conference in May 2005, the British philosopher Galen Strawson claimed that intentionality must always imply consciousness, even for analytical philosophers. Thus, he argued against the idea,

also in analytical philosophy, that there is a broader type of intentionality which does not presuppose consciousness, for instance in living organisms. Just like the Peircean idea that mind is a broader concept than consciousness, and that only certain sophisticated types of mind are equipped with consciousness.

Such an idea Strawson rejected. His reason was one basic argument, the so-called 'stopping problem'. The idea is that if we take the chain of causes connecting an object with the intention intending it, then we lack criteria for where to 'stop' in the causal chain. To him, only the conscious experience can yield such a criterion. In Peircean notions, it is the question of where to 'stop' in the causal chain connecting index and object.

Strawson's idea is that it is only consciousness which makes it possible to 'stop' at the right place in that causal chain. If consciousness is not there to stop it, then any effect in the universe may be taken as a sign of any one of its many causes – and intentionality would be everywhere. This is evidently not an attractive consequence – even if Peirce sometimes spoke as if he was willing to accept something like it. But if we should not accept this consequence, how should we solve the 'stopping problem' for indices without ending in the opposite extreme and embracing Strawson and consciousness as the only source of proper intentionality?

The problem is evidently a major issue for the development of biosemiotics, because the idea here precisely is that biological processes down to biochemical detail can be understood as signs, that is, can be understood as intentional. But how do we see what biological signs below the organism level are 'signs for'? – in some sense they could be said to be signs for any segment in the homoeostatic process they are part of. When von Uexküll's famous tick waiting in its tree smells butylic acid, this chemical releaser may be said to function as a sign for 'mammal', as a sign for 'jump', as a sign for 'blood', as a sign for 'food' or 'reproduction' – because all these phenomena enter into the action chain involving the tick smelling the chemical and acting correspondingly.

But what is the right place to stop? Strawson implies that we cannot determine any stopping place from merely behavioral criteria. How do we frame the stopping problem in order to grant that a sign process is going on, no matter whether the tick has any inner feeling of what's going on? We shall return to the index stopping problem in the biosemiotics section below as well as in the discussion of Eco's idea of indexicality as subject to negotiation.

43 Sometimes also known as seme, pheme, and delome; or sumisign, dicisign, and suadisign.

44 It goes without saying that this reference may point to any kind of fictituous reality. Another discussion which we shall not go into here is the dicisign's character of assertion which we will here take as the default case. Of course, it is possible to consider the satiated predicate with its assertive force placed in brackets so that the expression in itself has no reference, just like the addition of assertive force to the expression is not the only speech act possibility. The same satiated predicate can be used in other speech acts than assertion; questions, imperatives, wishes, etc. In his Gamma graphs, Peirce outlines the basics of theory to cover such cases, and so qualifies as an early father of speech act theory.

45 As part of his elegant definition of the whole third trichotomy: rhemes are signs making explicit their information, dicisigns are signs making explicit their object, and arguments are signs making explicit their interpretant.

46 Human beings are mortal, Socrates is a human being, Socrates is mortal - this old example implicitly refers to a law due to which a continuum of other human beings than Socrates may be inserted in the latter part of the premiss and the conclusion - hence its law-like generality.

47 The three distinctions recur in Peirce's abstraction theory where the three types of distinctions dissociation, prescission, and discrimination correspond to Scotus' real, formal, and logical distinctions, respectively. They distinguish between *res* and *res, res* and *realitas, res* and *ens rationis*, respectively. We shall return to Peirce's abstraction theory below, Chap. 11.

48 '... grâce à laquelle *homme* est suffisament indéterminé pour que, par une intellection unique, il soit conçu quidditativement en tout homme' (Scotus, quoted by de Libera, 345).

49 Diodorus Cronus and Philo the Logician or Philo of Megara formed part of the so-called 'Dialectical school' at the turn of the fourth and third centuries BC – a school which was among the initiators of propositional logic. The latter was a pupil of the former, and we only know their work from secondary sources. Diodorus outlined the famous 'Master argument' to prove the existence of actuality only

(briefly: every proposition about the past is necessary; the impossible does not follow from the possible; it seems like possibilities exist which are not nor will ever become true. From this follows, Diodorus argues, that nothing possible exists which is not or will not become true. Much about the details of the argument is still unknown.) Thus, when Peirce in his later years toys with Diodoran implication, the realism as to possibilities which he tries to articulate does not have any support from the Master argument (As to Peirce and the Master argument, see Øhrstrøm 2001). Thus, it is rather Diodoran implication which interests Peirce. It seems like Diodoran implication involved time: a conditional is Diodoran true if it is never the case that P is true and Q is false – it must be Philonian true not only now but at all times. But still it does not seem to require any connection between P and Q such as Peirce seems to indicate (Bobzien 2004).

[50] Two years later, he says similarly ‘For my part, I am a Philonian; but I do not think that justice has ever been done to the Diodoran side of the question. The Diodoran vaguely feels that there is something wrong about the statement that the proposition “If it is lightening, it will thunder,” can be made true merely by its not lightening.’ (‘Types of Reasoning’ (1898), Peirce 1976, 169).

[51] This would not be satisfied by what we know of Diodorus’ view of implication either, however, cf. above.

[52] Strict implication avoids the problems quoted, but involves other problems: [] $p \rightarrow q$ (it is necessary that p implies q) gives rise to problems like the following. ‘If there are lions in South America, then $5+2=7$’. As $5+2=7$ in any possible world, the statement read as strict implication is true. In Peirce’s later years, he makes some attempts to extend the Beta and Gamma logic graph systems to cover real possibilities, cf. Stjernfelt 2006.

[53] To substantiate the existence of such ontological extravaganza in Peirce, see the discussion of ‘The Law of Mind’ (which is the earliest locus in Peirce where this tendency is given free rein) in the Appendix. The general problem is that it is not clear how many of the phenomena subsumed under Thirdness are taken to characterize the whole of that category. Take for instance the symbol concept. It is easily taken for a pure sign concept only appearing at some point of biological evolution, and in Chap. 11, we shall indeed discuss it in that way. Peirce’s cosmology, however, gives him the interesting idea that nature itself may take habits, and from here to assuming that such habits possess the full character of symbols is a short step which Peirce often does not hesitate to take. See, e.g. the conclusion to his magnificent discussion of his sign theory in ‘New Elements’ (1904): ‘A symbol is an embryonic reality endowed with power of growth into the very truth, the very entelechy of reality. This appears mystical and mysterious simply because we insist on remaining blind to what is plain, that there can be no reality which has not the life of a symbol.’ (NEM IV, 262). There are indeed interesting ideas in Peirce’s cosmology: ‘If we are to explain the universe, we must assume that there was in the beginning a state of things in which there was nothing, no reaction and no quality, no matter, no consciousness, no space and no time, just nothing at all. Not determinately nothing. For that which is determinately not *A* supposes the being of *A* in some mode. Utter indetermination. But a symbol alone is indeterminate. Therefore, Nothing, the indeterminate of the absolute beginning, is a symbol.’ (‘New Elements’, 1904, EPII, 322). But I do not think we can take them as proofs for the omnipresence of symbols in the universe. Such ideas seem to be related to Peirce’s revaluation of Hegel – whom he almost despised during the 70s and 80s – in the 90s. Hegel’s speculative dialectics and its jumps between Nothing and Everything inspires such parts of Peirce’s cosmology: ‘With the emergence of Time, the first book, as it might be called, of an Historical *Encyclopädie*, what Hegel terms “Logic” is brought to a close and we pass to what he calls the “Naturphilosophie” ’ (‘Abstracts of 8 lectures’, undated, NEM IV, 138) Such claims are by no means rare in the mature Peirce, take for instance this from 1898 (RLOT, NEM IV 345): ‘. . . I myself entertain that every true universal, every continuum, is a living and conscious being . . .’ Such claims imply there are not only Thirdness, habits, and signs, but also symbols, life, and consciousness already in prebiological nature and we must assume gravitation, prime example of Thirdness in Peirce, suits this description of a conscious, personalized symbol. Still, I hesitate to venture into these waters where all distinctions between real possibilities, Thirdness as such, continuity, symbols, entelechy, life, evolution, etc. vanish. I prefer to remain seeing such claims as mystical and mysterious until further notice, which is why I only treat Peirce’s cosmology tangentially in this book. A related issue to which we return in the biosemiotics section is the question whether there are indeed signs and life in prebiological nature, and if so, then in what sense.

It should immediately be said that this absolute idealism tendency is not alone in Peirce - there is also a counter-tendency, so as for instance: 'My reply is that while I hold all logical or intellectual, interpretants to be habits, I by no means say that all habits are such interpretants. It is only *self-controlled* habits that are so, and not all of them, either' ('Pragmatism' 1907, EPII, 431). Logical and final interpretants are the same, so here Peirce delimits final interpretants to self-controlled – that is, consciously controlled, primarily human – habits. See Chap. 11 for our development of this idea.

54 An attempt at construing a Peircean notion of causation is Hulswit (2002) with the sound aim of elaborate '... an anatomy of (Peircean) events, according to which it belongs to the nature of an event, not only *that* it creates a new event (Secondness) but also that it creates a new event *of a certain type* (Thirdness)' (215). Unluckily, Hulswit choses the same path as the absolute idealist tendency in Peirce, claiming that all causations are inherently teleological: '... Peirce forcefully defends the position that all natural processes are to some extent teleological, and can only be explained on the basis of final causes, which are general principles. The empirical fact that, as far as I know all (physical, biological, psychological, and sociological) processes show a definite tendency toward a general end state – whether it be described as chaotic or as involving a higher level complex order – cannot be explained without referring to certain general principles that somehow direct the process.' (213). The error here is to assume 'end states' of all processes. This begs the question, for only teleological processes could be said to have an 'end state'. To impose an end in a physical process is artificially splitting up its continuous development. Thus, Hulswit's identification of general principles (such as laws) with final causes does not hold (even if teleological processes form a subset of general principles). We would restrict Peircean teleology and mind to be biological properties primarily, thus occurring only locally in certain specifiable processes, cf. the biosemiotics section below.

55 They do not imply, however, that all concepts of relations etc. have real counterparts. So the relations between relations between relations... need not be a real would-be and basic relational realism needs not fall victim to Ockham's razor.

56 Thus, there is also an important link from real possibilities and to Peirce's own method, highlighted by Risto Hilpinen (1995, 274) who calls it 'the method of ideal examples' as opposed to 'attempts to characterize concepts by necessary and sufficient conditions: the latter method has often prompted philosophers to be excessively concerned about 'difficult' and 'borderline' cases which has sometimes hampered fruitful philosophical theorizing.' Being obsessed with category borderlines, we may add, is precisely an extensionalist vice, while the continuist will be rather obsessed with 'ideal examples' (prototypes, if you wish, or 'ideale Spezies' (Husserl)) and their bundles of interconnected real possibilities.

57 Given Peirce's 'extreme realism', is it not the case that it severely conflicts with his idea of truth as that which the scientific community asymptotically approaches in the long run – just like Hendricks (2001, 101) says, interpreting the latter claim as anti-realist. It is true that this claim has often been interpreted as opting for a consensus or coherence theory of truth, but I think it is equally compatible with a correspondence theory of truth. The idea of ultimate agreement among investigators is, in fact, an *idea*, it is what they would agree upon if they behaved as impeccable scientists and reasoners. So truth is not mixed up with what actual communities of scholars may agree upon or not. On the other hand, the idea is fallibilist: the vast majority of our current scientific views *are* probably true, but a few of them may lack credibility. So overall, science as a whole corresponds to reality, apart from some points. But we have no other means of discerning those points than further research.

58 It must be noted, however, that it is well-known that similarity within categories is psychologically perceived in a not completely symmetrical way. Rosch's and Lakoff's studies show that there less typical category members are perceived as more similar to the more typical members than vice versa. i.e., penguins and sparrows display a range of similarities, but the penguin is seen as more like a sparrow than a sparrow is like a penguin.

59 One could add that the opposite seems to be the case regarding transitivity. Similarity is not transitive: if A is similar to B and B is similar to C, then A is similar to C, is generally not the case, not even if we restrict similarity to deal with one single type of quality. One red nuance may be similar to another which in turn is similar to a third nuance – without it being the case that the first and third are similar. Take on the other hand transitivity in signification: if A signifies B, and B signifies C, then A signifies C. This is, if not generally realized in discourse, then in general a possible, that is, valid construction

in the system in question, at least given certain contexts. Peirce seems to admit this principle under the Scholastic notion 'nota notae': 'The logical principle Nota notae est nota rei ipsius, that is, the predicate of the predicate is the predicate of the subject, which is laid down in several places by Aristotle as the general principle of syllogism.' (a planned continuation of 'Syllabus', 1903, 3.590)

[60] 'Das Zeichen hat mit dem Bezeichneten inhaltlich zumeist nichts gemein, es kann ihm Heterogenes ebesowohl bezeichnen, als ihm Homogenes. Das Bild hingegen bezieht sich auf die Sache d u r c h Ä h n l i c h k e i t, und fehlt sie, so ist auch von einem Bilde nicht mehr die Rede. (...) Es wäre eine deskriptiv unrichtige Auffassung der Sachlage, wenn man denken würde, der ganze Unterschied bestehe darin, dass dieselbe Intention, die einmal an die Erscheinung eines dem gemeinten Objekt ä h n l i c h e n Objektes genknüpft ist, ein andermal an die Erscheinung eines ihm u n ä h n l i c h e n Objekts geknüpft sei. Denn auch das Zeichen kann dem Bezeichneten ähnlich sein, ja vollkommen ähnlich. Die Zeichenvorstellung wird dadurch aber nicht zur Bildvorstellung. Die Photographie des Zeichens *A* fassen wir ohne weiteres als Bild dieses Zeichens auf. Gebrauchen wir aber das Zeichen *A* als Zeichen für das Zeichen *A*, wie wenn wir schreiben: *A ist ein römisches Schriftzeichen*, so fassen wir *A* trotz bildmässiger Ähnlichkeit nicht das Bild, sondern eben als Zeichen.

Also die objektive Tatsache der Ähnlichkeit zwischen Erscheinendem und Gemeintem bestimmt keinen Unterschied. Gleichwohl ist sie für den Fall der Bildvorstellung nicht belanglos. Dies zeigt sich in der möglichen Erfüllung; und es ist ja nur die Erinnerung an diese Möglichkeit, welche uns die 'objektive' Ähnlichkeit hier heranziehen ließ. Die Bildvorstellung hat offenbar die Eigentümlichkeit, daß, wo immer ihr Erfüllung zuteil wird, ihr als 'Bild' erscheinender Gegenstand sich mit dem in erfüllenden Akte g e g e b e n e n Gegenstand durch Ähnlichkeit identifiziert. Indem wir dies als Eigentümlichkeit der Bildvorstellung bezeichnet haben, ist schon gesagt, daß hier die E r f ü l l u n g d e s Ä h n l i c h e n d u r c h Ä h n l i c h e s d e n C h a r a k t e r d e r E r f ü l l u n g s s y n t h e s i s a l s e i n e r i m a g i n a t i v e n i n n e r l i c h b e s t i m m t.' (LU II, 54–55)

[61] 'An icon is a representamen which fulfills the function of a representamen by virtue of a character which it possesses in itself, and would possess just the same though its object did not exist.' (5.73, Lectures on Pragmatism, 1903). In general, the relation of Peirce's sign descriptions to his pragmatism has been undervalued cf. e.g. 'It appears to me that the essential function of a sign is to render inefficient relations efficient, – not to set them into action, but to establish a habit or general rule whereby they will act on occasion.' (8.332, Letter to Lady Welby, Oct 12, 1904). Peirce's sign distinctions are made with reference to a basically functionalist (in a pragmatic sense of the word) sign conception.

[62] The animal known as a walking stick evidently functions as a sign fooling hungry birds, but there's no conscious, let alone human intention behind this functioning.

[63] Thus, it could be argued, Hilbert the father of formalism was no formalist (Hempel, in Benacerraf and Putnam (1983) makes this point). He insisted not on formalism but on finitism, that is, the formal means of proof should be finite, even if the mathematical content of the theorems proved might be infinite, and thus it was his attempt at solving the same infinity problem as the intuitionists, not to make formalisms the only goal of science.

[64] Of course, one should not overrate the 'turn' fact. Many foolish ideas have presented themselves as 'turns'.

[65] A related quarrel loomed large in early Cognitive Science under the label of the 'imagery debate'. Do mental pictures form real parts of thought processes, and may the pictural properties of such figures play effective roles in reasoning? Or are such pictures merely secondary to a basically symbolical, propositional thought process having pictures as epiphenomena or taking them as objects? Stephen Kosslyn's and Zenon Pylyshyn's old debate is even resurfacing these years. For a Peircean point of view, however, iconicity and propositionality is not in any way mutually exclusive, cf. Chap. 4 on diagrams.

[66] Actually, this similarity problem involves Gödel's well-known incompleteness theorem and entails that the set of possible A's may not be delimited beforehand because of being not computable. Of course, such insights do not interest the crusader of difference.

[67] Peirce makes exactly the same observation, in fact, when stating his pre-Popperian argument against the necessity of inductive reference.

[68] The reason why Goodman does not think this is the case is probably because he sees the idea of a 'common property' as something which may easily be formalized in a discrete symbolic algebra. If phenomenon 1 has the property a and phenomenon 2 has the property a, then it takes no larger amount of insight to state that $a = a$. But if quality spaces are continuous, this proposition forms a mere surface referring to continuous transformations which need not be simple.
[69] Thus, there must be an internal contradiction in the theory; the idea of a 'sémiotique naturelle' being impossible without the possibility to identify similar phenomena as similar. I hasten to say that this theory has in other respects, in particular dealing with narrative issues, proved fertile cf. Chap. 17–18.
[70] One can even find cooperations between the Nietzschean version and the extreme formalist version of anti-iconism, as for instance in the Italian Hjelmslevo-Nietzscheanism of Alessandro Zinna, a direct heir to the Ecoists' Hjelmslevo-Marxism of the seventies. When first you get the strange idea it is easy to see how this coupling works: Hjelmslev's sign-function destabilizes the sign to the extent that it is not a part of the language system, and his definition of form of expression and form of content, respectively, as the giving form to amorphous substance yields a place for Nietzschean active nihilists to undertake these creative acts of producing not only signs, but also expression and content forms.
[71] Even if Eco does not know the Erlanger program and thinks that geometrical similarity is only invariance to scalar transformations (and not, for instance, projections, rotations, mirrorings, topological 'stretchings' etc.).
[72] Already Peirce, by the way, raged against the mysticist claims that the Gizeh pyramids documented ancient Egyptian knowledge of the expansion of π, of e, the precise position of the North Star etc. – claims so well parodied in *Foucault's Pendulum*.
[73] The argument here is structurally analogous to Kant's argument for the indispensability of an *Urteilsraft*, an ability to judge. How can a rule be applied to a particular case? You may not propose another rule to take care of the application, for then you run into an infinite regress: how can this second rule be applied, then, and so on. The same goes for the attempt to get rid of similarity by means of conventions: how do you apply the conventions to similar cases – by means of yet other conventions...?
[74] We shall not go into Eco's arguments here except to notice that his treatment of the token-type problem is at best superficial. He obviously thinks that expression types are 'ruled by a *ratio facilis*', that is, they are easy to produce and reproduce, even if he admits that tokens of content types may be more difficult to handle. But even expression tokens constitute a very complicated problem, the bundle of transformations securing the various kind of similarities between various subtypes of a type has been investigated by Douglas Hofstadter (1986) and myself (1992).
[75] In short, one could sum up, it conceives of the thing as if it was an animal. René Thom's idea of the semantics of the substantive as a generalized animal here receives an indirect support. It is a pity that the consequences of this theory are not more thoroughly discussed – does this ascription of generality and organic teleology precede categorization *as such*? If so, is it not a necessary corollary that animal species categories are prototypical categories, so that other empirical concepts receive their (deceptive?) natural-kind character from this implicit biology of meaning? (No wonder, if this be the case, that philosophers always chose animals when talking about natural kinds, from Kant's dog over 'the cat is on the mat' and to Eco's own platypus...).
[76] Nolan 1994, 230.
[77] In a variety of sources, we find indications of such a distinction, for instance Groupe μ's insistence on the autonomous existence of visual types as distinct from their virtual linguistic denomination (1992), Jakobson's distinction between privative and qualitative oppositions, and Medin and Barsalou's (1987) distinction between prototype and boundary classifications, respectively, in categorical perception – see Chap. 10.
[78] Eco's promising negotiation theory of semantics ought to be compared to Hintikka's game-theoretical semantics.
[79] Rather late in the development of his theory, Peirce saw the icon as being not general, even if the icon can be used as an icon of a continuity of objects which has the same quality: '...neither icon nor index possesses generality' ('A Guess at the Riddle', c. 1890, 1.372). Peirce's development of the notion of hypoicons, including diagrams, introduces an important change to this idea, making general icons not only possible, but central to his epistemology (cf. Chap. 4). The existence of general

pictures has often been ridiculed in philosophy (Locke's famous triangle), but an important undercurrent continues a position cognate to Peirce's, see e.g. H.H. Price's overlooked *Thinking and Experience* (written in 1953 in the midst of Wittgenstein language game ecstasy in Britain) where the importance of 'images as general symbols' is emphasized. Price makes a strong case for the central role of general images in the psychology of thought. Regarding diagrams as external representations, James Robert Brown's *Philosophy of Mathematics* (1999) makes, disguised as a phil-of-math introduction, a strong case for the possibility of pictures to 'provide solid evidence, too, evidence which is as rigorous as any traditional verbal/symbolical proof' (192). Brown's involvement with computer aided mathematical imagery suggests that information technology may function as a partisan in the renaissance of the acknowledgment of general images.

[80] Peirce does not identify the copula with the verb to be, rather it is the very assertion of the predicate+subject structure. Here, he anticipates speech act theory by admitting questions, imperatives and the like as proposition types on a par with assertions.

[81] This problem of the reducibility of relations entails an enormous and still unfinished discussion. Löwenheim and Quine have, each in their way, tried to show that all logical propositions may be reduced to dyads, but the general validity of their results are still subject to doubt, cf. Burch 1991. As to actants in linguistics, René Thom (1980) has argued there is a formal, topological reason for the apparent four-actant limit in sentences.

[82] Cf. Tom Short's (1983) important observation regarding 'How to Make our Ideas Clear': less clear ideas are not prevented from participating in making more general ideas clear. It suffices that the role played by the less clear ideas in the clearer picture is, in fact, clear. Clarity is not necessarily built up from below using crystal clear basic elements but may reside on certain levels of organization, cf. the 'bottomlessness' of mereology in contrast to set theory.

[83] Maybe this would also provide an answer to the old Husserl's famous problem in *Ursprung der Geometrie* (1939) about how original evidence might be transferred through history by means of symbols not containing this evidence...

[84] The whole problem of the synthesis of the various profiles of the object is another, even if it possesses interesting analogues to the diagrammatic conception of the object. One profile of the object already contains hidden similarities in a Peircean conception.

[85] See Roberts 1973, for a groundbreaking treatment of Peirce's logic graphs. Among recent diagrammatic scholars investigating Peirce's logic graphs could be mentioned John Sowa, Allwein/Barwise, Sun-Joo Shin and Ahti-Veikko Pietarinen. Shin claims the heuristic virtues of existential graphs and aims at rearticulating Peirce's basic graph operations for practical purposes, while Pietarinen argues for a philosophically deeper relevance of the 'endoporeutic' (outside-in) interpretation direction of Peirce's graphs, the idea that the outermost layer of a diagram must be interpreted before going on to its interior. To Pietarinen, Shin's reformulation of Peirce's conventions attempting to render the graphs compositional goes against the outside-in reading of symbolic calculi and he points to Hintikkan game-theoretical, dialogical intepretations of logic in order to further develop Peirce's graphs. The outside-in reading, to Pietarinen, is important because it makes the interpretation depend on the context implicitly provided by the phemic sheet on which the graphs are drawn – the phemic sheet corresponding not to the universe as such, but to the universe of discouse, depending on tacit understandings between the dialogue partners (Pietarinen 2006, 120–22). Maybe Pietarinen's observation can be generalized to Peirce's diagram doctrine as such: the outside-in reading is preferred because taking the diagram as a gestalt informed by a context. In specific diagram interpretation, the outside-in reading will, of course, mix with inside-out readings in an ongoing trial-and-error process, but it is right for Pietarinen to insist that diagrams generally may not be presupposed to be compositional; compositionality rather forming a restricted subclass of special diagrams. Everyday examples of diagrams like a map of a country or a school timetable are evidently interpreted outside-in rather than inside-out.

[86] Pure icons only exist as a limit category in Peirce – concrete signs being, as a rule, composite. 'Hypoicon' is Peirce's notion, then, for signs whose mode of object reference is primarily iconic.

[87] The force of this idea in metaphor analysis is obvious – and it is recognized, albeit in non-Peircean clothing, by the cognitive semantics tradition mentioned above.

[88] In the development of Peirce's thought, the idea of a general diagrammatology thus precedes his construction of existential graphs rather then the opposite. It is his diagrammatology and his category phenomenology which permits him putting them to use in graphical logic representations as a special case.

[89] This fact is elaborated ingeniously in Hintikka 1997.

[90] In addition to this basic, operational icon criterion, however, Peirce also has a continuous idea of measuring different degrees of iconicity of representations. Thus, he sees his own logic graphs as more iconic than symbolic representations because a variable is here presented by one continuous line of identity as against the repeated occurrence of a number of *x*s with the same reference in a symbolic representation. While the former preserves the unity of the variable, the latter represents it in a shattered way untrue to the unity of the reference of the variable. Thus, in his Beta graphs, he has two different ways of expressing identity between variables – one identifying them by means of a continuous 'identity-line', another identifying them by attaching the same letter ('selective') to them. The former Peirce sees as *more iconical* than the latter (even if the latter may be hueristically superior). This points to another, *optimal* iconicity concept in Peirce in addition to the basic operational iconicity. See Stjernfelt 2006.

[91] It is, for instance, not sufficient to rebaptize objects *a, b, c*... in order to perform a formalization if a rational transformation syntax is missing. By this criterion, hence, the infertility of some classical formalization attempts in semiotics becomes understandable; e.g. Hjelmslev's ambitious algebra of glossematics (1975) which did not permit transformational possibilities of any larger interest.

[92] I prefer to count such sign use as diagrammatic, notwithstanding some Peirce's more strict definitions demanding the presence of explicit intentional diagrams. This definition conflicts with other descriptions of diagram use, e.g. his characterization of mental imagery experiments as diagrammatic or his ideas of grammar as a type of diagrams, and is closer to his pragmatic *in actu*-requirement for sign use. I follow the latter tendency in calling icon experimentation involving rule-bound manipulation of icon parts diagrammatic.

[93] This points to the fact that the organization of perception includes highly elaborated diagrammatic capacities without explicit conscious representation.

[94] Lakoff and Johnson's metaphor theory thus involves that structure is mapped from one domain onto another; Fauconnier and Turner's generalized 'blending' theory (comprising also non-oriented mappings) involve a schematic, so-called 'generic' space granting the coherence of the blending output.

[95] For a thorough investigation of Kant's schematism, see Frovin Jørgensen 2005.

[96] Diagrams as 'the main if not the only way we acquire new knowledge of relationships' have been acknowledged as a Peircean doctrine by Johansen 1993, 99.

[97] It must be added, though, that Peirce's attitude towards Kant's famous distinction is not unanimous. In his early and middle period, Peirce simply sees analyticity as identical to deductive necessity, while syntheticity covers ab- and inductive probability (cf. for instance Peirce's discussion with Dr. Carus in CP 6.595 (1893), see Otte 1997 353f), thus pertaining to ideal and real realms, respectively. Consequently, mathematics is taken to be analytic – in contrast, of course, to Kant. Later, the issue becomes more muddled, not less because Peirce now only rarely refers to the analytic/synthetic concepts explicitly. Here, theorematic reasoning – corresponding to the experiment attitude in diagram manipulation – is generally seen as synthetic. We shall return to the issue in more detail in Chap. 8.

[98] Correction in the quote made from Robin (293), 59; NEM IV has 'represent*ed* existential or experiential peculiarities'.

[99] The distinction between pure and applied diagrams roughly corresponds to Kant's distinction between a priori and a posteriori schemata.

[100] Barwise and Etchemendy highlight this important feature in diagrammatic modeling: '5. Every possibility (involving represented objects, properties, and relations) is representable. That is, there is no possible situations that are represented as impossible. 6. Every representation indicates a genuine possibility.' (1995, 215).

[101] But doesn't this example run counter to Peirce's observation that the grammar of natural language is diagrammatic? No, because the contents of the words 'round' and 'square' are not defined by grammar. The diagrammaticality of (parts of) natural language syntax rather lies in its instantiation of some basic logic and ontological categories (argument structure, subject/predicate structure, etc.) It is important to

remember – cf. our painting example above – that concrete signs may possess both diagrammatic and non-diagrammatic aspects, just like they may be composed of differently defined diagrams, the relation between which need not in itself be diagrammatic. Some of the strength of natural language probably lies in precisely this: it freely unites diagrams on different levels (expression, grammar, lexical semantics of the different word classes, narratology), the relative independence of which constitute language's plasticity and its ability to talk about many things, including impossible objects.

102 Of course, this requires that the diagram is consistent. But the very syntax of a diagram forces it to be consistent: it is impossible to draw a square circle. This does not imply, however, that it may not be in many cases rather or extremely difficult to determine whether a given diagram is in fact consistent. For instance, an equation – a subspecies of algebraic diagrams – may hide an inconsistency very difficult to ascertain at first glance but which requires lot of work to determine: if you can derive a contradiction from it (the reductio ad absurdum method), then it is false (if we do not admit intuitionist logic etc.). The seminal difference is that you cannot derive from the mere grammar of the symbolic expression 'a square circle' an analogous contradiction; in order to do so, you have to attempt to make a diagram of its content.

103 Yet, this distinction is in many cases impossible to draw beforehand, so to speak – cf. for instance the fact that a certain amount of empirical data shows up to yield a Gaussian distribution: on a first glance, this result may be conceived of as an empirical law, but it might hide a deeper law, yet uncovered, which would rather make the distribution a logical result of general mathematical principles.

104 This icon-index distinction in Peirce of course refers back to Kant's contention that existence (haecceity, referred to by an index) is no predicate (quality, referred to by an icon), just like it refers forward to Kripkean reference theory's rigid designators (as a certain class of indices).

105 Many basic proofs in mathematics may be represented in more or less immediately accessible visual diagrams, see e.g. Nelsen 1993.

106 See Misfeldt 2006 for an empirical investigation emphasizing the change between different representations of the same object in mathematical thought.

107 Our description of the workings of such experiment processes might give the idea that they are psychological and thus dependent upon a person's psychological grasping of the diagrams. This is not, however, the case. It is a crucial part of Peirce's pragmatism that it shares a basic anti-psychologism with Husserlian phenomenology (cf. Chap. 6). Pragmatism insists that it is possible actions on diagrams which count – but such actions need not be performed by psychological means (cf. Peirce's notion of mind being much broader than that of psyche). Diagram operations are, by their very nature, purely formal and do not owe their validity to the psychology of those performing them. If the description of such processes may in some cases sound as if informed by psychology, this is only for the sake of understanding. When talking about the 'imaginary moment' as a phase in diagram manipulation, this of course refers to the psychology of the manipulator, but the decisive thing is that this moment is made possible by structural iconicity between diagram and object – not by the psychology of he or she who contemplates that iconicity. The case is parallel to when Peirce himself refers to human minds as bearer of signs, but immediately adds that this is only a pedagogical 'sob to Cerberus' to make his own conception easier understandable (Letter to Lady Welby 23 December 1908, EPII, 478). The validity of diagrammatical representations in general depends just as little on psychology as does the special case of logical formalisms.

108 Peirce makes a distinction making this understandable – between corollaries and theorems. The former are propositions directly read from a diagram; the latter propositions only to be found after some more or less 'ingenious' experiment. The distinction is valid, but can not be sharp: there is a continuum between, say, measuring a distance on a map; measuring the same distance with corrections according to the map projection used; constructing that projection; proving that the geometry of the surface of a sphere is isomorphous to a non-Euclidean geometry... We return to this distinction below.

109 It must be added that the truth of Kekulé's discovery story is a matter of ongoing debate in the history of science and has not yet been definitively settled. The story is only recorded by Kekulé himself in 1890, in a celebration speech 25 years after the discovery was published. The case is even more complicated from the fact that the German Chemists' Society at a conference in 1886 published a mock-periodical in which they appear not as the deutsche but as the 'durstige chemische Gesellschaft'. In this

joke, the Benzene ring is depicted with the Carbon atoms as six apes grapping each others' arms and legs (playing on the similarity between 'Affe' and 'Affinität' (ape and affinity) in German). Hence, it is argued, Kekulé's 1890 memory might have been influenced by this recent joke. Thus, the story was an object of doubt rather early, and already in 1927, Kekulé's son stated in an article that according to his childhood recollections, his father had often told the story many years before it appeared in print without ever referring to any apes – thus adding to its probability (Sponsel and Rathsmann-Sponsel 2000). Also the source of the 'Uroboros' motive – the snake biting its own tail – is discussed, and maybe traced to a sign on a pharmacy door which Kekulé remembered (ibid.). On the other hand, it has been pointed out that the German chemist Josef Loschmidt was a forerunner of Kekulé because he had, already in 1861, described a long series of organic molecules as involving ring-shaped Carbon structures (albeit not the simplest one, Benzene). Kekulé knew Loschmidt's work as is evident from his dismissive references to it already in the same year where he refuses the shapes given by Loschmidt to have any connection to real molecule shapes (Bader and Parker 2001). This thus forms a strong argument that Kekulé's 1863–65 discovery may have been influenced by his reading of Loschmidt some years earlier – but, on the other hand, it does not prove this influence may not have appeared in the guise of the half-dreamt snake ring of the original anecdote.

[110] Ernst Cassirer's concept of 'symbolic pregnance' may be interpreted as referring to such cases of 'spontaneous proto-diagrams'.

[111] Another example would be maps with high direction sensitivity but no metric, e.g. maps of the starry sky as seen from the earth; distances on this map measured in minutes and seconds of arc do not refer to real distances between stars in the universe, while directions do refer to real orientations in space.

[112] Hintikka's distinction is most thoroughly presented in the papers collected in the volume *Lingua Universalis vs. Calculus Ratiocinator* (1997).

[113] The drawbacks of this tradition have large overlaps with those ridiculed by Barry Smith's concept of 'fantology' (2005).

[114] In the logic-as-universal-language tradition, Peirce's distinction will be invisible, because any chain of reasoning will here be represented as a valid, finished symbol string *post hoc*, so that the theorematic, constructive part may merely be added as further premisses among others to the inference at issue. Peirce's distinction, however, becomes crucial because the viewpoint of the logic-as-calculus stance rather envisages the issues *ante hoc* – logic deals with the solution of problems and thus displays a continuity with heuristics and theory of science which is absent in the language tradition. When seen from the problem rather than from the solution aspect, Peirce's distinction suddenly becomes pertinent: given a set of premisses, it is of huge importance how and what to construct on their basis in order to reach a desired result.

[115] It is not so easy, however, that the universal-language and the calculus tradition are simply anti-iconic and iconic, respectively. Hintikka, e.g. places Hilbert and Goodman firmly in the calculus tradition. Hilbert is, according to Hintikka, no formalist, rather a fore-runner of model theory and, as discussed later, much less anti-iconist than often assumed, while Goodman the staunch anti-iconist is a calculus supporter because of his plurality of languages. On the other hand, the position of being universal-language and iconist at the same time is also possible – cf. the younger Wittgenstein with his picture theory of language maintained at the same time as the ineffability of semantics, famously making it impossible actually to point out any particular examples of the logical atoms claimed to found the theory of *Tractatus*. So even if Hintikka's calculus-language distinction is indeed orthogonal to the iconist-anti-iconist distinction, the combination of universal-language and anti-iconism is strong (and stronger than the combination of universal-language and iconism) in both analytic and continental traditions.

[116] See some preliminary remarks in May and Stjernfelt 1996.

[117] The German mathematician Moritz Pasch explicitly noticed this geometrical error and proposed a pure geometry in terms of purely formal manipulation of symbols with no regard to their intuitive signification, an idea that was fully developed by his famous pupil David Hilbert's formalism.

[118] The concept of 'symbol' has a history so confused that it almost ought to be completely discarded; in any case, any use of it should be explicit about the precise signification intended, cf. Sørensen 1963.

In formalism, symbols are arbitrary, simple signs to be manipulated syntactically; in Peirce they are not necessarily simple, and dependent on iconic meaning and indexical reference. On the symbol concept in the Kantian tradition, see my 'Die Vermittlung zwischen Anschauung und Denken' (2000).

[119] Greaves, in turn, quotes from a paper by Michael Hallett who provided translation and italics.

[120] Greaves has even found an amazing quote by Hilbert's close collaborator Paul Bernays in an unpublished lecture from 1921, where the stroke counting ability is directly expressed in terms of basic iconicity suggesting, as Greaves says, 'a distinctly Peircean explanation': 'The philosopher is inclined to speak of this representation [between sign and number] as a relation of meaning. However, one should note that, in contrast to the usual relation between word and meaning, there is [in this example] the essential difference that *the object doing the representing contains the essential properties of the object to be represented.* Thus the relations which are to be investigated between the objects represented *are to be found in* the objects doing the representing, and thus can be established through consideration of these.' (190–191; translated by Michael Hallett, emphases by Greaves). The intuition necessary for the metamathematical finitism is hence (strongly restricted, it must be admitted) iconicity. Still remains, of course, the Hilbertian distinction between these finite calculi and the potentially infinite objects they may be taken to refer to.

[121] But even if we grant the basic iconicity of any 'symbolic' calculus, a Peircean approach will still be faced with the problem of evidence in cases where the 'imaginary moment' is precluded or where it simply refuses to appear, cf. for instance the discussion of the computer proof of the four-color map theorem of topology which – because of its enormous size – is hard to understand as an ordinary proof which a skilled reader may adorn with interpretations from beginning to end. In proofs of this type, the trust is put in the infallibility of the computer: each step in the proof is logically valid; ergo the whole proof is valid, even if nobody has ever *observed* its truth in Peircean evidence or in Husserlian 'kategoriale Anschauung'.

[122] 3.467, from 'Grand Logic,' 1893.

[123] Life as such seems formally to involve simple diagrams known as 'categorical perception,' see Chaps. 9–12.

[124] '... il n'y a science qu'à partir du moment où on peut plonger le réel dans le virtuel' (Thom 1988, 69)

[125] Of course, this counterfactuality is easier to hide in experimental sciences where diagram experiments may, in many cases, be verified by similar experiments on the object itself. When this possibility falls away, counterfactual speculation prevails, cf. for instance cosmology or issues like the origin of life and origin of language. It is interesting to note, however, that the insight in the connection between counterfactual constructions and scientificity is taken up in non-experimental sciences in recent years. For instance historiography, so long trapped in a positivist determination to record only what actually happened, now seems (through inspiration from, among others, chaos theory and the formal concept of phase space in general qualitative dynamics) to realize that the actual event is only made intelligible through its juxtaposition with a rational idea of what would have happened if some central factors in the initial conditions of the situation were changed (Ferguson 1997).

[126] Nöth uses the term when describing Piaget's structuralism, Todorov's narratology, Jakobson's classification of sign systems as well as in the context of more general issues such as the semiotic concept of 'code' and the description of the transferral of semiotic structures into media such as theatre or gesture.

[127] By 'semiotics' we shall here take those disciplines which in various ways investigate signs and their meaning; we should not delimit ourselves to currents only which explicitly makes use of the term 'semiotics' or related expressions.

[128] We shall here remain so general in order not too quickly to tie ourselves to one specific rendering of these concepts so as for instance function as modelled by set theory as a 1-1 relation between the elements of one set to those of another. We shall return to different versions of the function concept below.

[129] The following presentations are not, of course, thorough introductions to the relevant theories' assumptions and methods, but are centered around the transformation concepts in play. Some of the theories I have discussed in more detail elsewhere.

[130] Cf. the detailed criticism in Scubla 1998 where the different attempts at formalization of the 'canonical formula of myth' is presented: the equivalence between dual opposition pairs and the mediations between them; Boolean logic, group theory, the Klein bottle as well as Jean Petitot's catastrophy theory

interpretation of it. I cannot here run through these vast discussions, a discussion of the last example can be found in Stjernfelt 1992a.

[131] 'Or, la notion de transformation est inhérente à l'analyse structurale. Je dirais même que toute les erreurs, tous les abus commis sur ou avec la notion de structure proviennent du fait que leurs auteurs n'ont pas compris qu'il est impossible de la concevoir séparée de la notion de transformation. La structure ne se réduit pas au système: ensemble composé d'éléments et des relations qui les unissent. Pour qu'on puisse parler de structure, il faut qu'entre les éléments et les relations de plusieurs ensembles apparaissent des rapports invariants, tels qu'on puisse passer d'un ensemble à l'autre au moyen d'une transformation' (Lévi-Strauss 1988, 159).

[132] 'Quant à la méthode comparative, elle ne consiste pas, je l'ai souvent dit, à comparer d'abort et à généraliser ensuite. Contrairement à ce qu'on croit souvent, c'est la généralisation qui fonde et rend possible la comparaison' (ibid. 179).

[133] D'Arcy Thompson thus is the heir to a sort of structuralism widely disregarded in the French structuralism of the 60s, namely that of Goethe (see Cassirer 1945). Goethe's theory of the 'Urpflanze', the arch plant, and its metamorphoses of course transgresses – as Jean Petitot remarks – Kant's prohibition against intellectual intuition, as when he intuitively seeks to grasp the schema for vegetative morphogenesis from an inner power which modifies the arche-part of the plant, the leaf, into assuming different functions. Yet his idea that such a schema might be possible constitutes a source for later and more rational attempts at formalizing such schemata (cf. the unproblematic talk about 'Baupläne' for each of the large groups of animals). The concept of metamorphosis in Goethe thus forms a first outline of a biological concept of transformation.

[134] If we look back on Lévi-Strauss' inspiration from d'Arcy Thompson, it is general: he does not, generally, use continuous transformations, but discontinuous ones only (as in the equivalence between semantic opposition pairs), even if some of the models in which he tries to formalize these transformations are indeed continuous (topology).

[135] The discussion in this section is based on Kline 1972.

[136] Many more could be mentioned, thus the concept of 'categorical perception' in phonetics, generalized in Stevan Harnad (where a discontinuous partition of the continuum of perception allows that a particular, concrete perception is transformed to the category of a typical one); Eleanor Rosch's prototype semantics (where a similar argument is made on the semantical level); Gestalt Theory (where stable structures in perception may be reached by transformation from concretely perceived instantiations, often even with apparently insufficient input in cases where a pregnant form is completed).

[137] I have done that in Stjernfelt 1992a.

[138] Thus, the central unfolding of a function germ in Catastrophe Theory constitutes a transformation having the character of an eidetic variation making clear which possibilities lie in the function germ in question.

[139] I have discussed the principal aspects of this idea in Stjernfelt 1992c.

[140] It may be added that the programmatic binarism of this part of the theory remains problematic.

[141] As Jean Petitot argued in his catastrophe theory modeling of the basic structures of that theory, cf. Petitot 1985.

[142] See Bundgaard et al. 2006.

[139] All these ideas in Husserl have a mathematical background, namely his never published dissertation on 'Beiträge zur Variationsrechnung' from 1882 with references to Felix Klein. We return to eidetic variation in Chap. 8.

[140] This is a central argument in Jean Petitot's philosophy which is strongly presented in the introduction to *Naturalizing Phenomenology* (Petitot et al. 1999).

[141] See Chaps. 6 and 12.

[142] Jaakko Hintikka (1983) has highlighted the importance of this distinction: 'theorematic' reasoning requires the introduction of a new set of quantified objects and hence, in the diagrammatic representation of symbolic logic, a further quantifier.

[143] State changes must be understood as temporal transformation invested with material ontological categories pertinent for the field investigated. In physics, invariance refers to the constance of matter and energy, in other sciences other invariants are picked out: political power is redistributed in political

science; intersubjectivity in sociology; meaning in semiotics; gene pool and life in biology (cf. the centrality of 'survival'). These transformations in real time only become understandable when they are embedded in a more comprehensive virtual space accessible by eidetic variation, cf. René Thom's idea: understanding is the nesting of the real within the virtual.

[144] Cf. the Goodman-Eco discussion above.

[145] Thus, transformation as procedure is crucial to the enlightenment project. A decisive component here is the giving up of any idea of metaphysically unprenetrable distinctions between domains – which may be transgressed by the transformation of structure between them (cf. Newton's merging together of supra- and sublunar worlds or the search in cognitive science for neural consciousness correlates (NCCs)). Such transformations, of course, have trial-and-error character (to avoid dogmatic assumptions of wrong transformations), and one might assume that transformation is what realizes the otherwise incredible step from rule to case (the necessity of judgment in Kant following from the fact that no rule is possible to guide the use of the rule, for this should be governed by a further rule, etc.) – by a trial-and-error transformation of rule-bound structure onto the particular case.

[146] The references to English versions of *Logische Untersuchungen* quotes are to Findlay's translation, Husserl 1970.

[147] Herbert Spiegelberg's pioneering 1956 article has not given rise to much further work. Here, Spiegelberg finds the following four basic agreements between the phenomenology of the two: the intuitive approach to the immediately given without preconceived theories; the disregard of questions of reality or unreality in that investigation; the radical difference between phenomenology and psychology; and the foundational role of phenomenology for philosophy and logic (182). This indeed forms a basic agreement, and as Spiegelberg says, makes it legitimate to talk both of them being phenomenologists. Spiegelberg also lists a series of points in which Peirce differs from Husserl, most notably his emphasis of the discovery of categories and the absence of intentionality as a main concern – and he concludes a brief review of the two's mutual knowledge of the other that the similarities primarily are based in their common situation: two mathematicians turned philosophers who attempt to found philosophy as a rigorous science on phenomena given in experience – in short, as two historical parallels, not unlike Newton and Leibniz. Still we will argue that considerably more similarities may be studied than the four mentioned in Spiegelberg's paper – while his basic conclusion as to mutual influence remains correct, even in the light of the further information which has appeared since his paper (NEM, *Hua* XXII, etc.).

A few further papers have commented upon the relation between the two. The fine article by Dougherty 1983 on Peirce's phenomenological defence of deduction concludes that Peirce's phenomenological approach is what allows him to articulate the 'applicability of the ideal triad to the real world' due to his different notions of abstraction (cf. Chap. 11), and thus reconcile formal and empirical justifications of deduction. By doing so, he has reached a result '... remarkably similar to that of Husserl' (ibid.), namely founding phenomenology as a method to attain the ideal realm of meanings.

Leila Haaparanta 2001 continues Dougherty's observation and emphasizes the commonality between the two in bracketing existence presuppositions and utilizing related abstraction methods in their research. Inspired by Hintikka, she points to a commonality in Peirce and Husserl in the analogy to geometrical method in both of them; in Peirce explicitly, of course, in Husserl, in the idea of working 'backwards' from experience – cf. Pietarinen 2006 on the importance of Peirce's backwards, 'endoporeutic' interpretation rule for his logic diagrams.

[152] '... durch ihre besondere Originalität, Einfachkeit und Eleganz imponiert. Sie erscheint, zumal nach der Vereinfachung, die *Schröder* ihr zuteil werden läßt, als eminent brauchbar, und es wäre demgemäß ein größerer Reichtum an sie anschließender und ausgerechneter Aufgaben erwünscht gewesen' (43).

[153] Peirce thus explicitly refers to Hegel for the notion, and this takes place in a period where Perice highly admires Hegel and *Phänomenologie des Geistes* especially – as against his earlier anti-Hegelianism of the 70s and 80s. Still, the temporal coincidence with Peirce's acquaintance with LU is striking – just as the similarities in their way of using the concept, whereas Peirce's 'phenomenology' has little to do with the Hegelian emergence of the Geist through history.

[154] Spiegelberg 1956 points to the fact that Peirce probably gave up 'phenomenology' for ethics-of-terminology reasons between Oct 3, 1904 when he writes to James about his need to find a new term and

Oct 12 the same year when he writes to Lady Welby, now using the term 'ideoscopy'. Presumably, he felt that the Hegelian use of the word were too far from his own and that Hegel must be given primacy.

[155] Rollinger 1999 (58) relates an amusing anecdote of categorial intuition: in Husserl's copy of Brentano's *Vom Ursprung sittlicher Erkenntnis*, 'in the margin next to the passage where it is claimed that perceivability cannot be the same as existence since the non-real exists and yet cannot be perceived (p. 62), it is written "categorial perception!"' – Kategoriale Wahrnehmung!

[156] Measured against this central role in Husserlian epistemology, categorial intuition has hardly received the interest it deserves. Among the most important contributions to its clarification count the following: Bernet 1988, 1988a; Cobb-Stevens 1990; Lohmar 1987, 1990, 1998; Seebohm 1990; Sokolowski 1964, 1974, 1982; Ströker 1978, to which I shall refer in the following. Most of these accounts aim at a reconstruction of Husserl's views (and are very useful, even fertile in so doing) and do not attempt to go into comparisons with other accounts of the problem. I suspect, however, that such comparisons might bring us closer to the problem itself.

[157] Husserl's own examples of the higher-order objects grasped in categorial intution in the 6.*LI* include the identity of an object, the relation of part to whole, relations, collections, the 'ideierende Abstraktion' and its intiution of essences, the determinate and indeterminate grasping of single objects ('das A', 'ein A'). Sokolowski 1982 presents a thorough analysis of the steps from an unanalyzed experience to its categorial articulation in subject and predicate. Lohmar 1998 articulates a general 3-step structure for categorial intuition: 'Gesamtwahrnehmung, Sonderwahrnehmungen, kategoriale Synthesis'.

[158] Even if 'Wesensschau' is a later term, it can be seen as a successor concept to categorial intuition, cf. Hintikka 2003, 187.

[159] This is, of course, what makes transcendental phenomenology possible as a project: the idea of taking the essences of consciousness as an object of eidetic study. It seems possible, however, to retain the idea of the possibility of fulfilment of (certain) intentions aimed at eidetic phenomena as well as phenomenology as an anti-psychologist science of consciousness, without assuming the constitutive nature of the latter. This would preclude (or, at least, bracket) the possibility of a transcendental phenomenology, but preserve eidetic phenomenology.

[160] Later, in his drafts of a new version of the sixth *LI*, Husserl took the position that they *always* involve such meaning, cf. Melle 2002.

[161] 'Le paradoxe est qu'une forme catégoriale signitive ne peut être remplie que par une intuition déjà elle-même catégorisée.'

[162] We find in the three degenerate act types (imaginative, indexical, and signitive, respectively) a not coincidential parallel to Peirce's three different ways of signifying an object (icon, index, symbol). This has been noted by Münch 1993, 218.

[163] As is evident, this variation procedure is modeled upon function analysis in mathematics, even in Husserl's terminology.

[164] As to the history of the concept of ontology, see Øhrstrøm et al. 2005.

[165] This point and its relation to the transcendence issue is not always clearly emphasized; a strong exception is Willard 1982, 397.

[166] In fact, if categoriality were identified with formal logic only, then some version of logical positivism might be the outcome. But it is not necessary to identify categoriality nor the propositional stance with language. Rather, language is one (prominent, to be sure) instrument developed on the basis of the cognitive potentials of abstraction and categoriality. A very broad definition of categoriality – comprising all higher-level acts founded on perception – is proposed in Smith (2000). Similarly, Bernet (1988) points to the fact that categoriality in general 'is about the intuitive givenness of ideal objects' (33).

[167] The intimate connection between categorial intuition and this 'theory of theories' is highlighted in Cobb-Stevens 1990.

[168] In Peirce, this problem, of course, is solved by taking general meaning to have a continuum of merely possible (but vague) referents as its extension and his idea that generality corresponds to the possibility of choosing any one of these – close to Husserl's idea of 'no matter which one'. When Husserl talks about 'representation' here, it must not be read symbolically as in normal Anglo-Saxon usage – in that case, intuitive representations would lose their direct object contact (cf. Roy 1999, 111 – 147).

[169] Here, Husserl is on a par with Peirce for whom the diagram is not the particular drawing on the page nor the reader's perception of it. Peirce thus distinguishes between the diagram token – the particular drawing on the page, corresponding to Husserl's 'Anhalt' – and the diagram type which we are able to grasp through a reading of that token, governed by a symbolic sign (which, in Peirce's terminology, implies generality).

[170] See Chaps. 8 and 12.

[171] Lohmar (1987, 1990, 1998) highlights this example in order to correct an error which Husserl himself later detects (1939). Husserl's idea in the LU was that the categorial act of collection by means of the 'and' operator could only reach fulfilment by a 'reflection upon the act itself', because it is the very act that constitutes the collection of entities envisaged (all possible entities whatsoever may be so collected). As an alternative to this strange idea where the performance of an act becomes the intuition fulfilling that same act, Lohmar points to fulfilment as *Deckungssynthesis* – covering synthesis – between partial intentions. Thus, collection would be so to speak a zero-*Deckungseinheit* and is probably involved in all more complicated acts because it simply co-localizes its entities in one and the same categorial place.

[172] This corresponds to Peirce's semiotics where the classification of objects in types does not require symbols, while the opposite is the case.

[173] We have already remarked that a strong case can be made that the set of possible typefaces or writing styles is so large that it cannot be exhausted by computational algorithms (cf. Hofstadter 1986; Stjernfelt 1992c). Thus, the very concept of writing style cannot be the result of a variation procedure limited to computational strength. Thus it points to the fact that the variation implied in grasping essences does not always – if ever – proceed to completion through all possible variants. Rather we intuit the fact that such a variation may go on indefinitely.

[174] We can ask, then, what is implied in the *mere* signification of the word 'steam engine', before the analogical fulfilment by the general imagination of the model? 'Steam engine' is a composite noun, that is, it means an engine somehow concerned with steam. Depending on the underlying schema chosen, such an engine could work by steam, produce stem, fight against steam, etc. The syntax of noun composition only tell us it is an engine in some way concerned with steam (thus founded upon simpler signitive acts aimed at 'steam', and 'engine', respectively; but these are both acts concerned with general objects and thus already presupposing categoriality in the form of generalizing abstraction). The same structure characterizes Descartes' famous 'chiliogon'-example which Husserl uses to argue against the representative use of diagrams in geometry. True, we can not imagine such a figure in its complete shape, and thus the understanding of the P-S structure of the word, literally 'thousand-edge', rests on our categorial understanding of the syntactical structure as well as our categorial understanding of each of the composite elements: a figure with a thousand edges. A further step in the *Erfüllungsreihe* may now prompt us to try and construct the figure in imagination. We still cannot imagine it as a figure, but we may imagine the procedure to construct it (take a rectangle and subdivide the sides until you get a number close to thousand, then add or subtract sides until you get thousand) – just like Husserl himself, when describing eidetic variation, emphasizes that the imagination can not cover all cases, but we can imagine the imagination act to go on ... An contrast case is the well-known 'round square' with its impossibility of intuively construing such an object. All such composite expressions prompt an *Erfüllungsreihe* prescribed both by their syntactical structure, by their founding acts. The compound noun problem forms a center of the discussion of grammatical 'blending' in Turner and Fauconnier's cognitive semantics (Bundgaard et al. 2006; Fauconnier and Turner 2002).

[175] However, a problem remains concerning the purely empirical species concepts. Husserl distinguishes three cases: sensuous abstraction giving sensuous concepts, categorial abstraction yielding pure categorial concepts, and a large group of *mixed* concepts (with the examples 'color', 'house', 'judgment', 'wish' for the first ones, 'unity', 'plurality', 'connection', 'concept' for the second ones, and 'coloredness', 'virtue', 'parallel axiom' as examples for the mixed group.) The latter two can of course be seen as direct parallels to Peirce's pure and applied diagrams, respectively. Peirce also admits the existence of concepts not (yet?) analyzable in diagrams and mentions 'murder' as an example. Still, even concepts as these are not without diagrammatic content. Both Husserl's 'house' and Peirce's 'murder' are founded concepts; both presupposes a schema of wilful, human action oriented towards a goal and towards eliminating certain factors opposing that goal (life of some person in the former case; cold, rain, theft,

etc. in the latter. The reason why Peirce will not take 'murder' as a diagram is that its semantics contains no rational relations. But this only implies that it is a 'stiffened' diagram: it is not possible to perform any information-yielding experiments on it. But it is still a diagram in so far it is a schematic relation able to subsume instantiations by variation. Thus, I believe that the field of pure sensuous concepts is probably narrower than both Husserl and Peirce suppose; they seem to be restricted to primitive sense qualities and even then, the categorial apparatus of variation is necessary for isolating them from phenomenological experience.

[176] Thus, I perfectly agree in his insistence that it is a mistake for Husserl to claim that the 'vague morphologies' principally resist mathematization. It is interesting to note that this was not unanimously Husserl's contention in the first version of the LU where he still claims the ideal that 'Die vagen Gebilde der Anschauung mittels exakter Begriffe möglichst deutlich zu charakterisieren, ist überhaupt eine phänomenologische Aufgabe...', even if he immediately admits that this task '... lange nicht genug angegriffen und auch in Beziehung auf die vorliegende Untersuchungen nicht gelöst ist' (3. LU, Hua XIX, 249). The second version replaces this contention with a longer argumentation to emphasize that 'Offenbar sind die Wesensgestaltungen aller anschauliche Gegebenheiten als solcher prinzipiell nicht unter 'exakte' oder 'Ideal-Begriffe', wie es die mathematischen sind, zu bringen' (ibid.). (Eng. 'The descriptive concepts of all pure description, i.e. of description adapted to intuition immediately and with truth and so of all phenomenological description, differ in principle from those which dominate objective science. To clear up these matters is a phenomenological task never yet seriously undertaken and not carried out in relation to our present distinction'; 3rd LU, Vol. 2, 451).

[177] Of course, discontinuous operation procedures are possible, so as e.g. the stepwise construal of $(5^3)^4$. But discontinuity is dependent on continuity, Peirce would argue: the only way we are able to synthesize the single components of a discontinuous procedure into one state-of-affairs is by embedding them in a continuous space. Thus, discontinuous phenomena are always embedded into continuous ones, and discontinuous objects or calculi presuppose (explicitly or not), continuity.

[178] Holenstein 1972, 288. Husserl's argument is built on §22 in LU 3, where he notes that not every whole requires a specific moment of unity, only the 'zerstückbare' require that. All really uniting consists of relations of foundation, and unity is a categorial predicate, on the one hand, at the same time as this unity is given directly, sensuously in Gestalts.

[179] This fact is, of course, what gives rise to the attempts at finding a vocabulary of simple schemata, so as e.g. the 'kinaesthetic image schemas' of Lakoff 1987 or the schemata in Peruzzi 1999. The precise amount of such schemata which the human mind is able to process is no doubt governed by empirical psychological regularities – but an a priori, phenomenological corollary is that *any* possible mind will have to do with some finite vocabulary or other of such simple diagram atoms. This will, in fact, be the phenomenological equivalent of the impossibility of 'intellectual intuition': if we possessed such a faculty, infinitely complicated diagrams would be possible to process in one glance.

[180] 'Mereology', from Greek *meros*, part. As a matter of fact, it might have been called 'merology', but after the Polish logician Lesniewski the form quoted has become ubiquitous. Lesniewski, in his works from the 20s and 30s, considered mereology to be one out of three basic branches of philosophy, 'protothetic', 'ontology', and 'mereology', respectively. Protothetic is a doctrine of propositions and their interrelations and it forms the logical basis of his theory; ontology is based on a distributive rendering of class membership, so that distributive class expressions are identical with general nouns applicable to individuals. Mereology, then, considers collective class expressions understood as being composed of parts. By the distinguishing of the two latter branches of philosophy, Russell's paradox is avoided, and mereology is weaker than set theory because it only admits one relation of inclusion (part of), opposed to the two in set theory (membership relation and subset relation, which in Lesniewski's thought are separated as belonging to ontology and mereology, respectively). Mereology in this sense of the word thus has the advantage of being 'bottomless', the compositional foundation of a class upon the existence of ultimate *Ur-elemente* being avoided. This implies that mereology is 'phenomenological' in so far as it may describe a given level of phenomenal existence without recourse to a bottom level of atomistic ontology, a crucial aspect of the semiotic perspectives in mereology. This implies the possible affinity of mereology to strongly nominalist positions claiming mereology to be without any ontological presuppositions whatsoever (as in Lesniewski's case, and after him, Goodman) – even if this is no

necessity, and mereology may as well be connected to neutral (the younger Husserl) or strong realist positions (Smith). Lesniewski was influenced by the part-whole reflections of the Brentanian tradition: Husserl's third and fourth *Logische Untersuchungen*, in turn influenced by Brentano's *Deskriptive Psychologie* (1890), and Carl Stumpf's *Über den psychologischen Ursprung der Raumvorstellung* (1873). For further accounts for these developments, see Smith 1982, 1994. In this context, we shall stick to mereological aspects of decidedly semiotic currents of thought.

Even if none of the authors discussed here use the word, I have chosen it as shorthand for 'doctrine of whole and parts' and similar complicated expressions.

[181] Of course, these two alternatives display, each of them, a host of subvariants of more and less sophisticated types, ranging from a completely compositional logicist theory in the one end and to almost mystical insistances on the autonomy of wholes in the other. Moreover, there is not necessarily a contradiction between the two; it is perfectly possible to imagine compromises, so as e.g. an emphasis on the primacy of the holist level giving rise to motivated, iconic syntaxes governing its parts – combined with a recognition of the possibility for these syntaxes of assuming, once established, an autonomous status involving local compositionality.

[182] Cf. Hua XII which contains among other writings 'Zur Logik der Zeichen (Semiotik)'.

[183] Husserl's disciple from the 1910s, Roman Ingarden, has probably developed the most ambitious further detailing of dependency theory, distinguishing no less than four different ontological species of dependency: autonomy/heteronomy, originality/derivation, self-sufficiency/connectivity, independence/dependency which may further combine into eight types of signs (Ingarden 1965–74, 123). It seems evident that dependency is in need of further subdistinctions (a head is an independent part of the body, as an autonomous part, but it is a special part because the body ceases to function as a body if it is in fact cut off). Ingarden's suggestions have not, however, yet given rise to a fertile further development.

[184] Fassen wir irgendein Paar von Teilen eines Ganzen ins Auge, so bestehen folgenden Möglichkeiten:

1. Zwischen beiden Teilen besteht ein Verhältnis der Fundierung.
2. es besteht diese Verhältnis nicht. Im ersteren Falle kann die Fundierung
    a) eine **gegenseitige**,
    b) eine **einseitige** sein... (LU II, 264–65)

[185] After Holenstein 1976, 58.

[186] Jakobson 1985, 281.

[187] Jakobson 1971b, 713.

[188] Jakobson 1985, 203.

[189] Ibid. 189. Here, he claims that the third Logical Investigation is '... one of the milestones for the initial advance of structural linguistics ...'.

[190] Like in 'Parts and Wholes in Language' where he begins: 'In the second part of Edmund Husserl's *Logische Untersuchungen* – still one of the most inspiring contributions to the phenomenology of language – two studies devoted to 'Wholes and Parts' introduce the philosopher's meditations on 'the Idea of Pure Grammar'. In spite of manifold aspects of interdependence between wholes and parts in language, linguists have been prone to disregard this mutual relationship.' (Jakobson 1971b, 280), and further 'The structure of the verbal code is perhaps the most striking and intricate example of whole-part relations that are built hierarchically' (282–83).

[191] Jakobson 1971a, 314.

[192] It is a strange fact that Jakobson does not explicitly refer to the *Logische Untersuchungen* in his definitions of the zero-sign or the markedness/unmarkedness distinction. Holenstein 1975, 1976 does not mention any such direct influence either.

[193] 'Eine der wesentlichen Eigenschaften der phonologischen Korrelation besteht darin, dass die beiden Glieder eines Korrelationspaares nicht gleichberechtigt sind: das eine Glied besitzt das betreffende Merkmal, das andere besitzt es nicht; das erste wird als *merkmalhaltig* bezeichnet, das zweite – als *merkmallos*...' Jakobson 1971b, 3, with reference to Prince Troubetzkoy.

[194] Jakobson 1971b, 213.

[195] Even politically correct language politics, eager to avoid the asymmetry inherent in the markedness-unmarkedness distinction, can not avoid this, cf. long marked forms like 'African-American' vs. short unmarked forms like 'African' or 'American'.

[196] For a comparative discussion of the bases of Hjelmslev's theory in relation to Brøndal's, see Jørgensen and Stjernfelt 1987.
[197] The only reference to Husserl found in Hjelmslev is in the early *Principes*, but even if it refers to the fourth investigation, it is pejorative: '... la théorie étrange du philosophe HUSSERL' (40). None of two recent comprehensive investigations on Hjelmslev mentions any possible relation to Husserl (Gregersen 1992; Rasmussen 1992).
[198] Diderichsen returns over and over again to the fact that linguistics in general tends to focus upon 'three main types of grammatical connexion', and at several occasions he mentions in that context Husserl's mereological analyses from *Logische Untersuchungen* as strikingly similar to structural linguistics (Diderichsen 1966, 107 [1947]; 137 [1948]; 207 [1952]) but he yields no indication as to the possible relationship between Husserl and Hjelmslev.
[199] Marty, whose 1908 *Untersuchungen zur Grundlegung der allgemeinen Grammatik und Sprachphilosophie* refers to the *Logische Untersuchungen*. Hjelmslev refers to Marty at several occasions, but not directly in connection to the dependency algebra.
[200] So, it seems like mid-twentieth century mereology has left behind some interesting ruins – or drafts for impressive new constructions? – in Ingarden's and Hjelmslev's complicated mereologies, hopefully to be developed by future research.
[201] Here, Diderichsen is admirably clear in his early insistence that formal glossematic description is impossible without a prior phenomenological sensibility for identities and differences (Diderichsen 1966, 123 (1948)).
[202] This consequence is still visible in Greimasian semiotics, having inherited the whole of its metatheoretical apparatus from the *Prolegomena*. In the semiotics of the Paris school, the orthodoxy teaches that denominations of theoretical as well as metatheoretical terms are completely arbitrary; still they are invariably chosen so they are relatively easily understandable with reference to ordinary language or linguistic tradition.
[203] For instance Ronald Langacker at the 'Wholes and their Parts' conference in Bolzano, where a first draft of this chapter was given as a paper.
[204] The shortest presentation is probably: 'For mathematical reasoning consists in constructing a diagram according to a general precept, in observing certain relations between parts of that diagram not explicitly required by the precept, showing that these relations will hold for all such diagrams, and in formulating this conclusion in general terms. All valid necessary reasoning is in fact thus diagrammatic' ('Lessons from the History of Science', 1896, 1.54). The object of mathematics will be pure diagrams of any kind, while ordinary reasoning as well as the empirical sciences will use diagrams applied in being constrained by existing relations as well.
[205] Maybe this very general notion of diagram – parts with unspecified interrelations being open to abstraction to classes of interrelations – makes Peirce's schema an early forerunner of category theory. Finally, I imagine Peirce's very wide concept of diagram may not only generalize Kant's idea of a schema, but also serve as a unifying concept for the various schema-like ideas that prevail in present-day cognitive semantics (image schemas, force dynamics, landmark-trajectory, frames, scripts etc.). If so, then the semantic processes charted in this tradition – metaphorical mapping, blending, conceptual extension etc. – would be understandable as specific variants of diagram manipulation.
[206] Austria avoided the wave of German idealism (partly due to politically motivated prohibitions) and thus maintained a realist Aristotelian curriculum in philosophy all the way to 1918. This contributed to the Austrian commitment to ontological issues when such issues were long since surrounded by epistemological (on the Continent) or logical (the Anglo-Saxon world) scepticism, cf. Smith 1994.
[207] Smith 1992, 1996b.
[208] Peer Bundgaard (2004a, 2004b) has pointed to the fact that Husserl actually outlines *two* different a priori theories in that investigation, namely of grammar and semantics respectively.
[209] Cf. Barry Smith 1992
[210] It is interesting to note that parts of semiotics have undertaken similar investigations, even if in less ontological clothing, cf. Greimas' fertile analyses of 'challenge' and 'anger' in Greimas 1982. In Chap. 18, we shall attempt a related analysis of 'espionage'.
[211] See Caldwell 2004 and Grassl and Smith (eds.) 1986.

[212] In the original, the first question was: *'Do the empirical theories with the help of which we seek to approximate a good or true picture of reality rest on any non-empirical presuppositions?'* We must assume this version is a typo since it 'turns the wrong way'. Extreme empiricists would hardly embrace non-empirical presuppositions. In order to give meaning, we have 'inverted' the question to fit the answer.

[213] Cf. Cassirer's interest in mediate concepts in, for instance, biology, cf. next chapter. It must be added that Carnap was not unaware of the coordination problem between these two tiers – the discussion of this problem was central in early positivism. Still, he thought to have solved it with the distinction between analytical and descriptive parts of the logical-linguistic framework and thus lost sight of it as a special problem (Friedman, 82n).

[214] Friedman's ideas have an independent correlate in the French Neo-Kantian Jean Petitot and his idea of an epistemological 'golden rule' requiring an intermediate level of objectivity-constituting concepts between mathematical models used in science on the one hand and empirical data on the other (see Petitot 1992; Stjernfelt 1992a). Petitot is even, from time to time, collaborating with Barry Smith.

[215] A considerable wave of interest is also palpable in analytical philosophy, cf. the Boghossian and Peacocke volume *New Essays on the A Priori*. I think it is fair to say, though, that these purely philosophical discussions are less interested in history of science – and, maybe for the same reason, less interested in the synthetical a priori.

[216] What does one do when trying to find a – conceivable or imaginable – counterexample to a synap sentence? One *varies* the states of affairs mentioned, performs a diagram experiment, in short.

[217] This and the following are my translations. 'Das Eidos, das *reine Wesen*, kann sich intuitiv in Erfahrungsgegebenheiten, in solchen der Wahrnehmung, Erinnerung usw., exemplifizieren, ebensogut aber *auch in bloßen Phantasiegegebennheiten*. Demgemäß können wir, ein Wesen selbst und *originär* zu erfassen, von entsprechenden Anschauungen ausgehen, *ebensowohl aber auch von nicht erfahrenden, nicht-daseinserfassenden, vielmehr 'bloß einbildenden' Anschauungen.*

Erzeugen wir in der freien Phantasie irgendwelche Raumgestaltungen, Melodien, soziale Vorgänge, u. dgl., oder fingieren wir Akte des Erfahrens, des Gefallens oder Mißfallens, des Wollens u.dgl., so können wir daran durch 'Ideation' mannigfache reine Wesen originär erschauen und ev. sogar adäquat: sei es die Wesen von räumlicher Gestalt, von Melodie, sozialem Vorgang, usw. *überhaupt*, sei es von Gestalt, Melodie, usw. des betreffenden besonderen *Typus*.' (16–17)

[218] '1. erzeugences Durchlaufen der Mannigfaltigkeit der Variationen;

2. einheitliche Verknüpfung in fortwährender Deckung;

3. herausschauende aktive Identifizierung des Kongruierenden gegenüber den Differenzen.' (EU 419)

[219] Husserl's theory is that empirical generals – in an analogy to the pure generals – form successive layers, beginning with 'concrete generals' based on repetition of the same experience ('gleiche' – as it can not, of course, be numerically identical experiences, it must be 'same' experiences, give and take some margin), while generals based on similarities only form species comprising several such concrete generals. The same two steps are repeated when it comes to partial sameness or similarity: here generals comprising abstract aspects, and, in turn, species of such aspects, are constituted. Common to all these empirical generals on different levels, however, is that they refer merely to 'real possibilities' as distinct from the 'pure possibilities' covered by the variation manifold referred to by pure concepts. It is crucial to keep in mind, here, that the distinction between empirical and pure does not – as in Kant – coincide with the distinction between material and formal. This forms the very basis of the synthetical a priori in Husserl: the pure concepts may refer to a priori structures, both in formal and material domains.

[220] 'Sie beruht auf der Abwandlung einer erfahrenen oder phantasierten Gegenständllichkeit zum beliebigen Exempel, das zugleich den Charakter des leitenden 'Vorbildes' erhält, des Ausgangsgliedes für die Erzeugung einer offen endlosen Mannigfaltigkeit von Varianten, also auf einer *Variation*. M.a.W. wir lassen uns vom Faktum als Vorbild für seine Umgestaltung in reiner Phantasie leiten' (EU, 410–11).

[221] '. . . das besagt nicht, daß ein *wirkliches* Fortgehen in die Unendlichkeit erforderlich wäre, eine wirkliche Erzeugung aller Varianten . . .' (EU, 412).

[222] 'Es kommt vielmehr darauf an, daß die Variation als Prozeß der Variantenbildung selbst eine *Beliebigkeitsgestalt* habe, daß der Prozeß im Bewußtsein beliebiger Fortbildung von Varianten vollzogen sei.' (EU, 413).

[223] 'Zu jeder Variationsmannigfaltigkeit gehört wesentlich dieses merkwürdige und so überaus wichtige Bewußtsein des 'und so weiter nach Belieben'. Dadurch allein ist gegeben, was wir eine 'offen unendliche' Mannigfaltigkeit nennen; evidenterweise ist sie dieselbe, ob wir langehin erzeugend oder beliebig Passendes heranziehend fortschreiten...' (EU, 413).
[224] A 'composite photograph' was – and is – the name of a technique where different variants of a type are superposed by subjecting the film to the light from each of the variants for exactly the same amount of time – then differences are levelled out, and a type appears. The technique is nowadays used to display the 'favourite woman' of a decade by the composite photograph of a series of actresses and models. See Hookway 2002 for Peirce's use of the metaphor.
[225] '... die logische Forderung der Individualität als eines Gegenstandes, als eines identischen Substrates für Prädikate, bzw. für objektive Wahrheiten (die unter dem Satz vom Widerspruch stehen), ist nicht erfüllt durch die Vereinzelung eines Begriffsumfanges, sondern steht unter Bedingungen der Zeit. Das sagt, daß wir für die individuelle Vereinzelung unter der Forderung einer Möglichkeit einstimmiger Ausweisung in einem *kontinuierlichen* Zusammenhang wirklicher und möglicher (an die wirklichen anschließbarer) Anschauungen stehen' (431, our italics).
[226] von Hayek's (1978) idea that the whole of our perception apparatus is geared to the extraction of abstract features of the environment and that abstractions (Peircean prescissions) are thus phenomenologically primary as compared to particulars might support Peirce's controversial idea of the direct perception of generalities.
[227] See Chap. 4.
[228] Earlier, an exception to this rule made it possible to let a broken coastline signify that nothing beyond that point had yet been explored, a sign pertaining not to geographical ontology, but rather to that of discovery.
[229] He even proposes a third interpretation: that what is synthetic is only what is theorematical in the sharpened sense of deriving from the introduction of some new axiom (and not theorematical in the sense of deriving form some well-known theorem which has not been mentioned in the problem at issue). According to Levy, these two 'degrees' of theorematicity must generally be distinguished – between, e.g. the introduction of an auxiliary line in a geometrical proof where the existence of such a line is granted beforehand by the axioms of geometry even if the line has not in any way been mentioned in context of the present problem, on the one hand, and e.g. the introduction of the Power Set Axiom in Peirce's proof for the claim that there is no greatest multitude, on the other. Even if this distinction is indeed valuable, to let the analytic/synthetic distinction follow it would be to historicize it to unrecognizability.
[230] In both special sciences and everyday reasoning, diagrams and mathematics may be *implicit*, though – just like is the case in syllogisms, in logical arguments more general, or in diagrams underlying grammar according to Peirce. There is no need that we should in every single case be able actually to draw out the diagram in question (even if it generally makes explicit what was only implicit and thus may open up a new set of possibilities of diagram experiments). Just as being the case in many symbols (covering both terms, propositions, and arguments in Peirce) – in most cases the content of those symbols may rest implicit and not made the explicit object of a diagram. In this respect, diagrams may play a role parallel to the variation procedure in the eidetic variation which, in Husserl's terminology might remain a result of 'passive synthesis' before yielding the resulting concept as an object of active intention; Husserl's intentionality distinction between passive and active may here be covered by Peirce's semiotic distinction between implicit and explicit. Such implications, however, will not be part of reasoning proper which, in Peirce, comprises only inferences which are explicitly controllable and subject to an ongoing scrutiny (cf. Chap. 11).
[231] Another such problem is: given fallibilistic apriorism, it is possible for individuals or cultures to be at fault regarding parts of a given synthetic a priori conceptual cluster – how does this idea fit with the clain that synthetic a priori laws are generally tacitly known and govern common sense actions in different domains? One answer is that the common sense world has a basic pragmatic quality which to some extent sets it apart from scientific accounts for the same phenomena, cf. for instance the mainly Aristotelian 'naive physics' sufficient for most everyday tasks but at odds with real physics. See Petitot and Smith 1990.

[232] In Stepanov 1971, cf. Sebeok 1991, 7. Around the same time, C.H. Waddington concluded his *Towards a Theoretical Biology* (Vol. III, 1972, 289) with the idea that language should become 'a paradigm for the theory of General Biology', and René Thom founded his biolinguistics in Thom 1972.
[233] The latter half of the *Kritik der Urteilskraft*, 1790.
[234] For instance René Thom (1972), or Emmeche and Hoffmeyer (1991).
[235] Hoffmeyer has recently summed up his own work in the field in *Biosemiotik* (2005).
[236] Translations from *Kritik der Urteilskraft* are my own. 'Ich behaupte aber, dass in jeder besonderen Naturlehre nur so viel *eigentliche* Wissenschaft angetroffen werden könne, als darin *Mathematik* ist...' (Kant 1786, A IX).
'...eine reine Naturlehre über bestimmte Naturdinge (Körperlehre und Seelenlehre) ist nur vermittelst der Mathematik möglich...' (ibid. A X)
[237] 'Dass seine Form nicht nach blossen Naturgesetzen möglich sei, d.i. solchen, welche von uns durch den Verstand allein, auf Gegenstände der Sinne angewandt, erkannt werden können; sondern dass selbst ihr empirischens Erkenntnis, ihrer Ursache und Wirkung nach, Begriffe der Vernunft voraussetze' (Kant 1974 [1790], 232).
[238] '...ein Ding existiert als Naturzweck, wenn es von sich selbst [...] Ursache und Wirkung ist' (Kant 1974 [1790], 233).
[239] That is to say, it could not be mathematized with the mathematics available to Kant. But there is no reason to accept Kant's identification of mechanical explanation with mathematical formalization of a given domain; the latter is far richer than the former. Whether more advanced mathematics might do (parts of) the job is a possibility open for an historicized version of Kant's apriorism, permitting the synthetic a priori field to develop. As we have already mentioned, Petitot has a strong argument for the possibility of qualitative dynamics to mathematically describe phenomenological aspects of non-mechanical reality aspects – 'pheno-physics'.
[240] 'dieser Begriff fürht nun notwendig auf die Idee der gesamten Natur als einer Systems nach der Regel der Zwecke, welcher Idee nun aller Mechanism der Natur nach Principien der Vernunft (...) untergeordnet werden muss' (op. cit. 242).
[241] 'Auch Schönheit der Natur, d.i. ihre Zusammensstimmung mit dem freien Spiele unsere Erkenntnisvermögen in der Auffassung und Beurteilung ihrer Erscheinung kann auf die Art als objektive Zweckmässigkeit der Natur in ihrem Ganzen als System, worin der Mensch ein Glied ist, betrachtet werden' (op. cit. 243).
[242] One could say that Kant admits the anthropomorphic character of the concept of organism and related notions – but at the same time he offers a naturalistic (!) explanation of it: it because we ourselves are animals and thus teleological beings that we are able to discern organic properties in other objects. Thus, the tragedy of teleology is not that it is wrong or misleading, it is, rather, that we are unable to understand its 'inside'.
[243] In this interpretation of Kant, it must be emphasized, we do not subscribe to subjectivism, nor to the idea of a prioric regularities as a privilege of some transcendental subject. They are, rather, objective conditions valid for, but not necessarily known by, any possible subject whatever, cf. Chap. 8.
[244] On Cassirer's Uexküll inspiration, see Stjernfelt (in press b).
[245] My translation. 'Sie setze an die Stelle des Zweckbegriffs den Ordnungs- und Systembegriff, und sie charakterisiert das Leben dadurch, dass sie ihm die Eigenschaft des Systems zuschreibt' (Cassirer 1957/1991, 221).
[246] Of course, the problem here deals with the idea that semiotic concepts might contribute to the very basis of biology, not, for instance, to its import into more restricted subarea like zoo-semiotics where their use is much more unproblematic, at least for higher animals with elaborate mental representations of the surroundings. Already phyto- and myco-semiotics investigating plants and fungi, take the problem much deeper into the metaphysics of biology.
[247] The semiotic interpretation of the behaviour of E. Coli presented here has been developed through discussions with the biologist Claus Emmeche.
[248] The illustrations are from Alberts et al. 1994.
[249] Such as the fine introduction in Alberts et al. 1994, 773–85.
[250] The fact that man and bacterium alike may be fooled by exactly the same chemical supports such a hypothesis. The qualia of sweetness, somehow actualized by our gustatory system, is no doubt lacking

in the bacterium and bears witness to a much more sophisticated process of tasting in higher animals – but still exploiting the same rather simple mechanisms. Categorial perception by means of 'active sites' probably plays a central role in the important transition from uni- to multicellular animals studied by Lynn Margulis. According to her endosymbiont hypothesis, already the eucaryote cell forms a symbiosis between simpler species: mitochondria, chloroplast, cell kern, and maybe the undilipodia facilitating cell movement like cilia or the flagella as in E. Coli. Further symbiosis at a higher level requires the coordinated cooperation between bacterial colonies able to communicate, and the interesting intermediate forms behaving like single cells in some phases and environments only to change to more coordinated behaviour in other environments are possible only through elaborate intercell communication. During unfavorable environment constraints, e.g. the outer bacteria of the colony are able to secrete chemical signs in order to communicate the danger to the more inner layers, causing them to take defensive action and change into a more intert, spore-like form. Simultaneously, the whole colony changes shape forming a sort of fruit stalk able to let these spores spread to faraway, hopefully more favorable locations. To that extent, bacterial colonies share certain of the coordinated behaviours in much more sophisticated multicellular organisms where intercell sign use may be harder to observe directly.

[251] Even if biologists like de Loof and Broeck (1995) maintain a short definition not far from what follows here: life as the total sum of all acts of communication performed by a compartment (an organized, bounded whole) from the lowest to the highest level at a given moment *t*.

[252] Rosen (1989) argues from Gödel's incompleteness theorem for the introduction of semantic notions in biology. Gödel's theorem may be expressed as stating that for a given, sufficiently rich system, syntactic truth (that is, theorems provable in the system) is but a subset of semantic truth (true theorems not provable in the system). Rosen calls 'complex systems' material systems which embody this property, in opposition to mechanic systems, and sees living beings as prototypical examples of such complex systems. As is evident, this argument is structurally analogous to the anti-functionalist Gödel argument in AI (from Lucas to Penrose).

[253] A priori analyses of various domains' conceptual structures still often have an ill reputation due to empiricist prejudices or to the widespread myth that a priori analyses necessarily implies subjectivism; see Chap. 8.

[254] We also find this crucial idea in Kauffman as a property in 'autonomous agents', e.g. his (2000).

[255] In Stjernfelt 1992b, I proposed this property as a prerequisite to all formation of signs. The notion of 'categorical perception' is taken from experimental phonetics (Liberman, Studdert-Kennedy etc.) where it refers to the ability of language speakers to immediately identify and categorize different variants of spoken phonemes, cf. Petitot 1985a.The fact that the set of macromolecules possessing the same active site can probably not be mapped by any effective procedure supports the Robert Rosen-like idea that the semantics inherent in this primitive categorization is subjected to undecidability limitations. The set of macromolecules triggering a specific reaction in the organism is (just like the whole of organic chemistry) potentially open-ended and thus can not be defined extensionally. I did not know, in 1992, that Giorgio Prodi had some years before put forward almost exactly the same proposal (Prodi 1988).

[256] This crucial difference is noted already by Pattee 1977 ('our *conceptual* distinction between a molecule and its message or interpretation depends on the *physical* distinction between strong and weak forces', 265) and more recently by Igambardiev's (1992) distinction between information and energetic levels (128).

[257] This crucial size differences between the sign and its signification probably entails that semiotic systems must constitute a part of what Bak (1997) calls self-organized critical systems. The idea here is that a long range of different systems from physics and geology over biology and well into the human and social sciences (sandpiles, earthquakes, traffic jams etc.) display a characteristical relation between the size of events (avalanches in a sand pile, for instance) and their frequency, so that a power law holds between event size and event frequency. Bak argues that brains must be at the self-organized critical limit by a negative argument: if brains were subcritical, too few events would influence their behavior, rendering them insensible; if they were supercritical, too many events would influence them and make them chaotic (Bak 1997, 176–77). This brain argument may be generalized to all semiotic systems: they must react with insignificant changes to most small impulses but with significant changes to few small impulses; the latter class now constitute signs of various 'sizes' dependent on the size of events they trigger.

[258] I borrow this use of the term from Merrell (1997).
[259] These ideas resume deliberations in Thom 1975, 128f.
[260] And even science – according to René Thom's fertile definition of science as the nesting of actual events into a space of virtuality – finds its roots here.
[261] As discussed in Chap. 4, the concept of *schema* goes back to Kant's first Critique (as a notion for the possible meeting point for concepts and intuitions) and plays a central role for realist semiotics from Cassirer to cognitive semantics. Santa Fe complexity theory school has recently realized the necessity of the presence of *schemata* in complex, adaptive systems (Gell-Mann 1995; Martin 1994) and found coarse graining and perception of regularities among their chief properties.
[262] As to consciousness involving qualia, experienced selfhood, etc., we have no behavioral criteria for determining its occurrence during evolution, but a first hypothesis might be that it appears as an integrating factor in organisms with central nervous systems, facilitating more efficient perception-action cycles.
[263] Peirce's pan-psychism of course implies that even inanimate matter would possess some kind of dull 'feeling'. A recent argument for the existence of proto-consciousness in living beings has been put forward by Penrose (1990, 1994): the remarkable fact that most life reacts similarly to the same basic types of anesthetic chemicals. The question of when consciousness shows up in the course of evolution, however interesting it is, is not decisive to the discussion of the relevance of semiotics for biology, so we shall leave it in this context.
[264] Thus, the code duality (analogue/digital, cell architecture/genome) discussed by my colleagues in the Copenhagen biosemiotic school is probably no logical necessity for life as such, even if it seems to be for the reproduction of it.
[265] Elsewhere, Peirce restricts himself to claiming that the universe is representable (because of appearing in the three categories) – this would be a more modest, Scotean claim that symbols and arguments may chart the universe because they mirror structures in it and thus have a *fundamentum in re* in such proto-semiotic *realitates*, rather than claiming that the universe *consists* of symbols.
[266] A further support for this thesis comes from Santa Fe theoretical biologist Stuart Kauffman and his theory of 'autonomous agents'. The idea here is that they derive from 'autocatalytic sets' of chemical compounds giving rise to chemical cycles. When they become able to stabilize and reproduce, we get 'autonomous agents' as a general notion for (not only biological) organisms. Furthermore, by the introduction of Per Bak's 'self-organized criticality' into this framework, Kauffman makes the guess that the biosphere as a whole – maybe even the whole universe – evolves at the maximum speed possible into what he calls the 'adjacent possible', defined by consisting of yet unexisting chemical compounds at one-reaction-distance from existing compounds. If this radical version of a naturalized Enlightenment optimism is really correct, it would probably imply that biological evolution takes the shortest way possible in the evolution of still more intelligent beings. Kauffman specifies in this context an interesting assumption in our context: even if the biosphere (and maybe cosmos as a whole, he speculates) thus must be supercritical with respect to the autocatalytic power of its chemical networks, the single autonomous agent (organism, cell, or whatever) must necessarily be subcritical inside its skin – if not, autocatalytic bursts of still new compounds (following Bak's power law distribution of events at the critical limit) would rapidly destroy it. This hypothesis – speculative as it is – would imply interesting semiotic consequences; the Kantian notion of the finitude of the subject (here, the agent) in relation to the thing itself (here, the biosphere) would have a naturalistic counterpart, as would the Hegelian idea of a 'ruse of reason' where the individual agent unknowingly contributes to the common best. Semiotically speaking, this asymmetry between the agent and its environment would arithmetically necessitate categorization (by a finite mind of an infinity of possibilities) and hence schematic signs as general and underdetermined means of coping with an overdetermined environment, cf. our remarks on the economy of sign use above.
[267] See, for instance, Emmeche et al. 1997 and 2000.
[268] Thom 1990, 43: 'Efficient causality and final causality may be subsumed under the heading of formal causality. It should be observed that in Aristotelian theory final causality is in fact always conditional. An 'act,' described as a process by an analytical model, may be stopped short if some impediment intervenes during its course. So the act may fail because of interaction with an accidental factor. When so considered, Aristotelian finality is perfectly compatible with our modern view of (local) determinism.'

[269] The fact that biosemiotic vocabulary is indispensable also for higher ontological levels does not in any way imply that higher domains may be *reduced* to biology. von Uexküll's dark *Staatsbiologie* – in fact no less than a fascist biologist doctrine of state – should be sufficient as a warning against such shortcircuits.

[270] It might be argued that this privilege of biology might be a coincidence due to our biological being. Kauffman (2000), for instance, toys with the possibility that his equivalent concept to biosemiotic organisms, 'coevolving autonomous agents' may be applied on a cosmological scale (such as the organization of galaxies) as well as on a microphysical scale (the organization of elementary particles and below). Of course, there may be several levels of organization between the fermion-boson level and Planck scale; we do not know and might never come to know.

[271] A lineage involving scholars like Goethe, Geoffroy Saint-Hilaire, von Baer, d'Arcy Thompson, Spemann, Driesch, Waddington, Brian Goodwin, René Thom, Robert Rosen, Stuart Kauffman, inter alia.

[272] René Thom has attempted to rescue the term vitalism, claiming that it is reductionism which is a dogmatist view, trying to repudiate the reality of the organism, evident for all to see. His insistence on the term, however, does not include the usual assumptions of the existence of extra-physical forces or the like, so his idea of saving the term, however sympathetical, should probably be left behind. Nevertheless, in this connection he makes a surprising attack on Neo-Darwinism, hinted at above: it is Neo-Darwinism which is the metaphysical theory, not at all so empiricist as it often sees itself, and what is more, in contrast to the epigenetic tradition's emphasis on ontogenesis, the Neo-Darwinist emphasis on phylogenesis gives it a *holism* on the species level, contrary to its own reductionist claims, in so far as it is the genome of a species which is considered the main object of research. (Thom 1990, 600)

[273] 'Umwelt' is often quoted as von Uexküll's own terminological neologism; in fact, it was already used by Goethe and originates from the Danish-German Enlightenment-Early Romanticism poet Jens Baggesen who coined it in an ode to Napoleon in the year 1800.

[274] cf. von Uexküll 1973, 220.

[275] The inclusion of von Uexküll in the current repertoire of semiotics is thus largely due to Sebeok. But despite his exile status in relation to biology departments, there is a series of other local 'rediscoveries'. Most notably, probably, his status as ancestor to present-day ethology via Lorenz and Tinbergen. But also his role in the biological parts of Cassirer's theory of knowledge should be mentioned (Cassirer (1991) quotes with approval the 'nicht-stoffliche Ordnung' of the *Theoretische Biologie* as a central concept for an anti-vitalist biology – see Stjernfelt (in press a), in René Thom's catastrophe theory (1975, 1990) with his insistence of the centripetal vs. centrifugal construction principles of machines and organisms, respectively (despite the non-mechanist teleology inherent in both), and, most elaborated, in Merleau-Ponty of around 1957–60: in his lecture notes for that period (1995), dealing with the concept of nature, we find a large chapter on the phenomenological importance of the *Umwelt* concept: its introduction of 'plans naturels' making possible the 'commencement de culture' and its gradual evolution, being 'moins en moins orienté vers un but et, de plus en plus, interprétation de symboles.' (231) – see Chap. 10. Finally, the 'embodiment' wave of recent cognitive studies also displays a growing von Uexküll interest.

[276] The illustration is from von Uexküll 1973, 105. Such circles are most often, in von Uexküll, connected to individual purposes of the organism: drinking, preying, eating, mating, fighting, etc. Collective functional circles of more complex kind, relevant to a flock or herd Umwelt must be presupposed in the semiotically interesting cases of social animals; we shall not go into this issue here.

[277] A Kantian idea, to be sure, prefigured in Kant's idea of causal circles in 'Kritik der teleologischen Urteilskraft' – see the previous chapter.

[278] This goes for animals, even down to unicellular animals, but not for plants in so far they are generally seen as possessing a truncated *Umwelt* all right (von Uexküll calls it a *Wohnhülle*, a dwelling place), but no functional circle because of their stationary life and lack of nervous system, making them unable to move and hence unable to perform *Wirken*. This distinction seems somewhat problematic (heliotropes or carnivorous plants e.g. seem to form spectacular exceptions and not the only ones) but we shall leave further discussion of it here.

[279] An example given by several occasions, e.g. von Uexküll 1973, 1982, 56–57.

[280] Prominent figures in a Neo-Kantian semiotic family tree would include Peirce, Cassirer, and in our days Jean Petitot and the American cognitive semantics traditions; the differentia specifica will be the assignment of a central role to schemas or schema-like concepts in the respective semiotic theory. For a discussion of Neo-Kantian semiotics and Peirce, Cassirer, and Heidegger, see Stjernfelt 2000. Taken in a more wide sense, of course, Neo-Kantianism as meaning merely semiotic reflections inspired by Kant, it would include the doctrines of Nietzsche, Heidegger, brands of *Lebensphilosophie* as well as evolutionary epistemology.

[281] The German quote can be found in von Uexküll 1973, 9.

[282] This attempt at the organism's auto-constitution of space, however, is mysterious: a single segment of the (in itself spatially defined) functional circle is taken to be constitutive of space... So the same problem remains here: if the organism is taken to be a subject in the constitutive sense of the word, then it is posited anterior to time and space, and its body remains part of the empirical world just like any other object – if the organism, on the other hand, is taken to be the empirical animal, then its bodily sensations are seminal to the construction of its surrounding space-time – but not of space-time as such.

[283] The same is the case in von Uexküll's famous tick example. The tick *Umwelt* as it is described here, has little to do with the simplicity we must suppose pertains to the real tick *Umwelt* but is ripe with concepts from the scientist's *Umwelt*: the branch, the fall, the mammal etc.; and the example lives from our intuitive grasping of a mapping from the scientist's *Umwelt* to the tick *Umwelt* and back. But this mapping presupposes a space in which both these *Umwelten* partially coincide, namely an objective space from which scientist and tick *Umwelt* select each their sets of relevant signs. If no such space is supposed, the mapping between scientist and tick *Umwelt* can not be presupposed, and the *Umwelt* hypothesis will fall (consequently, the idea of an ontological 'pluriverse' of *Umwelten* is contradictory).

[284] Merleau-Ponty finds a development in von Uexküll from an early Kantianism in which the *Umgebung* behind the various *Umwelten* plays the role as Ding-an-sich and to a later Schellingianism, in which Nature becomes a name for the integration of all *Umwelten*. (Merleau-Ponty 1995, 232)

[285] Translations from *Theoretische Biologie* are mine. 'Das Material, aus dem sich eine fremde Umwelt aufbaut, besteht unter allen Umständen aus unserer objektivierten Qualität, weil uns andere Qualitäten gar nicht zugänglich sind. Das einzige Unterschied zu unserer Umwelt besteht darin, dass sie geringer an Zahl sind. [...] Ein Ort, der für uns mehr nach links liegt als ein anderer, liegt auch in der fremden Umwelt, wenn beide Orte als Merkmale in ihr vorhanden sind, ebenfalls weiter nach links, auch wenn die Anzahl von örtlichen Merkmalen die sie voneinander trennen, geringer ist als in unserer Welt.' (von Uexküll 1973, 104)

[286] An analogous argument as to the possibility of inter-species *Umwelt* mappings due to 'general-purpose sensory and effector mechanisms' can be found in Cariani 1998, 252–53. The possibility of mappings within and between different functional circles in the same species must is in itself count as a core contribution to the semiotic generativity in higher animals and humans especially: they constitute metaphors, blendings, metonymies etc. and thus form part of the biological basis of cognitive semantics' 'literary mind', cf. Turner 1996.

[287] This does not, of course, rule out the issue of the experienced qualia consciousness and its 'inside'. It is perfectly possible, even highly probabale, that higher animals (that is, animals possessing a central nervous system) entertain inner states of this kind, and the question whether lower animals or even plants share related properties remains undecided. Yet, this question must be kept apart from the functional *Umwelt* definition, because the existence of a functional *Umwelt* in a species does not imply the presence of qualia consciousness. As to research into the qualia consciousness issue, however, we have not, at present, any scientifically valid method to ascertain nor investigate such states.

[288] Thus it will no longer be the case, as von Uexküll quotes the radical conservative sociologist Werner Sombart: 'No 'forest' exists as an objectively prescribed environment. There exists only a forester-, hunter-, botanist-, walker, nature-enthusiast-, wood gatherer-, berry-picker- and a fairytale-forest in which Hansel and Gretel lose their way.' (29). The fact that the forest is defined exactly with reference to these and related groups (and not e.g. football players, state secretaries, unemployed, Irishmen, or other collective subjects) is an objective fact which we – so Sombart – are able to registrate and which thus forms part of the characterization of the forest, not its dissolution. The Nietzschean skepticism in this quote is related to other radical conservative and irrationalist vitalist features in von Uexküll, for instance his pessimism toward civilization, not to mention his Nazist leanings.

[289] 'Sobald die Umrisse der Körper als Merkmale auftreten, ändert sich das Bild der Merkwelt von Grund aus, denn nun beginnt das Nebeneinander im Raum eine ausschlaggebende Rolle zu spielen. [...] Erst wenn räumliche Unterschiede im Merkmal selbst auftreten, kann man von einer höheren Stufe der Merkwelt sprechen.' (189)

'Voraussetzung für die Existenz von Gegenständen in der Merkwelt eines Tieres bildet die Fähigkeit des Tieres, eigene Funktionsregel für sein Handeln zu bilden.' (191)

[290] 'Bref, c'est la thème de mélodie, beaucoup plus que l'idée d'une nature-sujet ou d'une chose supra-sensible qui exprime le mieux l'intuition de l'animal selon Uexküll. Le sujet animal est sa réalisation, trans-spatiale et trans-temporelle. Le thème de la mélodie animale n'est pas en dehors de sa réalisation manifeste, c'est un thématisme variable que l'animal ne cherche pas à réaliser par la copie d'un modèle, mais qui hante ses réalisations particulières, sans que ces thèmes soient le but de l'organisme.' (Merleau-Ponty 1995, 233)

[291] Probably the ability to distinguish pure from false tones (relative to a given scale) is innate in humans: already 4-months old babies express disgust when faced with false tones (Wilson 1998, 166).

[292] The place of these categorical perception examples in functional circles even anticipates a central issue in the recent decades' category research: Eleanor Rosch's observation that there is a 'basic level' of categorization relevant to corresponding basic motor routines involving those objects. Functional circle categorical perception is defined with respect to the *Wirken* and hence on a motor routine basic level.

[293] The categorical perception tradition in cognitive psychology has its roots in phonetics but has since then spread onto other senses (e.g. color perception) and species (higher animals categorical perception). In Stevan Harnad's central 1987 anthology around the subject, most papers still focus upon perception issues, but one seminal paper (Medin and Barsalou) undertakes a detailed comparison of categorical perception research with semantic categorization and finds much more similarities than expected. Many brands of structuralism tacitly assume the same parallel (e.g. in Hjelmslev's thoroughgoing formal parallelism between expression and content). The same continuity between perceptive and semantic categorization is often tacitly assumed in the cognitive linguistics and cognitive semantics traditions in the recent decades. Categorical perception in the Uexküllian *Umwelt* connection is a strong argument for this continuity to the extent that the perceptual categorizations in primitive animals are directly linked to semantic categorizations (namely of the corresponding *Wirkmale*, specific actions). Thus, we here immediately use the categorical perception concept about animal categorization. Generally, no sharp distinction seems to delimit perceptual and semantical categories which rather form parts of a continuum.

[294] Cf. Günter Ehret and Patricia K. Kuhl in chaps. 10 and 12, respectively, in Harnad 1987.

[295] It is striking how much Harnad's model mirrors features in Peirce's classical account for the generality of signs, where token and type denotes the particularity and generality of the sign itself, while the type due to its typicality in turn may be used as a symbol to signify some other general object.

[296] von Uexküll even, in a *lebensphilosophical* gesture typical for his time, makes the specificity of the human *Umwelt* a tragic problem for our species. It is strange that he can, on the one hand, claim that his theory of meaning 'culminates in explicating this connection' (the relation between a human need and the corresponding utensilia, such as coffee and coffee-cup), at the same time as he on the other hand can state that 'It should suffice to point out that with all our utensils we have built bridges between ourselves and nature. In so doing, we have come no closer to nature; in fact we have removed ourselves from her. [...] In the city we are exclusively surrounded by artifacts [...] The much-lauded technology has lost all feeling for nature: Indeed, it presumes to solve the most profound questions of life – such as the relationship between man and god-like nature – with totally inadequate mathematics.' (66–67) It is by no means an evident corollary to his theory that man should possess this tragic distance to nature; quite on the contrary. If Uexküll's concepts are removed from his creationism and *lebensphilosophische* pessimism of culture, it may be integrated into evolutionism such that man's especially malleable *Umwelt* becomes a consequence of the ever more complex *Umwelten* of higher animals, cf. Jesper Hoffmeyer's idea of a gradient of growing 'semiotic freedom' during evolution.

[297] Even if the *Bedeutungslehre* is sceptical against neutral objects in animal functional circles, it admits the possibility of distinguishing properties in the object with greater or lesser relevance for the animal. This, Uexküll argues, is what gives rise to the Aristotelian essentia/accidentia distinction, referring to a rank-ordering of properties by importance according to the functional circle in question

(28). This distinction may receive a formal interpretation in the light of categorical perception, implying the crucial strengthening of cross-boundary distinctions (relevant for the essentia) as well as the weakening of the intra-boundary properties (which are hence accidental, according to the categorization in question).

[298] 'Bei diesem Vorgang ist die Melodie der Richtungszeichen das Gestaltende, die uns aber nur in der 'Gestalt' zum Bewusstsein kommt. Die gestaltende Melodie nannte Kant *ein 'Schema'*, und die verborgene Kunst des Gestaltens in unserem Gemüt nannte er den 'Schematismus'.' (118)

[299] Here, we disagree with Chebanov's otherwise exciting paper (1999) and its claims that the concept of *Umwelt* does not entail semiosis. A similar discussion which we mentioned earlier was raised at the conference World and Mind in Copenhagen May 2005 where the British philosopher Galen Strawson argued against the existence of non-conscious intentionality from a so-called 'stopping problem'. If we take the causal chain involved in a perception or other intention, then the intention selects its object in that chain. Not so for non-conscious processes, Strawson argued: here the object intented can not be distinguished from other segments of the chain such as the light falling upon the object or my retina or my visual cortex. Obviously, it would be absurd to say that I see my neurons working in my visual cortex. When consciousness is absent as stopping criterion, so Strawson, there is no criterion for determining the object of intention and hence not for intentionality at all. Strawson's argument begs the question, however, in presupposing intention to be translatable into a causal chain which is, of course, not intentional. If intention, in an Uexküllian manner, is defined by the functional circle, any further segment of that circle is functionally intented. Even if microscopically continuous, the perception and action signs cut up the circle in characteristic schematic segments, due to granularity, and in the circle, a segment may serve as a sign for a later segment. Butylic acid serves as a sign for falling, serving as a sign for warm skin (or for crawling back on a branch), serving as a sign for biting... Thus, the stopping problem is posed in a wrong way as pertaining to linear rays of intention. It overlooks that such lines basically constitute segments of the functional meaning circle – what is intended is intended with respect to its virtual inclusion in this circle.

[300] Here, we recognize two different sets of semiotic oppositions: contrary and contradictory relations (in e.g. Greimas) vs. marked/unmarked (Jakobson), the former privileging boundaries over prototypes, the latter prototypes over boundaries.

[301] Niklas Luhmann's autopoiesis theory seems to be an attempt at generalizing this insight in sociology, seeing societies as consisting of systems with each their defined *Umwelten.*

[302] Thus, the centrality of some version of CP in any functional circle definition points to it as being a formal a priori prerequisite of biosemiotics – adding to the arguments connecting biosemiotics to formal ontology in Chap. 9.

[303] It should be noted that this recursive definition of symbols does not correspond to Peirce's original account. In Deacon, indices are made out of icons plus icons holding between icons, and symbols, in turn, out of indices plus indices holding between indices. In Peirce, however, the three are irreducible, and the icon-index structure rather forms the internal anatomy of the symbol without being sufficient for its compositional definition.

[304] So as for instance that the increase in brain size necessitating more protein made early man turn to a more carnivorous behavior. While the *Männerbund* went hunting, the mothers nursing the children were waiting for protein to be brought home. The hunting man's gene pool was threatened, however, by his woman's possible unfaithfulness during hunt, and she and her child was correlatively threatened by protein undernourishment if he did not return. This situation calls for stabilization by marriage which in turn requires stable institutions guaranteed by language – in turn calling for (further) development of symbol use. This hypothesis is interesting indeed, but it includes many specific issues and premisses which are not our concern in this context.

[305] Baldwin was primarily known as one of the pioneers of child psychology. Incidentally, he was also one of Peirce's acquaintances and invited Peirce to contribute his many entries to Baldwin's influential *Dictionary of Philosophy and Psychology* (1901). Sometimes, Peirce expressed ideas not far from Baldwinian evolution: 'For we must remember that the organism has not made the mind, but is only adapted to it. It has become adapted to it by an evolutionary process so that it is not far from correct to say that it is the mind that has made the organism.' ('Abstract of 8 lectures', undated, NEM IV, 141)

[306] After *The Symbolic Species*, Deacon has refined the Baldwinian hypothesis assuming two phases often interchanging in complex evolution, involving the 'masking' and 'unmasking' of selection pressure on a specific gene (Deacon 2003). A gene may be masked by a learned behaviour which makes the effect of that gene superfluous. Deacon's example is early primates being introduced to fruit in the diet with the result that the gene governing Vitamin C synthesis in early primates (as in most other mammals) was masked against selection with the result that the gene degenerated. Now, primates were forced to stick to a Vitamin C rich diet in order to survive, a fact which unmasked a completely different gene, namely in the set of genes governing the anatomy of the retina. Now, colour vision all of a sudden became an advantage in order to get Vitamin C from red, yellow, and blue fruit on the green foliage background, and via two duplications of a retinal gene, primates became able to see colours. Genetic degeneration facilitated by masking is thus an important factor in evolution because it loosens genetic control over certain phenotypical features and allows self-organizing possibilities in the organism to experiment with new properties. Thus, Deacon conjectures that early symbol use in hominids masked genetically constrained stereotypical communication and thus allowed for the brain to self-organize in new ways giving rise to the growth of interconnectedness of the human brain – in turn making more complicated symbol use possible and thus unmasking other cerebral genes and exposing them to Baldwinian selection pressure from linguistically skilled group members.
[307] It may even cover lower animals as well, cf. E. coli's ability to swim upstream in a saccharine gradient which in Peirce's terms must be classified as symbolic with respect to its *esse in futuro* – see Chap. 9. See Harnad 1987 for a long range of investigations of categorical perception tied to behavior – hence forming symbols – in many different species. The issue of the natural extension of Peirce's symbol concept is complex – as already discussed in Chap. 2. There is no agreement upon it, and Peirce's own idea on this point seems to find symbols even in inorganic nature – an idea nowadays supported by John Deely. Thomas Sebeok tended to make semiotics co-extensive with biology, hence only finding symbols in biological nature. Deacon, in turn, heightens the distinctive limit to make symbols a human prerequisite. On this point, we support Sebeok's middle way.
[308] A further problem is that Deacon's reconstruction of the icon-index-symbol triad as referred above makes it compositional, so that higher sign types are presumed reducible to combinations of the lower. But if pure icons – so Peirce – are mere possibilities, taken by themselves, then the actuality dimension of indices can not be created by the composition of ever so many icons; correlatively, the general, future dimension of symbols can not be the result of a composition of ever so many actual, indexical moments. Peirce's description faces the opposite direction: symbols are wholes, and icons and indices are moments of the symbol's anatomy.
[309] Deacon refers at length to the famous *Kanzi* case where a young bonobo on its mother's back learned the symbolic language which scientists were trying to teach its mother. There is little doubt that *Kanzi* is a symbol user, both in Deacon's and (less surprisingly) in Peirce's sense of the word.
[310] See, e.g. 'Minute Logic' (1902):
'Reasoning, properly speaking, cannot be unconsciously performed. A mental operation may be precisely like reasoning in every other respect except that it is performed unconsciously. But that one circumstance will deprive it of the title of reasoning. For reasoning is deliberate, voluntary, critical, controlled, all of which it can only be if it is done consciously. An unconscious act is involuntary: an involuntary act is not subject to control; an uncontrollable act is not deliberate nor subject to criticism in the sense of approval or blame. A performance which cannot be called good or bad differs most essentially from reasoning.' (2.182)
[311] Both must, furthermore, be distinguished from *induction* dealing with a series of related, empirical phenomena and proposing a probable law uniting them. Induction is often by empiricists confused with abstraction, but like Husserl (2nd LU), Peirce keeps these problems apart, and neither of the abstraction types have anything to do with extracting regularities from a set of examples.
[312] The central quote is the following:
'The terms 'precision' and 'abstraction,' which were formerly applied to every kind of separation, are now limited, not merely to mental separation, but to that which arises from *attention to* one element and *neglect of* the other. Exclusive attention consists in a definite concepton or *supposition* of one part of an object, without any supposition of the other. *Abstraction* or precision ought to be

carefully distinguished from two other modes of mental separation, which may be termed *discrimination* and *dissociation*. Discrimination has to do merely with the senses of the terms, and only draws a distinction in meaning. Dissociation is that separation which, in the absence of a constant association, is permitted by the law of association of images. It is the consciousness of one thing, without the necessary simultaneous consciousness of the other. Abstraction or precision, therefore, supposes a greater separation than discrimination, but a less separation than dissociation. Thus I can discriminate red from blue, space from color, and color from space, but not red from color. I can prescind red from blue, and space from color (as is manifest from the fact that I actually believe there is an uncolored space between my face and the wall); but I cannot prescind color from space, nor red from color. I can dissociate red from blue, but not space from color, color from space, nor red from color.

Precision is not a reciprocal process. It is frequently the case, that, while A cannot be prescinded from B, B can be prescinded from A. (...)'

313 We may note that in this argumentation, the three separation modes are tied to three different modes of presentation: imagining, supposing, and representating, respectively.

314 Peirce's definition of the distinction abstraction types connects them intimately to a part-whole dependency calculus. This idea is exactly parallel to the connection between Husserl's 2nd and 3rd LU with the anti-empiricist abstraction theory of the 2nd (abstraction is not inductive generalization, abstraction is a special idealizing focusing act related to an object's properties) and the mereology of the 3rd (the properties thus grasped should be seen as different parts and moments of the object, and a calculus is possible to map these parts' internal relationships). This connection between abstraction and mereology is a highly original idea crucial for the possibility of a realist understanding of the cognition of abstract objects.

315 'In general, prescission is always accomplished by imagining ourselves in situations in which certain elements of fact cannot be ascertained.' ('Supplement', 1893, 2.428) Prescission thus is Peirce's version of Duns Scotus' famed 'formal distinction'; it refers to a distinction made by the mind, but with a *fundamentum in re*. As to the color-extension issue, Husserl and Peirce analyze it differently. To Husserl, they are mutually dependent, to Peirce, colour is dependent upon extension (cf. Chap. 6–8); we tend to support the latter.

316 A simple example is the train of thought as follows: *a white particular thing* - (P) - *white things as such* - (A) - *whiteness*, with P for prescission and A for abstraction. A more complicated example is hinted at in set theory and may be reconstructed as follows: *elements* - (P) - *belonging together* - (A) - *a set* - (P) - *bigger/smaller* - (A) - *multitude* - (P) - *relation to other multitudes* - (A) - *cardinal number*... This example is reconstructed from 'Consequences of Critical Common-Sensism' (1905), 5.534. The unlimited character of this abstraction process does not entail it is infinite.

317 So as for instance ('Consequences of Critical Common-Sensism' (1905), 5.534). The famous quote stems from the third interlude in Molière's last play, *Le malade imaginaire* which introduces a grotesque ceremony of doctors dancing and singing medical latin. Here, a medicine student answers a doctor's question as follows:

'BACHELIERUS/ Mihi a docto doctere/ Domandatur causam et rationem quare/ Opium facit dormire./ A quoi respondeo,/ Quia est in eo/ Virtus dormitiva,/ Cujus est natura/ Sensus assoupire. CHORUS/ Bene, bene, bene, bene respondere./ Dignus, dignus est entrare/ In nostro docto corpore.' (Molière, 660); in my translation: 'Bachelor/ Me the learned doctor/ asks about the cause and reason why/ Opium puts to sleep./ To this I answer/ That there is in it/ A sleep-inducing power/ Whose nature it is/ To weaken the senses.// CHORUS/ Good, good, good, good answer./ Honorable, honorable is it to enter/ into our learned society.'

318 Peirce's theory of hypostatic abstraction thus forms a strong argument against the current fad in rhetorics where it is claimed that abstract noun use is just a showoff strategy trying to impress the reader with difficult wording (the opium argument of our time), while texts which express 'the same' in more concrete terms are praised as more honest and easier to read. In a Danish context, the two writing styles are even hastily connected to males and females, respectively, rhetoricians taking the party of the latter. If this claim were true, it would do the feminist cause a questionable service, as abstractions are indeed necessary for thought to occur.

[319] Roberts, 64.

[320] Peirce has a primitive mereological intuition in so far he refuses to distinguish between parthood and elementhood (as in set theory) and sees those two as shadings of one and the same basic relation.

[321] Cf. e.g. '... I do not think that we need have any further scruple in admitting that abstractions may be real, - indeed, a good deal less open to suspicion of fiction than are the primary substances.' ('Pragmatism Lectures' (1903), Peirce 1997, 136.) This idea forms a very important phenomenological principle in Peirce: the objects to which we have direct access are neither completely abstract nor concrete; they are at different intermediate levels. Thus, both abstract objects and the concrete object are constructions reached by abstraction, and the ladder of levels is virtually bottomless; we have no guarantee that it terminates 'downwards' in some elementary, atom-like entities. As collections are also abstractions, this consideration also goes for scalar properties. This idea is basically a mereological idea (whole-part relations are pertinent on all levels of observation or reflection) and fits nicely with Peirce's proto-mereological refusal of distinguishing element-of and part-of relations (like it was later formalized in Lesniewski).

[322] T.L. Short (1983) traces roots of this idea in Peirce's thought to the famous 'How to Make our Ideas Clear' paper from 1878 where Short emphasizes the crucial idea that 'we *can* use ideas that are *less* clear to make other ideas *more* clear' (290). Thus, the fact that the higher, more abstract terms may be clearer than their concrete basis is a crucial insight in order to avoid infinite regresses and appreciate the role played by abstraction in Peirce.

[323] Helmut Pape: '... the 'abstract in concrete form' brought about by a 'realistic hypostatization of relations' is a deductively valid form of reasoning which he at other places calls 'hypostatic abstraction' and which is now called class abstraction.' (Pape 1997, 171) Pape explains in a note: 'It is obviously deductively valid to conclude that, if there is a red rose, the class of red things has at least one member, namely, this rose.' (182n). In the very same volume we read T.L. Short: '... neither is the inference to it logically necessary. Rather, that inference could be deductively valid only with the additional, logically contingent premiss that the regularity in question has an explanation (...) Absent that assumption, the inference is not deductive but an extreme case of what Peirce called 'abduction' ...' (297).

[324] This should, of course, be taken to refer to reflective self-consciousness. Pre-reflective self-consciousness (cf. Zahavi 2000) as a moment of any conscious experience is presupposed by reflective self-consciousness, and may probably, unlike the latter, be found in large parts of the animal kingdom and maybe even in lower organisms.

[325] The quote talks about 'races', but we should not take this as an indication that any idea of 'racism' could be found in Peirce's thought; he is merely using the word as coextensive with 'culture' or 'social or geographical group of people', such as was a commonplace at the time.

[326] The Husserlian idea of a pure grammar – to some extent shared by Peirce – may make the Chomsky-Deacon conflict around grammatical inneism irrelevant. If there are indeed a priori rules for grammar, then we should expect evolution (biological as well as linguistic evolution) to conform to them in a gradual approximation, making the riddle of possible innate chunks of universal grammar easily understandable, because no empirical selection pressure will be needed for their articulation. This would correspond to the fact that we have learned elementary arithmetics and performs that easily without anybody wondering about the specific selection pressures giving rise to an 'arithmetic module' in the brain. Terrence Deacon seems on the same track when toying with an 'extraterrestrial' grammar valid for any speaking subjects whatever...

[327] The issue of the 'semiotic missing link' is currently the object of intense investigation. A famous suggestion parallel to Deacon's is Michael Tomasello's idea that what separates man from other higher primates is the ability of *shared attention* during learning. Mother and child, for instance, may direct their attention to one and the same object, initiated by the child understanding the mothers gaze direction or pointing gesture, while young apes would rather look at the pointing finger than at the object indicated. There is not, however, any common scientific language or set of criteria making us able to decide between such proposals – or even to decide whether they are mutually exclusive, reconcilable – or even maybe expressing related or identical ideas in different vocabularies. It would not be strange, for instance, if shared attention necessarily involved hypostatic abstraction – the explicit control necessary for the latter being provided by the common discursive field opened by the former. Here, further investigation, both empirical and conceptual, is necessary.

[328] A problem here, though, is that there is hardly any definite boundary separating metaphorical projection from concept extension (both refer to the application of a concept on hitherto uncovered domains). The difference lies in the domains rather than in the extension process.
[329] The grasp of this problem depends on the understanding of 'a priori reasoning'. If it is to be taken in the Kantian tradition where a priori refers to validity prior to any empirical knowledge as well as to evident givenness for any knowing subject whatsoever, it is clear that the CS tradition is right in refusing to indulge in a priori reasoning, cf. Chap. 8.
[330] See Chap. 10.
[331] Doing so, Merleau-Ponty follows the example set by Ernst Cassirer – philosophizing by sifting the results of the special sciences - rather than that of Heidegger and his irrationalist claim that the 'Wissenschaft denkt nicht'. There are more references to Cassirer than to Heidegger in *Phénoménologie de la perception.*
[332] Despite the idea that the *Umwelt* is a subjective construction on part of the organism, von Uexküll's position is not necessarily subjectivist in the sense that it is a construction taking place in the organism's consciousness. von Uexküll's methodology for *Umweltforschung* was, in fact, a sort of behaviorism: an animal's *Umwelt* should be investigated by observation of its anatomy and its interaction with its surroundings, rather than through any sort of empathy.
[333] Merleau-Ponty thus does not hesitate to solve von Uexküll's problem with the possibility of 'neutral objects': higher animals with central nervous systems do, in fact, access such objects, cf. below.
[334] This idea anticipates René Thom's idea that environment objects are prototypically conceived of as a sort of generalized 'animals' so that nouns in general refer to such animal-objects.
[335] In this idea, Merleau-Ponty's concept of symbol approaches the Peircean symbol, being defined by its *esse in futuro* and its ability to anticipate future action by incarnating habits. At the same time, moreover, Merleau-Ponty anticipates Terrence Deacon's fertile idea of symbolicity as a crucial threshold in evolution, cf. the previous chapter.
[336] Translations from *La nature* are mine. 'Le corps appartient à une dynamique de comportement. Le comportement est enlisé dans la corporéite.' (239)
[337] This importance of animal appearance not only goes for its relation to species mates and to predators, it must be assumed, but also in relation to other species with which it competes, enters into symbioses, etc.
[338] 'La phénoménologie dénonce l'attitude naturelle et, en même temps, fait plus qu'aucune autre philosophie pour la réhabiliter.' (104)
[339] 'La chose m'apparaît comme fonction des mouvements de mon corps.' (106)
[340] Do animals hunting in bands not possess a sort of shared attention? Maybe not at any conscious level as known from human learning, but still several stages during the evolution from inter-animality over inter-corporality and to intersubjectivity seem to await proper conceptual determination.
[341] Without explicitly talking about the a priori of regional ontologies, Kauffman thus makes a parallel argument to our charting of a conceptual cluster of biosemiotics in Chap. 9.
[342] It is an interesting and strange implication of Kauffman's work (which it shares with the organism definitions of much romantic *Naturphilosophie*) that this cluster of concepts defining an autonomous agent may not stay a privilege for biology as we know it (geomorphology in geology, galaxy clusters in astronomy, companies in economics, states in international politics, etc. may be other candidates for autonomous agents in Kauffman's definition) – cf. our related discussion in Chap. 9.
[343] Referring to the putting to use of a biological property for another purpose than that for which it originally evolved.
[344] This idea forms Kauffman's suggestion for a fourth thermodynamical law - the universe invades the 'adjacent possible' with the highest possible speed. It remains unclear, though, against what this 'highest speed' could be measured.
[345] What Kauffman calls 'clusters of concepts', Husserl would probably call a regional ontology of a priori concepts for any possible biology.
[346] When I have presented this idea in lectures, it has been challenged by distinguished semioticians, like Susan Petrilli and Lucia Santaella, on Peircean grounds. They claim that Peirce's metaphysical continuity prohibits such thresholds from having any ontological value. I would argue this is a fallacious

argument. It is perfectly possible to maintain a basic continuist ontology and at the same time search for discontinuities segmenting this continuum. This lies already in one of Peirce's basic arguments for the primacy of continuity: the continuum problemlessly includes discontinuities, vice versa not so (and all attempts at 'building' the continuum from discrete entities fail). Moreover, even when no apparent discontinuity is at stake, clear differences are possible (take e.g. the continuum of color which does not imply that there is no difference between, say, yellow and green, having each their prototype (cf. Berlin and Kay's classic investigation of color names)). A sharp demarcation between any two colors will possess a certain arbitrarity but still it will possess a *fundamentum in re*. It is important not to let continuity assume a role of thought prohibition, a new darkness in which all cows become grey. This would violate Peirce's basic motto: Do not block the way of inquiry.

The biosemiotic *Scala Naturae* may be conceived of on the basis of Gould's idea of 'punctuated equilibrium'. Even if evolution is basically continuous, it may be segregated into long periods with little change, separated by short periods with huge changes. Thus, the development of man's unique semiotic possibilities has taken place over a very short timespan on an evolutionary scale, and it may be possible that the same thing holds for the passing of some of the other biosemiotic thresholds.

[347] Some of the thresholds proposed may cross each other, e.g. some of the candidates for 'missing link' thresholds, or the 'intercellular' threshold and the 'Uexküll threshold'.

[348] Mikkel Bogh (2002) argues that the introduction of poststructuralist theories in art history has had double consequences. On the one hand, it heightened the level of reflection compared to the strongly biographical traditional art history – on the other hand, it tended to leave the interest in the formal and phenomenological properties of the artwork to oblivion in favor of mere context.

[349] For a further discussion of negativism as avant garde kitsch, see Stjernfelt 'The Vulgar Metaphysics of Transgression', in *Text und Kontext* 2001 or Thomsen and Stjernfelt 2005.

[350] For a closer gaze, a certain tension thus prevails between two of the basic ideas of the historical avant garde: anti-figuration and anti-convention. Both of them are dubious as ontological constraints on art (why must an artwork not portray something; why could a conventional work not possess artistic value?), but it is interesting in this context that anti-conventionalism naturally implies that the anti-conventional artwork can not rely on purely anti-figurative mechanisms of meaning only (meaning must be either motivated or conventional, or both). Anti-convention must thus base itself on a similarity (but maybe one not seen before!). This lies already in the notion of 'abstract' art: if a picture is abstract, it does not imply it is not figurative, only that its similarity is so general that many different objects fall under it. The famous white square on white (to which we return below) refers to many possible concrete squares, herein lies its abstraction, and the game of providing the abstraction with the concrete objects it subsumes is left over to the viewer. The fact that things are hardly thus conceived by the main line of the art history of our time has its reason elsewhere, in another reason for the scepticism towards similarity, namely that similarity exists as a possibility before the finished artwork and thus offends the idea of the artist as an inspired deity, creating objects never before seen nor possible. The artwork's basis in similarity makes, from this point of view, the artist into a copist; even if the status of a sharp-sighted observer, seeing what nobody before him has been able to see, should not be so poor as a substitute for radical 'creativity'.

[351] In this, the metaphor definition is in agreement with the actual tradition of cognitive semantics where general metaphors and blendings are widespread in ordinary language, even if in an unspectacular way because understanding them is so easy - here the spectacular metaphors and blendings of poetry owe their effects to a sophisticated but derivative use of ordinary linguistic and conceptual material. This forms the cognitive equivalent to the core idea of 'deviation' poetics highlighted by the Russian formalists such as Shklovski.

[352] At the conference 'Thinking With Diagrams 98', Alan Blackwell and Yuri Engelhardt gave an overview over the many existing attempts at diagram taxonomies which all seem insufficient in some way or another (Blackwell and Engelhardt 'A Taxonomy of Diagram Taxonomies', paper given at 'Thinking With Diagrams 98', University of Wales, Aberystwyth). As mentioned Peirce has one sketchy attempt at a tripartition of diagrams into maps, algebra, and graphs (May and Stjernfelt 1996).

[353] That is, they can be used as diagrams and are regularly so used in ordinary picture observation. It must immediately be admitted that even if visual pictures are thus icons and diagrams among others,

their visual specificity may be pointed out referring to the special features of the visual system, cf. e.g. the already classical *Traité du signe visuel* (Groupe μ 1992) where the visual sign is characterized by its possible utilization of a long series of different optical transformations. These transformations are relevant here, because they are visually specific versions of the general rule-governed manipulability of diagrams. Visual diagrams are thus characterized by certain classes of manipulability, and it may be expected that a rational taxonomy of visual pictures may be constructed from the classification of such transformations.

[354] It is possible to argue for a distinction between two types of pictures, those which make possible the construction of a space for the perceiving body, and those without this possibility, cf. next chapter.

[355] A further investigation of this device follows in Chap. 16.

[356] Erik Fischer *C.W. Eckersberg*, Copenhagen 1993. I reviewed the book in the Copenhagen daily *Information* (Oct. 6, 1993) where some of the reflections in this paper were originally born.

[357] John 17.6–19 (*King James Bible* 1611):

6 'I have manifested thy name unto the men which thou gavest me out of the world: thine they were, and thou gavest them me; and they have kept thy word.
7 Now they have known that all things whatsoever thou hast given me are of thee.
8 For I have given unto them the words which thou gavest me; and they have received them, and have known surely that I came out from thee, and they have believed that thou didst send me.
9 I pray for them: I pray not for the world, but for them which thou hast given me; for they are thine.
10 And all mine are thine, and thine are mine; and I am glorified in them.
11 And now I am no more in the world, but these are in the world, and I come to thee. Holy Father, keep through thine own name those whom thou hast given me, that they may be one, as we are.
12 While I was with them in the world, I kept them in thy name: those that thou gavest me I have kept, and none of them is lost, but the son of perdition; that the scripture might be fulfilled.
13 And now come I to thee; and these things I speak in the world, that they might have my joy fulfilled in themselves.
14 I have given them thy word; and the world hath hated them, because they are not of the world, even as I am not of the world.
15 I pray not that thou shouldest take them out of the world, but that thou shouldest keep them from the evil.
16 They are not of the world, even as I am not of the world.
17 Sanctify them through thy truth: thy word is truth.
18 As thou hast sent me into the world, even so have I also sent them into the world.
19 And for their sakes I sanctify myself, that they also might be sanctified through the truth.'

[358] It is a well-known fact, dating at least from Kant's *Kritik der Urteilskraft* (1790), that artworks and biological organisms share this property, making both adequate objects for 'teleological judgment'. For both kinds of objects, however, we should not expect every single detail or feature to serve the whole teleologically; this would be a misplaced and dogmatic use of the concept of artwork or organism, respectively. Rather, artworks make use of completely ordinary structures of perception, thought, objects, which they take over for their own purposes (as argued in Bundgaard 2004) – much like biological organisms are built from ordinary physical structures.

[359] Cf. Groupe μ's criticism of this tendency, op.cit.

[360] See Esmann 1995.

[361] The color white naturally plays a special role in relation to the 'absolute' character of the painting because of the neutral and 'all-encompassing' character of this color (consisting of all visual wavelengths) which gives it a more absolute character than the individual spectral colors. Another experiment might thus substitute for the two white nuances two red nuances, e.g., with a larger degree of concretion as a contrastive result.

[362] An imperfect similarity relation, of course, because the infinitely thin lines of ideal geometry by their very nature cannot be drawn.

[363] Something like this naturally also occurred to Rudolf Arnheim cf. his *Visual Thinking* (Berkeley 1967). Sensitive phenomenologists might feel an intellectualist offense by this emphasis on diagrams, but on a closer look it is by no means strange to the phenomenological tradition. As discussed in

Chaps. 6–8, the experimentative manipulation on a diagram is directly related to the 'eidetic variations' leading to phenomenological 'Wesenserschauung', cf. Husserl's *Erfahrung und Urteil*, and the emphasis on the body in phenomenology from the later Husserl, Merleau-Ponty and onwards (cf. 'la chair' in *Le visible et l'invisible*, Paris 1964) may easily be reconciliated with diagrammatical reasoning. The diagrammatical experiment forms a central part of Peirce's pragmatism, the experiment involves the imagination of something *one could do*, and this imaginable doing has a natural corporal basis: 'It is not a historical fact that the best thinking has been done by words, or aural images. It has been performed by means of visual images and muscular imaginations.' (Peirce: 'Sketch of a new philosophy', undated, NEM IV, 375). In many pictures, there is a large class of diagram experiments which are immediately bodily; in landscape pictures, the viewer involuntarily imagines himself wandering into it, cf. Anne Fastrup's research into Diderot's 'Salons' (Fastrup, in press).
A remarkable consequence of pictures' diagram character is the refusal of the widespread semiotic idea that a picture is a sign to the extent that it implies a code which must be decoded directly, in an analogy to a denotative proposition. Diagrams leave, thanks to their more or less open manipulability a considerable initiative to the observer, and different schools of picture viewing may probably be described by their favorite diagram manipulation types. The systematic and rule-bound character of diagrams prevents, on the other hand, the complete handing over of initiative to the observer, as known from the different forms of scepticist aesthetics, from reception analysis and to (de)constructionism.

[364] It could be said that the analysis of pictures must give up its dogma of the concreteness of the picture and thus make an epistemological self-criticism analogous to historiography. Here, the Rankean idea of the description of concrete events 'wie es eigentlich gewesen' implied the rejection of all speculations of the type 'what if…'. But a semiotic analysis of these ideas demonstrates that it is impossible to describe even the most concrete event without the use of universals, types, generalities, and as soon as reference is made to general objects, a 'what if?' is implied, cf. the pragmatic maxim's 'would-bes'. If you say, for instance, that the rearmament of the enemy released the war, then you claim by the same token, that if the enemy had not rearmed, all other things being equal, then the war would not have come - a claim which would normally be rejected out of hand as counterfactual speculation. If you should refrain from indulging in such things, you should positivistically limit yourself to registrate events without reflecting on their type, nor on relations of cause and effect between them. I do not think any historiography has been written which conforms to such ideas. Younger historians (such as Niall Ferguson (ed.) *Virtual History*, London 1997) are now realizing that counterfactual speculations are inevitable - and if you think you can do without them, then you are fooling yourself. The equivalent in analysis of pictures, I would claim, is that if you think you can see and analyze pictures without the counterfactual variations of diagrams, then you fool yourself.

[365] Translations from *Hua* XXXIII are my own.'Wenn hier zwischen Sache und Bild unterschieden wird, so merken wir bald, dass der Begriff des Bildes hier ein doppelter ist. Der abgebildeten Sache steht nämlich ein Doppeltes gegenüber: 1) Das Bild als physisches Ding, als diese bemalte und eingerahmte Leinwand, als dieses bedruckte Papier usw. In diesem Sinn sagen wir, das *Bild* ist verbogen, zerrissen, oder das Bild hängt an der Wand usw. 2) Das Bild als das durch die bestimmte Farben- und Formgebunt so und so erscheinende *Bildobjekt*. Darunter verstehen wir nicht das abgebildete Objekt, das *Bildsujet* sondern das genaue Analogon des Phantasiebildes, nämlich das erscheinende Objekt, das für das Bild*sujet* Repräsentant ist.' (*Hua* XXIII, 19)

[366] 'Drei Objekte haben wir. 1) Das physische Bild, das Ding aus Leinwand, aus Marmor usw. 2) Das repräsentierende oder abbildende Objekt, und 3) das repräsentierte oder abgebildete Objekt.' (ibid.)

[367] 'Was im Inhalt des Bildobjektes repräsentativ fungiert, das ist in eigentümlicher Weise ausgezeichnet: *Es stellt dar*, es *vergegenwärtigt*, verbildlicht, *veranschaulicht*. Das Sujet blickt uns gleichsam durch *diese* Züge an. Diese Züge treten erst in Einzelbeachtung hervor und scheiden sich erst in ihr von den anderen Zügen des Bildobjektes: von Momenten, Teilen, Bestimmtheiten, die entweder ausgeprägt den gegensätzlichen Charakter, den des Widerstreits mit entsprechenden Bestimmtheiten des gemeinten Sujets, haben, oder denen weder der eine noch der andere Charakter anhaftet. Solche charakterlosen Züge verbildlichen nichts, es bleibt aber unbestimmt, wie sich darin das wirkliche Objekt [dar]stellt.' (30–31)

[368] '... das Bild muss sich *klar* von der Wirklichkeit scheiden, d.h. rein intuitiv, ohne alle Beihilfe von indirekten Gedanken.' (41)

[369] 'Which relation pertains between picture and object? Which relation pertains between pictorial object and object?'

[370] As to the issue of similarity in icon definitions in general, see Chaps. 2–3.

[371] See Groupe μ 1992 and the chapter on the esquisse. Here they distinguish between the visual sign, the type it signifies, and the object it refers to – where the type is the invariance kept constant in the transformation between the sign and its object. The type thus corresponds to Husserl's pictorial object, and Husserl's two-step transformation is analyzed as a one-step transformation with invariance – which may be viewed as different ways of representing the same state-of-affairs. The type in Groupe μ may be grasped separately with transformation relations to the two poles at each end so that it subsumes both the sign and the object as instantiations. In this idea it seems to be the case that the pictorial object is more general (or, at least, not less general) than the picture and the object.

[372] *'Vielmehr: "Bewusstsein" besteht durch und durch aus Bewusstsein, und schon Empfindung so wie Phantasma ist "Bewusstsein"'* (265)

[373] Noema of course being the object pole in the central analysis of perception as intentional act in the *Ideen.*

[374] I translate 'setzende' by 'assertive'; it refers to an act's character of existence claim as to its object; 'nicht-setzende' correlatively refers to acts not involving such claim.

[375] The context might be Meinong's doctrine of non-existent objects where 'assumptions' play an important role: they are non-assertive propositions (or, propositions minus existence claims) and form a much wider category than fantasy, comprising e.g. assumptive linguistic utterances deprived of fantasy imagery.

[376] 'Ich folge den Aufführungen eines Theaterstückes, oder ich betrachte ein Gemälde. Hier unterscheiden sich, ähnlich wie bei der reproduktiven Phantasie, die im Bilde dargestellten (bildlich produzierten) Vorstellungen, Wahrnehmungen, Urteile, Gefühle, etc., von den in mir, dem Zuschauer, erregten und aktuellen. Aktuell ist die Darstellung selbst.' (383)

[377] 'Aber nur wo ein Bildbewusstsein sich gründet in einem Wahrnehmungszusammenhang (wo eine Perzeption eben in einem perzeptiven Zusammenhang steht, aber sich ihm nicht in Einstemmigkeit einfügt) oder wo ein Bildbewusstsein sich in einem Erinnerungszusammenhang ebenso einfügt, haben wir Bildbewusstsein mit Widerstreit verknüpft; was alles genauer zu fassen wäre. Ich muss mich hüten vor unangemessener Verallgemerinerung: als ob Bild und Widerstreit notwendig zusammengehörte.' ('Zur Lehre vom Bildbewusstsein und Fiktumbewusstsein', c. 1912, 494)

[378] We may add that Husserl now generally sees *illusion* as a category where the *Widerstreit* is lacking, cf. his overall wax doll – human being example (fx. 39; 490). This is analogous to that phase in picture contemplation where the *Widerstreit* is suspended (in the old theory, however, the *Widerstreit* could be called back). It is the close relation between these two forms, of course, which lies at the root of the mathematical and theory-of-science disbelief in pictures as medium of thought.

[379] 'Auseinanderhalten wird man müssen die *Fälle wirklichen Widerstreits*, wie im Fall schauspielerischer Darstellung, und die Fälle *potentiellen Widerstreits*, der an Umgebungsintentionen hängt, die dem anschaulichen Gegenstand anhaften, die aber erst 'aufgewickelt' werden müssen, um zu wirklichen Widerstreiten zu führen.' ('Zur Lehre von den Anschauungen und ihren Modis', c. 1918,510)

[380] Fantasy (incuding pictures) remains in this period fundamentally 'modification'. This idea may puzzle, because it does not refer to any originary sensation of which the modified is a modification. Rather, modification refers to the fact that neither fantasy nor picture is present in 'anschauliche Fülle'. Husserl takes this concept of 'fullness' from the British psychologist Alexander Bain, and it plays a central role in his transcendental period's definition of perception as adequate act. The analogous problem (which phenomenological character has the givenness of actual objects, different from phantasized, ideal, and other objects?) is solved in a different way in Peirce: as we saw, he introduces the scholastic term 'haecceity', thisness from Duns Scotus as a concept for this special insistence characterizing the actually existing object different from merely potential or phantasized objects. In both cases, Husserl and Peirce, this difference is grounded on a 'purified' psychological concept, but while Husserl hints at the possibility of continuous grading (objects may be given with more or less fullness), Peirce's solution points to an all-or-nothing solution with a certain tension to his overall metaphysical continuism.

[381] Taken to the LU letter, a fairy-tale remains a purely 'signitive' intention, so it is striking that Husserl without further ado may involve this example in the discussion of the species of imagination. This points to the fact that the definition of signitive intentions only with difficulties may free itself completely from imaginative components – just like it was argued in the chapter on categorial intuition above.
[382] Ingarden enrolled as Husserl's student in Göttingen in May 1912, only to follow him (with a few breaks) closely in the years to come. He followed him from Göttingen to Freiburg and from around 1916 he was very close to Husserl as his personal assistant. In the summer of 1918, Ingarden returned to the Polish state then in the process of resurrection, but he remained in close contact with Husserl until his death in 1938, despite their well-known disagreement as to the issue of idealism/realism, cf. Ingarden's memories in his edition of Husserl's letters to him (Husserl 1968).
[383] Peirce has an analogous point when he points to the possible growth of information in diagrammatical reasoning by making explicit what was only implicit.
[384] Beilage VII (appr. Sept. 1906) introduces an interesting distinction between several possible types of *Widerstreit.* In addition to the empirical contrast between appearing pictorial object and its physical base, a contrast between pictorial object and sujet (that the people in a photograph are gray, but not reality) is proposed, as well as a second contrast in pictorial context (hallucinations in the midst of the visual field, e.g.), and a third contrast with our knowledge about the type of object depicted so that the object depicted becomes self-contradictory, a so-called logically mediated skin. Husserl does not provide examples on the latter, but it must be taken to refer to fallacious perceptions like the well-known 'impossible objects' in psychology of perception. Husserl's idea here seems to be to refine the second contrast, that between pictorial object and sujet, be pointing out different possibilities of the structure of this contrast.
[385] 'Ist nicht bei aller auch reinen Phantasie Widerstreit? Mit Wahrnehmungen, mit Erinnerungen, mit Antizipationen. Ich inhibiere alle Weltsetzung. Aber hat nicht jede Phantasie irgend eine Stelle, irgend etwas, das sie bestreitet?' (593)
[386] As a matter of fact, most often, the object of a picture is not a particular depicted at an unextended point of time. Even a photograph depicts its object during a short but extended timespan, and paintings very often involves a generalization aimed at typical properties of its object over a certain period.
[387] See Chap. 6.
[388] In Ingarden's theory of literature four strata are distinguished, each possessing a formal level – the aesthetic intention may select either of them as its object, as well as any of their mutual interrelations. See Chap. 17 on Ingarden.
[389] '*Der Inhalt des Gegenstandes selbst ist nicht ästhetisch bedeutungslos.* Es ist nicht gleich, ob es ein Kaiser ist oder nicht, ob ein bedeutendes Schicksal oder ein alltägliches etc. Handelt es sich da um ain Anklingen von Gemütswirkungen (Ehrfurcht, Ergebenheit)? Aber auch Anderes: Jede Gegenständlichkeit, die existentiale Freude motiviert, oder phantasiert, *quasi*-Freude. In sich ist diese Freude nicht ästhetisch. Aber das ästhetische Gefallen, das an der Erscheinungsweise hängt, kann sich mit dieser Freude (als einer Aktualität) verbinden, und das Ganze hat den Charakter einer erhöhten ästhetischen Freude. Stilleben. Das Wechselspiel aktueller Freuden oder *quasi*-Freuden (an der Natur: Freude über die fruchttragenden Obstbäume, Felder, etc.) und Leiden und sonstigen aktuellen Stellungnahmen ist aber selbst ein Hauptstück wirklich ästhetischer Freude. Also auch das gehört zur 'Erscheinungsweise'.'
[390] As a matter of fact, these determinations must also hold for perception proper (and hence for natural beauty, beauty in architecture, etc.) which does not need pictorality nor fiction, but merely a certain weakening of the distinction between appearances and the sujet pole.
[391] The connection of such indeterminacy spots to the problem solving of diagrams may be indicated by the ideas of Otto Selz (1913) talking about the role of 'completion on the basis of a schema' in his treatise of the 'Laws of the ordered thought process'. Selz argues for the 'Komplexergänzung', the completion of complexes as against 'Konstellationsergänzung', the completion of associative constellations of though in James, the latter being judged too simple. Selz's laws of complex completion are relevant here: a part or a schema of a complex has a tendency to call forth the reproduction of the whole complex.
[392] Of course, the question must be asked whether the *Unbestimmtheitsstellen* do not possess a foundation in ordinary perception. When, in perception, we spontaneously endow the objects with back sides, cf. the analysis of perception of the *Ideen*, then we add *general* backsides to those objects (we do not know nor see

where the cracks are on the back side of a wooden box), and the single perception always possesses a certain granularity beyond which we know nothing further. We do assume, that is right, in the perceptual *Einstellung*, that all unclarities might in principle be made clear by a further observation of the object (wa walk around it, take a closer look, etc.) – but no real series of perceptions may ever exhaust the properties of the object. Thus, even in the most meticulous observation a degree of underdetermination, that is, ideality, remains. Peirce radicalizes this idea to the claim that the object *is* not in fact particular, but still possesses a certain generality (its future, e.g. is not in all details decided), and that, conversely, the complete determination in the individual object is an idea governing perception. We shall not go that far but merely stay by the claim that the existence of *Unbestimmtheitsstellen* in fiction and language are so to speak prepared by the existence of spots of indeterminacy already in ordinary perception which makes possible the strange fact that we do not find it strange at all to meet e.g. persons with vast dimensions of indeterminacy in fiction.

[393] 'Danach kann Phantasie eigentlich kein Individuum als solches wiedergeben, sie gibt 'etwas', das in der Form eines Individuellen auszugestalten ist und nur nach seinem hinsichtlich der Individualität unbestimmten Sinnesgehalt anschaulich wird.' (nr. 19, 1922/23, 552).

[394] 'Möglichkeiten, individuelle reine Möglichkeiten sind fliessend' (Beilage 553).

'Unendlich viele mögliche Wirklichkeiten, die miteinander unverträglich sind etc.' (Beilage LXII, c. 1918, 566)

'*Prädikat* ist Identisches, das in vielen, in 'Mannigfaltigem' identisch sein kann, das in wandelbaren und beliebigen Möglichkeiten identisch vorkommen und das Konkretum bestimmen kann.' (nr. 19, 1922/23, 564)

'Und ebenso kan ich irgendein Subjekt mir denken, das das ansetzt auf Grund seiner Tatsachen.' (570).

'... wenn ich exemplarisch den Umfang als reinen Umfang durchlaufe, fällt es mir nicht ein, auf die bezüglichen subjektiven individualisierenden Differenzen mich einzustellen. Vergegenwärtige ich mir einen einzelnen Zentauren, so meine ich 'ihn' nicht als gerade von mir vorgestellten. Sollen wir sagen, dass das Singuläre eines reinen Umfangs (eines Umfangs reiner individueller Möglichkeiten), das exemplarisch Einzelne selbst schon ein Allgemeines <ist>, ein Identisches, das seinerseits einen reinen Möglichkeitsumfang hat?' (Beilage LXIII, c. 1920/21, 568).

[395] 'In den Unbestimmtheitshorizont, den für jedermann die gegebene Welt und gegebene Zeit hat, stellt der Dichter, ihn bestimmend, ausfüllend, Gestalten hinein.

Leser, Dichter in einer Welt und Zeit. Zwei Extreme:

a) Die gegebene Welt und Zeit kann eine so voll bestimmte sein, als für uns jetzt unsere Umwelt ist (nicht die wirkliche Welt). Z.B. das heutige Berlin, so bestimmt, wie es für uns und selbst für die Berliner selbst ist.

b) Extremer Gegenfall. Es war einmal, irgendwo, in irgendeinem Fabelland, in irgendeiner Zeit, in irgendeiner Welt mit ganz anderen animalischen Wesen, sogar anderen Naturgesetze etc.

Zwischen diesen beiden Extremen bewegt sich alle Kunst.' (Beilage LIX, 1916 el. 1918, s540)

[396] 'Oder vielleicht doch noch ein wenig. Wir haben den Menschen in Deckung, die beiden überschieben sich, aber geben weit klaffende Unterscheide. Der Unterschied ist so gross, dass wir nicht mehr uns gestört fühlen durch das Bewusstsein der Verfälschung bei der Neigung zur Identifizierung, sondern es fehlt die Neigung zur Identifizierung und bei dem grossen Abstand finden wir die Darstellung lächerlich.' (143)

[397] '... one *flows* into the other and is yet again felt as different.' (our italics)

[398] 'Warum wirkt die Natur, eine Landschaft als 'B i l d'? Ein fernes Dorf. Die Häuser 'kleine Häuser'. Diese kleinen Häuser haben a) eine geänderte Grösse gegenüber den Häusern, wie wir sie gewöhnlich sehen, b) eine geringere Stereoskopie, geänderte Färbungen etc. Sie werden ähnlich als Bilder aufgefasst wie Spielzeughäuser. Ebenso Menschen: Puppen.' (144)

[399] As found for instance in Nelson Goodman, cf. Chap. 3. In a Danish context, this idea has been influentially inherited in Søren Kjørup (2000, originally 1971). The conventionality hypothesis may take the form that the ability to grasp similarities is acquired, or the form that similarities do not play any role and must be substituted by conventions which are then what is acquired. See the discussion in Chap. 2.

[400] The fact that exactly landscape (pictures) furnish the immediate example of this duplicity in perception is probably due to the fact that fundamental spatial orientation connects the sense of vision

with the possibility of bodily movements in space. That this is the case, is supported by the metaphor the positivist Otto Neurath uses in the foreword to his amazing political-pedagogical book of diagrams *Modern Man in the Making* (1939) where he argues that modern man may get at more direct understanding of the political relations of modenity by means of diagrams depicting basic social relations. Neurath developed a set of so-called 'isotype' diagrams (especially making use of maps, column diagrams and the cake diagrams which he himself invented), and he introduces them as follows: 'An attempt has been made to evolve for this purpose a special picture-text style which sould enable anybody to walk through the modern world that is beginning to appear about us and see it as he may see a landscape with its hills and its plains, woods and meadows.' (7) The immediate accessability of landscape – in both the concrete and abstract meanings of this word – are mapped onto the diagram wherein complicated relations may be viewed with the same directness.

401 We do not have to wait for a *Abwehrvorgang* to appear such as Freud believed.

402 'Ich kann wohl auch mich in das Bild 'hineinphantasieren'. Das kann aber nur sagen, dass ich den Bildraum über mich und meinen Umgebungsraum ausdehne und mich selbst unter Ausschluss der wirklichen Dinge, die ich sehe, mit ins Bild aufnehme, wodurch ich meine Aktualität ausschalte; ich werde dann selbst zum modifizierten Ich, zum setzungslosen. Dann ist meine Teilnahme die Teilnahme eines bildlichen Zuschauers (sie gehört zum Bildobjekt), nicht eines sympathisierenden vor dem Bild.'

403 '...Neubildung von solchen Synthesen und Analysen, welche aus den einheitlichen Erscheinungen Teilerscheinungen herauslösen mit den Teilmeinungen in Synthesis gebracht usw. Im Expliziten treten also neue Stoffe (Kerne) auf. Also *Explikation des in vorgegebenen Erscheinungen verborgenen*.' (316).

404 We thus receive a category to grasp the specificity of a project like the Flemish painter Coninxloo's landscape painting which is probably the first to block the access of the observer by orienting the gaze directly into brushes, bushes, and foliage and thus makes possible a distinct forest feeling for the observer lost in the wilderness.

405 A further taxonomy of rather demanding character is constituted by the spectrum of imaginary action possibilities which the single picture offers the observer. It is well known that books and magazines exist which should, as Sade put it, be read with one hand only (the other hand presumably being elsewhere occupied): pornography is a pragmatic type of picture immediately suggesting which (more or less) imaginary action the observer may indulge in. It seems evident that enormous spectrum of possible actions open up which we can not make an even preliminary mapping of here – of how the single type of picture may be integrated into narratives, fictive or not. It is noteworthy, though, that these possibilities of action – thanks to the purposive structure of actions – mostly will be pertinent in figurative pictures. We have none or only marginally abstract pornography, and something similar holds for cookbooks, literary illustration, comics and a series of other picture genres correlated with actions possibilities, phantazised or real.

406 'Solche dargestellten Stimmungen, Gefühle etc. setzen nicht voraus eine Mitdarstellung des Beschauers, obschon er in eigener Weise in Aktion tritt. Genauer, sicher gehöre ich mit *dieser* Stimmung nicht ins Bild hinein.'

407 Many other types of experiments may suit a given type of picture, cf. Chap. 13.

408 Of course, the linguistic term has other prerequisites alongside with the visual type: knowledge of its use, bodily experience with that use, the object's relation to other objects, etc. – see the discussion of Eco's Cognitive Type and Nuclear Content in Chap. 3.

409 Cf. Chap. 3.

410 Translations from the *Traité* are my own. '... on peut supprimer certaines lignes pour ne laisser substister que celles jugées importantes. On obtient ainsi l'*esquisse*, qui fournit du sujet une vision synthetique, et livre souvent des tracés régulateurs utiles pour l'interpretation correcte de l'oeuvre finale.' (310).

411 See Hintikka 1997; also Chap. 4.

412 Peirce uses various other notions for abduction during his theoretical development: hypothesis, hypothetic inference, retroduction. Ab-, de-, and induction, being subspecies of arguments, are subsubspecies of symbols in Peirce's systematic sign doctrine.

413 Michael Hoffmann (1999, 2002) has collected and discussed the various arguments pertaining to the concept of abduction in Peirce and his most important interpreters.

[414] As Hoffmann (2002, 257) comments, this forms an early version of Hempel's induction as the confirmation of a hypothesis by making it open for falsification.

[415] In that respect, it is involved in analogy. Peirce a few times mentions *analogy* as a further form of reasoning after Aristotle's "paradeigma", composed from ab- and induction ("History of science", 1896, 1.65), but he never develops a it thoroughly, and the supposed workings of analogy seems to be explainable by the abductive trial-and-error application of diagram structure on different materials.

[416] Hoffmann 2002 argues strongly and at length for this distinction between the logical form of abduction and its character of 'creative perception'.

[417] A basic hint lies, of course, in Peirce's possibility realism. Among all hypotheses which are possible in the weak sense of 'not known not to be true', hypotheses supported by known real possibilities (already recorded laws, tendencies, patterns, etc.) stand out as the hypotheses most promising to test.

[418] Peirce's discussions of the three different types of induction in the years after the turn of the century (in 'On the Logic of Drawing History from Ancient Documents,' (1901, EPII, 75; 7.164–231); in L75 (1902); in 'G,' (1905, 2.755–759); in L224 to William James, (1909, EPII, 492), etc., and their respective subtypes merit a whole separate discussion. Here, we shall stick to a short presentation aimed at the particular problem at issue.

[419] As always, other possible hypotheses abound: the appearance of death symbolism is an accident or coincidence, it owes it existence to the fact that the poem is pieced together of two autonomous parts, it owes it existence to the special location depicted, in which death signs naturally occur ... etc.

[420] Cf. Freud 1948.

[421] In what sense are these observations deductions from the idea of the poem as a piece of *memento mori*? They are obviously not ordinary deductions; it does not follow with necessity that a poem about death contains these images. But they are deductions in the sense of 'probable deductions' which differ from induction as a probable inference. Probable deductions conclude with necessity about probabilities. Given this piece of knowledge of cultural history, an ideal type death poem could be said to exist, of which there is, with necessity, a certain probability that a death poem will instantiate certain features. If no necessary consequences flowed from the fact that the poem is a death poem, on the contrary, the claim would be empty.

[422] This is one of the implications of Ingarden's idea of a level of 'schematized aspects' of the literary text: the objectivities presented in the text are always seen as schemas, that is, with lots of 'Unbestimmtheitsstellen', that is, undetermined parts. The reader's filling-in of such parts prototypically takes place by 'crude induction' – if a man is introduced, he will be grasped as two-legged, until the contrary may be suggested – cf. next chapter.

[423] See Umberto Eco's discussion volume of the same title (1992). Eco here proposes a 'Popperian' principle which allows not to find the *best* interpretations, but to ascertaining which interpretations are 'bad' (Eco 1992, 52). It is an Augustinian idea: 'any interpretation given of a certain portion of the text can be accepted if it is confirmed by, and must be rejected if it is challenged by, another portion of the same text' (Eco 1992, 65). Eco picks the example of the British-Italian scholar Gabriele Rossetti's reading of Dante. He sets out to prove that Dante was a Rosicrucian (and, as a corollary, that the Rosicrucian symbol of a cross in a rose with a pelican ripping its own flesh from its breast underneath is present in his work). Rossetti finds a cross and a pelican at the same spot in Dante (not so strange as the self-sacrificing bird is an old symbol of Christ), he also finds a rose, but unfortunately not at the same place. His supporting hypothesis now goes that the parts of the symbol have been so dispersed in Dante's work for secrecy reasons. Eco, of course, points out, that this is a blatant case of overinterpretation, made possible only by this extra, unfounded hypothesis. In relation to the early existence of the Rosicrucians, Eco says that Rossetti adds the absurd argument of *Post hoc, ergo ante hoc*: because Rosicrucian symbols, dating after Dante, seemingly can be found in his text, the order must be even *older* than him. In our discussion, the theoretical assumption that the Rosicrucians are earlier than Dante governs the whole hypothesis in the symbol reading – instead of it being explicitly admitted as a hypothesis which may be confirmed or refuted depending on the findings of the Dante exegesis.

[424] Which is, of course, completely different from the deconstruction fad of the 80s assuming that all texts necessarily contain deep contradictions.

[425] As is evident, both deductions contain the word 'probably' – but this does not mean they are not necessary inferences, quite on the contrary. Probable inferences – inductions – build on observations, while deductions are based on idealized models. The idealized model of a catholic priest has a tonsure, just like the idealized model of a priest ties him to the city he lives in. This does not entail that all empirical priest are tonsured nor stick to their hometowns, but it implies that prototypical priests do so. Thus, the word 'probably' here refers to the relation between the ideal model and its empirical extension, before any empirical observation.
[426] Ingarden's whole œuvre thus forms a late flower of what is often called 'Göttingen' phenomenology from the period when Husserl lectured there. Among Ingarden's other chief influences thus count other early realist phenomenologists from the Göttingen circle or the München phenomenologists they were associated with: Adolf Reinach, Alexander Pfänder, Alexius Meinong, Johannes Daubert, etc.
[427] 'Idealism' here to be taken in the sense of the construal of the real world by the subject – not in the sense of claiming the real existence of ideal objects; in the former sense, Ingarden is an anti-idealist, in the latter, he – just like other phenomenologists, including Peirce – is an idealist. This turn has been the object of much discussion, ranging from the assertion that it took place somewhere between *Logische Untersuchungen* (1901) and *Ideen* (1913), with 1906–07 as peak candidate, to the claim that it never took place and that the internal continuity in Husserl's work remains far greater than the discontinuities. Ingarden was Husserl's pupil from around 1912–18 and obviously in this period found Husserl's philosophy as well as his teaching basically realist (in concert with the other Göttingen pupils), and he claims the internal tensions between realism and idealism in Husserl's thought only tipped to the side of idealism in the early twenties, only to be made public as late as in *Cartesianische Meditationen* and *Formale und transcendentale Logik*. In any case, Ingarden's experience of such a turn was what prompted his early publications 'Essentiale Fragen' (1925) and 'Bemerkungen zum Problem "Idealismus-Realismus" ' (1929) as well as *Das literarische Kunstwerk*. Indeed, Ingarden's enormous *Der Streit* (1965–74) forms the late culmination of this *Auseinandersetzung*. In Ingarden's letters to Husserl, the issue is discussed openly in detail (Husserl 1968).
[428] *Vom Erkennen* and *Der Streit* are preceded by Polish versions (from 1937 and 1947–48 and 1960–61, respectively). As I do not read Polish and as the German versions are Ingarden's own, revised, translations, I stick to them and the corresponding English translations.
[429] In the Austrian tradition, the notion of *Sachverhalte* often has this duplicity or ambiguity: referring both to the content of propositions, on the one hand, and to the objectivity referred to by those propositions, on the other.
[430] As always in the phenomenological tradition, 'object' here does not merely refer to physical things, but includes everything 'regardless of objectivity category and material essence. Thus it refers to things as well as persons, but also to all possible occurrences, states, acts performed by persons, etc.' (219)
[431] Thus, the beloved ambiguity and paradox of New Criticism, as well as the self-contradiction revered by deconstruction, are already grasped by Ingarden's theory – albeit not as indispensable properties of the literary work, but rather as possibilities among others.
[432] Ingarden gives a reference to Pfänder for the notion of 'quasi-judgment' as an intermediate with respect to Pfänder's distinction between two functions of the copula; assertive and predicative, respectively. Furthermore, the two extremes between which the *quasi* is localised are described with reference to Russell (the assertive judgment) and Husserl (neutrality modification), respectively (176). We may add that the very terminology (and the basic idea) also comes from Husserl (cf. Chap. 15) who uses 'quasi' intensely in the 1910s when Ingarden studied by him in Göttingen.
[433] In *Logik der Dichtung*, Käte Hamburger protests against the definition of the literary work by quasi-judgments. Quasi-judgments possess no special formal characteristics making it possible to isolate them – and Hamburger is on the search for such formal linguistic criteria (as the 'erlebte Rede' (style indirect libre), the historical present tense, and, especially, the epic past tense 'Tomorrow it was the first day...'). Hamburger's valuable work on these forms notwithstanding, they still do not constitute formal criteria of fiction – they are neither necessary nor sufficient, for there are many fictive works without any of these forms. So her use of these forms as weapon against Ingarden does not work – her argument that Ingarden's idea is circular because it only says a judgment is a quasi-judgment because it appears

in a novel would hit herself to the same degree if we were to take her formal studies as definitory for literariness or fiction. Maybe a misunderstanding is at stake here – Ingarden does not claim that what he finds is not circular, it is no attempt at *defining* the literary work by means of properties in its constituents, it is rather a description of the status of those constituents.

434 Ingarden's dwelling with these metaphors indirectly refers to his idealism-realism agenda and his ongoing discussion with and of Husserl's gradual turn to idealism. Ingarden regards *Ideen* as ambiguous as regards idealism. The idealism-realism issue had not explicitly been the object of Husserl's published writings until the *Ideen*, and Husserl's Göttingen students obviously assumed – according to Ingarden – *Logische Untersuchungen* to be realist. They were thus surprised to see the much more explicitly idealist *Ideen I*, especially because Husserl's lectures was in a '... wenn man so sagen darf, realistischen Ton gehalten.' (Ingarden 1998 (1968), 407) – thus Husserl's lectures at the same time (summer semester 1913) on 'Natur und Geist' – being close to the more realist approach of the *Ideen II*. (ibid.)

In the *Ideen I*, on the one hand, the claim is repeated that the object is 'transcendent' in relation to the act, and the epistemology for ideal objects is not, unlike that of real objects, revised in the *Ideen*, so that Husserl still appears as a realist in these respects. On the other hand, the whole noesis-noema doctrine of the act is seen by Ingarden as idealist, reducing the object world of space-time to: '... ein Sein, das das Bewußtsein in seine Erfahrungen setzt, das prinzipiell nur als Identisches von motivierten Erscheinungsmannigfaltigkeiten anschaubar und bestimmbar – *darüber hinaus* aber ein Nichts ist.' (*Ideen I*, 93, quoted in Ingarden 1998, 211 (1956)) – a quote to which Ingarden returns.

This criticism of Husserl's position is constant in Ingarden's many writings about the subject. He writes – as early as in his 1918 letter to Husserl about the re-edition of the VI. investigation -: 'Wir kommen am Ende zu der Gleichung: Ding = ein eigentümlich gebautes Noema-Bewußtsein.' (Ingarden 1998, 5). So the relation between the single aspects-noemas and the thing-noema constituted by them is what recurs in Ingarden's idea of the states of affairs forming a 'net' around the represented objectivity which thus to some extent corresponds to the noema. Unlike Husserl's noema, this objectivity is, of course, distinct from the correlated real or ideal object itself (if it exists).

435 Here, an important difference in relation to Peirce's metaphysics resides in the fact that Peirce would never allow particulars of any sort to be completely determined, inter alia because of the indeterminacy of their future states. To Peirce, thus, there are *real Unbestimmtheitsstellen* mirroring the case of semiotic representations. Indeterminacy, in any case, must be far more extensive in representations.

436 Wellek says (1991, 379): 'I find myself in wide agreement with his views and acknowledge learning from him, on many of these questions, more than from anybody else.' – and still 'I find it difficult to isolate this stratum of schematic aspects from that of represented objects...' (383). The polemical dialogue between the two is recorded in later editions of *Das literarische Kunstwerk* and in Wellek 1991.

437 'All so-called sensory qualities – but not only these – are self-presenting' (196). Just like categorical entities in Husserl may also permit intuitive access, it is not only sensory qualities which satisfy self-presentation. We saw that occurrence states also did it to some extent (due to what – due to the schematic character of their event type?) – and we must assume that (certain simple?) ideal objects also display self-presentation (cf. Chap. 6). This becomes important for the stratum of schematized aspects.

438 Ingarden's account for these qualities has hardly left the first, descriptive phase. It is evident that reading great literature, qualities like the mentioned are often experienced, but their status does not seem perfectly clear in Ingarden's theory. They seem to be objective qualities dependent on temporal situations depicted in the object level – but why is it that they are presented as extraordinary, almost quasi-religious breaks with everyday life (cf. the word 'revelation' used as common denominator for them)? It is true that literature often describes such qualities, but literature also has the power of redescribing aspects of everyday life with an intensity which may make us experience it as with new eyes. So their 'revelation' character ought to be described more meticulously in order to avoid identifying them rightaway with vitalist ecstasies or intensities. Recall, for a corrective point of view, that the doctrine of Russian formalism with its avant garde aesthetics applauded 'deviation' exactly for its potential for making us see the trivial things anew – to see again the stoneness of the stone, etc. If constructed properly, Ingarden's 'metaphysical qualities' should include this sort of aesthetic experience as well, far from the

vitalist shocks seeming to play center stage (in his examples, at least). What about the boring, the sick, the tedious, the ordinary, the vexatious, the brooding, the slowly threatening, the contingent... it takes a true master to make such metaphysical qualities appear with the same intensity as Ingarden's more vitalist examples.

439 I owe many insights in these paragraphs to my discussions with Thomas Illum Hansen who also works with the ideas of integrating cogntive semantics insights into an Ingardenian framework (Hansen 2001, 2004).

440 This would also throw new light on Ingarden's attempt at distinguishing the literary work of art from the scientific literary work. Of course his principal distinction goes between literary quasi-judgments and scientific judgments – where the latter serve the depitction of transcendent (real or ideal) objects. More debateable is his idea that in a scientific work one must focus upon the objects represented solely, and that the '... situation of the aspects which may be held in readiness in a scientific work is analogous, unless the aspects perform a particular cognitive role in the work.' (1968, 157). The addition is significant, for in the scientific work, the aspects held in readiness may perform a *cognitive* role akin to the aesthetic role played in the literary work – in both cases the function of making the objects represented *intuitive* unites the role of the aspects. Diagrams may directly play the role of such aspects. While we agree with Ingarden in the basic distinction (against all sorts of deconstructivism blurring fiction and reality), we find a greater degree of relatedness betweeen the two.

441 Cf. Hayek 1978.

442 With respect to this, a myth probably exists in phenomenology about the definite concreteness of what is perceptually given. But the concreteness of ordinary perceptual experience does not lie in it being constantly highly detailed, but rather in the fact that we know that we could always, in a given situation, choose to focus more attentively on this or that feature and thus bring forth further detail – which is not possible, of course, in purely textually or diagrammatically presented objectivities.

443 This does not, however, imply that literary quality necessarily lies in the making explicit and thematic such thought experiments in themselves. As Nils Gunder Hansen (2004) argues, Iser's idea that the reader's explicit grasping of literary conventions and machinery involved therein constitutes the access to complicated literature and its aesthetic values, is wrong. The diagram experiments here discussed are undertaken primarily in order to co-constitute the level of represented objectivities, not to abstract from it and seeking the work's *raison-d'être* only in its ingenious use of literary devices.

444 What about literary genres like poetry or essays, do they share the basic definition by quasi-judgments? Another of Ingarden's famous opponents, Käte Hamburger, has a counterargument along these lines. She famously claims that poetry is emphatically *not* fictive, but rather a special genre, to be classified along with the 'Ich-Erzählung', the 'I-tale', referring to the subjective experience reality of the lyrical ego. We shall not go into this discussion here, but Hamburger is obviously right that it seems questionable to define the notoriously fuzzy concept of literature as a whole on the basis of fictivity. Ruttkowski 1978, 291–92 details this argument: poetry may be fictive or not, just like epic literature (containing both adventure novels and autobiographies). Similarly, other counterexamples can be listed: fictional texts which are not literature (myth; scientific or other thought experiments). So most likely, it is not the case that literature and fictionality is coextensional, and Ingarden's theory does not hold for the whole wide field covered by the umbrella concept of literature but only for that large prototypical subset of it which develops fictive universes.

445 A further point to be discussed is *formal* ontology. Ingarden mentions the idea in a short note (302) and says that as long as a literary work deals with *objects* at all, the laws of formal ontology must be observed. Given the broad phenomenological idea of objects (covering events, states, processes, qualities, being ideal, real, intentional; physical, psychic etc.) it may be more difficult than expected to transgress formal ontology, so assumably, there is hardly any specific literary truth effect connected to adherence to formal ontology.

446 This taxonomy of iconicities in literature only pertains to the level of represented objectivities and thus does not prevent further, indirect iconicities to play a role in the literary work which are orthogonal to the types here mentioned and may freely combine with all of them: That of experienced events being seen through a narrator or a character; that of an indirect, symbolical or allegorical reference; that of letting linguistic devices function iconically (word order, etc.), that of adding illustrations (maps, graphs,

diagrams, etc.) to the text or directly shape the text after such icons as in figure poems, etc. Johansen (1996, 2003) chart some of these iconicity possibilities after Peircean mode of representation (image, diagram, metaphor, respectively) which is, of course, independent of the five types discussed here, so a combination will yield further subtypes to be investigated.

447 Maybe these types of truth could ease the tension between Ingarden and Hamburger. The latter attacks fiction theories of literature with reference to the non-fictionality of poetry and first-person narration (and a related case could be made for the literary essay) – but even if 'quasi-judgments' are basically fiction, the two concepts are not completely coextensive. Quasi-judgments come in many types (they form a continuum between assertion and mere predication or assumption) and the most assertion-like quasi-judgments, so to speak, only marginally differ from full-fledged assertions. The personal truths experienced in poetry reading or the journalistic or philosophical truth in essayism could thus be seen as a mixture of assertion-like quasi-judgments with pseudo-truths of the types here listed.

To return to the discussion of interpretation in chapter 16, Ingarden's doctrine may also permit us to distinguish between the often mixed-up notions of reading, analysis and interpretation. Reading is prototypically Ingardenian concretization – it is the appropriation of the work which performs the different procedures of filling-in and of actualization of schematized aspects held-in-readiness. Analysis, then, is the operation which – typically on the basis of a reading – attempts to get an overview of the text's organic structure, which parts relate to what other parts in which way. Analysis, then, makes explicit a lot of things unthematically participating in reading – at the same time as it is, typically, ascetic regarding the filling-in which is artificially kept back. Interpretation, finally, is the appropriation of the text which is primarily aimed at the pseudo-truth of the work – in some or all of the five senses here mentioned. As is evident, many empirical examples of papers about specific literary texts indulge in combinations of all three procedures; probably the prototypical academic, informed treatment of literary texts will do so.

448 Clausewitz's *Vom Kriege* is even remarkably sparse as to observations on the role of espionage in warfare; all is a 1-page chapter about 'Nachrichten im Krieg' containing little exceeding common sense: 'Ein grosser Teil der Nachrichten, die man im Kriege bekommet, ist widersprechend, ein noch größerer ist falsch und bei weitem der größte einer ziemlichen Ungewißheit unterworfen.' (48)

449 In the case of espionage, we meet such structures in the recurrent, transhistorical claims about the nature, essence, principles, or problems of espionage in spy literature. In *Spys and Spymasters*, e.g., we read about twentieth century espionage that 'Though considerate advances had been made in technology, the basic principles and problems of intelligence remained unchanged.' (144). In the same vein, we are told that as to the human element of espionage '... nothing had changed since the days of Joshua.' (146). Such general ideas are subsequently applied in the analyses of specific subjects, as when the espionage satellites of the twentieth century are seen as evolutionary heirs to the balloons of the eighteenth century They, in turn, had the function '... to take one stage further the instruction Moses gave to his spies: "Go up into the mountain, and see the land!" ' (166).

450 Smith has thus founded a center for philosophy and geography and conceives of political geography as an exemplary case for a priori studies, e.g., of border types. The idea of such a relation between reality and semantics remains, though, controversial. The present chapter has thus been turned down by several distinguished scholarly journals, not because of its quality (at least, so they claimed), but because of the fact that it included real-world issues in the discussion of a literary genre and thus was deemed unfit for literary studies.

451 In an arbitrarily selected dictionary, *Nudansk Ordbog*, Copenhagen: Politiken 1977. This procedure in inspired by Greimas's investigation of the concepts 'challenge' and 'anger', in Greimas 1982.

452 A prominent example is the allies' large-scale deception operation before D-day in order to make Hitler believe the Dunkirk area to be the invasion spot, including not only a planning of a feinted invasion there but also the planning of a feinted feint, a more northerly invasion supposed to take place from Scotland, thus adding further credibility to the Dunkirk possibility.

453 I believe this is not generally acknowledged, and among many literary scholars, Greimas even counts as an especially malignant reductionist. This rests, however, upon a misinterpretation of Greimas' 'narrative schema' as an assumedly identical deep structure underlying all concrete texts. This idea

overlooks a crucial moment in all decent structuralisms: the concept of *transformation*, cf. Chap. 5. The schema is a prototype only which must be transformed in order to grasp the single text's specificity. The specific features of the single text is grasped only by understanding – not only the schema – but the specific transformation (and its motivations and implications) resulting in just that text. Moreover, the schema may develop with the addition of further assumptions which make new aspects of the fight appear. The schema is not a causal regularity, it is a teleological regularity, and hence it may bifurcate at every possible joint, failing to satisfy the telos in question. The simple phases of it are not causally determined by earlier phases; rather, later phases presuppose earlier phases.

454 Here sociological criteria enter: espionage does not seem to have been anomalous in GDR, for instance, measured on what is known about the number of informants in the people employed by the Stasi, and generally espionage is considerably less controversial in pre-democratic or totalitarian states. But even here, the anomalous character is preserved in the secrecy of procedures.

455 Carl Schmitt's personal career is highly controversial, involving extreme right wing positions and active support for the Nazi regime in the 30s. Despite Schmitt's dubious – to say the least – political positions, it is possible to discuss his more general philosophical and scientific points of view on a democratic basis. Schmitt's notion of sovereignty is explicitly copied from theological concepts; cf. the hypothesis of *Politische Theologie* that the concepts of modern political theory are constituted by secularized theology. A corollary is that fundamental political and politological issues inherit structures from theology; the political wars of the largely atheist twentieth century support this idea, cf. also Vögelin's notion of religious politics. It is easy to recognize the problems of incarnation and of theodicée in relation to espionage: how may democratic ideas become flesh? How may democracy be morally good when its own secret services are not?

456 Schmitt does not explicitly claim this, but the idea clearly appears, e.g. in *Der Begriff des Politischen*, (34n) where the famous dictum of Clausewitz is interpreted with the conclusion that politics is determined by the Friend-or-Foe logic of war.

457 The latter expression is, surprisingly, rather new and dates back only to American discussions in the beginning of the Cold War.

458 We presuppose, of course, that the Prussians did not have the possibility of incarcerating the two and keep them as prisoners of war. We may note *en passant* that according to John Keegan, it was the Prussians' victory in the Franco-German war which made Clausewitz an international hero in military academies worldwide. This development formed part of the radical brutalization of war during the twentieth century supported by Clausewitz's idea of the war as tending to the utmost release of violence and aiming at the total defeat of the enemy.

459 Brandt 1983, 129.

460 Analogous cases occurred in USA during the same period – president Kennedy's affair with Judy Exner whom he shared with mafioso Sam Giancana, just like his affairs with the East German girl Ellen Rometsch and several upper class whores with connections to the Profumo case. These affairs were only made silent due to intensive emergency work by Robert Kennedy and J. Edgar Hoover (according to Hersh 1997).

461 With Jens-Martin Eriksen, I wrote two books about the Balkan wars of the 90s and the actual predicaments of the region (2003, 2004).

462 According to West, 296–97.

463 Famous is the alleged refusal of the USA to perpetuate the services in the period between the World Wars, with reference to the fact that 'gentlemen do not read each other's mail'.

464 The military function is the second function out of the three in Dumézil's theory of three main functions in Indo-European culture: *justice/magic, war, fertility.)*

465 From *Ornicar!*, Vol. 19, Paris 1979.

466 We may remark the British military historian John Keegan's empirical claim that the Napoleonic revolutionary armies with their general conscription played a decisive role in the democratization of Europe.

467 Of course, military intelligence is subject to the same control, but the recurrent and delicate political tension between state security and the human rights of the same state's citizens is structurally relevant for domestic services primarily.

[468] Cf. the Swedish conflict researcher Ola Tunander's work on the Palme case.
[469] When senator Moynihan (cf. below), after having served under Kennedy, Johnson, Nixon, and Ford, was elected to Senate, he got admission to his own FBI file of 561 pages, naming him as a communist.
[470] In one of the rare cases of principal reflection on these issues – the last chapter of former CIA-boss Allen Dulles's book *The Craft of Intelligence* (1963) – he claims that the president himself controls the services, that Dulles himself has supported a proposal for a civil control commission, that all his own knowledge of the services gives reason for trust: 'After more than a decade of service, I can testify that I have never known a group of men and women more devoted to the defense of our country and its way of life than those who are working in the Central Intelligence Agency.' (264). Apart from the fact that a natural scepticism easily awakens faced with claims like these from the leader's own lips, it remains correct that assumptions like the ones quoted is all that you have to rest your head on. On the other hand, Dulles adds immediately afterwards, as the very last two sentences of the book that 'The last thing we can afford to do today is to put our intelligence in chains. Its protective and informative role is indispensable in an era of unique and continuing danger.' (264). Dulles thus summarizes in a few lines all possible points of view: control, trust, limitation of control . . .
[471] A recent Danish example is the schoolmaster Kristian Kjær Nielsen who recently (in the Danish daily *Information* March 10, 1999) told about how, as a member of the Danish Neo-Nazi Party DNSB in the seventies, he spied on party comrades. The information he collected was delivered anonymously by P.O. Box in Copenhagen, and the spy had never any clear idea as to who his commissioners were. Obvious candidates included Israeli, West German, and Danish intelligence services, just like Jewish organizations for the tracing of World War II criminals are possible.
[472] Moynihan relates, not without comical effects, how general Butler, one of the main responsibles for the American atomic strategy, visited the Soviet Union for the first time in 1988 and got a shock. Everything is falling apart, and the chauffeur in the official limousine transporting him breaks off the gear stick. After all these years, Butler in a split second realizes that he has been dealing all along with a caricature (78–79). Moynihan himself tells about a parallel experience regarding the Sandinistas in Nicaragua where he, as an official guest, is witness to the Secretary of Interior trying in vain to order beans at a restaurant – all at the same time as the illegal Iran-Contra scandal develops on a CIA automat idea that the Sandinist state should be a strong and dangerous enemy (208–12). Instead, Moynihan's proposal would be that a '. . . reasonable American response to the new Communist government in Managua would have been a statement of condolence.' (207)
[473] Ed. Nigel West, London 1993.
[474] We may note that the postmodernist sceptic Jean Baudrillard took his most salient examples for his radical idea of the 'disappearance of the real' from the world of secret services. Who was responsible for the Italian terror bombs of the seventies and eighties? – many different groupings claimed responsibility, maybe it was instructed by one political wing in order to discredit the other, maybe by the police in order to discredit both, maybe by foreign interests in destabilizing Italy . . . reality vanishes behind such interpretations and their effects.
[475] I pick these informations from Haswell 1975, 48–49.
[476] The first organized use of this effect was probably Bismarck's famous espionage chief Wilhelm Stieber who was the father of many classical espionage inventions. He erected the so-called 'Greenhouse' in Berlin, an especially sophisticated and depraved whorehouse, with the intention of its use in blackmail of its customer circle involved in international politics.
[477] An explicit example supporting this idea being of course Graham Greene the Catholic.
[478] Maybe this fact gives part of an explanation of the often-noticed but relatively unexplained partition of the film and literature public into two segments: a masculine segment preferring the detectives, front pigs, spies, thrill, and objective action of B-movies, while a feminine segment wants doctors, artists, love, children, passion in melodramas with full possibilities of heavy psychological identification. Nobody could possibly overlook this, least of all Hollywood, still it seems anathema for academic investigation in most of literature and film departments.
[479] 'Metaphysics consist in the results of the absolute acceptance of logical principles not merely as regulatively valid, but as truths of being.' ('The Logic of Mathematics: An Attempt to Develop my Categories from Within', c. 1896, CP 1.487)

[480] Peirce presents several different systems of sciences, but the most developed ones stem from his mature period, the same period in which the concept of continuity is central. In Peirce's large application for the Carnegie foundation from 1902 (NEM IV, 17), the system is schematically presented; in Robin 693 (op.cit. 189 ff. undated), the classification is commented upon; in CP 1.203–83 (1902) it is the object of a thorough argumentation. Mathematics is 'engaged solely in tracing out the consequences of hypothesis' (CP 1.247) – Peirce's interpretation of his father Benjamin Peirce's definition of mathematics as the science which draws necessary conclusions. Its hypothetical character is important: '*Mathematics* is the study of what is or is not logically possible, without undertaking to ascertain what actually exists.' ('Reason's Conscience', undated, NEM IV, 191). Consequently, it is left to the more empirical sciences to ascertain which of its hypotheses may have which applications. Conversely, mathematics is the most general science, whose results (even if they may not be unambiguous, cf. the Non-Euclidean geometries) all the more specific sciences must apply. The philosophical sciences – in Peirce phenomenology (or phaneroscopy), normative sciences (including logic equaling semiotics) and finally metaphysics, in that order – hence depend on mathematics; it is, of course, their own decision which parts of mathematics they build on. It is decided by what may appear in experience (phenomenology), by which forms valid inferences may take (logic) and by which general properties reality possesses (metaphysics), the latter of these sciences specifying the former, not vice versa. Only after these sciences, the special sciences follow under the headline of 'idioscopy'. In the series of philosophical sciences, continuity has a central role, both in the sense of first philosophy and as a sort of connecting glue between the single domains.

[481] The very first philosopher on this point is probably Bolzano (*Paradoxien des Unendlichen*, 1850) to whom Peirce only refers in his late period. Peirce refers one of his main ideas (that the equivalence of sets with respect to size should be measured by a one-to-one pairing of elements), but at the same time he claims he has not read his work ('Bedrock', 1906, 6.175, and 'Some Amazing Mazes', 1908, 4.651).

[482] Peirce claims at several occasions that Dedekind's influential book – where the real numbers are defined by 'Dedekind cuts' – should be influenced by papers sent to him by Peirce. Cf. the following lecture note from 1903: 'My own first contribution to this branch was made in the autumn of 1867' (Robin 316a, NEM III, 129), and further 'My second contribution was 1881 [...] My third [...] That is all I have printed. My work has been, I believe, completely independent of Cantor. I never knew anything definite about him until 1884. I have seen it stated in some book that I modified the statements of Dedekind. But the truth is that Dedekind's *Was sind und was sollen die Zahlen* first appeared in [1888]. It contains not a single idea which was not in my paper of [1881?], of which an extra copy was sent to him and I do not doubt influenced his work.' (ibid. 130; the square brackets around the years are Eisele's). The unspecified reference to 1881 probably refers to 'On the Logic of Number' (CP 3.252–88), cf. what Peirce says a couple of years later in a footnote ('Pragmaticism, Prag. [4]' c. 1905, CP 5.526n): 'C.S. Peirce's paper *On the Logic of Number*, published in the spring of 1881 contained, though not in a perspicuous form, the leading results of Dedekind's classic of 1887.' We shall not take that text into accout here, as it – with respect to continuity – rests upon insuffucient definitions: 'A continuous system is one in which every quantity greater than another is also greater than some intermediate quantity greater than that other.' (3.256) – which is valid already for non-continuous objects, such as certain arrangements (after size, e.g.) of rational numbers.

[483] Cantor's decisive articles are the six-part 'Über unendlich lineare Punktmannigfaltigkeiten' (1879–84) which widely transgresses the subject indicated by the title to provide a philosophical argumentation for the basis of his theory of sets, as well as the two-part 'Beiträge zur Begründung der transfiniten Mengenlehre' (1895–97) which provides a rigorous mathematical presentation.
According to Parker 1992, Peirce only knew the first of these papers through a short excerpt in French translation (of part 6, §16), 'Sur divers théorèmes de la théorie des ensembles de points situés dans un espace continu a *n* dimensions' (1883), just like he only knew the second part of the second paper in addition to a final third paper 'Mitteilungen zur Lehre vom Transfiniten' (1887–88) which consists of a collection of letters discussing and polemizising with various authors. The background for these claims about Peirce's selective reading of Cantor is not provided in Parker 1992.
Most references to Cantor in Peirce's work are later than 1897, but e.g. CP 6.164 (The 'Continuity' entry from *Century Dictionary* 1889) or the reference in 1892 'The Law of Mind' points to the fact that Peirce at this point must have known considerably more about Cantor than §16 (1883a); in the notes of the CP a

reference points to §10 of (1879–84). In Peirce's draft for a letter to Cantor dating from 1900 (NEM III, 767) he says, 'Dear Sir: Before reading your wonderfully beautiful and profound studies (in fact, I have only just read your memoir in the Math. Annalen XLVI and XLIX, and have not had time to digest them, and have only looked a little at the memoir in Vol. XXI, because I could not catch the idea of your ordinal numbers)...' According to this passage, Peirce knows *both* parts of Cantor 1895–97, while the reference to a paper in the same journal's Vol. 21 refers to 'Grundlagen einer allgemeinen Mannigfaltigkeitslehre' (1883) which is 'Nr. 5' (by far the largest) of the six parts of Cantor 1879–84, containing among other things the first articulation of the continuum hypothesis. Thus it does not seem correct to claim that Peirce only knew the French version of Nr. 6, §16. When Peirce says about his knowledge of these texts that he has 'only looked a little', we may not be so sure he is speaking the whole truth; firstly, Peirce claims elsewhere that he has anticipated Cantor and hence it would not be clever of him to admit much too close an acquaintance with his early writings (even if Peirce's decisive argument for his autonomy 'On the Logic of Number' (CP 3.252–88) dates from 1881); secondly, he adds that he has not really understood ordinal numbers which is no doubt correct. The references to Cantor earlier than 1897, thus in 'The Law of Mind', seem to refer to a knowledge of part 5 of Cantor 1879–84.

484 The centrality of time, however, in Peirce's continuity argument is evident as early as in the *Monist* papers from 1892 (cf. below) so Potter and Shields' idea of a fourth period based on the 1908 note may be subject to discussion.

485 Thereafter Peirce adds that this ambitious and romantic philosophical program might have certain roots in his Concord childhood during the period in which Emersonian transcendentalism dominated the intellectual climate with ideas 'caught from Schelling, and Schelling from Plotinus, from Boehm, or from god knows what minds stricken with the monstrous mysticism of the East.' He is not, however, 'conscious of having contracted any of that virus', but he still holds it probable that 'some cultured bacilli, some benignant form of the disease was implanted in my soul, unawares, and that now, after long incubation, it comes to the surface, modified by mathematical conceptions and by training in physical investigations.' Thus Peirce perfectly realizes the apparently strange fact that his evolutionary cosmology introduces rather mystical ideas in the cool pragmatism. In spite of parallel formulas regarding mind and matter, Peirce in these lectures rejects monism (6.24) in favor of an idealism making matter a derivative phenomenon; later (6.73) he presents himself as monist, probably motivated by the continuity between mind and matter which is a recurring idea in him. Even if Peirce is thus no neutral monist, it seems fair to call his idealism a monism, because it refers to the metaphysical interpretation of the concept of continuity, due to which no absolute metaphysical distinctions are allowed.

486 As Parker remarks (16, 52), the concept of continuity is here invoked at two epistemologically and metaphysically decisive points: the beginning point of science and the ground of being of reality; both of these solid and problematic entities are dissolved by means of continua establishing their connection to the actual level of knowledge at any time, backwards (knowledge) as well as forwards (being). This allows the young Peirce to cut down his category table in 'New List' (EPI, 6; 1.555) from five (minus Substance and Being) to the canonical three, at the same time as he gets rid of the Kantian problem of a *Ding an sich*.

487 This argument remains constant in the mature Peirce, cf. 4.641 ('The Amazing Mazes. First Curiosity, 1908): 'The argument which seems to me to prove, not only that there is such a conception of continuity as I contend for, but that it is realized in the universe, is that if it was not so, nobody could have any memory.'

488 The difference remains, of course, that Husserl's question refers to ideal objects only, defined by that the access to them is in principle always possible. Peirce's argument refers to memory in general where this principle does not hold.

489 Ketner and Putnam's preface to *Reasoning and the Logic of Things* refers to this affinity between Peirce and non-standard analysis.

490 The first occurrence of this intuition is probably Bernard Bolzano in his *Paradoxien des Unendlichen* from 1850 which Peirce only later realizes, cf. 'Notes on the Theory of Multitude', (NEM III, 1088) from 1903.

491 Even if equivalent to Dedekind and Cantor's definitions, Peirce's 1881 description of infinite sets in terms of the syllogism of transposed quantities differs in being an intensional description, pointing

to a seminal difference between Cantorian set theory and Peirce's approach. Peirce later highlighted the difference in a letter to *Science* in 1900 (3.563), and Randall Dipert (1997) convincingly argues that Peirce's criticism forms an 'extremely subtle' point. Peirce protests that Dedekind's notion of proper part ('echter Theil') remains undefined, and Dipert argues that this is because there is no procedure given to determine whether a given candidate is, in fact, a proper part of an infinte set. In order to tell whether a set A is a subset of another set B, we must have sufficient knowledge af that set in order to determine whether each of its elements are in fact elements of B as well. But we may only possess such knowledge of A if we know each of its individual elements by some procedure. Dipert acknowledges the constructivist leanings of this idea – but Peirce's obsession does not lie, as in the later constructivists after Brouwer, with the impossibility of admitting objects defined by infinite procedures, but rather with the distinctness of the individual elements. This is why Peirce insists on an intensional definition of subsets in the syllogism of transposed quantities, Dipert argues: 'One cannot *legislate* an ability to form proper subsets of a given purported collection. That is, of any purportedly collective entity, one cannot simply assert that one can, say, partition it. Rather, the ability to do so is a consequence of, and hence required for, its being a collection in the first place. A conception of the members of the collection as individuated entities comes first, and of the collections of such entities, the Axiom of Choice will in fact be *descriptively* true.' (65). Peirce only later, in the 1900 letter to Cantor, make explicit this idea, but its germ is present already in the definition of infinity by means of the syllogism of transposed quantities where the crucial subset (the subset of women in the French population) is intensionally defined. Dipert rightly notices the connection from Peirce's intensionality claim to his realism as to rules and laws.

492 In 'The Logic of Quantity' (1893, 4.126) Peirce says (almost) the exact opposite: 'The multiplicity of points upon a surface must be admitted, as it seems to me, to be the square of the points of the line and so with higher dimensions.' If you do not add the rule of calculation claiming that Aleph-n times Aleph-n = Aleph-n, then this claim is fallacious. This only mentioned to indicate how vacillating Peirce may be in this field, and how difficult it may be to reconstruct his changing positions.

493 Peirce later regrets his condemnation of Kant's continuity definitions and his adherence to Cantor's definition of the continuum as a set of points. In L77 (NEM III, 780), a letter to Paul Carus from 1899, he says that Kant in the first Critique does not claim that space is infinitely divisible but only that it may *not* be so divided in order to reach any least part. This definition is not, of course, more correct than the first one, but it is closer to the spirit of Peirce's idea that points are not parts of the line but limit values on it. Even later, Peirce returns to another and better idea in Kant: that continuity implies that all parts have parts of the same sort.

494 Elsewhere, Peirce has a definition of the concept of 'perfect' deviating from Cantor's ('Notes on the Theory of Multitude', 1903?, NEM III, 1092). Peirce here comments upon Schönfliess' 'Die Entwickelung der Lehre von dem Punktmannigfaltigkeiten' in Heft VIII of the *Jahresbericht der Deutschen Mathematiker-Vereinigung* 1900, presenting Cantor's theory. In Peirce's note to page 32 he claims that 'A type (or a collection) is called secluded or concluded abgeschlossen if every fundamental series in it has its limit within it.

It is called *close* (in sich dicht) if every element is a capital element.

It is called *perfect* (perfekt) if it is both secluded and close. (That is, if you take away an element from a perfect type it will be *close* just the same but will cease to be *concluded*. But if you add an element it will still be *concluded* but will cease to be *close*. Hence we may translate *in sich dicht* by *dense* or *condensed* or *concise but as a whole*, and *abgeschlossen* by *complete*.)'

Here, perfect is defined as being close plus dense, where perfect earlier was only close, but had to be completed by density in order to yield continuity. The definitions of continuity are hence equivalent, even if the definition of perfection has changed.

495 If you take e.g. a closed set of rationals on the number line, then the derivative set in this sense is the same closed set (because every rational is a limit element, cf. the Dedekind cut). A set of this kind is irreducible by this procedure, while on the other hand a convergence series is reducible to the number towards which it converges.

496 The same Aristotelian idea plays a central role in René Thom's *Esquisse d'une sémiophysique*.

497 If we go back to Cantor's definition of a perfect point, it is not far from Peirce's Aristotelicity. He defines (Cantor 1872) a 'limit point' as one of which it holds that in any neighborhood around the

point there will be an infinity of points belonging to the set, and he furthermore defines a derived set P' as that which consists of the limit points of a given set. If the derived set is identical to the set itself, then the set is perfect (Cantor 1879–84, 193), but this is only the case if it is not open (an open set's derived set is larger han itself, because it contains, in addition to the set itself, the points which might close it (which is what we today would call its limit points, NB), its end points. Already in his article about *Continuity* in *The Century Dictionary* (1889, CP 6.164), Peirce refers Cantor's definition, here with acclamation.

[498] Cf. Maddy 1990, 110, referring Cantor's definition of perfection as follows: 'A closed set of reals is one that contains all its accumulation points; a perfect set is closed, and every one of its points is an accumulation point', corresponding to Peirce's version: the former requirement grants Aristotelicity, the latter Kanticity.

[499] A truism, it must be said, which Peirce does not continue claiming. In his late philosophy with its finalist interpretation of continuity he no longer finds it strange that the future should be able to influence upon the present now (albeit only in general terms, in opposition to the past); this is the very definition of finality.

[500] It is from arguments like these that Peirce is later able to claim that his doctrine does not exclude the possibility of telepathy (even if it does not, on the other hand, positively imply nor argue for such a thing . . .).

[501] The infinitesimalist time theory very logically – but counterintuitively – leads Peirce to claim that the energy of a perceived idea in the present must be infinite, while an idea in the past has a finite and relative energy. This idea really claims a primacy of perception, cf. Peirce's theory of 'perceptual judgment' where generality is directly perceived.

[502] Cf. Holenstein 1972 on associationism in Husserl.

[503] An idea reminding of Kant's philosophy of history, according to which we can never know whether history has a goal – but we ought to act as if it had.

[504] In this, Peirce continues Kant, for whom the finitude of man prevents us from reaching the *noumenon*, which is why we must be satisfied with *phenomena*. But this delimitation so to speak incarcerates us in generalia (not in particulars, as one might think); all we can do is to predicate a substance with general predicates (but maybe even only the simpler general predicates . . .). Elsewhere, Peirce says: '. . . I had long before declared that absolute individuals were entia rationis, and not realities. A concept determinate in all respects is as fictitious as a concept definite in all respects.' (A letter to Signor Calderoni, 1905, 8.208); the idea of an object determinate through and through is rather a regulative idea which requires of us not to remain satisfied with any partial knowledge of the object. But we can not even reach any thoroughgoing knowledge of *realized* generals, Peirce here adds: they are necessarily vague.
Towards the end of the article, Peirce's speculations speed up, and in the course of the next few paragraphs we learn that the association of ideas, just like personality, has a teleological aim, because the future reference is a crucial element in both (just like in Peirce's symbol category in general). Consequently, the idea of an evolutionary universe is coextensive with the idea of a personal Creator, Peirce all of a sudden concludes – as against determinism which is always unable to make sense of any personal contact with the deity. The question why we have not long since received full clarity about this personal God is answered with reference to our unability to grasp facts right before our noses (6.162). These surprising teleological and even theological speculations are brief and do not influence the preceding overall argument. We include them here in order to show the range of Peirce's trust in the continuity concept. It may be added, that Peirce's famous personal epiphany probably took place around this time (cf. the mystic musings in the last lines of 'Man's Glassy Essence', EPI, 350–51; 6.271); theological reflections connected to the continuity concept are not generic in Peirce's later writings.

[505] The beginning of the text in question with its rich use of these terms is so sudden because it constitutes page 23 of a manuscript with its first pages missing.

[506] Later, around 1909, Peirce realizes that not only CH but also its generalized version, the general continuum hypothesis (GCH) may not hold: '. . . it has never been exactly proved that there are no multitudes between two successive abnumerable multitudes, nor, which is more important, that there is no multitude greater than all the abnumerable multitudes.' ('Some Amazing Mazes, Fourth Curiosity', 4.656).

[507] Discussing Peirce's different ideas about CH, it is important to keep in mind that he omits Cantor's doctrine of ordinal numbers. In fact, Cantor's Aleph series is constructed not as a power set series, but as the series of powers of the successive sets of ordinal transfinite numbers. These two, in Cantor, are different procedures for constructing larger cardinal numbers. Thus, Cantor's CH basically asks the question whether the cardinal series of ordinal number set powers is the same series as the series obtained by power set constructions taking Aleph-0 as its point of departure. As far as I can judge, Peirce never understood Cantor's concept of ordinal transfinite numbers and hence did not grasp this aspect of CH. This is probably why he – until late in his life – immediately identifies the Aleph series with the series obtained by successive power set constructions from Aleph-0. But this identification, in Cantorian set theory, is the whole issue of CH.
[508] While Peirce insists on the existence of infinitesimals, that is, numbers smaller than any finite real number but larger than zero, then for the infinitesimal x we may take its inverse $1/x$. This number must be larger than any finite real number, but still smaller than infinity. Since infinitesimals may assume any infinitely small size, some of these must make $1/x$ an integer larger than any given finite integer but still smaller than Aleph-0.
[509] Cantor has already found that Aleph-2 must be the power of the set of all real functions (if you take the concept of functions as referring to the formation of n-tuples between the elements of a number of sets, then the set of real functions will of course be of the same power as the set of subsets of the reals). Elsewhere, Peirce does not refrain from giving examples of Aleph-2 or even Aleph-3 sets.
[510] Peirce here writes 'Exp γ, $(\text{Exp})^2\gamma$, $(\text{Exp}\gamma)^3\gamma$'; supposedly, he intends the version given.
[511] What may have lead Peirce to believe in this proof may be found in the alternative version found in R&LOT. Here a decisive step in the argument is the following 'But now let us consider a collection containing an individual for every individual of a collection of collections comprising af collection for every abnumeral multitude. That is, this collection shall consist of all finite multitudes together with all possible collections of those multitudes, together with all possible collections of collections of those multitudes, together with all possible collections of collections of collections of those multitudes, and so on *ad infinitum*. This collection is evidently of a multitude as great as that of all possible collections of its members. But we have just seen that this cannot be true of any collection whose individuals are distinct...' (159) The way of constructing the pseudo-set S here differs from the way just presented. When a set of power Aleph-n + 1 is made a part of S, then at the same time it is supposed to give an inventory of the subsets of Aleph-n. Aleph-1, e.g., constitutes a set of subsets of Aleph-0, and it is of course correct that this is in a certain sense the case (because the decimal expansion of every single real number constitutes a set of figures which corresponds to a subset of the natural numbers). This trick now allows Peirce to claim that just like any Aleph is constituted by the subsets of the former Aleph, then by the simple sum of all Aleph-n's you get the subsets of S which will then have the same power as all of its subsets. But here Peirce does two things at the same time: he conceives of the elements of an Aleph as simple entities and as sets. For all these sets are naturally subsets of S but do they *exhaust* its subsets? Some subsets 'lack', namely those that appears by the pairing of e.g. each of the elements in Aleph-0 with each of the elements of Aleph-1. But they *are* included, Peirce will say (we must assume), because Aleph-0 can make no additions to Aleph-1. But by saying so, he no longer regards the elements as sets, but merely as elements. If we pair, e.g. 1 and $\sqrt{2}$ (1.4142...) then Peirce may say that we may just construct the element 1.14142... which is already a member of Aleph-1. That is correct, but not for the set $\{1, \sqrt{2}\}$. In that way he conceives of S's elements as sets, when that is called for, and as elements when that is needed, and this is why he can abstain from taking the exponent of the whole sum of Alephs: he so to speak divides the whole series with 2^n – and so it is no wonder he may obtain a result like $n = 2^n$.
[512] Cf. Thom's mirror idea that the concept of 'point' might yet contain secrets able to solve the continuum riddle.
[513] This lecture, as it appears, is a construct on the part of the editors; it is composed by Robin 751 (as a sort of preface) plus Robin 439 which probably is the lecture proper, because its contents correspond to what is referred to in the other lectures. In our context, this is of little importance.
[514] These signs give, by the embedding of the former in the latter, rise to the three classes of signs which Peirce in his third tricotomy calls 'rhema', 'dicisign', and 'argument', respectively, – also 'seme', 'pheme', and 'delome'. Closest to ordinary language is the names 'term', 'proposition', and 'argument'.

This analysis, moreover, constitutes the background for Peirce's own construction of existential graphs, providing the verbal, predicative, iconic part of the graphs; the demonstrative-pronoun, indexical parts; and the logical connective parts, respectively. ('Detached Parts Continued', 1898, NEM IV, 339). The first form the diagram icon proper, the second form propositions connecting that icon with indices; the third connect such propositions into inference chains and provide the possibility of controlled experiment with the diagram icon.

[515] This idea refers to Putnam's classic paper 'Mathematics without foundations' (1967, reprinted in Benacerraf and Putnam 1983) where he claims the idea that mathematics has a modal logic character so that every proposition about mathematical objects may be read as a proposition about a possible construction: 'In short, if one fastens on the first picture, (the "object" picture), then mathematics is wholly extensional, but presupposes a vast totality of eternal objects; while if one fastens on the second picture (the "modal" picture), then mathematics has *no* special object on its own, but simply tells us what follows from what.' (300).

[516] In contrast to Peirce's metaphysics, this interpretation is too constructivist. Peirce follows Cantor all the way as regards actual infinity; this is indeed the very metaphysical motivation for his position. Continuity is merely potential, but it may, in turn, be actualized as 'real possibilities' in the general phenomena of reality (that is, without any finitistic restriction is imposed on its actualized parts, which, on the other hand, seems to be Putnam's classical position).

[517] It is interesting here to note that also John von Neumann's version of set theory allows for the construction of a multitude which is the union set of all transfinite numbers and which hence exceeds their powers (his 'proper class'). In his case this class lies beyond not only the series of Aleph-n's, but beyond any ensuing series of Aleph-Alephs etc. of classical set theory.

[518] This is obviously no strong argument for Peirce's particular doctrine of a supermultitudinous set *S* (which, interpreted in standard set theory, would have the cardinality of Aleph-omega-0; all other cardinalities might as well be invoked here). But this undecidability could, more generally, be taken as an indication of the possibility that no formalism may exhaust the continuum and hence that the continuum might be 'larger' than any formalization attempt – thus supporting Peirce's modal interpretation of it.

[519] Unfortunately many of his notes on Husserl are written in the vanishing brand of Austrian stenography known as Gabelsberger shorthand and has not yet been transcribed and published.

[520] The intuition necessary for symbolic calculi – assumed to be infallible – should amount to no more than the ability of distinguishing 0 from 1 on the line, to count the length of finite symbol strings in number of units, to manipulate symbols and symbol groups on the line according to certain rules, etc. In short, this presupposes the intuition of the line, segmented in distinct places, and of the characterization of those places measured in (at least) two different possibilities. It presupposes, moreover, the parsing of the row of symbols into meaningful groups (dictated by the given symbol language), and, most complicated, the substitution for certain symbol or symbol groups by others (determined by the axioms, expressed in the same symbol language).

[521] This argument is very often stated in a strangely naive manner: because physics is unthinkable without mathematics, the two sciences are taken to be inseperable (this argument is also found in the structuralist Resnik, cf. Resnik 1997) without considering the equally possible conclusion that a unilateral dependency is at stake (so that mathematics is independent of physics even if physics depend on mathematics). This would, of course, be Peirce's position.

[522] A fine example of this is the application of (parts of) mathematical group theory in everyday perception, such as Cassirer 1944 argues. Here it is shown how a series of arch problems of the psychology of perception can be explained by reference to the group of transformations characterizing a group in group theory. Object constancy (why do we see objects having the same shape when it changes contour in different perspectives and distances?), color constancy (why do we see the object having the same color under different light conditions?) can be interpreted as referring to an invariance not affected by transformations. As is known from the Erlanger program, different geometries can be defined by each their group of transformations, so that the more tranformations are allowed, the more 'empty' and general the corresponding invariant object will become – culminating in topology, as both Peirce and Cassirer realize (cf. Chap. 5). But the variety of geometries is not – as it is often taken to be – an argument for the lack of objectivity of geometry (and, consequently, for the subjectivity of perception), but quite on the contrary an argument for the strength of the mind, possessing a differentiated series of objective

tools for the conception of space, some of which are already in play in ordinary perception. In this sense, it becomes understandable how already everyday perception is ripe with sophisticated knowledge – as well as how it is possible to isolate and sophisticate the evidence of mathematical basic objects on the basis of everyday perceptions without assuming Benacerafferian causal-reference access to such objects. In this manner, this train of thought – parallel to Peirce – runs counter to the materialist-empiristic idea that perception 'first' consists of an uncoordinated aggregate of sensory stimuli which 'later' is put in order by the subject. Corresponding to Peirce's claim for the existence of infinitesimals, perception in this perspective always treats continuous, organized phenomena – and the idea of an uncoordinated set of pure sense impressions will be relegated to a limit case never actually realized: it is an idea, not reality. Quite on the contrary, the object is always constituted as an invariance measured on different sets of transformations, and perception is never limited to a present now without any extension. This is also equivalent to Peirce to whom pure secondness, pure thisness, pure haeccity, remains a mere limit case – any event is only minimally graspable to the extent to which it is referred to a domain in some qualitative continuum, to which it is embedded in a more extended temporal process, to which it forms part of a general, lawlike space.

[523] The resume in the following paragraph of developments in the CH discussion from Cohen's 1964 till now is indebted to Chap. 4 in Maddy 1990.

[524] Gödel, in the period from 1938 to 1947, also changes his viewpoint in this direction (Maddy 134–35) and in the same way Moschovakis 1980 (610), as quoted by Maddy, takes it as an unsuitable constraint on the concept of '*arbitrary* set', because there is no a priori reason to suppose that every subset should be definable (definability by first order formulas is equivalent to being a finite union of open intervals with named endpoints).

[525] '...donc, elle est un fait "objectivement" scientifique, et à ce titre, elle doit être "objectivement" expliquée. Si l'on n'a pas à sa disposition une théorie immanentiste du continu dans notre physiologie neuronale, qu'on m'explique la genèse de l'illusion!' (Thom 1992, 140)

[526] Harthong, in fact, proposes four new antinomies for modern mathematical rationality (not unrelated to Kant's): randomness/determinism, continuity/discontinuity, simplicity/complexity, and finally a finite/infinite number of basic rational principles. He argues that all of these antinomies spring from two basic phenomena: 1) knowledge being approximative (corresponding to Kant's idea of human finitude, and 2) the existence of several different size scales. The latter he views as a mathematical constraint (not only a property in empirical nature, 354), because a scalar shift may let a discrete and complicated description give way for a continuous and simple description (the arch example of course being thermodynamics in Boltzmann). Naturally it is the fact that nature does present us with vast differences in size which makes us aware of this antinomy: '...les antinomies de la rationalité mathématique se manifestent sur de très grandes échelles de grandeurs; elles n'existeraient pas si les grandeurs existant dans la nature étaient toutes du même ordre.' (363) To Harthong, the possibility of ever making CH decidable vanishes for exactly this reason: 'Le débat reste métaphysique tant que la plus petite des deux échelles de grandeurs reste *inaccessible* á la perception (...) Les lois "utiles" sont des lois phénoménologiques qui marchent à l'échelle où nous observons.' – he pragmatically concludes, polemically aimed against reductionist brain science and its assumptions that the final neuronal explanation of the mind is close at hand.

[527] In Salanskis' Continuity volume, Robinson's non-standard analysis plays a constant role because it reintroduces infinitesimals – even if it takes place on the base of orthodox set theory, as non-standard points in a halo around the standard points of **R**, appearing in external and internal sets, respectively. Yet, this 'galaxy' of non-standard points have, as Salanskis says (199) the effect of expressing the 'richness' of the continuum which is not present in standard analysis – this especially so in the so-called Harthong-Reeb version of non-standard analysis.

[528] '...ce qui apparaît ainsi dans l'ensemble de ces travaux de non standardistes reprenant la tradition borélienne, c'est que le continu est quelque chose comme un "effet" qui résulte d'un *excès*. Si l'approche de l'aspect "désensemblisation" du destin du continu de Cantor-Dedekind met en vedette l'*incommensurabilité*, la prise en compte du regain d'attention envers le constructif et sa place dans le monde des entités mathématiques nous fait rencontrer le thème philosophique de l'*excès*. Le lien entre les deux termes, cela dit, nous semble clair: l'excès engendre de l'incommensurabilité, et la distribution de l'incommensurabilité sur la droite entière lui ajoute la *qualité* constitutive du continu (...)' (210)

529 Petitot's contribution to Salanskis 1992 is complemented by the programmatical paper 'Pour un platonisme transcendantal' (1994).

530 Namely this: if anything exists as a psychological primitive, this phenomenon is eo ipso also natural (pace naturalism in psychology) and has reality which is not merely psychological, se the paragraph on Maddy.

531 'Le platonisme transcendantal est "négatif", et non pas "positif" comme le platonisme ontologique vulgaire. Il permet *d'inverser* les affirmations philosophiques de l'antiplatonisme *et de transformer les phénomènes d'incomplétude et d'indécidabilité en arguments en faveur du platonisme.*' (254)

532 The axiom of determinacy states that if there is a game between two players, each in turn selecting successive nested subsets of the unit interval, then there is a strategy for one of the players to let the selecting process terminate in a number belonging to a given subset of the unit interval.

533 'Evidemment, demeure entière l'énigme de l'intuition pure, c'est-à-dire de la façon dont nous sommes affectés par l'exteriorité. Ce problème n'est pas d'abord mathématique mais *physique et cognitif.*' (261)

534 Among many possible quotes from his work see e.g. *Physique du sens* from 1992 where the introduction claims that the morpho-mathematical foundation of structuralist models necessitates a critique of the formalist position regarding language and meaning: 'Ce dernier point pourra paraître irrecevable aux linguistes et sémioticiens qui croient encore, suite aux dogmes du positivisme logique, que la conception métamathématique hilbertienne (c'est-à-dire le rapport entre syntaxe et sémantique que l'on trouve en théorie des modèles peut être transférée à l'analyse *des rapports de dépendance* constitutifs des structures.' (XXII). The fact that this critique does not aim at the semiotic domain only is evident from, e.g., 53 where, following Cavaillès, it is claimed that in any doctrine of objectivity, formal logic must be connected to an ontology: 'Ses concepts doivent acquérir un contenu transcendantal – c'est-à-dire concertant l'*objet* de la connaissance et non pas simplement sa *forme discursive* – et, selon Kant, ils le peuvent si les catégories prennent pour "matière" le divers pur (spatio-temporel) de l'Esthétique transcendantale (schématisme).' Also mathematics has a content to be schematized (cf. the schema 55).

535 For a prototypical articulation of this point of view, see the influential American philosopher of mathematics Saunders MacLane's (the father of category theory) antiplatonist doctrine *Mathematics. Form and Function* (1986), 385: 'These results [the proof of the independence of CH, AC, and other axioms from set theory], and others too numerous to mention, show that many interesting Mathematical questions cannot be settled on the basis of the Zermelo-Fraenkel axioms for set theory. (...) This variety and the undecideability results indicate that set theory is indeterminate in principle: There is no unique and definitive list of axioms for sets; the intuitive idea of a set as a collection can lead to wildly different and mutually inconsistent formulations. On the elementary level, there are options such as ZFC, ZC, ZBQC or intuitionistic set theory; on the higher level, the method of forcing provides many alternative models with divergent properties. The platonic notion that there is somewhere *the* ideal realm of sets, not yet fully described is a glorious illusion.' MacLane later turns the argumentation directly against Gödel and his 'realist view' (449).

536 This implies, of course, problems to which we shall later return: the metaphysics (1) which operate on the threshold between the many and the one Platonic world, how is that to be situated in relation to that metaphysics (2) which constructs the four-part scheme above and consequently must be more extensive than mathematics?

537 I think an account in a Peircean spirit would claim that the hierarchy of prescission and abstraction (Chap. 11) allows us to reason with infinite procedures, because they are here taken as objects and subjected to imaginary completions.

538 An example: 'Pure Mathematics is the study of pure hypothesis regardless of any analogies they may have in our universe.' ('Lectures on Pragmatism' 1903, NEM IV, 149)

BIBLIOGRAPHY

Alberts B, Bray D, Lewis J, Raff M, Roberts K, Watson JD (1994) Molecular biology of the cell, 3rd edn. Garland Publ., New York

Allwein G, Barwise J (1996) Logical reasoning with diagrams. Oxford University Press, New York

Almeder R (1973) Peirce's pragmatism and scotistic realism. Transactions of the Charles S. Peirce Society 9(1):3–23

Anderson DR (1986) The evolution of Peirce's concept of abduction. Transactions of the Charles S. Peirce Society 22(2):145–164

Armstrong D (1989) Universals. An opinionated introduction, Westview Press, Boulder etc.

Armstrong D (undated) Four disputes about properties, www.uni-konstanz.de/FuF/Philo/Philosophie/Spohn/workshop/armstrong/armstrong.pdf

Arnheim R (1969) Visual thinking. University of California Press, Berkeley

Bader A, Parker L (2001) Joseph Loschmidt, physicist and chemist. Physics Today Online, March 2001; http://www.physicstoday.org/pt/vol-54/iss-3/p45.html

Bak P (1997) How nature works. Oxford University Press, Oxford

Baldwin JM (ed) (1901) Dictionary of philosophy and psychology, vol 1–4. The MacMillan Co., New York

Baldwin JM (1902) Development and evolution. MacMillan Company, New York

Barwise J, Etchemendy J (1995) Heterogeneous logic. In Glasgow, 1995

Bateson G (1979) Mind and nature. Fontana, London

Bell JL (preprint) Dissenting Voices. Divergent Conceptions of the Continuum in 19th and Early 20th Century Mathematics and Philosophy

Benacerraf P, Putnam H (eds) (1983) Philosophy of mathematics. Cambridge University Press, Cambridge

Benoist J (1997) Phénoménologie, sémantique, ontologie. P.U.F., Paris

Berg HC (1988) A physicist looks at bacterial chemotaxis. Cold Spring Harbor Symposia on Quantitative Biology, 53:1–9

Berlin B, Kay P (1969) Basic color terms: their universality and evolution. University of California Press, Berkeley

Bernet R (1988) Perception, categorial intuition and truth in Husserl's sixth 'logical investigation'. In: Sallis J, Moneta G, Taminiaux J (eds) The collegium phenomenologicum. Kluwer, Dordrecht, pp33–45

Bernet R (1988a) Husserl's theory of signs revisited. In: Sokolowski R (ed) Studies in philosophy and the history of philosophy, vol 18. Catholic University of America Press, Washington, pp1–24

Bernet R, Kern I, Marbach E (1989) Edmund Husserl. Darstellung seines denkens. Felix Meiner, Hamburg

Blackwell A, Engelhardt Y, (1998) A taxonomy of diagram taxonomies. Paper at the conference *Thinking With Diagrams 98*, University of Wales, Aberystwyth

Bobzien S (2004) Dialectical school. Stanford encyclopedia of philosophy, http://plato.stanford.edu/entries/dialectical-school/

Bogh M (2002) Formalitet og figurativitet. Fænomenologiske perspektiver i nyere kunstteori [Formality and figurativity. Phenomenological perspectives in recent art theory]. In: Christensen HA, Michelsen A, Wamberg J (eds) Kunstteori. Positioner i nyere kunstdebat. Borgen, København, pp215–250

Boghossian P, Peacocke C (eds) (2000) New essays on the a priori. Clarendon Press, Oxford

Boler J (1963) Charles Peirce and scholastic realism. University of Washington Press, Seattle

Boler J (2004) Peirce and medieval thought. In Misak 2004, pp58–86
Brandt PA (1983) Sandheden, sætningen og døden [Truth, sentence, and death]. Sjakalen, Århus
Brentano F (1982) Deskriptive psychologie. Meiner, Hamburg
Brook A (1997) Kant and the mind. Cambridge University Press, Cambridge
Brøgger A (preprint) Billedernes retorik [The rhetoric of pictures]. PhD thesis
Brown JR (1999) Philosophy of mathematics. An introduction to the world of proofs and pictures. Routledge, London
Brunning J, Forster P (eds) (1997) The rule of reason. The philosophy of Charles Sanders Peirce. University of Toronto Press, Toronto
Bundgaard P (2004) Kunst. Semiotiske beskrivelser af æstetisk betydning og oplevelse [Art. Semiotic descriptions of aesthetic meaning and experience]. Haase, Copenhagen
Bundgaard P (2004a) The ideal scaffolding of language: Husserl's fourth logical investigation in the light of cognitive linguistics. Phenomenology & the cognitive sciences 3:49–80
Bundgaard P (2004b) Configuration sémantique et combinaison syntaxique dans la IVe Recherche logique de Husserl. Recherches Husserliennes 21:3–33
Bundgaard P, Egholm J, Skov M (eds) (2003) Kognitiv semiotik. Haase, København
Bundgaard P, Østergaard S, Stjernfelt F (2006) Water proof fire stations? Conceptual schemata and cognitive operations involved in compound constructions. *Semiotica* 161(1/4):363–393
Burch RW (1991) A Peircean reduction thesis, Texas Tech University Press, Lubbock
Caldwell B (2004) Hayek's challenge. An intellectual biography of F.A. Hayek. University of Chicago Press, Chicago
Cantor (1872) Über die Ausdehnung eines Satzes aus der Theorie der trigonometrischen Reihen, in Cantor 1966, 92–102
Cantor (1879–84) Über unendlich lineare Punktmannigfaltigkeiten in Cantor 1966, 139–246
Cantor G (1966) Gesammelte Abhandlungen mathematischen und philosophischen Inhalts. In: Zermelo E (ed) Inhalts. Olms, Hildesheim
Cariani P (1998) Life's journey through the semiosphere. Semiotica 120(3/4)
le Carré J (1965[1963]) The spy who came in from the cold. Gollancz, London
Cassirer E (1944) The concept of group and the theory of perception. Philosophy and Phenomenological Research 5:1–35 (original French version 1938)
Cassirer E (1945) Structuralism in modern linguistics. Word I(II), New York
Cassirer E (1991[1957]) Das Erkenntnisproblem in der Philosophie und Wissenschaft der neueren Zeit. Von Hegels Tod bis zur Gegenwart (1832–1932), vol IV. Georg Olms, Hildesheim
Chaitin G (1987) Algorithmic information theory. Cambridge University Press Cambridge, MA
Chalmers D (1996) The conscious mind. Oxford University Press, Oxford
von Clausewitz C (1963[1832]) Vom Kriege. Rowohlt, Hamburg
Cobb-Stevens R (1990) Being and categorial intuition. Review of Metaphysics 44:43–66
Cohen PJ (1966) Set theory and the continuum hypothesis, WA Benjamin New York
Cowan GA, Pines D, Meltzer D (1994) Complexity. Metaphors, models, and reality. Addison-Wesley, Reading, MA
Dauben J (1995) Peirce and history of science. In Ketner 1995, pp146–195
Deacon T (1997) The symbolic species. W.W. Norton & Co, New York
Deacon T (1999) The trouble with memes (and what to do about it), (unpublished working paper, Boston)
Deacon T (2003) Multilevel selection in a complex adaptive system: the problem of language origins. In Weber and Depew 2003, pp81–106
Deely J (1992) Semiotics and biosemiotics: are sign-science and life-science coextensive?. In Sebeok and Umiker-Sebeok 1992, pp45–76
Deely J (2001) Four ages of understanding. The first postmodern survey of philosophy from ancient times to the turn of the twenty-first century. Toronto University Press, Toronto
Deighton L (1986[1983]) Game, set, and match. Hutchinson, London
Diderichsen P (1966) Helhed og struktur. ['Whole and Structure'] Gad, Copenhagen

Dipert R (1995) Peirce's underestimated place in the history of logic: a response to Quine. In Ketner 1995, pp32–58

Dipert R (1997) Peirce's philosophical conception of sets. In Houser et al (eds) 1997

Dipert R (2004) Peirce's deductive logic: its development, influence, and philosophical significance. In Misak 2004, pp287–324

Dougherty CJ (1983) Peirce's phenomenological defense of deduction. In Freeman 1983, pp167–177

Drummond J (1988) Realism *versus* Anti-Realism: a Husserlian Contribution. In: Sokolowski R (ed) Studies in philosophy and the history of philosophy, vol 18. Catholic University of America Press, Washington, DC, pp87–106

Dulles A (1963) The craft of intelligence. Harper and Row, New York

Dumézil G (1979) Interview in *Ornicar!* vol 19. Paris

Eco U (1976) A theory of semiotics. Indiana University Press, Bloomington

Eco U (1989) Foucault's pendulum. Secker and Warburg, London

Eco U (1992) Interpretation and overinterpretation. Cambridge University Press, Cambridge, MA

Eco U (1999) Kant and the platypus. Secker, London

Emmeche C (1997) Aspects of complexity in life and science. Philosophica 59(1):41–68

Emmeche C (2001) Does a robot have an Umwelt? Reflections on the qualitative biosemiotics of Jakob von Uexküll. Semiotica 134(1/4):653–693

Emmeche C (2002) The chicken and the Orphean egg: on the function of meaning and the meaning of function. Sign systems studies 30(1):15–32

Emmeche C (2004) Causal processes, semiosis, and consciousness. In: Seibt J (ed) Process theories: crossdisciplinary studies in dynamic categories. Kluwer, Dordrecht, pp313–333

Emmeche C (2004b) A-life, organism and body: the semiotics of emergent levels. In: Bedeau M, Husbands P, Hutton T, Kumar S, Suzuki, H (eds) Workshop and tutorial proceedings. Ninth international conference on the simulation and synthesis of living systems (Alife IX), Boston, MA, pp117–124

Emmeche C, Hoffmeyer J (1991) From language to nature – the semiotic metaphor in biology. Semiotica 84(1/2):1–42

Emmeche C, Hoffmeyer J (1992) Code-duality and the semiotics of nature. In: Anderson M, Merrell F (eds) Semiotic modeling. Gruyter, Berlin

Emmeche C, Køppe S, Stjernfelt F (1997) Explaining emergence: towards an ontology of levels. Journal for general philosophy of science 28(1):83–119

Emmeche C, Køppe S, Stjernfelt F (2000) Levels, emergence, and three versions of downward causation. In: Andersen PB et al (eds) Downward causation. Aarhus University Press, Copenhagen, pp13–34

Emmeche C, Kull K, Stjernfelt F (2002) Reading Hoffmeyer, rethinking biology. University of Tartu Press, Tartu

Engel-Tiercelin C (1992) Vagueness and the unity of C.S. Peirce's Realism. Transactions of the Charles S. Peirce Society XXVIII(1):51–82

Eriksen J-M, Stjernfelt F (2003) Hadets anatomi. Rejser i Bosnien og Serbien ['Anatomy of hatred. Travels in Bosnia and Serbia'] Lindhardt og Ringhof, København

Eriksen J-M, Stjernfelt F (2004) Krigens scenografi. Nye rejser i Bosnien og Serbien ['Scenography of war. New travels in Bosnia and Serbia'] Lindhardt og Ringhof, København

Esmann J (1995) Sort i sort - da Gud reinkarnerede i modernismen. Kandinsky, Malevich, Reinhardt ['Black in Black - when God reincarnated in modernism']. Kritik 119, Copenhagen

Falk EH (1981) The poetics of Roman Ingarden. University of North Carolina Press, Chapel Hill

Fastrup A (2007) Sensibilitetens bevægelse - Denis Diderots fysiologiske æstetik ('The Movement of Sensibility – the Physiological Aesthetics of Denis Diderot') Museum Tusculanum, Copenhagen

Fauconnier G, Turner M (2002) The way we think. Conceptual blending and the mind's hidden complexities. Basic Books, New York

Ferguson N (ed) (1997) Virtual history. Alternatives and counterfactuals. Picador, London

Fisch M (1967) Peirce's progress from nominalism towards realism, in Fisch 1986, pp184–200

Fisch M (1986) (eds. K.L. Ketner and C.J.W. Kloesel) Peirce, Semeiotic, and Pragmatism. Indiana University Press, Bloomington

Fischer E (1993) C.W. Eckersberg. Edition Bløndal, Copenhagen
Franck D (1999) Auswirkungen der uexküllschen Umweltlehre auf die moderne Verhaltensbiologie. Fol Baer VII, Tartu pp81–91
Freeman E (ed) (1983) The relevance of Charles Peirce. The Hegeler Institute, La Salle, IL
Freud S (1948[1912]) Ratschläge für den Arzt bei der psychoanalytischen Behandlung. In: Zentralblatt für Psychoanalyse, vol II. s483–489; in Freud A et al (eds) (1948) Gesammelte Werke vol VIII. S.Fischer Verlag, Frankfurt/M, s376–387
Friedman M (2000) The parting of the ways: Carnap, Cassirer, and Heidegger. Open Court, Chicago
Friedman M (2001) Dynamics of reason. CSLI Publications, Stanford
Gadamer H-G (1960) Wahrheit und Methode. J.C.B. Mohr, Tübingen
Gardies J-L (1985[1975]) Rational grammar. Analytica, München
Gell-Mann M (1995) Complex adaptive systems. In: Morowitz H, Singer J (eds) The mind, the brain, and complex adaptive systems, Santa Fe Institute Proc. vol XXII. Addison Wesley, Reading, MA
Glasgow J et al (eds) (1995) Diagrammatic reasoning. Cognitive and computational perspectives, AAAI Press/MIT Press, Menlo Park, CA
Gödel K (1986) Collected works I, (eds Feferman et al), Oxford University Press New York
Godfrey-Smith P (1996) Complexity and the function of mind in nature. Cambridge University Press, New York
Goel V (1995) Sketches of thought. MIT Press, Cambridge, MA
Goodman N (1976) Languages of art. Hackett Publishing Co, Indianapolis
Gould SJ (1972) Punctuated equilibria: an alternative to phyletic gradualism. In: Schopf TJM (ed) Models in paleobiology. Freeman, Cooper & Co.,San Francisco, pp82–115
Grassl W, Smith B (1986) A theory of Austria. In: Nyíri JC (ed) From Bolzano to Wittgenstein: the tradition of Austrian philosophy, Hölder-Pichler-Tempsky and Reidel, Vienna and Dordrecht, pp11–30
Grassl W, Smith B (1986) (eds) Austrian economics. Historical and philosophical background. New York, New York University Press
Greaves M (2002) The philosophical status of diagrams. CSLI Publications, Stanford
Gregersen F (1992) Sociolingvistikkens (u)mulighed. ['The (Im)Possibility of Socio Linguistics'] Tiderne Skifter, Copenhagen
Greimas AJ (1976) Maupassant. Seuil, Paris
Greimas AJ (1979) Sémiotique. Seuil, Paris
Greimas AJ (1982) Du sens II. Seuil, Paris
Groupe μ (1992) Traité du signe visuel. Seuil, Paris
Haack S (1992) Extreme scholastic realism: its relevance to philosophy of science today. Transactions of the Charles S. Peirce Society XXVIII(1):19–50
Haack S (1993) Peirce and logicism: notes towards an exposition. Transactions of the Charles S. Peirce Society XXIX(1):33–56
Haack S (1997) The first rule of reason. In Brunning and Forster 1997, pp241–261
Haaparanta L (2001) On Peirce's methodology of logic and philosophy. In: Ylikoski P, Kiikkeri M (eds) Explanatory connections. Electronic essays dedicated to matti sintonen (http://www.valt.Helsinki.fi/kfil/matti/)
Haas, WP (1964) The conception of Law and the Unity of Peirces Philosophy. Fribourg, Switzerland: University Press
Hadamard J (1954[1945]) The psychology of invention in the mathematical field. Dover, Mineola, NY
Hamburger K (1968) Die Logik der Dichtung. Klett, Stuttgart
Hansen NG (2004) Dialog mellem læsere ['Dialogues between readers']. Kritik 172:15–24, Copenhagen
Hansen TI (2001) Tidens øje – Rummets blik ['The eye of time – the gaze of space'] Odense Universitetsforlag, Odense
Hansen TI (2004) Fæno... hvad for noget? Om forholdet mellem litteratur og fænomenologi ['Pheno... what? On the relation between literature and phenomenology']. Kritik 172:5–14, Copenhagen
Harnad S (ed) (1987) Categorical perception. Cambridge University Press, Cambridge
Harnad S (1987) Category Induction and Representation. In Harnad 1987, pp1–52
Haswell J (1977) Spies and spymasters. Thames and Hudson, London

Hawkins BS Jr (1997) Peirce and Russell: the history of a neglected 'controversy'. In Houser et al (1997), pp111–146

Hayek FA (1978) The primacy of the abstract. In: New studies in philosophy, politics, economics, and the history of ideas. Routledge and Kegan Paul, London

Hayes P (1995) Section Introduction. In Glasgow 1995

Heijenoort J van (1997) Logic as calculus and logic as language. In Hintikka 1997, pp233–39.

Hendricks V (2001) The convergence of scientific knowledge. A view from the limit. Kluwer, Dordrecth

Hendricks V (2006) Mainstream and formal epistemology. Cambridge University Press, New York

Hersh SM (1997) The dark side of Camelot. Little Brown, Boston

Hilpinen R (1995) Peirce on language and reference. In Ketner 1995, pp272–303

Hintikka J (1983) C.S. Peirce's 'First Real Discovery' and its contemporary relevance. In: Freeman E (ed) The relevance of Charles Peirce. The Hegeler Institute, La Salle, IL, pp107–118

Hintikka J (1997) The place of C.S. Peirce in the history of logical theory. In: Brunning J, Forster P (eds) The rule of reason. University of Torotno Press, Toronto, pp13–33 (also in Hintikka 1997, pp140–61)

Hintikka J (1997a) Lingua Universalis vs. Calculus Ratiocinator. An ultimate presupposition of twentieth-century philosophy. Kluwer, Dordrecht

Hintikka J (1998) What is abduction? The fundamental problem of contemporary epistemology. Transactions of the Charles S. Peirce Society XXXIV(3):503–34

Hintikka J (2003) The notion of intuition in Husserl. Revue internationale de philosophie, no. 2, juin 2003, pp169–192

Hitchcock A (1955) The man who knew too much (screenplay John Michael Hayes), Hollywood

Hjelmslev L (1928) Principes de grammaire générale. Høst og Søn, Copenhagen

Hjelmslev L (1943) Omkring sprogteoriens grundlæggelse. Munksgaard, Copenhagen (Eng. version *Prolegomena to a Theory of Language*, 2nd edn. The University of Wisconsin Press, Madison, 1969)

Hjelmslev L (1972[1934]) Sprogsystem og sprogforandring ['Language System and Language Change'], Nordisk Sprog- Og Kulturforlag, Copenhagen

Hjelmslev L (1975) Resumé of a theory of language. Nordisk Sprog- Og Kulturforlag, Copenhagen (approx. 1943)

Hjelmslev L (1985 Nouveaux essays. PUF, Paris

Hoffmann M (1999) Problems with Peirce's concept of abduction. Foundations of science 4:271–305

Hoffmann M (April, 2002) Erkenntnisentwicklung. Ein semiotisch-pragmatischer Ansatz, Habilitationsschrift an der Technischen Universität Dresden

Hoffmeyer J (1996) Signs of meaning in the universe. Indiana University Press, Bloomington

Hoffmeyer J (2003) Semiotic aspects of biology: Biosemiotics. In: Posner R, Robering K and Sebeok T (eds) Semiotics. A handbook of the sign theoretic foundations of nature and culture. Gruyter, Berlin, pp2643–2666

Hoffmeyer J (2005) Biosemiotik. En afhandling om livets tegn og tegnenes liv ['Biosemiotics. A Treatise on the Signs of Life and the Life of Signs', English version in preparation], Ries, Copenhagen

Hofstadter D (1986) Metamagical Themas. Penguin, Harmondsworth

Holenstein E (1972) Phänomenologie der Assoziation. Martinus Nijhoff, Den Haag

Holenstein E (1975) Roman Jakobsons phänomenologischer Strukturalismus. Suhrkamp, Frankfurt/M

Holenstein E (1976) Linguistik Semiotik Hermeneutik. Suhrkamp, Frankfurt/M

Hookway C (2002) '... a sort of composite photograph': Pragmatism, ideas, and schematism. Transactions of the Charles S. Peirce Society XXXVIII(1/2):29–46

Houser N (1997) Introduction: Peirce as a logician. In Houser, Roberts, van Evra (eds) 1997, pp1–22

Houser N, Roberts DD, van Evra J (eds) (1997) Studies in the Logic of Charles Sanders Peirce. Indiana University Press, Bloomington

Hulswit M (2002) From cause to causation. A Peircean viewpoint. Kluwer Academic Publications, Dordrecht

Humphrey N (1992) A history of mind. Harper Collins, New York

Husserl E (1939) Die Frage nach dem Ursprung der Geometrie als intentional-historisches Problem. Revue internationale de philosophie 1:203–225

Husserl E (1939a) Entwurf einer 'Vorrede' zu den 'Logischen Untersuchungen' 1913. Tijdschrift voor philosophie 1(1, 2):106–33, 319–339

Husserl E (1968) Briefe an Ingarden mit Erläuterungen und Erinnerungen an Husserl (ed Ingarden R) Martinus Nijhoff, Den Haag

Husserl E (1970) Logical investigations (trans Findlay JN) Routledge and Kegan Paul, London

Husserl E (1970a) Philosophie der Aritmetik, Hua XII. Martinus Nijhoff, Den Haag

Husserl E (1971) Ideen zu einer reinen Phänomenologie und Phänomenologischen Philosophie, drittes Buch [Ideen III] Husserliana vol V. Max Niemeyer, Tübingen

Husserl E (1973) Ding und Raum, Hua XVI. Martinus Nijhoff, Den Haag

Husserl E (1973a) Experience and judgment. Routledge and Kegan Paul, London

Husserl E (1975) Logische Untersuchungen [LU] I, Hua XVIII. Martinus Nijhoff, Den Haag

Husserl E (1979) Aufsätze und Rezensionen (1890–1910), Hua XXII, Martinus Nijhoff, Dordrecht

Husserl E (1980) Phantasie, Bildbewusstsein, Erinnerung, [PBE] Hua XXIII. Martinus Nijhoff, Dordrecht

Husserl E (1980a[1913]) Ideen zu einer reinen Phänomenologie und Phänomenologischen Philosophie, [Ideen] Max Niemeyer, Tübingen

Husserl E (1984) Logische Untersuchungen| [LU] II, I.-II. Teil (Text nach *Hua* XIX/1-2) Felix Meiner, Hamburg

Husserl E (1985[1939]) Erfahrung und Urteil [EU] Felix Meiner, Hamburg

Husserl E (1991) Ideen zu einer reinen Phänomenologie und Phänomenologischen Philosophie, zweites Buch [Ideen II] Husserliana vol IV. Max Niemeyer, Tübingen

Husserl E (1993[1937]) Das Gesetz der Fortpflanzung. in Die Krisis. Ergänzungsband, Husserliana vol. XXIX, Kluwer, Dordrecht, p317

Igambardiev AU (1992) Organization of biosystems: a semiotic approach. In Sebeok and Umiker-Sebeok 1992, pp125–44

Ingarden R (1925) Essentiale Frage. Ein Beitrag zu dem Wesensproblem. Jahrbuch für Philosophie und phänomenologische Forschung 7:125–304, Max Niemeyer, Band, Halle a.d. Saale

Ingarden R (1929) 'Bemerkungen zum Problem Idealismus-Realismus' Festschrift. Edmund Husserl zum 70. Geburtstag gewidmet. Max Niemeyer, Halle a.d. Saale, pp159–90

Ingarden R (1965[1931]) *Das literarische Kunstwerk* [LK], Max Niemeyer, Tübingen (Eng. version *The Literary Work of Art*, Northwestern University Press(1973), Evanston)

Ingarden R (1965–74) Der Streit um die Existenz der Welt I-III. Max Niemeyer, Tübingen

Ingarden R (1968) Vom Erkennen des literarischen Kunstwerk. Max Niemeyer, Tübingen (Eng. version *The Cognition of the Literary Work of Art*, Northwestern University Press(1973), Evanston)

Ingarden R (1969) Erlebnis, Kunstwerk and Wert. Wissenschaftliche Buchgesellschaft, Darmstadt

Ingarden R (1970) Innføring i Edmund Husserls Fenomenologi. Tanum, Oslo

Ingarden R (1976) Gegenstand und Aufgaben der Literaturwissenschaft. Max Niemeyer, Tübingen

Ingarden R (1975) On the Motives which led Husserl to Transcendental Idealism. Martinus Nijhoff, The Hague

Ingarden R (1998) Schriften zur Phänomenologie Husserls. Gesammelte Werke, Bd. 5 (ed Galewicz W), Max Niemeyer, Tübingen

Jakobson R (1971a) Selected writings I. Mouton, Berlin

Jakobson R (1971b) Selected writings II. Mouton, Berlin

Jakobson R (1985) Selected writings VII. Mouton, Berlin

Johansen JD (1993) Dialogic semiosis. Indiana University Press, Bloomington

Johansen JD (1996) Iconicity in literature. Semiotica 110(1/2):37–55

Johansen JD (2003) Iconizing literature. In: MüllerWG, Fischer O (eds) From sign to signing, John Benjamins, Amsterdam

Johansson A (2001) Modern topology and Peirce's theory of the continuum. Transactions of the Charles S. Peirce Society XXXVII(3):1–12

Johansson I (1998) Pattern as an ontological category. In: Guarino N (ed) Formal ontology in information systems, IOS Press, Amsterdam pp86–94

Jørgensen H, Stjernfelt F (1987) Substance, substrat, structure. Sur la controverse épistémologique qui a opposé Brøndal et Hjelmslev. Langages 86:79–94, Paris

Jørgensen KF (2005) Kant's schematism and the foundations of mathematics. Dissertation, Roskilde University Center (preprint)

Joswick H (1988) Peirce's mathematical model of interpretation. Transactions of the Charles S. Peirce Society XXIV:107–21

Kant I (1956[1781]) Kritik der reinen Vernunft. Felix Meiner, Hamburg

Kant I (1974[1790]) Kritik der Urteilskraft. Felix Meiner, Hamburg

Kant I (1993) Opus Postumum (ed Förster E), Cambridge University Press, Cambridge, MA

Kant I (1997[1786]) Metaphysische Anfangsgründe der Naturwissenschaft. Felix Meiner, Hamburg

Kauffman S (1993) The origin of order. Oxford University Press, New York

Kauffman S (2000) Investigations. Oxford University Press, Oxford

Keegan J (1993) A history of warfare. Vintage Books, London

Ketner KL (1985) How Hintikka misunderstood Peirce's account of theorematic reasoning. Transactions of the Charles S. Peirce Society XXI(3):407–18

Ketner KL (ed) (1995) Peirce and contemporary thought. Philosophical inquiries. Fordham University Press, New York

Ketner KL, Putnam H (1992) Introduction: The Consequences of Mathematics, in Peirce 1992

King James Bible 1611, http://www.biblegateway.com/

Kjørup S (2000) Kunstens filosofi: En indføring i æstetik ['Philosophy of Art: An Introduction to Aesthetics'], Roskilde Universitetsforlag, Roskilde

Kline M (1972) Mathematical thought, I-III. Oxford University Press, New York

Knudsen J (1999) Agent, men for hvem? ['Agent, but for whom?'], The Copenhagen daily *Information* 10(3)

Kreisel G (1982) Hilbert's programme. In Benacerraf P, Putnam H 1982

Kull K (1999) Biosemiotics in the twentieth century: a view from biology. Semiotica 127(1/4):385–414

Kusch M (1997) Husserl and Heidegger on Meaning. In Hintikka 1997, pp240–268.

Lachs J (2002) The insignificance of individuals. Transactions of the Charles S. Peirce Society XXXVIII(1/2):79–94

Lakoff G (1987) Women, fire, and dangerous things. Chicago University Press, Chicago

Lakoff G, Turner M (1989) More than cool reason. A field guide to poetic metaphor. University of Chicago Press, Chicago

Lakoff G, Johnson M (1999) Philosophy in the flesh. University of Chicago Press, Chicago

Lakoff G, Nuñez R (2001) Where mathematics comes from. How the embodied mind brings mathematics into being. Basic Books, New York

Lane R (1989) Peirce's triadic logic revisited. Transactions of the Charles S. Peirce Society XXV(2):284–311

Lane Robert (1997) Peirce's "Entanglement" with the Principles of Excluded Middle and Contradiction, in Transactions of the Cahrles S. Peirce Society, vol. XXXIII, no. 3, 680–703.

Lane Robert (1999) Peirce's Triadic Logic Revisited, in Transactions of the charles S. Peirce Society, vol. XXXV, no. 2, 284–311

Langton C (1997) Artificial life. An overview. The MIT Press, Cambridge, MA

Lévi-Strauss C (1974[1958]) Anthropologie structurale. Plon, Paris

Lévi-Strauss C (1971) L'homme nu. Mythologiques IV. Plon, Paris

Lévi-Strauss C (1973) Anthropologie structurale II. Plon, Paris

Lévi-Strauss C (1988) De près et de loin. Plon, Paris

Levy SH (1997) Peirce's theoremic/corollarial distinction and the interconnections between mathematics and logic. In Houser et al (1997), pp85–110

Libera A de (1996) La querelle des universaux. Seuil, Paris

Lohmar D (1987) Husserls Phänomenologie als Philosophie der Mathematik. Lohmar, Köln

Lohmar D (1990) Wo lag der Fehler der kategorialen Repräsentation? Zu Sinn und Rechweite einer Selbstkritik Husserls. Husserl Studies 7:179–197

Lohmar D (1998) Erfahrung und kategoriales Denken. Kluwer, Dordrecht

Lohmar D (2003) Husserl's concept of categorial intuition. In: Zahavi D, Stjernfelt F (eds) (2002) One hundred years of phenomenology. Kluwer, Dordrecht, pp125–46

de Loof A, Broeck J van den (1995) Communication: the key to defining 'life', 'death' and the force driving evolution, 'organic chemistry-based' versus 'artificial' life. Belg J Zool 125(1):45–76
MacLane S (1986) Mathematics. Function and form. Springer, New York
Maddy P (1990) Realism in mathematics. Oxford University Press, Oxford
Maddy P (1997) Naturalism in mathematics. Oxford University Press, Oxford
Marbach E (1980) 'Einleitung des Herausgebers' XXV-LXXXII. In Husserl 1980
Margulis L, Sagan D (1987) Microcosmos: four billion years of evolution from our microbial ancestors. HarperCollins, New York
Martin B (1994) The schema. In: Cowan M, Pines D, and Meltzer D (eds) Complexity. Santa Fe Institute Proc. vol XIX. Addison Wesley, Reading, MA
de Maupassant G (2000) Deux amis. In: Textes choisis. Gallimard, Paris.
Marty A (1976[1908]) Untersuchungen zur Grundlegung der allegemeinen Grammatik und Sprachphilosophie. Georg Olms Verlag, Hildesheim
May M (1995) Diagrammatisches Denken: Zur Deutung logischer Diagramme als Vorstellungsschemata bei Lakoff und Peirce. In Zeitschrift für Semiotik, Band 17, Heft 3–4.
May M (1999) Diagrammatic reasoning and levels of schematization. In: Skov M et al (eds) Iconicity. NSU, Copenhagen
May M, Stjernfelt F (1996) 'Måling, diagram, kunst'/'Measurement, Diagram, Art', Danish/English version in *Billeder fra det fjerne - videnskabelig visualisering. En antologi* (eds. Michelsen and Stjernfelt) Akademisk, Copenhagen, pp63–76, 191–204
McEwan I (1997[1987]) The innocent. Picador, London
Medin DL, Barsalou LW (1987) Categorization processes and categorical perception. In Harnad 1987
Melle U (2002) Husserl's revision of the sixth logical investigation. In Zahavi and Stjernfelt 2002
Merleau-Ponty M (1983[1945]) Phénoménologie de la perception. Gallimard, Paris
Merleau-Ponty M (1964) Le visible et l'invisible. Gallimard, Paris
Merleau-Ponty M (1995) La nature. Notes. Cours de Collège de France (ed Dominique Séglard), Seuil, Paris
Merrell F (1992) As signs grow, so life goes. In Sebeok and Umiker-Sebeok 1992
Merrell F (1997) Peirce, signs, and meaning. University of Toronto Press, Toronto
Mitchell WJT (Spring, 1981) Diagrammatology. Critical inquiry 7(3):622–33
Misak C (2004) The Cambridge companion to Peirce. Cambridge University Press, Cambridge, MA
Misfeldt M (2006) Mathematical writing. PhD dissertation, Danish University of Education (preprint)
Molière JP de (1962) Œuvres complètes. Seuil, Paris
Moschovakis Y (1980) Descriptive set theory. North-Holland Publishing Company, New York, NY
Moynihan DP (1998) Secrecy. The American experience. Yale University Press, New Haven
Mullin AA (1966) C.S. Peirce and E.G.A. Husserl on the Nature of Logic. Notre Dame Journal of Formal Logic 7(4):301–304
Münch D (1993) Intention und Zeichen. Suhrkamp, Frankf./M
Murphey M (1961) The development of Peirce's philosophy. Cambridge University Press, Harvard
Myrvold WC (1995) Peirce on Cantor's paradox and the continuum. Transactions of the Charles S. Peirce Society XXXI(3):508–41
Nelsen RB (1993) Proofs without words. Mathematical Association of America, Washington, DC
Nesher D (2001) Peircean epistemology of learning and the function of abduction as the logic of discovery. Transactions of the Charles S. Peirce Society XXXVII(1):23–58
Nesher D (2002) Peirce's essential discovery: Our senses as reasoning machines Can quasi-prove our perceptual judgments. Transactions of the Charles S. Peirce Society XXXVIII(1/2):175–208
Neurath O (1939) Modern man in the making. Alfred A. Knopf, New York
Niiniluoto I (1999) Defending abduction. Philosophy of Science 66 (Supplement: Proceedings of the 1998 Biennial Meetings of the Philosophy of Science):436–451
Noble NAB (1989) Peirce's definitions of continuity and the concept of possibility. Transactions of the Charles S. Peirce SocietyXXV(2):149–174
Nolan R (1994) Distinguishing perceptual from conceptual categories. In: Casati R, Smith B, White G (eds) Philosophy and the cognitive sciences. Holder-Pichler-Temsky, Vienna, pp221–231

Nöth W (2000) Handbuch der Semiotik. J.B. Metzler, Stuttgart

Øhrstrøm P (2001) Mesterargumentet i modlys ['The Master Argument in Backlight'] In: Thellefsen TL (ed) Tegn og betydning. Betydningsdannelse i filosofisk, biologisk og semiotisk perspektiv. Akademisk Forlag, Copenhagen pp201–221

Øhrstrøm P, Andersen J, Schärfe H (2005) What has happened to ontology?. In: Dau F, Mugnier M-L, Stumme C (eds) Common Semantics for sharing knowledge ICCS 2005, LNAI 3598, Springer Verlag, Berlin-Heidelberg, pp425–438

Øhrstrøm P, Hasle P (1995) Temporal logic, Kluwer, Dordrecht

Otte M (1997) Analysis and synthesis in mathematics from the perspective of Charles S. Peirce's philosophy. In: Otte M, Panza M (eds) Analysis and synthesis in mathematics. History and philosophy. Kluwer, Dordrecht, pp327–356

Panza M (1992) De la Continuité chez Aristote, sa Réponse à Zénon. In Salanskis and Sinaceur 1999, pp3–15

Pape, H (1997) The logical structure of idealism: C.S. Peirce's search for a logic of mental proccsses, In Brunning J, Forster P (eds.) (1997)

Pape H (1999) Abduction and the topology of human cognition. Transactions of the Charles S. Peirce Society XXXV(2):248–269

Pattee HH (1977) Dynamic and linguistic modes of complex systems. Int J General Systems 3 pp259–266

Pattee HH (1979) The complementarity principle and the origin of macromolecular information. BioSystems 11 pp217–226

Pattee HH (1988) Simulations, realizations, and theories of life. In: Langton C (ed) Artificial life. Addison-Wesley, Reading, MA pp63–77

Parker K (1992) The principle of continuity in Charles S. Peirce's phenomenology and semeiotic. PhD Dissertation, Vanderbilt University, printed from microfilm by University Microfilms International, Ann Arbor

Parker K (1998) The continuity of Peirce's thought. Vanderbilt University Press, Nashville

Peckhaus V (2004) Calculus Ratiocinator vs. Characteristica Universalis? The two traditions in logic, revisited. History and Philosophy of Logic 25, pp3–14

Peirce CS (1998[1931–58]) Collected papers [CP, references given by volume number and paragraph], I-VIII, (ed Hartshorne, Weiss, Burks) Thoemmes Press, London

Peirce CS (1976) New elements of mathematics [NEM], (ed Eisele C) I–IV, Mouton, The Hague

Peirce CS (1978) Contributions to the Nation [NATION] Texas Tech Press, Lubbock

Peirce CS (1992) Reasoning and the Logic of Things [RLOT], eds. Ketner K, Putnam H, Harvard University Press Cambridge, MA

Peirce CS (1992) The Essential Peirce, vol I. (1867–1893) [EPI] (eds Houser N, Kloesel C), Indiana University Press, Bloomington

Peirce CS (1997) Pragmatism as a Principle [PP], (ed Turrisi A), SUNY Press, Albany

Peirce CS (1998) The Essential Peirce, vol II. (1893–1913) [EPII] (eds Houser N, Kloesel C), Indiana University Press, Bloomington

Peirce CS (undated) Logic, Considered as Semeiotic [LCS], constructed from manuscript L 75 by Joseph Ransdell (http://members.door.net/arisbe/menu/library/bycsp/l75/ver1/l75v1-01.htm)

Peirce CS Manuscripts from Peirce's unpublished papers (cited by permission of the Houghton Library, Harvard University), manuscript numbers referring to Robin 1967 as [MS CSP XX], when discussed in the text referred to as Robin nr. XX

Penrose R (1990) The emperor's new mind. Oxford University Press, Oxford

Penrose R (1994) Shadows of the mind. Oxford University Press, Oxford

Peruzzi A (1999) An essay on the notion of schema. In: Albertazzi L (ed) Shapes of forms. Kluwer, Dordrecht, pp191–244

Petitot J, Varela FJ, Pachoud B, Roy J-M (eds) (1999) Naturalizing phenomenology. Issues in contemporary phenomenology and cognitive science. Stanford University Press, Stanford

Petitot J (1985) Morphogenèse du sens. PUF, Paris

Petitot J (1985a) Catastrophes de la parole. Maloine, Paris

Petitot J (1992) Physique du sens. Éditions du CNRS, Paris

Petitot J (1992a) Continu et objectivité. La Bimodalité Objective du Continu et le Platonisme Transcendantal, in Salanskis and Sinaceur 1992, pp239–266

Petitot J Pour un platonisme transcendantal. In: Panza M, Salanskis J-M (eds) L'objectivité mathématique. Masson, Paris

Petitot J (1999) Morphological eidetics for a phenomenology of perception. In Petitot et al (1999), pp330–371

Petitot J, Smith B (1990) New foundations for qualitative physics. In: Tiles JE, McKee GT, Dean CG (eds) Evolving knowledge in natural science and artificial intelligence. Pitman, London, pp231–249

Pfänder A (1921) Logik. Max Niemeyer, Halle

Pietarinen A-V (2006) Signs of Logic. Peircean Themes on the Philosophy of Language, Games, and Communication. Manuscript (in press)

Pihlström S (1998) Peircean scholastic realism and transcendental arguments. Transactions of the Charles S. Peirce Society XXXIV(2):382–413

Poli R (1998) Levels. Axiomathes 9(1–2):197–211

Potter V, Shields PB (1977) Peirce's definitions of continuity. Transactions of the Charles S. Peirce Society XIII(1):20–34

Price HH (1953) Thinking and experience. Hutchinson's University Library, New York

Prodi G (1988) Signs and codes in immunology. In: Sercarz et al (eds) The semiotics of cellular communication in the immune system. Springer, Berlin

Putnam H (1992) Comments on the lectures. In Peirce 1992

Putnam H (1975a[1967]) Mathematics without foundations. In Putnam 1975

Putnam H (1975) Mathematics, matter and method. Philosophical papers, vol 1. Cambridge University Press, Cambridge, MA

Quine WvO (1995) Peirce's Logic. In Ketner 1995, pp23–31

Ransdell J (1989) Is Peirce a phenomenologist?. http://members.door.net/arisbe/homepage/ransdell.htm

Rasmussen M (1992) Hjelmslevs sprogteori ['Hjelmslev's Theory of Language'], Odense Universitetsforlag, Odense

Resnik M (1997) Mathematics as a science of patterns. Clarendon Press, Oxford

Ricœur P (1983–85) Temps et récit I-III. Seuil, Paris

Roberts DD (1973) The Existential Graphs of Charles S. Peirce. Mouton, The Hague

Robin R (1967) Annotated catalogue of the papers of Charles S. Peirce. University of Massachusetts Press, Worcester, MA

Robin R (1997) Classical pragmatism and pragmatism's proof. In Brunning and Forster 1997, pp139–152

Rollinger R (1999) Husserl's position in the school of Brentano. Kluwer, Dordrecht

Rosch E (1978) Principles of categorization. In Rosch and Lloyd (eds) 1978

Rosch E, Lloyd B (eds) (1978) Cognition and categorization. Lawrence Erlbaum, Hillsdale, NJ

Rosen R (1989) The roles of necessity in biology. In: Casti J, Karlqvist A (eds) Newton to Aristotle. Towards a theory of models for living systems. Birkhäuser, Boston

Rosen R (1990) Life itself. Columbia University Press, New York

Rosenthal SB (1968) The 'Would-Be' present of C.S. Peirce. Transactions of the Charles S. Peirce Society IV(3):155–62

Rothbard M (January, 1957) In defense of 'Extreme Apriorism'. Southern Economic Journal 314–320 (www.mises.org/rothbard/extreme.pdf)

Roy J-M (1999) Saving intentional phenomena: intentionality, representation, and symbol. In Petitot et al (1999), pp111–147

Ruttkowski WV (1978) Typen und Schichten. Zur Einteilung des Menschen und seiner Produkte. Franke Verlag, Bern und München

Salanskis J-M, Sinaceur H (eds) (1992) Le labyrinthe du continu, Springer. Paris

Salanskis J-M (1992) Le Destin du Modèle de Cantor-Dedekind. In Salanskis J-M and Sinaceur H 1992, pp190–212

Schank RC, Abelson RP (1977) Scripts, plans, goals and understanding: an inquiry into human knowledge structures. L. Erlbaum, Hillsdale, NJ

Schmitt C (1963[1932]) Das Begriff des Politischen. Duncker & Humblot, Berlin

Schmitt C (1993[1922]) Politische Theologie. Duncker & Humblot, Berlin
Schmitt C (1970) Politische Theologie II. Duncker & Humblot, Berlin
Scubla L (1998) Lire Lévi-Strauss. Odile Jacob, Paris
Schuhmann K (1972) Forschungsnotizen über Husserls 'Entwurf einer 'Vorrede' zu den 'Logischen Untersuchungen'. Tijdschrift voor filosofie 34(3):513–24
Sebeok T (ed) (1986) Encyclopedic dictionary of semiotics I-III. Mouton de Gruyter, Berlin
Sebeok T (1991) A sign is just a sign. Indiana University Press, Bloomington
Sebeok T (1992) 'Tell me, where is fancy bred': the biosemiotic self. In Sebeok and Umiker-Sebeok
Sebeok T (2001) Global semiotics. Indiana University Press, Bloomington and Indianapolis
Sebeok T, Umiker-Sebeok J (eds) (1992) Biosemiotics. The Semiotic Web 1991. Mouton de Gruyter, Berlin
Seebohm T (1990) Kategoriale Anschauung. In: Seebohm et al (eds) Logik, Anschaulichkeit und Transparenz. Karl Alber, München, pp9–47
Selz O (1913) Über die Gesetze des geordneten Denkverlaufs. Verlag von W. Spemann, Stuttgart
Shin S-J (1994) The logical status of diagrams. Cambridge University Press, Cambridge
Shin S-J (2000) The iconic logic of Peirce's graphs. MIT Press, Cambridge, MA
Short TL (1983) Peirce and the incommensurability of theories. In Freeman 1983, pp119–131
Short TL (1997) Hypostatic abstraction in self-consciousness. In: Brunning J and Forster P (eds) The rule of reason. University of Toronto Press, Toronto
Short TL (2004) The development of Peirce's theory of signs. In Misak 2004, pp214–240
Skagestad P (2004) Peirce's semiotic model of the mind. In Misak 2004, pp241–256
Sloman A (1995) Musings on the roles of logical and non-logical representations in intelligence. In Glasgow 1995
Smith B (ed) (1982) Parts and moments. Philosophia, München
Smith B (1979) Roman Ingarden: ontological foundations for literary theory. In: Odmark J (ed) Language, literature & meaning I. Johns Benjamins, Amsterdam, pp373–390
Smith B (1980) Ingarden vs. Meinong on the logic of fiction. Philosophy and Phenomenological Research 16:93–105
Smith B (1983) Meinen und Vorstellen in der literarischen Gegenstandskonstitution. In: Wolandt G (ed) Kunst und Kunstforschung. Beiträge zur Ästhetik. Bouvier, Bonn, pp49–61
Smith B (1992) An essay on material necessity. In: Hanson P, Hunter B (eds) Return of the a priori (Canadian Journal of Philosophy, Supplementary vol 18)
Smith B (1994) Austrian philosophy. Open Court, Chicago
Smith B (1996a) The ecological niche, paper, SUNY (German version: Ontologie des Mesokosmos: Soziale Objekte und Umwelten, *Zeitschrift für philosophische Forschung*, 52 (1998), 521–540
Smith B (1996b) In defense of extreme (fallibilistic) apriorism. Journal of Libertarian Studies 12:179–192
Smith B (1998a) (with Roberto Casati and Achille Varzi) Ontological tools for geographic representation. In: Guarino N (ed) Formal ontology in information systems. IOS Press (Frontiers in Artificial Intelligence and Applications), Amsterdam, etc. pp77–85
Smith B (1998b) (with Achille Varzi) The niche. Nous 33(2):198–222
Smith B (2000) Logic and formal ontology. Manuscrito 23:275–323
Smith B (2001) Objects and their environments: from Aristotle to ecological ontology, paper, SUNY (to appear in A. Frank (ed.) The life and motion of socio-economic units. Taylor and Frank, London)
Smith B (2003) Ontology. In: Floridi L (ed) Blackwell Guide to the Philosophy of Computing and Information. Blackwell, Oxford, pp155–166
Smith B (2005) Against fantology. In: Reicher M, Marek J (eds) Experience and analysis. ÖBV & HPT, Vienna, pp153–170
Smith B, og Smith D (eds) (1995) The Cambridge companion to Husserl. Cambridge University Press, Cambridge
Sokolowski R (1964) The formation of Husserl's concept of constitution, Martinus Nijnott, The Hague
Sokolowski R (1974) Husserlian meditations. Northwestern University Press, Evanston
Sokolowski R (1982) Husserl's concept of categorial intuition. In: Mohanty JN (1982, ed) Phenomenology and the human sciences 1981. Norman, Oklahoma

Sokolowski R (2000) Introduction to phenomenology. Cambridge University Press, Cambridge, pp127–141

Sonesson G (1989) Pictorial concepts. Lund University Press, Lund

Sørensen BA (1963) Symbol und Symbolismus in den ästhetischen Theorien des 18. Jahrhunderts und der deutsche Romantik.Munksgaard, Copenhagen

Sowa J (1997) Matching logical structure to linguistic structure. In Houser, Roberts, Van Evra (eds) 1997

Sowa J (1999) Knowledge representation: logical, philosophical, and computational foundations. Brooks Cole Publishing Co, Pacific Grove, CA

Spiegelberg H (1956) Husserl's and Peirce's phenomenologies: coincidence or interaction. Philosophy and Phenomenological Research 17:164–185

Spiegelberg H (1969) The Phenomenological Movement. A Historical Introduction, 2 vols. Martinus Nijhoff, The Hague

Sponsel R, Rathsmann-Sponsel I (2000) Kekulés Traum. Über eine typisch-psychoanalytische Entgleisung Alexander Mitscherlichs über den bedeutenden Naturwissenschaftler und Chemiker August Kekulé (1829–1896), Mitschöpfer der Valenz, Vollender der Strukturtheorie und Entdecker der Bedeutung des Benzolrings. Alternative Analyse und Deutung aus allgemeiner und integrativer psychologisch-psychotherapeutischer Sicht. IP-GIPT. Erlangen: http://www.sgipt.org/th_schul/pa/kek/pak_kek0.htm

Stepanov JS (1971) Semiotika. Nauka, Moscow

Sterelny K, Griffiths PE (1999) Sex and death. An introduction to the philosophy of biology. The University of Chicago Press, Chicago

Stjernfelt F (1990) Baldr og verdensdramaet i den nordiske mytologi [The God Baldr and the drama of the world in old norse mythology]. Museum Tusculanum, Copenhagen

Stjernfelt F (1992a) Formens betydning. Katastrofeteori og semiotik [The meaning of form. Catastrophe theory and semiotics]. Akademisk Forlag, Copenhagen

Stjernfelt F (1992b) Categorical perception as a basic prerequisite to the formation of signs? in Sebeok and Umiker-Sebeok 1992, pp427–454

Stjernfelt F (1992c) Buchstabenformen, Kategorien und die Apriori Position. In: Gumbrecht and Pfeiffer *Schrift.* Wilhelm Fink, München, pp289–310

Stjernfelt F (1997) Rationalitetens himmel og andre essays ['The heaven of rationality and other essays']. Gyldendal, Copenhagen

Stjernfelt F (1997a) Wie ist Form möglich? The concept of nature in Kant: the third critique and the opus postumum. Prismer 14(1):43–58, Århus

Stjernfelt F (1997b) Den narrative kamp og dens beskrivelse. In: Rationalitetens himmel. Gyldendal, Copenhagen (French version: Récits de l'âgon: leur description linguistique. In: Michèle P (ed) Quand parlent les formes, vol 2. C.N.R.S., Paris, 1995)

Stjernfelt F (2000) Die Vermittlung zwischen Anschuung und Denken bei Kant, Cassirer und Peirce. In: *Zeitschrift für Semiotik,* Band 22, Heft 3–4, pp341–368

Stjernfelt F (2001) The vulgar metaphysics of transgression. Text und Kontext 23(1):144–155

Stjernfelt F (2003) Sebeotics at the threshold. Review of Ponzio and Petrilli *Thomas Sebeok. An Introduction.* Semiotica 147(1/4):485–494

Stjernfelt F (2004) Traffic jam, or objectivity and interdisciplinarity in the humanities. In: Jensen JHC (ed) The object of study in the humanities. Museum Tusculanum Press Copenhagen, pp69–78

Stjernfelt F (2006) Two iconicity notions in Peirce's diagrammatology. In: Proceedings from 6th international conference on conceptual structures. Springer Verlag, pp70–86

Stjernfelt F (2006a) Let us not get too far ahead of the story... A history of realist semiotics? Review of John Deely: Four ages of understanding, in Cybernetics and Human Knowing 13.1, pp91–104

Stjernfelt F (in press a) Simple animals and complex biology. The double von Uexküll inspiration in Cassirer's philosophy. In: Synthese

Stjernfelt F (with N Zeuthen) (in press b) The representation of consciousness in language and fiction. A cognitive theory of enunciation. In: Semiotica 165(1/4)

Stone A (2004) Heidegger and Carnap on the overcoming of metaphysics, unpublished (http://home.uchicago.edu/~abestone/papers.html)

Ströker E (1978) Husserls Evidenzprinzip. In: Zeitschrift für philosophische Forschung, Bd. 32:3–30

Stumpf C (1873) Über den psychologischen Ursprung der Raumvorstellung. Hirzel, Leipzig

Stumpf C (1907) *Zu Einteilung der Wissenschaften*, in *Philos.-histor. Abh.* 1906, 5.

Talmy L (2000) Toward a cognitive semantics, vol 1–2. MIT Press, Cambridge, MA

Thom R (1972) Stabilité Structurale et Morphogénèse. Ediscience, Paris

Thom R (1975) English version of Thom 1972: Structural Stability and Morphogenesis. Benjamin, Reading, MA

Thom R (1980) Modèles mathématiques de la morphogenèse. Chr. Bourgois, Paris

Thom R (1988) Ésquisse d'une sémiophysique, InterEditions, Paris

Thom R (1990) Causality and finality in theoretical biology: a possible picture. In: Casti J, Karlqvist A (eds) Newton to Aristotle. Towards a theory of models for living systems. Birkhäuser, Boston

Thom R (1992) L'Antériorité Ontologique du Continu sur le Discret. In Salanskis and Sinaceur, pp137–143

Thompson, d'Arcy W (1961 [1917/1942]) On growth and form (abridged version, Bonner JT (ed)). Cambridge University Press, Cambridge

Thomsen SU, Stjernfelt F (2005) Kritik af den negative opbyggelighed ['A critique of negativism']. Vindrose, Copenhagen

Tiercelin C (1993) Peirce's realistic approach to mathematics: or, can one be a realist without being a platonist. In: Edward CM (ed) Charles S. Peirce and the Philosophy of Science (papers from the Harvard Sesquicentennial Congress). University of Alabama Press, Tuscaloosa, pp30–48

Tiercelin C (1997) Peirce on norms, evolution and knowledge. Transactions of the C.S. Peirce Society 33(1):35–58

Tiercelin C (January, 1999) L'influence scotiste dans le projet peircien d'une métaphysique scientifique. *Revue des Sciences Philosophiques et Théologiques*, 83(1):117–134

Tiercelin C (2002) Dispositions and Essences, paper from the colloquium 'Dispositions et pouvoirs causaux', Paris X-ENS Ulm, Sept. 2002

Tiles JE (1988) Iconic thought and the scientific imagination. Transactions of the Charles S. Peirce Society 24(2):161–178

Tomasello M (1999) The cultural origins of human cognition. Harvard University Press, Cambridge, MA

Tunander O (1994) Den usynlige hånd og den hvide – en læsning af Palme-mordet [The invisible hand and the white hand – an interpretation of the Palme murder]. Kritik 110, Copenhagen

Tunander O (2004) The secret war against Sweden – US and British submarine deception in the 1980s. Frank Cass, London

Turner M (1996) The Literary Mind. Oxford University Press, New York

Uexküll Jv (1933) Staatsbiologie. Hanseatische Verlagsanstalt, Hamburg

Uexküll Jv (1973) Theoretische Biologie. Suhrkamp, Frankf/M (1920/28)

Uexküll, Jv (1982) The theory of meaning (transl. of Bedeutungslehre (1940)). Semiotica 42(1):1–24

Uexküll Tv (1982) Introduction: meaning and science in Jakob von Uexküll's concept of biology. Semiotica 42(1):1–24

Uldall H-J (1967) Outline of glossematics. Nordisk Sprog- og Kulturforlag, Copenhagen

Visetti Y-M (2004) Le continu en sémantique, in Texto! (online magazine), http:/www.revue-texto.net/Inedits/Visetti_Continu.html

Waddington CH (1968–1972) Towards a theoretical biology, vol 1–3. Aldine Publishing, Chicago

Wang ONC (1997) The Last Formalist, or W.J.T. Mitchell as Romantic Dinosaur, Interview with Mitchell by O.N.C. Wang, In: Wang ONC, Morillo J (series ed) Romantic circles praxis series. (http://www.rc.umd.edu/praxis/mitchell/mitch-about.html)

Weber B, Depew DJ (eds) (2003) Evolution and learning. MIT Press, Cambridge, MA

Weber M (1972 [1921]) Wirtschaft und Gesellschaft. J.C.B. Mohr, Tübingen

Wellek R (1991) A history of modern criticism 1750–1950, vol 7: German, Russian, and Eastern European Criticism, 1900–1950. Yale UP, New Haven

West N (ed) (1993) The Faber book of espionage. Faber, London

Willard D (1982) Wholes, parts, and the objectivity of knowledge. In Smith B (ed) 1982, p 379

Wilson EO (1998) Consilience. The unity of knowledge. Little, Brown, & Co., London

Zahavi D (1996) Husserl und die transzendentale Intersubjektivität. Kluwer, Dordrecht

Zahavi D (1999) Self-awareness and alterity. A phenomenological investigation. Northwestern University Press, Evanston

Zahavi D, Stjernfelt F (eds) (2002) One Hundred Years of Phenomenology. Kluwer, The Hague

Zelaniec W (1996) The recalcitrant synthetic a priori. ArTom, Lublin

Zeman JJ (1968) Peirce's graphs – the continuity interpretation. *Transactions of the Charles S. Peirce Society* 4(3):144–154

Zeman JJ (1983) Peirce on abstraction. In: Freeman E (ed) The relevance of Charles Peirce. The Hegeler Institute, La Salle, IL

Zeman JJ (1997) Peirce and Philo. In Houser et al (eds) (1997), Studies in the logic of Charles Sanders Peirce, Indiana University Press, Bloomington pp402–417

NAME INDEX

(E. Husserl and C. Peirce are not included)

SUBJECT INDEX

(much-used terms such as diagram, phenomenology, ontology, semiotics, etc. are not included)

SYNTHESE LIBRARY

1. J. M. Bochénski, *A Precis of Mathematical Logic.* Translated from French and German by O. Bird. 1959 ISBN 90-277-0073-7
2. P. Guiraud, *Problèmes et méthodes de la statistique linguistique.* 1959 ISBN 90-277-0025-7
3. H. Freudenthal (ed.), *The Concept and the Role of the Model in Mathematics and Natural and Social Sciences.* 1961 ISBN 90-277-0017-6
4. E. W. Beth, *Formal Methods.* An Introduction to Symbolic Logic and to the Study of Effective Operations in Arithmetic and Logic. 1962 ISBN 90-277-0069-9
5. B. H. Kazemier and D. Vuysje (eds.), *Logic and Language.* Studies dedicated to Professor Rudolf Carnap on the Occasion of His 70th Birthday. 1962 ISBN 90-277-0019-2
6. M. W. Wartofsky (ed.), *Proceedings of the Boston Colloquium for the Philosophy of Science, 1961–1962.* [Boston Studies in the Philosophy of Science, Vol. I] 1963 ISBN 90-277-0021-4
7. A. A. Zinov'ev, *Philosophical Problems of Many-valued Logic.* A revised edition, edited and translated (from Russian) by G. Küng and D.D. Comey. 1963 ISBN 90-277-0091-5
8. G. Gurvitch, *The Spectrum of Social Time.* Translated from French and edited by M. Korenbaum and P. Bosserman. 1964 ISBN 90-277-0006-0
9. P. Lorenzen, *Formal Logic.* Translated from German by F.J. Crosson. 1965 ISBN 90-277-0080-X
10. R. S. Cohen and M. W. Wartofsky (eds.), *Proceedings of the Boston Colloquium for the Philosophy of Science, 1962–1964.* In Honor of Philipp Frank. [Boston Studies in the Philosophy of Science, Vol. II] 1965 ISBN 90-277-9004-0
11. E. W. Beth, *Mathematical Thought.* An Introduction to the Philosophy of Mathematics. 1965 ISBN 90-277-0070-2
12. E. W. Beth and J. Piaget, *Mathematical Epistemology and Psychology.* Translated from French by W. Mays. 1966 ISBN 90-277-0071-0
13. G. Küng, *Ontology and the Logistic Analysis of Language.* An Enquiry into the Contemporary Views on Universals. Revised ed., translated from German. 1967 ISBN 90-277-0028-1
14. R. S. Cohen and M. W. Wartofsky (eds.), *Proceedings of the Boston Colloquium for the Philosophy of Sciences, 1964–1966.* In Memory of Norwood Russell Hanson. [Boston Studies in the Philosophy of Science, Vol. III] 1967 ISBN 90-277-0013-3
15. C. D. Broad, *Induction, Probability, and Causation.* Selected Papers. 1968 ISBN 90-277-0012-5
16. G. Patzig, *Aristotle's Theory of the Syllogism.* A Logical-philosophical Study of *Book A* of the *Prior Analytics.* Translated from German by J. Barnes. 1968 ISBN 90-277-0030-3
17. N. Rescher, *Topics in Philosophical Logic.* 1968 ISBN 90-277-0084-2
18. R. S. Cohen and M. W. Wartofsky (eds.), *Proceedings of the Boston Colloquium for the Philosophy of Science, 1966–1968, Part I.* [Boston Studies in the Philosophy of Science, Vol. IV] 1969 ISBN 90-277-0014-1
19. R. S. Cohen and M. W. Wartofsky (eds.), *Proceedings of the Boston Colloquium for the Philosophy of Science, 1966–1968, Part II.* [Boston Studies in the Philosophy of Science, Vol. V] 1969 ISBN 90-277-0015-X
20. J. W. Davis, D. J. Hockney and W. K. Wilson (eds.), *Philosophical Logic.* 1969 ISBN 90-277-0075-3
21. D. Davidson and J. Hintikka (eds.), *Words and Objections.* Essays on the Work of W. V. Quine. 1969, rev. ed. 1975 ISBN 90-277-0074-5; Pb 90-277-0602-6
22. P. Suppes, *Studies in the Methodology and Foundations of Science. Selected Papers from 1951 to 1969.* 1969 ISBN 90-277-0020-6
23. J. Hintikka, *Models for Modalities.* Selected Essays. 1969 ISBN 90-277-0078-8; Pb 90-277-0598-4

24. N. Rescher *et al.* (eds.), *Essays in Honor of Carl G. Hempel.* A Tribute on the Occasion of His 65th Birthday. 1969 ISBN 90-277-0085-0
25. P. V. Tavanec (ed.), *Problems of the Logic of Scientific Knowledge.* Translated from Russian. 1970 ISBN 90-277-0087-7
26. M. Swain (ed.), *Induction, Acceptance, and Rational Belief.* 1970 ISBN 90-277-0086-9
27. R. S. Cohen and R. J. Seeger (eds.), *Ernst Mach: Physicist and Philosopher.* [Boston Studies in the Philosophy of Science, Vol. VI]. 1970 ISBN 90-277-0016-8
28. J. Hintikka and P. Suppes, *Information and Inference.* 1970 ISBN 90-277-0155-5
29. K. Lambert, *Philosophical Problems in Logic.* Some Recent Developments. 1970 ISBN 90-277-0079-6
30. R. A. Eberle, *Nominalistic Systems.* 1970 ISBN 90-277-0161-X
31. P. Weingartner and G. Zecha (eds.), *Induction, Physics, and Ethics.* 1970 ISBN 90-277-0158-X
32. E. W. Beth, *Aspects of Modern Logic.* Translated from Dutch. 1970 ISBN 90-277-0173-3
33. R. Hilpinen (ed.), *Deontic Logic.* Introductory and Systematic Readings. 1971 *See also* No. 152. ISBN Pb (1981 rev.) 90-277-1302-2
34. J.-L. Krivine, *Introduction to Axiomatic Set Theory.* Translated from French. 1971 ISBN 90-277-0169-5; Pb 90-277-0411-2
35. J. D. Sneed, *The Logical Structure of Mathematical Physics.* 2nd rev. ed., 1979 ISBN 90-277-1056-2; Pb 90-277-1059-7
36. C. R. Kordig, *The Justification of Scientific Change.* 1971 ISBN 90-277-0181-4; Pb 90-277-0475-9
37. M. Čapek, *Bergson and Modern Physics.* A Reinterpretation and Re-evaluation. [Boston Studies in the Philosophy of Science, Vol. VII] 1971 ISBN 90-277-0186-5
38. N. R. Hanson, *What I Do Not Believe, and Other Essays*. Ed. by S. Toulmin and H. Woolf. 1971 ISBN 90-277-0191-1
39. R. C. Buck and R. S. Cohen (eds.), *PSA 1970.* Proceedings of the Second Biennial Meeting of the Philosophy of Science Association, Boston, Fall 1970. In Memory of Rudolf Carnap. [Boston Studies in the Philosophy of Science, Vol. VIII] 1971 ISBN 90-277-0187-3; Pb 90-277-0309-4
40. D. Davidson and G. Harman (eds.), *Semantics of Natural Language.* 1972 ISBN 90-277-0304-3; Pb 90-277-0310-8
41. Y. Bar-Hillel (ed.), *Pragmatics of Natural Languages.* 1971 ISBN 90-277-0194-6; Pb 90-277-0599-2
42. S. Stenlund, *Combinators, γ Terms and Proof Theory.* 1972 ISBN 90-277-0305-1
43. M. Strauss, *Modern Physics and Its Philosophy.* Selected Paper in the Logic, History, and Philosophy of Science. 1972 ISBN 90-277-0230-6
44. M. Bunge, *Method, Model and Matter.* 1973 ISBN 90-277-0252-7
45. M. Bunge, *Philosophy of Physics.* 1973 ISBN 90-277-0253-5
46. A. A. Zinov'ev, *Foundations of the Logical Theory of Scientific Knowledge (Complex Logic).* Revised and enlarged English edition with an appendix by G. A. Smirnov, E. A. Sidorenka, A. M. Fedina and L. A. Bobrova. [Boston Studies in the Philosophy of Science, Vol. IX] 1973 ISBN 90-277-0193-8; Pb 90-277-0324-8
47. L. Tondl, *Scientific Procedures.* A Contribution concerning the Methodological Problems of Scientific Concepts and Scientific Explanation. Translated from Czech by D. Short. Edited by R.S. Cohen and M.W. Wartofsky. [Boston Studies in the Philosophy of Science, Vol. X] 1973 ISBN 90-277-0147-4; Pb 90-277-0323-X
48. N. R. Hanson, *Constellations and Conjectures.* 1973 ISBN 90-277-0192-X

49. K. J. J. Hintikka, J. M. E. Moravcsik and P. Suppes (eds.), *Approaches to Natural Language.* 1973 ISBN 90-277-0220-9; Pb 90-277-0233-0
50. M. Bunge (ed.), *Exact Philosophy.* Problems, Tools and Goals. 1973 ISBN 90-277-0251-9
51. R. J. Bogdan and I. Niiniluoto (eds.), *Logic, Language and Probability.* 1973 ISBN 90-277-0312-4
52. G. Pearce and P. Maynard (eds.), *Conceptual Change.* 1973 ISBN 90-277-0287-X; Pb 90-277-0339-6
53. I. Niiniluoto and R. Tuomela, *Theoretical Concepts and Hypothetico-inductive Inference.* 1973 ISBN 90-277-0343-4
54. R. Fraissé, *Course of Mathematical Logic* – Volume 1: *Relation and Logical Formula.* Translated from French. 1973 ISBN 90-277-0268-3; Pb 90-277-0403-1
(For *Volume 2* see under No. 69).
55. A. Grünbaum, *Philosophical Problems of Space and Time.* Edited by R.S. Cohen and M.W. Wartofsky. 2nd enlarged ed. [Boston Studies in the Philosophy of Science, Vol. XII] 1973 ISBN 90-277-0357-4; Pb 90-277-0358-2
56. P. Suppes (ed.), *Space, Time and Geometry.* 1973 ISBN 90-277-0386-8; Pb 90-277-0442-2
57. H. Kelsen, *Essays in Legal and Moral Philosophy.* Selected and introduced by O. Weinberger. Translated from German by P. Heath. 1973 ISBN 90-277-0388-4
58. R. J. Seeger and R. S. Cohen (eds.), *Philosophical Foundations of Science.* [Boston Studies in the Philosophy of Science, Vol. XI] 1974 ISBN 90-277-0390-6; Pb 90-277-0376-0
59. R. S. Cohen and M. W. Wartofsky (eds.), *Logical and Epistemological Studies in Contemporary Physics.* [Boston Studies in the Philosophy of Science, Vol. XIII] 1973 ISBN 90-277-0391-4; Pb 90-277-0377-9
60. R. S. Cohen and M. W. Wartofsky (eds.), *Methodological and Historical Essays in the Natural and Social Sciences. Proceedings of the Boston Colloquium for the Philosophy of Science, 1969–1972.* [Boston Studies in the Philosophy of Science, Vol. XIV] 1974 ISBN 90-277-0392-2; Pb 90-277-0378-7
61. R. S. Cohen, J. J. Stachel and M. W. Wartofsky (eds.), *For Dirk Struik. Scientific, Historical and Political Essays.* [Boston Studies in the Philosophy of Science, Vol. XV] 1974 ISBN 90-277-0393-0; Pb 90-277-0379-5
62. K. Ajdukiewicz, *Pragmatic Logic.* Translated from Polish by O. Wojtasiewicz. 1974 ISBN 90-277-0326-4
63. S. Stenlund (ed.), *Logical Theory and Semantic Analysis.* Essays dedicated to Stig Kanger on His 50th Birthday. 1974 ISBN 90-277-0438-4
64. K. F. Schaffner and R. S. Cohen (eds.), *PSA 1972. Proceedings of the Third Biennial Meeting of the Philosophy of Science Association.* [Boston Studies in the Philosophy of Science, Vol. XX] 1974 ISBN 90-277-0408-2; Pb 90-277-0409-0
65. H. E. Kyburg, Jr., *The Logical Foundations of Statistical Inference.* 1974 ISBN 90-277-0330-2; Pb 90-277-0430-9
66. M. Grene, *The Understanding of Nature.* Essays in the Philosophy of Biology. [Boston Studies in the Philosophy of Science, Vol. XXIII] 1974 ISBN 90-277-0462-7; Pb 90-277-0463-5
67. J. M. Broekman, *Structuralism: Moscow, Prague, Paris.* Translated from German. 1974 ISBN 90-277-0478-3
68. N. Geschwind, *Selected Papers on Language and the Brain.* [Boston Studies in the Philosophy of Science, Vol. XVI] 1974 ISBN 90-277-0262-4; Pb 90-277-0263-2
69. R. Fraissé, *Course of Mathematical Logic* – Volume 2: *Model Theory.* Translated from French. 1974 ISBN 90-277-0269-1; Pb 90-277-0510-0
(For *Volume 1* see under No. 54)

70. A. Grzegorczyk, *An Outline of Mathematical Logic.* Fundamental Results and Notions explained with all Details. Translated from Polish. 1974 ISBN 90-277-0359-0; Pb 90-277-0447-3
71. F. von Kutschera, *Philosophy of Language.* 1975 ISBN 90-277-0591-7
72. J. Manninen and R. Tuomela (eds.), *Essays on Explanation and Understanding.* Studies in the Foundations of Humanities and Social Sciences. 1976 ISBN 90-277-0592-5
73. J. Hintikka (ed.), *Rudolf Carnap, Logical Empiricist.* Materials and Perspectives. 1975 ISBN 90-277-0583-6
74. M. Čapek (ed.), *The Concepts of Space and Time.* Their Structure and Their Development. [Boston Studies in the Philosophy of Science, Vol. XXII] 1976 ISBN 90-277-0355-8; Pb 90-277-0375-2
75. J. Hintikka and U. Remes, *The Method of Analysis.* Its Geometrical Origin and Its General Significance. [Boston Studies in the Philosophy of Science, Vol. XXV] 1974 ISBN 90-277-0532-1; Pb 90-277-0543-7
76. J. E. Murdoch and E. D. Sylla (eds.), *The Cultural Context of Medieval Learning.* [Boston Studies in the Philosophy of Science, Vol. XXVI] 1975 ISBN 90-277-0560-7; Pb 90-277-0587-9
77. S. Amsterdamski, *Between Experience and Metaphysics.* Philosophical Problems of the Evolution of Science. [Boston Studies in the Philosophy of Science, Vol. XXXV] 1975 ISBN 90-277-0568-2; Pb 90-277-0580-1
78. P. Suppes (ed.), *Logic and Probability in Quantum Mechanics.* 1976 ISBN 90-277-0570-4; Pb 90-277-1200-X
79. H. von Helmholtz: *Epistemological Writings. The Paul Hertz / Moritz Schlick Centenary Edition of 1921 with Notes and Commentary by the Editors.* Newly translated from German by M. F. Lowe. Edited, with an Introduction and Bibliography, by R. S. Cohen and Y. Elkana. [Boston Studies in the Philosophy of Science, Vol. XXXVII] 1975 ISBN 90-277-0290-X; Pb 90-277-0582-8
80. J. Agassi, *Science in Flux.* [Boston Studies in the Philosophy of Science, Vol. XXVIII] 1975 ISBN 90-277-0584-4; Pb 90-277-0612-2
81. S. G. Harding (ed.), *Can Theories Be Refuted?* Essays on the Duhem-Quine Thesis. 1976 ISBN 90-277-0629-8; Pb 90-277-0630-1
82. S. Nowak, *Methodology of Sociological Research.* General Problems. 1977 ISBN 90-277-0486-4
83. J. Piaget, J.-B. Grize, A. Szeminśska and V. Bang, *Epistemology and Psychology of Functions.* Translated from French. 1977 ISBN 90-277-0804-5
84. M. Grene and E. Mendelsohn (eds.), *Topics in the Philosophy of Biology.* [Boston Studies in the Philosophy of Science, Vol. XXVII] 1976 ISBN 90-277-0595-X; Pb 90-277-0596-8
85. E. Fischbein, *The Intuitive Sources of Probabilistic Thinking in Children.* 1975 ISBN 90-277-0626-3; Pb 90-277-1190-9
86. E. W. Adams, *The Logic of Conditionals.* An Application of Probability to Deductive Logic. 1975 ISBN 90-277-0631-X
87. M. Przełęcki and R. Wójcicki (eds.), *Twenty-Five Years of Logical Methodology in Poland.* Translated from Polish. 1976 ISBN 90-277-0601-8
88. J. Topolski, *The Methodology of History.* Translated from Polish by O. Wojtasiewicz. 1976 ISBN 90-277-0550-X
89. A. Kasher (ed.), *Language in Focus: Foundations, Methods and Systems.* Essays dedicated to Yehoshua Bar-Hillel. [Boston Studies in the Philosophy of Science, Vol. XLIII] 1976 ISBN 90-277-0644-1; Pb 90-277-0645-X

90. J. Hintikka, *The Intentions of Intentionality and Other New Models for Modalities.* 1975 ISBN 90-277-0633-6; Pb 90-277-0634-4
91. W. Stegmüller, *Collected Papers on Epistemology, Philosophy of Science and History of Philosophy.* 2 Volumes. 1977 Set ISBN 90-277-0767-7
92. D. M. Gabbay, *Investigations in Modal and Tense Logics with Applications to Problems in Philosophy and Linguistics.* 1976 ISBN 90-277-0656-5
93. R. J. Bogdan, *Local Induction.* 1976 ISBN 90-277-0649-2
94. S. Nowak, *Understanding and Prediction.* Essays in the Methodology of Social and Behavioral Theories. 1976 ISBN 90-277-0558-5; Pb 90-277-1199-2
95. P. Mittelstaedt, *Philosophical Problems of Modern Physics.* [Boston Studies in the Philosophy of Science, Vol. XVIII] 1976 ISBN 90-277-0285-3; Pb 90-277-0506-2
96. G. Holton and W. A. Blanpied (eds.), *Science and Its Public: The Changing Relationship.* [Boston Studies in the Philosophy of Science, Vol. XXXIII] 1976 ISBN 90-277-0657-3; Pb 90-277-0658-1
97. M. Brand and D. Walton (eds.), *Action Theory.* 1976 ISBN 90-277-0671-9
98. P. Gochet, *Outline of a Nominalist Theory of Propositions.* An Essay in the Theory of Meaning and in the Philosophy of Logic. 1980 ISBN 90-277-1031-7
99. R. S. Cohen, P. K. Feyerabend, and M. W. Wartofsky (eds.), *Essays in Memory of Imre Lakatos.* [Boston Studies in the Philosophy of Science, Vol. XXXIX] 1976 ISBN 90-277-0654-9; Pb 90-277-0655-7
100. R. S. Cohen and J. J. Stachel (eds.), *Selected Papers of Léon Rosenfield.* [Boston Studies in the Philosophy of Science, Vol. XXI] 1979 ISBN 90-277-0651-4; Pb 90-277-0652-2
101. R. S. Cohen, C. A. Hooker, A. C. Michalos and J. W. van Evra (eds.), *PSA 1974. Proceedings of the 1974 Biennial Meeting of the Philosophy of Science Association.* [Boston Studies in the Philosophy of Science, Vol. XXXII] 1976 ISBN 90-277-0647-6; Pb 90-277-0648-4
102. Y. Fried and J. Agassi, *Paranoia.* A Study in Diagnosis. [Boston Studies in the Philosophy of Science, Vol. L] 1976 ISBN 90-277-0704-9; Pb 90-277-0705-7
103. M. Przełęcki, K. Szaniawski and R. Wójcicki (eds.), *Formal Methods in the Methodology of Empirical Sciences.* 1976 ISBN 90-277-0698-0
104. J. M. Vickers, *Belief and Probability.* 1976 ISBN 90-277-0744-8
105. K. H. Wolff, *Surrender and Catch.* Experience and Inquiry Today. [Boston Studies in the Philosophy of Science, Vol. LI] 1976 ISBN 90-277-0758-8; Pb 90-277-0765-0
106. K. Kosík, *Dialectics of the Concrete.* A Study on Problems of Man and World. [Boston Studies in the Philosophy of Science, Vol. LII] 1976 ISBN 90-277-0761-8; Pb 90-277-0764-2
107. N. Goodman, *The Structure of Appearance.* 3rd ed. with an Introduction by G. Hellman. [Boston Studies in the Philosophy of Science, Vol. LIII] 1977 ISBN 90-277-0773-1; Pb 90-277-0774-X
108. K. Ajdukiewicz, *The Scientific World-Perspective and Other Essays, 1931-1963.* Translated from Polish. Edited and with an Introduction by J. Giedymin. 1978 ISBN 90-277-0527-5
109. R. L. Causey, *Unity of Science.* 1977 ISBN 90-277-0779-0
110. R. E. Grandy, *Advanced Logic for Applications.* 1977 ISBN 90-277-0781-2
111. R. P. McArthur, *Tense Logic.* 1976 ISBN 90-277-0697-2
112. L. Lindahl, *Position and Change.* A Study in Law and Logic. Translated from Swedish by P. Needham. 1977 ISBN 90-277-0787-1
113. R. Tuomela, *Dispositions.* 1978 ISBN 90-277-0810-X
114. H. A. Simon, *Models of Discovery and Other Topics in the Methods of Science.* [Boston Studies in the Philosophy of Science, Vol. LIV] 1977 ISBN 90-277-0812-6; Pb 90-277-0858-4

115. R. D. Rosenkrantz, *Inference, Method and Decision.* Towards a Bayesian Philosophy of Science. 1977 ISBN 90-277-0817-7; Pb 90-277-0818-5
116. R. Tuomela, *Human Action and Its Explanation.* A Study on the Philosophical Foundations of Psychology. 1977 ISBN 90-277-0824-X
117. M. Lazerowitz, *The Language of Philosophy.* Freud and Wittgenstein. [Boston Studies in the Philosophy of Science, Vol. LV] 1977 ISBN 90-277-0826-6; Pb 90-277-0862-2
118. Not published 119. J. Pelc (ed.), *Semiotics in Poland, 1894–1969.* Translated from Polish. 1979 ISBN 90-277-0811-8
120. I. Pörn, *Action Theory and Social Science.* Some Formal Models. 1977 ISBN 90-277-0846-0
121. J. Margolis, *Persons and Mind.* The Prospects of Nonreductive Materialism. [Boston Studies in the Philosophy of Science, Vol. LVII] 1977 ISBN 90-277-0854-1; Pb 90-277-0863-0
122. J. Hintikka, I. Niiniluoto, and E. Saarinen (eds.), *Essays on Mathematical and Philosophical Logic.* 1979 ISBN 90-277-0879-7
123. T. A. F. Kuipers, *Studies in Inductive Probability and Rational Expectation.* 1978 ISBN 90-277-0882-7
124. E. Saarinen, R. Hilpinen, I. Niiniluoto and M. P. Hintikka (eds.), *Essays in Honour of Jaakko Hintikka on the Occasion of His 50th Birthday.* 1979 ISBN 90-277-0916-5
125. G. Radnitzky and G. Andersson (eds.), *Progress and Rationality in Science.* [Boston Studies in the Philosophy of Science, Vol. LVIII] 1978 ISBN 90-277-0921-1; Pb 90-277-0922-X
126. P. Mittelstaedt, *Quantum Logic.* 1978 ISBN 90-277-0925-4
127. K. A. Bowen, *Model Theory for Modal Logic.* Kripke Models for Modal Predicate Calculi. 1979 ISBN 90-277-0929-7
128. H. A. Bursen, *Dismantling the Memory Machine.* A Philosophical Investigation of Machine Theories of Memory. 1978 ISBN 90-277-0933-5
129. M. W. Wartofsky, *Models.* Representation and the Scientific Understanding. [Boston Studies in the Philosophy of Science, Vol. XLVIII] 1979 ISBN 90-277-0736-7; Pb 90-277-0947-5
130. D. Ihde, *Technics and Praxis.* A Philosophy of Technology. [Boston Studies in the Philosophy of Science, Vol. XXIV] 1979 ISBN 90-277-0953-X; Pb 90-277-0954-8
131. J. J. Wiatr (ed.), *Polish Essays in the Methodology of the Social Sciences.* [Boston Studies in the Philosophy of Science, Vol. XXIX] 1979 ISBN 90-277-0723-5; Pb 90-277-0956-4
132. W. C. Salmon (ed.), *Hans Reichenbach: Logical Empiricist.* 1979 ISBN 90-277-0958-0
133. P. Bieri, R.-P. Horstmann and L. Krüger (eds.), *Transcendental Arguments in Science.* Essays in Epistemology. 1979 ISBN 90-277-0963-7; Pb 90-277-0964-5
134. M. Marković and G. Petrović (eds.), *Praxis.* Yugoslav Essays in the Philosophy and Methodology of the Social Sciences. [Boston Studies in the Philosophy of Science, Vol. XXXVI] 1979 ISBN 90-277-0727-8; Pb 90-277-0968-8
135. R. Wójcicki, *Topics in the Formal Methodology of Empirical Sciences.* Translated from Polish. 1979 ISBN 90-277-1004-X
136. G. Radnitzky and G. Andersson (eds.), *The Structure and Development of Science.* [Boston Studies in the Philosophy of Science, Vol. LIX] 1979 ISBN 90-277-0994-7; Pb 90-277-0995-5
137. J. C. Webb, *Mechanism, Mentalism and Metamathematics.* An Essay on Finitism. 1980 ISBN 90-277-1046-5
138. D. F. Gustafson and B. L. Tapscott (eds.), *Body, Mind and Method.* Essays in Honor of Virgil C. Aldrich. 1979 ISBN 90-277-1013-9
139. L. Nowak, *The Structure of Idealization.* Towards a Systematic Interpretation of the Marxian Idea of Science. 1980 ISBN 90-277-1014-7

140. C. Perelman, *The New Rhetoric and the Humanities.* Essays on Rhetoric and Its Applications. Translated from French and German. With an Introduction by H. Zyskind. 1979
ISBN 90-277-1018-X; Pb 90-277-1019-8
141. W. Rabinowicz, *Universalizability.* A Study in Morals and Metaphysics. 1979
ISBN 90-277-1020-2
142. C. Perelman, *Justice, Law and Argument.* Essays on Moral and Legal Reasoning. Translated from French and German. With an Introduction by H.J. Berman. 1980
ISBN 90-277-1089-9; Pb 90-277-1090-2
143. S. Kanger and S. Öhman (eds.), *Philosophy and Grammar.* Papers on the Occasion of the Quincentennial of Uppsala University. 1981 ISBN 90-277-1091-0
144. T. Pawlowski, *Concept Formation in the Humanities and the Social Sciences.* 1980
ISBN 90-277-1096-1
145. J. Hintikka, D. Gruender and E. Agazzi (eds.), *Theory Change, Ancient Axiomatics and Galileo's Methodology.* Proceedings of the 1978 Pisa Conference on the History and Philosophy of Science, Volume I. 1981 ISBN 90-277-1126-7
146. J. Hintikka, D. Gruender and E. Agazzi (eds.), *Probabilistic Thinking, Thermodynamics, and the Interaction of the History and Philosophy of Science.* Proceedings of the 1978 Pisa Conference on the History and Philosophy of Science, Volume II. 1981 ISBN 90-277-1127-5
147. U. Mönnich (ed.), *Aspects of Philosophical Logic.* Some Logical Forays into Central Notions of Linguistics and Philosophy. 1981 ISBN 90-277-1201-8
148. D. M. Gabbay, *Semantical Investigations in Heyting's Intuitionistic Logic.* 1981
ISBN 90-277-1202-6
149. E. Agazzi (ed.), *Modern Logic – A Survey.* Historical, Philosophical, and Mathematical Aspects of Modern Logic and Its Applications. 1981 ISBN 90-277-1137-2
150. A. F. Parker-Rhodes, *The Theory of Indistinguishables.* A Search for Explanatory Principles below the Level of Physics. 1981 ISBN 90-277-1214-X
151. J. C. Pitt, *Pictures, Images, and Conceptual Change.* An Analysis of Wilfrid Sellars' Philosophy of Science. 1981 ISBN 90-277-1276-X; Pb 90-277-1277-8
152. R. Hilpinen (ed.), *New Studies in Deontic Logic.* Norms, Actions, and the Foundations of Ethics. 1981 ISBN 90-277-1278-6; Pb 90-277-1346-4
153. C. Dilworth, *Scientific Progress.* A Study Concerning the Nature of the Relation between Successive Scientific Theories. 3rd rev. ed., 1994 ISBN 0-7923-2487-0; Pb 0-7923-2488-9
154. D. Woodruff Smith and R. McIntyre, *Husserl and Intentionality.* A Study of Mind, Meaning, and Language. 1982 ISBN 90-277-1392-8; Pb 90-277-1730-3
155. R. J. Nelson, *The Logic of Mind.* 2nd. ed., 1989 ISBN 90-277-2819-4; Pb 90-277-2822-4
156. J. F. A. K. van Benthem, *The Logic of Time.* A Model-Theoretic Investigation into the Varieties of Temporal Ontology, and Temporal Discourse. 1983; 2nd ed., 1991 ISBN 0-7923-1081-0
157. R. Swinburne (ed.), *Space, Time and Causality.* 1983 ISBN 90-277-1437-1
158. E. T. Jaynes, *Papers on Probability, Statistics and Statistical Physics.* Ed. by R. D. Rozenkrantz. 1983 ISBN 90-277-1448-7; Pb (1989) 0-7923-0213-3
159. T. Chapman, *Time: A Philosophical Analysis.* 1982 ISBN 90-277-1465-7
160. E. N. Zalta, *Abstract Objects.* An Introduction to Axiomatic Metaphysics. 1983
ISBN 90-277-1474-6
161. S. Harding and M. B. Hintikka (eds.), *Discovering Reality.* Feminist Perspectives on Epistemology, Metaphysics, Methodology, and Philosophy of Science. 1983
ISBN 90-277-1496-7; Pb 90-277-1538-6
162. M. A. Stewart (ed.), *Law, Morality and Rights.* 1983 ISBN 90-277-1519-X

163. D. Mayr and G. Süssmann (eds.), *Space, Time, and Mechanics.* Basic Structures of a Physical Theory. 1983 ISBN 90-277-1525-4
164. D. Gabbay and F. Guenthner (eds.), *Handbook of Philosophical Logic.* Vol. I: Elements of Classical Logic. 1983 ISBN 90-277-1542-4
165. D. Gabbay and F. Guenthner (eds.), *Handbook of Philosophical Logic.* Vol. II: Extensions of Classical Logic. 1984 ISBN 90-277-1604-8
166. D. Gabbay and F. Guenthner (eds.), *Handbook of Philosophical Logic.* Vol. III: Alternative to Classical Logic. 1986 ISBN 90-277-1605-6
167. D. Gabbay and F. Guenthner (eds.), *Handbook of Philosophical Logic.* Vol. IV: Topics in the Philosophy of Language. 1989 ISBN 90-277-1606-4
168. A. J. I. Jones, *Communication and Meaning.* An Essay in Applied Modal Logic. 1983 ISBN 90-277-1543-2
169. M. Fitting, *Proof Methods for Modal and Intuitionistic Logics.* 1983 ISBN 90-277-1573-4
170. J. Margolis, *Culture and Cultural Entities.* Toward a New Unity of Science. 1984 ISBN 90-277-1574-2
171. R. Tuomela, *A Theory of Social Action.* 1984 ISBN 90-277-1703-6
172. J. J. E. Gracia, E. Rabossi, E. Villanueva and M. Dascal (eds.), *Philosophical Analysis in Latin America.* 1984 ISBN 90-277-1749-4
173. P. Ziff, *Epistemic Analysis.* A Coherence Theory of Knowledge. 1984 ISBN 90-277-1751-7
174. P. Ziff, *Antiaesthetics.* An Appreciation of the Cow with the Subtile Nose. 1984 ISBN 90-277-1773-7
175. W. Balzer, D. A. Pearce, and H.-J. Schmidt (eds.), *Reduction in Science.* Structure, Examples, Philosophical Problems. 1984 ISBN 90-277-1811-3
176. A. Peczenik, L. Lindahl and B. van Roermund (eds.), *Theory of Legal Science.* Proceedings of the Conference on Legal Theory and Philosophy of Science (Lund, Sweden, December 1983). 1984 ISBN 90-277-1834-2
177. I. Niiniluoto, *Is Science Progressive?* 1984 ISBN 90-277-1835-0
178. B. K. Matilal and J. L. Shaw (eds.), *Analytical Philosophy in Comparative Perspective.* Exploratory Essays in Current Theories and Classical Indian Theories of Meaning and Reference. 1985 ISBN 90-277-1870-9
179. P. Kroes, *Time: Its Structure and Role in Physical Theories.* 1985 ISBN 90-277-1894-6
180. J. H. Fetzer, *Sociobiology and Epistemology.* 1985 ISBN 90-277-2005-3; Pb 90-277-2006-1
181. L. Haaparanta and J. Hintikka (eds.), *Frege Synthesized.* Essays on the Philosophical and Foundational Work of Gottlob Frege. 1986 ISBN 90-277-2126-2
182. M. Detlefsen, *Hilbert's Program.* An Essay on Mathematical Instrumentalism. 1986 ISBN 90-277-2151-3
183. J. L. Golden and J. J. Pilotta (eds.), *Practical Reasoning in Human Affairs.* Studies in Honor of Chaim Perelman. 1986 ISBN 90-277-2255-2
184. H. Zandvoort, *Models of Scientific Development and the Case of Nuclear Magnetic Resonance.* 1986 ISBN 90-277-2351-6
185. I. Niiniluoto, *Truthlikeness.* 1987 ISBN 90-277-2354-0
186. W. Balzer, C. U. Moulines and J. D. Sneed, *An Architectonic for Science.* The Structuralist Program. 1987 ISBN 90-277-2403-2
187. D. Pearce, *Roads to Commensurability.* 1987 ISBN 90-277-2414-8
188. L. M. Vaina (ed.), *Matters of Intelligence.* Conceptual Structures in Cognitive Neuroscience. 1987 ISBN 90-277-2460-1

189. H. Siegel, *Relativism Refuted.* A Critique of Contemporary Epistemological Relativism. 1987 ISBN 90-277-2469-5
190. W. Callebaut and R. Pinxten, *Evolutionary Epistemology.* A Multiparadigm Program, with a Complete Evolutionary Epistemology Bibliograph. 1987 ISBN 90-277-2582-9
191. J. Kmita, *Problems in Historical Epistemology.* 1988 ISBN 90-277-2199-8
192. J. H. Fetzer (ed.), *Probability and Causality.* Essays in Honor of Wesley C. Salmon, with an Annotated Bibliography. 1988 ISBN 90-277-2607-8; Pb 1-5560-8052-2
193. A. Donovan, L. Laudan and R. Laudan (eds.), *Scrutinizing Science.* Empirical Studies of Scientific Change. 1988 ISBN 90-277-2608-6
194. H.R. Otto and J.A. Tuedio (eds.), *Perspectives on Mind.* 1988 ISBN 90-277-2640-X
195. D. Batens and J.P. van Bendegem (eds.), *Theory and Experiment.* Recent Insights and New Perspectives on Their Relation. 1988 ISBN 90-277-2645-0
196. J. Österberg, *Self and Others.* A Study of Ethical Egoism. 1988 ISBN 90-277-2648-5
197. D.H. Helman (ed.), *Analogical Reasoning.* Perspectives of Artificial Intelligence, Cognitive Science, and Philosophy. 1988 ISBN 90-277-2711-2
198. J. Woleński, *Logic and Philosophy in the Lvov-Warsaw School.* 1989 ISBN 90-277-2749-X
199. R. Wójcicki, *Theory of Logical Calculi.* Basic Theory of Consequence Operations. 1988 ISBN 90-277-2785-6
200. J. Hintikka and M.B. Hintikka, *The Logic of Epistemology and the Epistemology of Logic.* Selected Essays. 1989 ISBN 0-7923-0040-8; Pb 0-7923-0041-6
201. E. Agazzi (ed.), *Probability in the Sciences.* 1988 ISBN 90-277-2808-9
202. M. Meyer (ed.), *From Metaphysics to Rhetoric.* 1989 ISBN 90-277-2814-3
203. R.L. Tieszen, *Mathematical Intuition.* Phenomenology and Mathematical Knowledge. 1989 ISBN 0-7923-0131-5
204. A. Melnick, *Space, Time, and Thought in Kant.* 1989 ISBN 0-7923-0135-8
205. D.W. Smith, *The Circle of Acquaintance.* Perception, Consciousness, and Empathy. 1989 ISBN 0-7923-0252-4
206. M.H. Salmon (ed.), *The Philosophy of Logical Mechanism.* Essays in Honor of Arthur W. Burks. With his Responses, and with a Bibliography of Burk's Work. 1990 ISBN 0-7923-0325-3
207. M. Kusch, *Language as Calculus vs. Language as Universal Medium.* A Study in Husserl, Heidegger, and Gadamer. 1989 ISBN 0-7923-0333-4
208. T.C. Meyering, *Historical Roots of Cognitive Science.* The Rise of a Cognitive Theory of Perception from Antiquity to the Nineteenth Century. 1989 ISBN 0-7923-0349-0
209. P. Kosso, *Observability and Observation in Physical Science.* 1989 ISBN 0-7923-0389-X
210. J. Kmita, *Essays on the Theory of Scientific Cognition.* 1990 ISBN 0-7923-0441-1
211. W. Sieg (ed.), *Acting and Reflecting.* The Interdisciplinary Turn in Philosophy. 1990 ISBN 0-7923-0512-4
212. J. Karpiński, *Causality in Sociological Research.* 1990 ISBN 0-7923-0546-9
213. H.A. Lewis (ed.), *Peter Geach: Philosophical Encounters.* 1991 ISBN 0-7923-0823-9
214. M. Ter Hark, *Beyond the Inner and the Outer.* Wittgenstein's Philosophy of Psychology. 1990 ISBN 0-7923-0850-6
215. M. Gosselin, *Nominalism and Contemporary Nominalism.* Ontological and Epistemological Implications of the Work of W.V.O. Quine and of N. Goodman. 1990 ISBN 0-7923-0904-9
216. J.H. Fetzer, D. Shatz and G. Schlesinger (eds.), *Definitions and Definability.* Philosophical Perspectives. 1991 ISBN 0-7923-1046-2
217. E. Agazzi and A. Cordero (eds.), *Philosophy and the Origin and Evolution of the Universe.* 1991 ISBN 0-7923-1322-4

218. M. Kusch, *Foucault's Strata and Fields.* An Investigation into Archaeological and Genealogical Science Studies. 1991 ISBN 0-7923-1462-X
219. C.J. Posy, *Kant's Philosophy of Mathematics.* Modern Essays. 1992 ISBN 0-7923-1495-6
220. G. Van de Vijver, *New Perspectives on Cybernetics.* Self-Organization, Autonomy and Connectionism. 1992 ISBN 0-7923-1519-7
221. J.C. Nyíri, *Tradition and Individuality.* Essays. 1992 ISBN 0-7923-1566-9
222. R. Howell, *Kant's Transcendental Deduction.* An Analysis of Main Themes in His Critical Philosophy. 1992 ISBN 0-7923-1571-5
223. A. García de la Sienra, *The Logical Foundations of the Marxian Theory of Value.* 1992 ISBN 0-7923-1778-5
224. D.S. Shwayder, *Statement and Referent.* An Inquiry into the Foundations of Our Conceptual Order. 1992 ISBN 0-7923-1803-X
225. M. Rosen, *Problems of the Hegelian Dialectic.* Dialectic Reconstructed as a Logic of Human Reality. 1993 ISBN 0-7923-2047-6
226. P. Suppes, *Models and Methods in the Philosophy of Science: Selected Essays.* 1993 ISBN 0-7923-2211-8
227. R. M. Dancy (ed.), *Kant and Critique: New Essays in Honor of W. H. Werkmeister.* 1993 ISBN 0-7923-2244-4
228. J. Woleński (ed.), *Philosophical Logic in Poland.* 1993 ISBN 0-7923-2293-2
229. M. De Rijke (ed.), *Diamonds and Defaults.* Studies in Pure and Applied Intensional Logic. 1993 ISBN 0-7923-2342-4
230. B.K. Matilal and A. Chakrabarti (eds.), *Knowing from Words.* Western and Indian Philosophical Analysis of Understanding and Testimony. 1994 ISBN 0-7923-2345-9
231. S.A. Kleiner, *The Logic of Discovery.* A Theory of the Rationality of Scientific Research. 1993 ISBN 0-7923-2371-8
232. R. Festa, *Optimum Inductive Methods.* A Study in Inductive Probability, Bayesian Statistics, and Verisimilitude. 1993 ISBN 0-7923-2460-9
233. P. Humphreys (ed.), *Patrick Suppes: Scientific Philosopher.* Vol. 1: Probability and Probabilistic Causality. 1994 ISBN 0-7923-2552-4
234. P. Humphreys (ed.), *Patrick Suppes: Scientific Philosopher.* Vol. 2: Philosophy of Physics, Theory Structure, and Measurement Theory. 1994 ISBN 0-7923-2553-2
235. P. Humphreys (ed.), *Patrick Suppes: Scientific Philosopher.* Vol. 3: Language, Logic, and Psychology. 1994 ISBN 0-7923-2862-0
   Set ISBN (Vols 233–235) 0-7923-2554-0
236. D. Prawitz and D. Westerståhl (eds.), *Logic and Philosophy of Science in Uppsala.* Papers from the 9th International Congress of Logic, Methodology, and Philosophy of Science. 1994 ISBN 0-7923-2702-0
237. L. Haaparanta (ed.), *Mind, Meaning and Mathematics.* Essays on the Philosophical Views of Husserl and Frege. 1994 ISBN 0-7923-2703-9
238. J. Hintikka (ed.), *Aspects of Metaphor.* 1994 ISBN 0-7923-2786-1
239. B. McGuinness and G. Oliveri (eds.), *The Philosophy of Michael Dummett.* With Replies from Michael Dummett. 1994 ISBN 0-7923-2804-3
240. D. Jamieson (ed.), *Language, Mind, and Art.* Essays in Appreciation and Analysis, In Honor of Paul Ziff. 1994 ISBN 0-7923-2810-8
241. G. Preyer, F. Siebelt and A. Ulfig (eds.), *Language, Mind and Epistemology.* On Donald Davidson's Philosophy. 1994 ISBN 0-7923-2811-6
242. P. Ehrlich (ed.), *Real Numbers, Generalizations of the Reals, and Theories of Continua.* 1994 ISBN 0-7923-2689-X

243. G. Debrock and M. Hulswit (eds.), *Living Doubt.* Essays concerning the epistemology of Charles Sanders Peirce. 1994 ISBN 0-7923-2898-1
244. J. Srzednicki, *To Know or Not to Know.* Beyond Realism and Anti-Realism. 1994 ISBN 0-7923-2909-0
245. R. Egidi (ed.), *Wittgenstein: Mind and Language.* 1995 ISBN 0-7923-3171-0
246. A. Hyslop, *Other Minds.* 1995 ISBN 0-7923-3245-8
247. L. Pólos and M. Masuch (eds.), *Applied Logic: How, What and Why.* Logical Approaches to Natural Language. 1995 ISBN 0-7923-3432-9
248. M. Krynicki, M. Mostowski and L.M. Szczerba (eds.), *Quantifiers: Logics, Models and Computation.* Volume One: Surveys. 1995 ISBN 0-7923-3448-5
249. M. Krynicki, M. Mostowski and L.M. Szczerba (eds.), *Quantifiers: Logics, Models and Computation.* Volume Two: Contributions. 1995 ISBN 0-7923-3449-3
Set ISBN (Vols 248 + 249) 0-7923-3450-7
250. R.A. Watson, *Representational Ideas from Plato to Patricia Churchland.* 1995 ISBN 0-7923-3453-1
251. J. Hintikka (ed.), *From Dedekind to Gödel.* Essays on the Development of the Foundations of Mathematics. 1995 ISBN 0-7923-3484-1
252. A. Wiśniewski, *The Posing of Questions.* Logical Foundations of Erotetic Inferences. 1995 ISBN 0-7923-3637-2
253. J. Peregrin, *Doing Worlds with Words.* Formal Semantics without Formal Metaphysics. 1995 ISBN 0-7923-3742-5
254. I.A. Kieseppä, *Truthlikeness for Multidimensional, Quantitative Cognitive Problems.* 1996 ISBN 0-7923-4005-1
255. P. Hugly and C. Sayward: *Intensionality and Truth.* An Essay on the Philosophy of A.N. Prior. 1996 ISBN 0-7923-4119-8
256. L. Hankinson Nelson and J. Nelson (eds.): *Feminism, Science, and the Philosophy of Science.* 1997 ISBN 0-7923-4162-7
257. P.I. Bystrov and V.N. Sadovsky (eds.): *Philosophical Logic and Logical Philosophy.* Essays in Honour of Vladimir A. Smirnov. 1996 ISBN 0-7923-4270-4
258. Å.E. Andersson and N-E. Sahlin (eds.): *The Complexity of Creativity.* 1996 ISBN 0-7923-4346-8
259. M.L. Dalla Chiara, K. Doets, D. Mundici and J. van Benthem (eds.): *Logic and Scientific Methods.* Volume One of the Tenth International Congress of Logic, Methodology and Philosophy of Science, Florence, August 1995. 1997 ISBN 0-7923-4383-2
260. M.L. Dalla Chiara, K. Doets, D. Mundici and J. van Benthem (eds.): *Structures and Norms in Science.* Volume Two of the Tenth International Congress of Logic, Methodology and Philosophy of Science, Florence, August 1995. 1997 ISBN 0-7923-4384-0
Set ISBN (Vols 259 + 260) 0-7923-4385-9
261. A. Chakrabarti: *Denying Existence.* The Logic, Epistemology and Pragmatics of Negative Existentials and Fictional Discourse. 1997 ISBN 0-7923-4388-3
262. A. Biletzki: *Talking Wolves.* Thomas Hobbes on the Language of Politics and the Politics of Language. 1997 ISBN 0-7923-4425-1
263. D. Nute (ed.): *Defeasible Deontic Logic.* 1997 ISBN 0-7923-4630-0
264. U. Meixner: *Axiomatic Formal Ontology.* 1997 ISBN 0-7923-4747-X
265. I. Brinck: *The Indexical 'I'.* The First Person in Thought and Language. 1997 ISBN 0-7923-4741-2
266. G. Hölmström-Hintikka and R. Tuomela (eds.): *Contemporary Action Theory.* Volume 1: Individual Action. 1997 ISBN 0-7923-4753-6; Set: 0-7923-4754-4

267. G. Hölmström-Hintikka and R. Tuomela (eds.): *Contemporary Action Theory.* Volume 2: Social Action. 1997 ISBN 0-7923-4752-8; Set: 0-7923-4754-4
268. B.-C. Park: *Phenomenological Aspects of Wittgenstein's Philosophy.* 1998 ISBN 0-7923-4813-3
269. J. Paśniczek: *The Logic of Intentional Objects.* A Meinongian Version of Classical Logic. 1998 Hb ISBN 0-7923-4880-X; Pb ISBN 0-7923-5578-4
270. P.W. Humphreys and J.H. Fetzer (eds.): *The New Theory of Reference.* Kripke, Marcus, and Its Origins. 1998 ISBN 0-7923-4898-2
271. K. Szaniawski, A. Chmielewski and J. Woleński (eds.): *On Science, Inference, Information and Decision Making.* Selected Essays in the Philosophy of Science. 1998 ISBN 0-7923-4922-9
272. G.H. von Wright: *In the Shadow of Descartes.* Essays in the Philosophy of Mind. 1998 ISBN 0-7923-4992-X
273. K. Kijania-Placek and J. Woleński (eds.): *The Lvov–Warsaw School and Contemporary Philosophy.* 1998 ISBN 0-7923-5105-3
274. D. Dedrick: *Naming the Rainbow.* Colour Language, Colour Science, and Culture. 1998 ISBN 0-7923-5239-4
275. L. Albertazzi (ed.): *Shapes of Forms.* From Gestalt Psychology and Phenomenology to Ontology and Mathematics. 1999 ISBN 0-7923-5246-7
276. P. Fletcher: *Truth, Proof and Infinity.* A Theory of Constructions and Constructive Reasoning. 1998 ISBN 0-7923-5262-9
277. M. Fitting and R.L. Mendelsohn (eds.): *First-Order Modal Logic.* 1998 Hb ISBN 0-7923-5334-X; Pb ISBN 0-7923-5335-8
278. J.N. Mohanty: *Logic, Truth and the Modalities from a Phenomenological Perspective.* 1999 ISBN 0-7923-5550-4
279. T. Placek: *Mathematical Intiutionism and Intersubjectivity.* A Critical Exposition of Arguments for Intuitionism. 1999 ISBN 0-7923-5630-6
280. A. Cantini, E. Casari and P. Minari (eds.): *Logic and Foundations of Mathematics.* 1999 ISBN 0-7923-5659-4 set ISBN 0-7923-5867-8
281. M.L. Dalla Chiara, R. Giuntini and F. Laudisa (eds.): *Language, Quantum, Music.* 1999 ISBN 0-7923-5727-2; set ISBN 0-7923-5867-8
282. R. Egidi (ed.): *In Search of a New Humanism.* The Philosophy of Georg Hendrik von Wright. 1999 ISBN 0-7923-5810-4
283. F. Vollmer: *Agent Causality.* 1999 ISBN 0-7923-5848-1
284. J. Peregrin (ed.): *Truth and Its Nature (if Any).* 1999 ISBN 0-7923-5865-1
285. M. De Caro (ed.): *Interpretations and Causes.* New Perspectives on Donald Davidson's Philosophy. 1999 ISBN 0-7923-5869-4
286. R. Murawski: *Recursive Functions and Metamathematics.* Problems of Completeness and Decidability, Gödel's Theorems. 1999 ISBN 0-7923-5904-6
287. T.A.F. Kuipers: *From Instrumentalism to Constructive Realism.* On Some Relations between Confirmation, Empirical Progress, and Truth Approximation. 2000 ISBN 0-7923-6086-9
288. G. Holmström-Hintikka (ed.): *Medieval Philosophy and Modern Times.* 2000 ISBN 0-7923-6102-4
289. E. Grosholz and H. Breger (eds.): *The Growth of Mathematical Knowledge.* 2000 ISBN 0-7923-6151-2

290. G. Sommaruga: *History and Philosophy of Constructive Type Theory*. 2000 ISBN 0-7923-6180-6
291. J. Gasser (ed.): *A Boole Anthology*. Recent and Classical Studies in the Logic of George Boole. 2000 ISBN 0-7923-6380-9
292. V.F. Hendricks, S.A. Pedersen and K.F. Jørgensen (eds.): *Proof Theory*. History and Philosophical Significance. 2000 ISBN 0-7923-6544-5
293. W.L. Craig: *The Tensed Theory of Time*. A Critical Examination. 2000 ISBN 0-7923-6634-4
294. W.L. Craig: *The Tenseless Theory of Time*. A Critical Examination. 2000 ISBN 0-7923-6635-2
295. L. Albertazzi (ed.): *The Dawn of Cognitive Science*. Early European Contributors. 2001 ISBN 0-7923-6799-5
296. G. Forrai: *Reference, Truth and Conceptual Schemes*. A Defense of Internal Realism. 2001 ISBN 0-7923-6885-1
297. V.F. Hendricks, S.A. Pedersen and K.F. Jørgensen (eds.): *Probability Theory*. Philosophy, Recent History and Relations to Science. 2001 ISBN 0-7923-6952-1
298. M. Esfeld: *Holism in Philosophy of Mind and Philosophy of Physics*. 2001 ISBN 0-7923-7003-1
299. E.C. Steinhart: *The Logic of Metaphor*. Analogous Parts of Possible Worlds. 2001 ISBN 0-7923-7004-X
300. P. Gärdenfors: *The Dynamics of Thought*. 2005 ISBN 1-4020-3398-2
301. T.A.F. Kuipers: *Structures in Science Heuristic Patterns Based on Cognitive Structures*. An Advanced Textbook in Neo-Classical Philosophy of Science. 2001 ISBN 0-7923-7117-8
302. G. Hon and S.S. Rakover (eds.): *Explanation*. Theoretical Approaches and Applications. 2001 ISBN 1-4020-0017-0
303. G. Holmström-Hintikka, S. Lindström and R. Sliwinski (eds.): *Collected Papers of Stig Kanger with Essays on his Life and Work*. Vol. I. 2001 ISBN 1-4020-0021-9; Pb ISBN 1-4020-0022-7
304. G. Holmström-Hintikka, S. Lindström and R. Sliwinski (eds.): *Collected Papers of Stig Kanger with Essays on his Life and Work*. Vol. II. 2001 ISBN 1-4020-0111-8; Pb ISBN 1-4020-0112-6
305. C.A. Anderson and M. Zelëny (eds.): *Logic, Meaning and Computation*. Essays in Memory of Alonzo Church. 2001 ISBN 1-4020-0141-X
306. P. Schuster, U. Berger and H. Osswald (eds.): *Reuniting the Antipodes – Constructive and Nonstandard Views of the Continuum*. 2001 ISBN 1-4020-0152-5
307. S.D. Zwart: *Refined Verisimilitude*. 2001 ISBN 1-4020-0268-8
308. A.-S. Maurin: *If Tropes*. 2002 ISBN 1-4020-0656-X
309. H. Eilstein (ed.): *A Collection of Polish Works on Philosophical Problems of Time and Spacetime*. 2002 ISBN 1-4020-0670-5
310. Y. Gauthier: *Internal Logic*. Foundations of Mathematics from Kronecker to Hilbert. 2002 ISBN 1-4020-0689-6
311. E. Ruttkamp: *A Model-Theoretic Realist Interpretation of Science*. 2002 ISBN 1-4020-0729-9
312. V. Rantala: *Explanatory Translation*. Beyond the Kuhnian Model of Conceptual Change. 2002 ISBN 1-4020-0827-9
313. L. Decock: *Trading Ontology for Ideology*. 2002 ISBN 1-4020-0865-1

314. O. Ezra: *The Withdrawal of Rights*. Rights from a Different Perspective. 2002
ISBN 1-4020-0886-4
315. P. Gärdenfors, J. Woleński and K. Kijania-Placek: *In the Scope of Logic, Methodology and Philosophy of Science*. Volume One of the 11th International Congress of Logic, Methodology and Philosophy of Science, Cracow, August 1999. 2002
ISBN 1-4020-0929-1; Pb 1-4020-0931-3
316. P. Gärdenfors, J. Woleński and K. Kijania-Placek: *In the Scope of Logic, Methodology and Philosophy of Science*. Volume Two of the 11th International Congress of Logic, Methodology and Philosophy of Science, Cracow, August 1999. 2002
ISBN 1-4020-0930-5; Pb 1-4020-0931-3
317. M.A. Changizi: *The Brain from 25,000 Feet*. High Level Explorations of Brain Complexity, Perception, Induction and Vagueness. 2003 ISBN 1-4020-1176-8
318. D.O. Dahlstrom (ed.): *Husserl's* Logical Investigations. 2003 ISBN 1-4020-1325-6
319. A. Biletzki: *(Over)Interpreting Wittgenstein*. 2003
ISBN Hb 1-4020-1326-4; Pb 1-4020-1327-2
320. A. Rojszczak, J. Cachro and G. Kurczewski (eds.): *Philosophical Dimensions of Logic and Science*. Selected Contributed Papers from the 11th International Congress of Logic, Methodology, and Philosophy of Science, Kraków, 1999. 2003 ISBN 1-4020-1645-X
321. M. Sintonen, P. Ylikoski and K. Miller (eds.): *Realism in Action*. Essays in the Philosophy of the Social Sciences. 2003 ISBN 1-4020-1667-0
322. V.F. Hendricks, K.F. Jørgensen and S.A. Pedersen (eds.): *Knowledge Contributors*. 2003
ISBN Hb 1-4020-1747-2; Pb 1-4020-1748-0
323. J. Hintikka, T. Czarnecki, K. Kijania-Placek, T. Placek and A. Rojszczak † (eds.): *Philosophy and Logic In Search of the Polish Tradition*. Essays in Honour of Jan Woleński on the Occasion of his 60th Birthday. 2003 ISBN 1-4020-1721-9
324. L.M. Vaina, S.A. Beardsley and S.K. Rushton (eds.): *Optic Flow and Beyond*. 2004
ISBN 1-4020-2091-0
325. D. Kolak (ed.): *I Am You*. The Metaphysical Foundations of Global Ethics. 2004
ISBN 1-4020-2999-3
326. V. Stepin: *Theoretical Knowledge*. 2005 ISBN 1-4020-3045-2
327. P. Mancosu, K.F. Jørgensen and S.A. Pedersen (eds.): *Visualization, Explanation and Reasoning Styles in Mathematics*. 2005 ISBN 1-4020-3334-6
328. A. Rojszczak (author) and J. Wolenski (ed.): *From the Act of Judging to the Sentence*. The Problem of Truth Bearers from Bolzano to Tarski. 2005 ISBN 1-4020-3396-6
329. A. Pietarinen: *Signs of Logic*. Peircean Themes on the Philosophy of Language, Games, and Communication. 2005 ISBN 1-4020-3728-7
330. A. Aliseda: *Abductive Reasoning*. Logical Investigations into Discovery and Explanation. 2005
ISBN 1-4020-3906-9
331. B. Feltz, M. Crommelinck and P. Goujon (eds.): *Self-organization and Emergence in Life Sciences*. 2005 ISBN 1-4020-3916-6
332. R. van Rooij: *Attitudes and Changing Contexts*. 2006 ISBN 1-4020-4176-4
333. L. Albertazzi: *Immanent Realism*. An Introduction to Brentano. 2006 ISBN 1-4020-4201-9
334. A. Keupink and S. Shieh (eds.): *The Limits of Logical Empiricism*. Selected Papers of Arthur Pap. 2006 ISBN 1-4020-4298-1
335. M. van Atten: *Brouwer meets Husserl*. On the Phenomenology of Choice Sequences. 2006
ISBN 1-4020-5086-0
336. F. Stjernfelt: *Diagrammatology*. An Investigation on the Borderlines of Phenomenology, Ontology, and Semiotics. 2007 ISBN 978-1-4020-5651-2

FSC
www.fsc.org
MIX
Papier | Fördert gute Waldnutzung
FSC® C083411

Kolibri 360 GmbH
Ferdinand-Jühlke-Straße 7
99095 Erfurt, Deutschland
produktsicherheit@kolibri360.de